Mathematical Foundations for
Chemistry, Physics, and
Engineering

Mathematical Foundations for Chemistry, Physics, and Engineering

Rita Kakkar

Department of Chemistry
University of Delhi
Delhi, India

WILEY

Published by John Wiley & Sons, Inc., Hoboken, New Jersey.

Library of Congress Cataloging-in-Publication Data has been applied for:

Names: Kakkar, Rita author | John Wiley & Sons publisher
Title: Mathematical Foundations for Chemistry, Physics, and Engineering / Rita Kakkar.
Description: Hoboken, New Jersey : John Wiley and Sons, [2027] | Publication date from CIP data view.
Identifiers: LCCN 2026030147 | ISBN 9781394368013 hardback | ISBN 9781394368020 epub | ISBN 9781394368037 epdf
Subjects: LCSH: Mathematical physics | Chemistry, Physical and theoretical—Mathematics | Engineering mathematics | Quantum chemistry—Mathematics
Classification: LCC QA401 .K35 2027
LC record available at https://lccn.loc.gov/2026030147

Cover Design: Wiley
Cover Image: Courtesy of Rita Kakkar

Dedicated to the memory of my parents,
who shaped my values and love for learning

Contents

Preface

Understand. Don't memorize. Learn principles, not formulas.

—R. Feynman

Having devoted almost a lifetime to teaching quantum mechanics and allied subjects—molecular spectroscopy, statistical mechanics, computer programming for chemists, and computational chemistry—I thought it was time to put my experiences on paper. The greatest challenge to teaching quantum mechanics is mathematics because students, particularly those from a chemistry background, have a mental block against anything mathematical. To top it, most quantum chemistry books and courses are filled with derivations without connection to physics. I had the privilege of being taught by some of the best teachers. As one of my teachers and later my senior colleague (an experimentalist) with whom I shared the spectroscopy paper once remarked, whatever math you teach, do not lose sight of the molecule. This is exactly what I have tried to do in this book—relating all abstract math to real problems. Each chapter begins with a mathematical problem and follows with applications.

That brings me to the second challenge. For a long time (including my own studentship), the first postgraduate physical chemistry course used to be an excellently drafted quantum chemistry course, divided into a mathematics portion, followed by quantum mechanics. When I studied or started teaching the course, finding appropriate reference books for the course was a challenge. I taught one topic from one book and another from another book, and most of these books were either pure math, mathematical physics, or engineering books. None of these was designed especially for chemistry students "keeping the molecule in mind." In chemistry books, mathematics was a quick interlude—just the rudimentary requirement.

It is this gap that I have tried to address in this book. The chapters are arranged in a sequence such that there is a connection between one chapter and the next. I started writing a book for the entire quantum chemistry course, but the mathematics portion itself ran into several pages. I also realized that you needed to know some topics before talking about some of these here. That emerged into a zero chapter, which, after a little expansion, emerged into a new book that should be out shortly. I feel that this book is for upper undergraduate and postgraduate students, while the next one is for undergraduate students. Together, these two books cover the essential math for science and engineering students, particularly chemistry.

I have written the book in a student-friendly style. I have sometimes left out the lengthy derivations for better flow, but there are always some students who want a little more—keep on asking why. For those, I have math text boxes and appendices. I have also included teacher notes sometimes (also helpful for students to self-learn) and a large number of worked examples. Besides some of the exercises, some of which are part of the assignments I gave to the students, I have added some conceptual questions for students to test their understanding. There are some advanced/ challenge-level questions that a general student can ignore. Only those interested in pursuing higher studies or interested in taking challenges should attempt to solve these.

I have intentionally kept the chapters short. On revising a chapter, if, at any point, I felt exhausted, I immediately stopped the chapter at a convenient point and placed the remaining part into a new chapter. To break the monotony, every chapter has its unique set of textboxes depending on the contents. For example, the Limits and Series chapter has a lot of historical references because of the large number of mathematicians involved.

This book (and its companion volume) covers nearly all the mathematics you will need for quantum chemistry, spectroscopy, statistical mechanics, and much of physical chemistry.

Besides Fact Boxes, each chapter has several Red Flags based on my own observations where students usually tend to make mistakes or are confused. Each chapter also has learning outcomes. Below, I describe each chapter in detail.

Chapter 1, Vectors and Coordinate Systems, introduces vectors, vector algebra, and coordinate systems used in physical sciences. Covers basis vectors, dot and cross products, direction cosines, and transformations between coordinate frames. Provides the foundation for later work with operators and differential equations. The basics are kept to a bare minimum, suitable for a PG-level book. More details can be found in my forthcoming book, *Mathematical Methods in Chemistry: From Basics to Applications.*

This chapter has a number of Fact Boxes to reinforce concepts. It lays the foundations for the next chapter on linear vector spaces.

Chapter 2, Linear Vector Spaces, develops the abstract framework of vector spaces, subspaces, linear independence, basis, and dimension. Includes inner products, norms, and orthogonality. Essential for understanding quantum states, function spaces, and orthogonal expansions.

Because of its relationship with quantum mechanics, the highlight of this chapter is the large number of connections to quantum mechanics text boxes, as well as fact boxes.

Chapter 3, Linear Transformations and Operators, explains linear transformations, operator representation, change of basis, and adjoint operators. Includes examples from physics and chemistry, emphasizing the role of linear operators in quantum mechanics and spectroscopy. The connection to quantum chemistry is reinforced through text boxes and fact boxes.

Besides these applications, the role of linear transformations in machine learning is explored. Of late, this has become important in chemistry, and its importance will continue to grow.

Chapter 4, Matrix Characteristics and Quantum-mechanical Operators, discusses matrix algebra, trace, determinant, rank, and properties of Hermitian, unitary, and normal matrices. It connects matrix characteristics to observable quantities, commutation relations, and symmetry operations in quantum mechanics.

Besides the usual chemistry and quantum chemistry connections, as well as fact boxes, this chapter addresses a very important question: Why at all do we need to find eigenvalues and eigenvectors? This is a big question in the minds of students. This is explained in easy geometric terms before discussing the machinery in the next chapter, which also addresses the question, "Why do we need to diagonalize matrices?" Without understanding these, it becomes just a meaningless mathematical exercise.

Chapter 5, Eigenvalue Problems, Orthogonality, and Completeness, covers characteristic equations, eigenvalues and eigenvectors, orthogonality, degeneracy, and completeness. Introduces applications to molecular vibrations, Hamiltonians, and normal modes.

This chapter develops the mathematical framework for determining eigenvalues and eigenvectors of matrices and linear operators. It also explores the conditions under which a matrix is diagonalizable to get ready for the next chapter.

Chapter 6, Diagonalization of Matrices, details diagonalization, similarity transformations, the spectral theorem for Hermitian operators, and applications to coupled systems. Includes examples from quantum mechanics, electronic structure, and matrix representations of observables.

Chapter 7, Group Theory and Symmetry, introduces symmetry elements, symmetry operations, point groups, representations, characters, and class structure. Emphasizes mathematical foundations with chemical applications kept minimal for clarity.

In this chapter, I have included several red flags from my own experience. Also, great emphasis is given on the assignment of point groups, for it is here that most students falter. This includes a decision tree and some shortcuts. Great care has been taken to cite only the relevant theorems. Emphasis is on examples rather than theory.

Chapter 8, Applications of Group Theory, applies group theory to vibrational analysis, selection rules, hybrid orbitals, and matrix elements. Includes construction of symmetry-adapted linear combinations and the use of character tables in spectroscopy.

The highlight of this chapter is the many shortcuts introduced to simplify calculations. Also, there are numerous examples of applications in chemistry—more than I originally planned.

Chapter 9, Limits, Series, and Power-series Expansions, reviews limits, continuity, Taylor and Maclaurin expansions, convergence tests, and series solutions, besides Fourier series. Provides tools needed for approximations, perturbation methods, and expansions used throughout the physical sciences.

Because of the large number of scientists involved, this chapter has several historical notes, besides the usual worked examples, red flags, concepts, and so on.

Chapter 10, Fourier Transforms, presents Fourier transforms and their properties using a consistent normalization. Includes convolution, modulation, symmetry, and common transform pairs. Applications to signal analysis, spectroscopy, and PDEs.

Longer derivations are included in Math Textboxes for the interested reader, and the narrative is kept clean, with one topic flowing into the next.

Besides the applications in quantum chemistry, spectroscopy, and solid-state physics, computer-based applications such as those in sound and image processing, data-driven discovery, and intelligent systems are discussed.

Chapter 11, Laplace Transforms, defines Laplace transforms, inverse transforms, and operational properties. Covers initial-value problems, convolution, and shifts. Includes rational functions, partial fractions, and applications to kinetics and differential equations.

The highlight of this chapter is the large number of worked examples, ranging from illustrations of principles to real-time applications, such as in chemical kinetics and electrochemistry.

Chapter 12, Linear Differential Equations, covers first- and second-order linear ODEs, constant-coefficient equations, reduction of order, and variation of parameters. Includes applications to chemical kinetics, oscillators, and operator methods. The Legendre series is discussed as an example.

This chapter naturally merges into the next chapter, where the important orthogonal series are studied in a unified way.

Chapter 13, Special Functions, introduces Legendre, associated Legendre, Hermite, and Laguerre functions, with definitions and key properties. Emphasizes their roles as solutions to physical differential equations.

Besides other text boxes, this chapter has a few teacher notes to help put the ideas across to the students. The ideas presented in this chapter are essential for quantum mechanics. A large number of tables present the properties of the polynomials.

Chapter 14, Partial Differential Equations, discusses classification, boundary conditions, separation of variables, heat, waves, and Laplace equations. Includes solutions in Cartesian, cylindrical, and spherical coordinates with applications to diffusion and quantum systems.

While this final chapter deliberately stops just short of quantum chemistry, it provides a heuristic understanding of how the Schrödinger equation naturally emerges as a partial differential equation governing physical systems. In addition, several other PDEs of fundamental importance in physics and engineering are included to place the methods in a broader scientific context.

This chapter thus provides the final mathematical steppingstone to quantum mechanics, preparing the reader to understand the Schrödinger equation as a natural culmination of the differential equations studied throughout the book.

A large number of appendices complete the book.

Appendices

Appendix A: Character Tables of Common Molecular Point Groups

Provides standard character tables (C_{2v}, C_{3v}, D_{4h}, D_{6h}, T_d, O_h, etc.) with basis functions and symmetry properties relevant to spectroscopy and quantum chemistry.

Appendix B: Fourier Transform Pairs and Properties (Unitary Convention)

Summarizes standard transform pairs, derivative and shift rules, convolution theorem, and common spectra, using the same Fourier convention as the chapter.

Appendix C: Laplace Transform Table (Compatible With Book Convention)

Lists Laplace transform pairs, inverse transforms, and operational properties used for solving initial-value problems.

Appendix D: Special Functions: Definitions, Orthogonality, and Recurrence Relations

Contains concise reference formulas for Legendre, Hermite, Laguerre, and Bessel functions.

Appendix E: Differential Operators in Cartesian, Cylindrical, and Spherical Coordinates

Provides gradient, divergence, curl, and Laplacian expressions for the coordinate systems used in quantum mechanics and PDEs.

Appendix F: Vector Calculus Identities

Lists essential vector identities, including gradient, divergence, curl, Laplacian, and triple-product identities.

Appendix G: Linear Algebra Identities and Matrix Decompositions

Includes key matrix relations, eigenvalue identities, and decompositions such as LU, QR, SVD, and Cholesky.

Appendix H: Numerical Methods Reference Sheet

Summarizes root-finding, numerical integration, and ODE algorithms (Euler, RK2, RK4) with stability notes.

Appendix I: Physical Constants and Conversions

Provides commonly used constants and unit conversions for spectroscopy, thermodynamics, and quantum chemistry.

Appendix J: Extended Proofs and Derivations

Reserved for longer mathematical derivations that support—but are not required within—the main chapters.

Appendix K: Notations and Conventions

Covers notations used in Chapters 2–6 and 12.

30 January 2026

Prof. Rita Kakkar
Retd. Senior Professor
Department of Chemistry
University of Delhi
Delhi, India

Acknowledgments

This book has grown out of many years of teaching, learning, and reflection, and I owe more than I can fully acknowledge to the people who accompanied me along this journey.

I remain deeply grateful to my teachers, from whom I first learned not only the beauty of mathematics and chemistry, but also the discipline of clear thinking and intellectual honesty. My association with the University of Delhi has been central to my academic life, and I thank the institution for providing a stimulating environment in which teaching and research could flourish side by side.

My interactions with students at different stages—undergraduate teaching during my years at Miranda House and postgraduate and doctoral teaching at the Department of Chemistry, University of Delhi—have been an enduring source of intellectual engagement. Their questions, discussions, and critical responses repeatedly compelled me to rethink explanations and clarify concepts that might otherwise have remained implicit. Much of the structure, emphasis, and clarity of this book emerged from these classroom and research interactions.

I thank the editorial and production teams at Wiley for their patience and professionalism during the preparation of this manuscript.

I also thank the anonymous reviewers for their careful reading of the proposal and for several helpful suggestions.

Above all, I acknowledge the quiet but unwavering support of my family. Their understanding, encouragement, and belief in the value of this work sustained me through the long and often demanding process of writing this book.

Any errors that remain are entirely my own.

About the Companion Website

This book is accompanied by an instructor companion website, protected via Wiley Reader:

www.wiley.com/go/Mathematicalfoundations

To access the website materials, click on the link under Related Resources on the book's page on wiley.com. You can also find this by searching http://www.wiley.com for the ISBN 9781394368013.

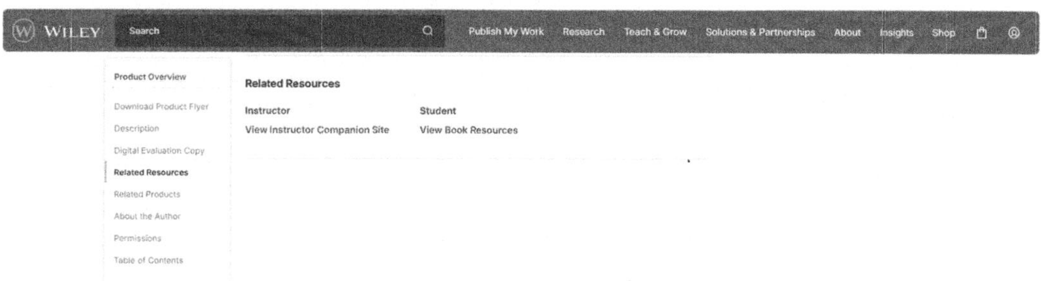

This website includes:

- Other - Instructor's solution manual

1

Vectors and Coordinate Systems

If one does not know to which port one is sailing, no wind is favorable.

—Seneca

Learning Outcomes

After completing this chapter, students should be able to:

1. **Distinguish** between scalars, vectors, and tensors, and explain their role in chemistry, physics, and engineering.
2. **Apply** vector operations (dot, cross, gradient, divergence, curl, Laplacian) to physical problems.
3. **Interpret** scalar and vector fields in terms of real-world applications (e.g., diffusion, electrostatics, molecular forces).
4. **Evaluate** line and surface integrals and **identify** conservative and nonconservative fields.
5. **Perform** coordinate transformations between Cartesian, spherical, and cylindrical systems, and **compute** Jacobians and vector operators in these systems.
6. **Relate** the mathematical framework of vectors and curvilinear coordinates to applications such as the hydrogen atom, rigid rotator, and fluid/mechanical systems.

1.1 Introduction

We are generally introduced to **vectors** in physics as quantities that have both magnitude and direction, in contrast to **scalar** quantities, which have only magnitude and no direction. Examples of vector quantities include velocity, dipole moment, force, and linear or angular momentum. The dipole moment is represented conventionally in chemistry as an arrow pointing from positive to negative charge, opposite to the physics convention.

Fact Box 1.1: Dipole Moment Conventions

Physicists point the arrow negative-to-positive (field direction); chemists reverse it, positive-to-negative (electron shift direction).

More formally, vectors are **tensors** of **rank** 1. Scalars are tensors of rank 0, since they can be represented solely by their magnitude (a zero-dimensional array). Vectors may be represented as one-dimensional arrays, while tensors of rank 2 may be represented as two-dimensional arrays, i.e., matrices. However, not all matrices are tensors. In three dimensions, the number of components of a tensor is rank 3. Thus, a scalar has one component, a vector has three components, and a rank-2 tensor has nine components.

Fact Box 1.2: What Makes a Tensor

A matrix is just an array of numbers. It becomes a tensor only when its components transform in a specific way under coordinate transformations. For a rank-2 tensor T in three dimensions, if we rotate coordinates, the components must transform as $T'_{ij} = \sum_{k,l} R_{ik} R_{jl} T_{kl}$, where R is the rotation matrix. This ensures physical laws work the same in all coordinate systems.

Application Box 1.1: Scalars and Vectors in Engineering

Stress is represented by tensors (not vectors), which illustrates why higher-rank tensors are needed in mechanics.

For mathematical manipulation, vectors must be expressed in a **coordinate system**. In many textbooks, scalar variables are written in italics and vectors in bold. In this chapter, for clarity and handwriting-friendliness, vectors are usually denoted with a right-pointing arrow above their symbol.

Fact Box 1.3: Vectors as Destiny

In physics, a vector combines two essential features: magnitude and direction. A force of 10 N is meaningless unless we also know the line along which it acts.

Interestingly, this mathematical necessity was understood intuitively by ancient philosophers long before vector algebra was formalized. The Stoic Philosopher Seneca famously wrote: "If a man knows not to which port he sails, no wind is favorable."

In this metaphor, the wind represents **magnitude**—the energy and potential to move. The port represents **direction**. Without the vector alignment of the two, the strongest wind (magnitude) is useless, or even dangerous.

Thus, the classroom aphorism—"A vector is not just magnitude; it is intention"—captures both the scientific and the human dimension of vectors.

In the Cartesian coordinate system, a vector is represented in terms of its projections on the three coordinate axes (Figure 1.1). The quantities a_x, a_y, and a_z are the **components** of the vector, and this three-dimensional space is called Euclidean space.

According to the Pythagorean theorem, the **magnitude** of a vector $\vec{a}$ is

$$|\vec{a}| = \sqrt{a_x^2 + a_y^2 + a_z^2}. \tag{1.1}$$

The magnitude or norm of a vector can also be denoted by $||\vec{a}||$. If we define three **unit vectors** in the Cartesian directions as $\hat{i}, \hat{j}, \hat{k}$, we may write

$$\vec{a} = a_x\hat{i} + a_y\hat{j} + a_z\hat{k}. \tag{1.2}$$

For brevity, a vector with components (a_x, a_y, a_z) may also be written directly in this shorthand form.

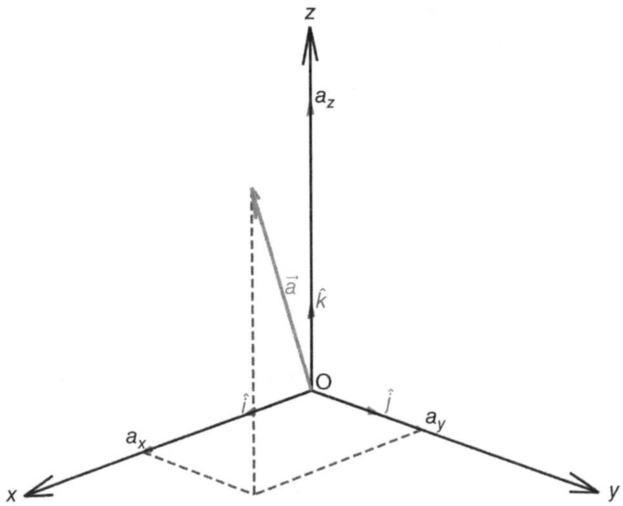

Figure 1.1 Vector components and unit vectors in Cartesian coordinates.

Example 1.1: Velocity and Speed

In a molecular dynamics (MD) simulation, a neon atom has an instantaneous **velocity vector**

$$\vec{v} = (1200, -900, 300)\ \text{m s}^{-1}.$$

a. Explain the difference between **velocity** and **speed** in the context of MD trajectories.
b. Calculate the atom's **speed**.
c. Would two atoms with identical speeds necessarily have identical kinetic energies? Explain.

Solution:

a. Velocity is a **vector**, specifying the magnitude and direction of motion.
Speed is the **magnitude** of the velocity and is a **scalar**, independent of direction.

b.

$$|\vec{v}| = \sqrt{1200^2 + (-900)^2 + 300^2} = 1530\ \text{m s}^{-1}.$$

c. Yes. The expression for kinetic energy is

$$E = \frac{1}{2}m|\vec{v}|^2,$$

so, atoms of the **same species** with the same speed have the same kinetic energy, regardless of direction.

Reminder: Dot and Cross Products

Fact Box 1.4: Work as a Dot Product

The dot product of force and displacement directly gives mechanical work.

Two fundamental operations with vectors are the **dot product** and the **cross product**.

The **dot product** (also called the scalar product) of two vectors $\vec{a}$ and $\vec{b}$ is

$$\vec{a} \cdot \vec{b} = |\vec{a}|\,|\vec{b}| \cos\theta,$$

where θ is the angle between them. The result is a scalar and represents the projection of one vector on another. A common application is the work done by a force:

$$W = \vec{F} \cdot \vec{r}.$$

! Red Flag 1.1: Units in Dot Product

The dot product result carries physical units—for example, work (joules) = force (Newton)· displacement (meter).

The **cross product** (or vector product) of two vectors $\vec{a}$ and $\vec{b}$ is

$$\vec{a} \times \vec{b} = |\vec{a}|\,|\vec{b}| \sin\theta\, \hat{n},$$

where $\hat{n}$ is a unit vector perpendicular to the plane of $\vec{a}$ and $\vec{b}$, with direction given by the **right-hand rule** (Figure 1.2).

In terms of the components, the vector product $\vec{a} \times \vec{b}$ can be written in the form of a determinant,

$$\vec{a} \times \vec{b} = \begin{vmatrix} \hat{i} & \hat{j} & \hat{k} \\ a_x & a_y & a_z \\ b_x & b_y & b_z \end{vmatrix} = \hat{i}(a_y b_z - a_z b_y) + \hat{j}(a_z b_x - a_x b_z) + \hat{k}(a_x b_y - a_y b_x).$$

Applications include **torque**,

$$\vec{\tau} = \vec{r} \times \vec{F},$$

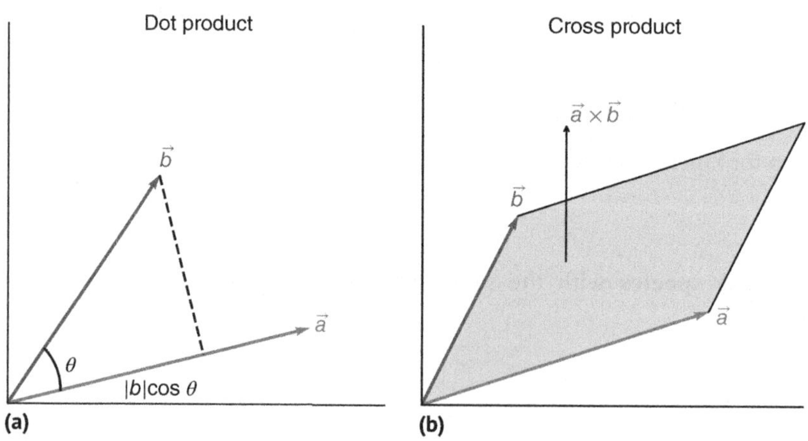

Figure 1.2 Geometric interpretation of dot and cross products: (a) the dot product projects one vector onto another; (b) the cross product yields a vector perpendicular to the original two, with magnitude equal to the area of the parallelogram they span.

and **angular momentum**,

$$\vec{L} = \vec{r} \times \vec{p}.$$

The Triple Scalar Product

The triple scalar product of three vectors $\vec{a}, \vec{b}, \vec{c}$ is defined as $\vec{a} \cdot \left(\vec{b} \times \vec{c}\right)$. It may also be written as $\left(\vec{a} \times \vec{b}\right) \cdot \vec{c}$, so the placement of brackets is immaterial. In determinant notation, this is represented as follows:

$$\vec{a} \cdot \vec{b} \times \vec{c} = \begin{vmatrix} a_x & a_y & a_z \\ b_x & b_y & b_z \\ c_x & c_y & c_z \end{vmatrix}$$

These brief reminders suffice.

Application Box 1.2: Torque in Molecular Rotation

The torque on a diatomic molecule due to an external field is given by a cross product, $\vec{\tau} = \vec{r} \times \vec{F}$.

Red Flag 1.2: Cross Product Sign

The cross product is **not commutative**: $\vec{a} \times \vec{b} = -\left(\vec{b} \times \vec{a}\right)$.

1.2 Vector Spaces and Linear Independence

So far, we have considered vectors mainly as arrows in Euclidean three-dimensional space. At a more general level, vectors are defined as elements of a **vector space**, which is a collection of objects that can be added together and multiplied by scalars, while still remaining within the space. The Euclidean space with its Cartesian vectors is only one example. In later chapters we shall see more abstract vector spaces, such as Hilbert spaces of functions, which play a central role in quantum mechanics (QM).

An important concept in vector spaces is **linear independence**. A set of vectors $\vec{a}_1, \vec{a}_2, \dots, \vec{a}_n$ is said to be linearly independent if no vector in the set can be expressed as a linear combination of the others. If such a relation does exist, the vectors are linearly dependent. Geometrically, two dependent vectors lie along the same line, three dependent vectors lie in the same plane, and so on.

Linear independence can be **tested using determinants**. In three dimensions, if the determinant formed by taking three vectors as columns (or rows) is zero, the vectors are linearly dependent; otherwise, they are independent.

Fact Box 1.5: Linear Independence in Quantum Chemistry

Atomic orbitals in a basis set must be linearly independent. If they are not, computations become unstable and redundant.

Example 1.2: Linear Independence

Determine whether the following sets of vectors are linearly independent:

a. (1, 2, 3), (0, 1, −1), (2, −1, 0)
b. (2, 3, 5), (4, 6, 10), (1, −2, 1)

Solution:

a. Construct the determinant

$$\begin{vmatrix} 1 & 0 & 2 \\ 2 & 1 & -1 \\ 3 & -1 & 0 \end{vmatrix} = (1)(0-1) - (0)(0+3) + (2)(-2-3) = -1 - 0 - 10 = -11 \neq 0.$$

Hence, the vectors are linearly independent.

b. Construct the determinant

$$\begin{vmatrix} 2 & 4 & 1 \\ 3 & 6 & -2 \\ 5 & 10 & 1 \end{vmatrix} = (2)(6+20) - (4)(3+10) + (1)(30-30) = 52 - 52 + 0 = 0.$$

Therefore, these vectors are linearly dependent. In fact, the second column vector is twice the first, so the third can be expressed as a linear combination of the first two.

Red Flag 1.3: Linear Dependence Test

A zero determinant means the vectors are dependent, not that they vanish.

The idea of linear independence extends naturally to higher and even infinite-dimensional spaces. In fact, **functions themselves can be regarded as vectors**. This is the basis of QM, where the solutions of the Schrödinger equation (wavefunctions) form a vector space known as a **Hilbert space**. Just as geometric vectors must be linearly independent to span three-dimensional space, basis functions such as atomic orbitals must be linearly independent to span the Hilbert space of solutions of the Schrödinger equation. We return to this formal treatment of vector spaces in Chapter 2.

1.3 Vector Calculus and Fields

In many physical and chemical problems, we deal not just with vectors but with fields. A **scalar field** assigns a scalar quantity to every point in space, for example, the temperature distribution $T(x, y, z)$ in a furnace, or the electrostatic potential $\varphi(x, y, z)$ in a molecule. A **vector field** assigns a vector quantity to every point, such as the velocity field of a fluid or the electric field in space.

To analyze such fields, we make use of the **del operator**, denoted by ∇:

$$\nabla = \hat{i}\frac{\partial}{\partial x} + \hat{j}\frac{\partial}{\partial y} + \hat{k}\frac{\partial}{\partial z}.$$

When applied to scalar or vector functions, this operator generates the fundamental quantities of vector calculus: the gradient, divergence, curl, and Laplacian.

1.3.1 Gradient

The **gradient** of a scalar field $\varphi(x, y, z)$ is defined as

$$\nabla\varphi = \left(\frac{\partial\varphi}{\partial x}\right)\hat{i} + \left(\frac{\partial\varphi}{\partial y}\right)\hat{j} + \left(\frac{\partial\varphi}{\partial z}\right)\hat{k}.$$

The gradient is itself a vector and points in the direction of the steepest increase of the scalar field. Its magnitude gives the rate of this increase.

Fact Box 1.6: Gradient and Steepest Ascent

When hiking on a mountain, the **gradient** of the terrain tells you about the direction in which the slope rises most steeply. In chemistry, this same idea applies to **potential energy surfaces**: the gradient of the potential gives the **force** acting on atoms. Just as a hiker follows the gradient uphill, molecules follow the gradient downhill to reach stable configurations.

Chemical examples:

- A **concentration gradient** drives diffusion of reactants in solution.
- A **potential energy surface** in quantum chemistry has gradients that correspond to forces acting on nuclei during molecular motion.

Example 1.3: Gradient

Find a unit vector normal to the surface $x^2 + y^2 - z = 0$ at the point $P(1, 1, 1)$.

Solution:

We define

$$\varphi(x, y, z) = x^2 + y^2 - z.$$

The gradient of φ is

$$\nabla\varphi = (2x,\ 2y,\ -1).$$

At the point $P(1, 1, 1)$,

$$\nabla\varphi = (2,\ 2,\ -1).$$

By definition, the gradient vector $\nabla\varphi$ is **perpendicular (normal) to the surface** $\varphi(x, y, z) = 0$.

To normalize the vector, we divide by the magnitude, given by Equation (1.1). Therefore, the required unit normal vector is

$$\frac{1}{\sqrt{2^2+2^2+(-1)^2}}(2,2,-1)=\frac{1}{3}(2,2,-1).$$

Fact Box 1.7: Gradient and Normal Vectors

The gradient of a scalar field,

$$\nabla\varphi(x,y,z),$$

is always **perpendicular (normal) to the level surface** defined by $\varphi(x, y, z) = \text{constant}$.

- ◊ **Geometric picture:** On a topographic map, the contour lines mark constant altitude. The gradient of the height function points directly uphill—always perpendicular to the contour line.
- ◊ **In 3D surfaces:** For a surface like $x^2 + y^2 - z = 0$, the gradient at any point gives the **normal vector** to that surface.

This is why, in Example 1.3, we could immediately identify $\nabla\varphi = (2, 2, -1)$ as the normal direction at point $P(1, 1, 1)$. Normalizing it gave the **unit normal vector**.

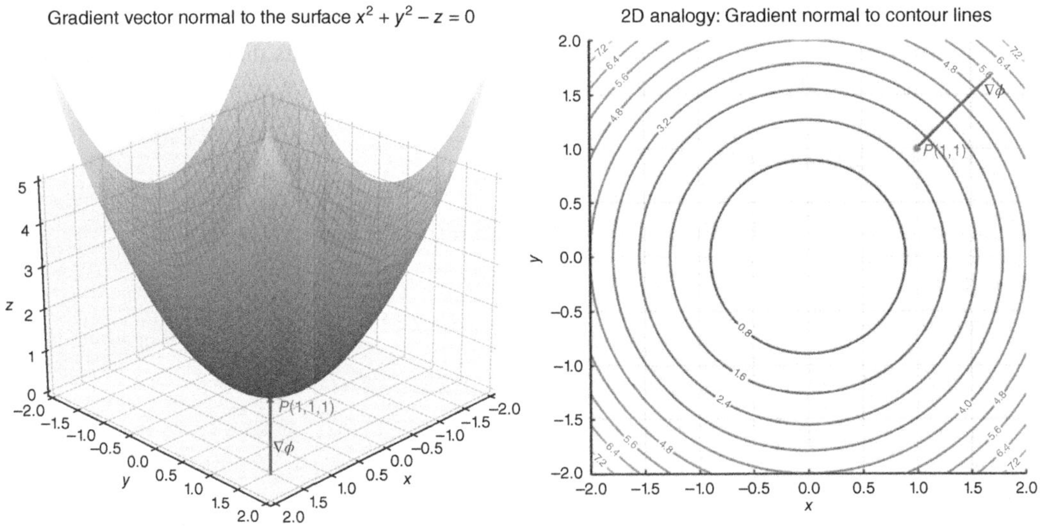

Figure for Example 1.3:
Left: The paraboloid surface $x^2 + y^2 = z$ is shown in three dimensions:

- The point $P(1, 1, 1)$ is marked.
- The gradient $\nabla\varphi$ at P, showing the **normal direction** to the surface.

Right: The **2D analogy figure** for Example 1.3. A **contour map** of $\varphi(x, y) = x^2 + y^2$:

- Point $P(1, 1)$ marked.
- The **gradient** $\nabla\varphi$, pointing perpendicular to the contour line through P.

This pairs neatly with the 3D surface figure—together they reinforce that the **gradient is normal to level curves/surfaces.**

Fact Box 1.8: Gradient in Chemistry

A stable geometry occurs where $\nabla V = 0$ — zero force on every nucleus, the condition geometry optimizations search for.

Red Flag 1.4: Gradient Direction

The gradient points in the direction of steepest *ascent*. The negative gradient points toward the steepest *descent* (and often toward a local minimum).

Force is the negative gradient of the potential energy, V, i.e., $\vec{F} = -\nabla V$.

Example 1.4: Potential Energy and Force

Suppose a particle experiences a potential energy $V(x, y, z) = k_x x^2 + k_y y^2 + k_z z^2$. Find the force acting on the particle.

Solution:

The force is the negative gradient:

$$\vec{F} = -\nabla V = (-2k_x x,\ -2k_y y,\ -2k_z z).$$

1.3.2 Divergence

The **divergence** of a vector field $\vec{A} = (A_x, A_y, A_z)$ is defined as

$$\nabla \cdot \vec{A} = \frac{\partial A_x}{\partial x} + \frac{\partial A_y}{\partial y} + \frac{\partial A_z}{\partial z}.$$

It is a scalar quantity that measures the net **flux** of the field "flowing out" of an infinitesimal volume.

Fact Box 1.9: Divergence and Conservation

The **divergence** of a vector field tells us whether something is being created or destroyed at a point. In fluid dynamics, positive divergence means a **source** (like water bubbling from a spring), while negative divergence means a **sink** (like water draining in a plughole). In electrostatics, Gauss' law connects the divergence of the electric field directly to the presence of **charges**.

Applications:

- In electrostatics, Gauss' law relates the divergence of the electric field to charge density.
- In diffusion processes, the divergence of the flux gives the rate of change of concentration.

Application Box 1.3: Divergence in Fluid Dynamics

Zero divergence indicates incompressible flow.

Red Flag 1.5: Misinterpreting Divergence

Zero divergence ≠ zero vector field.

Example 1.5: Divergence in Fluid Dynamics

Suppose the velocity field in a reactor is $\vec{v} = (x, -2y, z)$. Find the divergence at point (1, −2, 1).

Solution:

$$\nabla \cdot \vec{v} = \frac{\partial x}{\partial x} + \frac{\partial(-2y)}{\partial y} + \frac{\partial z}{\partial z} = 1 - 2 + 1 = 0.$$

Thus, the flow is incompressible at this point.

1.3.3 Curl

The **curl** of a vector field $\vec{A}$ is defined as

$$\nabla \times \vec{A} = \begin{vmatrix} \hat{i} & \hat{j} & \hat{k} \\ \frac{\partial}{\partial x} & \frac{\partial}{\partial y} & \frac{\partial}{\partial z} \\ A_x & A_y & A_z \end{vmatrix}.$$

The curl is itself a vector that measures the tendency of the field to circulate about a point.

Fact Box 1.10: Curl and Rotations

The **curl** measures the local rotation of a field. An everyday analogy is a **paddlewheel dropped in a stream**: if the wheel spins, the velocity field has curl at that point. In quantum mechanics, curl connects to **angular momentum operators**; in electromagnetism, the curl of the magnetic vector potential gives the magnetic field.

Faraday's law: curl of $\vec{E}$ relates to time-varying magnetic fields.

⚠ Red Flag 1.6: Curl of Gradient

Always $\nabla \times \nabla\varphi = \vec{0}$.

Applications:

- In electromagnetism, the curl of the electric field is related to the rate of change of the magnetic field (Faraday's law).
- In QM, angular momentum operators involve curl-like operations (Figure 1.3).

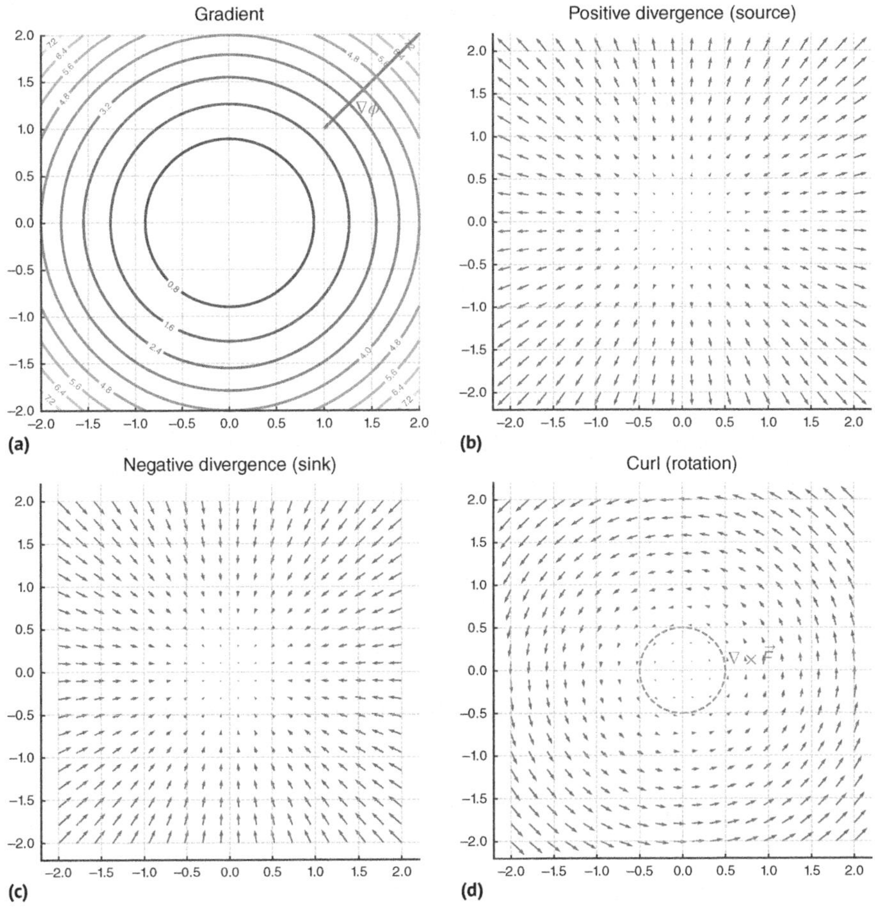

Figure 1.3 Visualizing gradient, divergence, and curl.

Figure 1.3 shows:

a. **Gradient of a Scalar Field:** The gradient $\nabla\varphi$ points in the direction of steepest ascent of the scalar field φ.
b. **Divergence (Positive–Source):** Positive divergence: a "source" where field lines originate. Net outward flux is positive.
c. **Divergence (Negative–Sink):** Negative divergence: a "sink" where field lines terminate. Net outward flux is negative.
d. **Curl (Rotation):** Nonzero curl: the field has a rotational component. The curl vector is perpendicular to the plane of rotation.

Fact Box 1.11: Differential Forms—One Operator to Rule Them All

- **Core Idea:**
  Gradient, curl, and divergence are not three separate operations. They are all special cases of a single operator, the **exterior derivative** d.
- **Dictionary:**

Vector Calculus	Differential Forms
Scalar field φ	0-form φ
Gradient $\nabla\varphi$	$d\varphi$ (derivative of a 0-form)
Curl $\nabla \times \vec{A}$	$d\vec{A}$ (derivative of a 1-form)
Divergence $\nabla \cdot \vec{A}$	$d\vec{A}$ (derivative of a 2-form)

- **Key Consequence:**
  The famous vector identities

$$\nabla \times (\nabla\varphi) = 0, \qquad \nabla \cdot \left(\nabla \times \vec{A}\right) = 0$$

become one universal truth:

$$d^2\omega = 0 \text{ (the boundary of a boundary is empty).}$$

Unified Theorem:
Gauss, Green, and Stokes theorems are all cases of the **generalized Stokes theorem**:

$$\int_M d\omega = \int_{\partial M} \omega.$$

Why Bother? This elegant formalism shows the deep unity of calculus and provides the natural language for modern physics, from electromagnetism to differential geometry.

If $\nabla \times \vec{F} = \vec{0}$, the field $\vec{F}$ is said to be a **conservative field**, i.e., it can be derived from a **scalar potential**.

Example 1.6: Conservative Fields

Determine whether the vector field $\vec{F} = (yz, xz, xy)$ is conservative.

Solution:

$$\nabla \times \vec{F} = \begin{vmatrix} \hat{i} & \hat{j} & \hat{k} \\ \partial_x & \partial_y & \partial_z \\ yz & xz & xy \end{vmatrix}$$

$$= \hat{i}(\partial(xy)/\partial y - \partial(xz)/\partial z) - \hat{j}(\partial(xy)/\partial x - \partial(yz)/\partial z) + \hat{k}(\partial(xz)/\partial x - \partial(yz)/\partial y)$$

$$= \hat{i}(x - x) - \hat{j}(y - y) + \hat{k}(z - z) = (0, 0, 0) = 0.$$

Since the curl vanishes, the field is conservative.

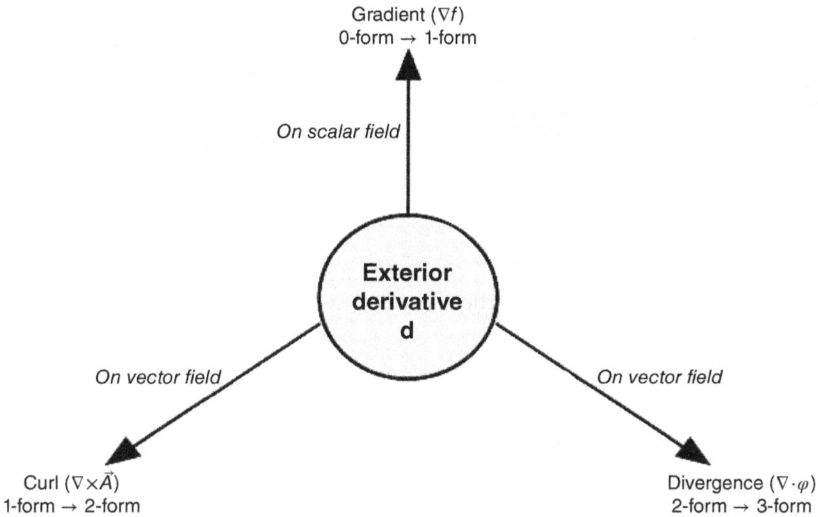

Gradient, curl, and divergence are "shadows" of a single operator—the exterior derivative d.

1.3.4 Laplacian

The **Laplacian** is a scalar operator obtained by applying divergence to the gradient:

$$\nabla \cdot \nabla\varphi = \nabla^2\varphi = \frac{\partial^2\varphi}{\partial x^2} + \frac{\partial^2\varphi}{\partial y^2} + \frac{\partial^2\varphi}{\partial z^2}.$$

It measures how a function differs from its average value around a point.

Fact Box 1.12: Laplacian in Quantum Mechanics

The Laplacian operator, ∇^2, may seem like a dry second derivative. But in the Schrödinger equation it directly represents the **kinetic energy of particles**. In heat conduction, it describes how temperature smooths out over time, while in electrostatics it appears in **Laplace's equation**. Thus, the Laplacian links energy, flow, and equilibrium across physics and chemistry.

Applications:

- In QM, the **kinetic energy operator** in the Schrödinger equation involves the Laplacian.
- In diffusion theory, the Laplacian governs the spreading of concentration or temperature fields.

Application Box 1.4: Laplacian in Quantum Mechanics

Sharply curved wavefunctions (large $\nabla^2\psi$) correspond to high kinetic energy — this is why confining a particle to a smaller region always raises its kinetic energy.

Red Flag 1.7: Laplacian Sign in QM

The Schrödinger equation has $-\frac{\hbar^2}{2m}\nabla^2$, not $+$.

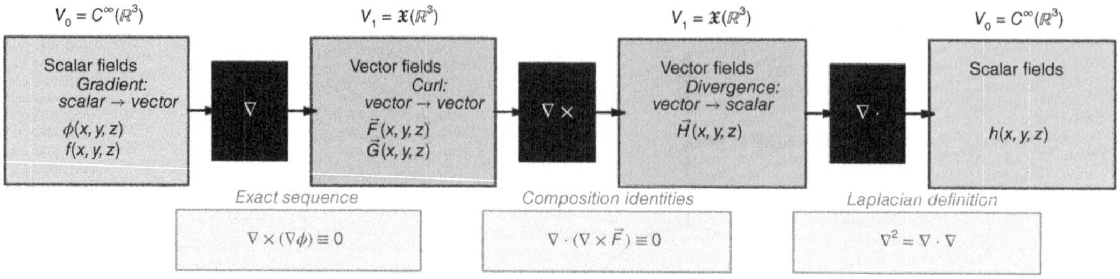

Figure 1.4 Vector operators as linear maps between function spaces.
This figure shows:

- **Gradient** ∇ maps scalar fields → vector fields.
- **Curl** $\nabla\times$ maps vector fields → vector fields.
- **Divergence** $\nabla\cdot$ maps vector fields → scalar fields.
- **Fundamental identities:** $\nabla \times \nabla\varphi = \vec{0},\ \nabla \cdot (\nabla \times \vec{F}) = 0, \nabla^2 = \nabla \cdot \nabla.$

Figure 1.4 shows the vector operators as linear maps between function spaces.

This completes the essentials of vector calculus for scalar and vector fields. These operators—gradient, divergence, curl, and Laplacian—will reappear throughout chemistry and physics, from diffusion and electrostatics to QM and spectroscopy.

1.4 Line and Surface Integrals (Conservative Fields)

Often, physical quantities are accumulated along a path or across a surface. This gives rise to **line integrals** and **surface integrals**, which are fundamental tools in chemistry, physics, and engineering.

The **line integral** of a vector field $\vec{F}$ along a curve C is defined as

$$\int_C \vec{F} \cdot d\vec{r},$$

where $d\vec{r}$ is an infinitesimal displacement along the curve. Physically, this represents the work done by a force field in moving a particle along the path C.

If the vector field is the gradient of a scalar function, $\vec{F} = \nabla\varphi$, then the integral depends only on the end points:

$$\int_A^B \nabla\varphi \cdot d\vec{r} = \varphi(B) - \varphi(A).$$

Such a field is called **conservative**, and the function φ is its **potential**. For conservative forces, the work done is **path independent** (Figure 1.5).

Applications:

- In mechanics, the work–energy theorem follows directly from line integrals of force.
- In electrostatics, the electric field is conservative, and the potential difference between two points is given by a line integral of the field.
- In thermodynamics, state functions such as internal energy, enthalpy, and entropy are defined through path-independent integrals.

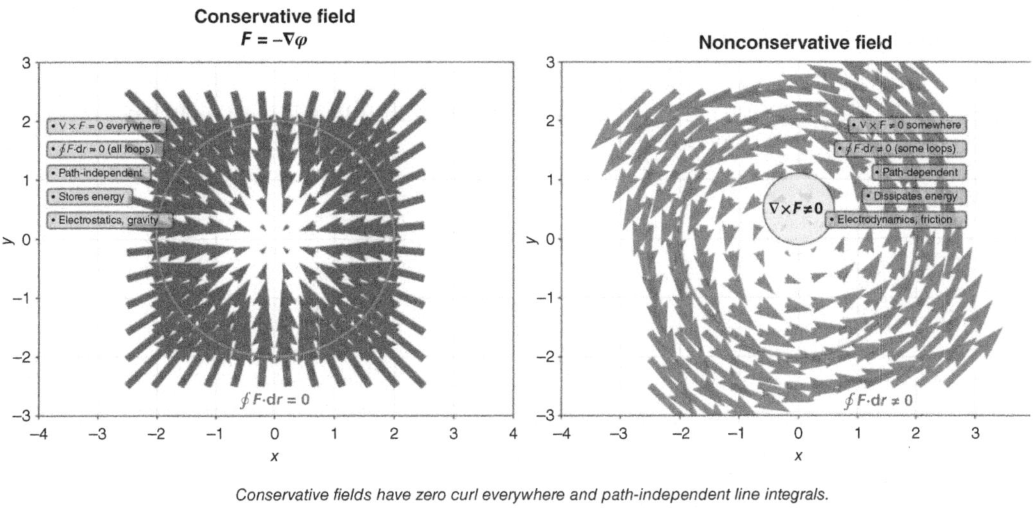

Figure 1.5 Conservative and nonconservative vector fields.

Fact Box 1.13: Path Independence

Conservative fields → path independence.

Fact Box 1.14: Equivalent Conditions for Conservative Fields

For a vector field $\vec{F}$ in a simply connected region, the following are equivalent:

1. $\nabla \times \vec{F} = \vec{0}$ everywhere (irrotational).
2. $\oint_C \vec{F} \cdot d\vec{r} = 0$ for any closed path.
3. $\int_A^B \vec{F} \cdot d\vec{r}$ is path-independent.
4. $\vec{F} = -\nabla\varphi$ for some scalar potential φ.

In simply connected regions, any one condition implies all the others.

Application Box 1.5: Thermodynamic State Functions

Energy, enthalpy, and entropy are path-independent integrals, unlike heat and work.

Red Flag 1.8: Path Independence

Only conservative forces qualify; students often generalize incorrectly.

Example 1.7: Work–Energy Theorem

Show that the work done on a particle by a resultant force during a displacement from A to B is equal to the change in its kinetic energy.

Solution:

The work is

$$W = \int_A^B \vec{F} \cdot d\vec{r}.$$

Using Newton's law, $\vec{F} = m\frac{d\vec{v}}{dt}$, and $d\vec{r} = \vec{v}\,dt$, we obtain

$$W = \int_A^B m\frac{d\vec{v}}{dt} \cdot \vec{v}\,dt = \frac{1}{2}m(v_B^2 - v_A^2).$$

Thus, the work equals the change in kinetic energy, which is the **work–energy theorem**.

Surface integrals are also widely used. The **flux** of a vector field $\vec{A}$ through a surface S is given by

$$\iint_S \vec{A} \cdot d\vec{S},$$

where $d\vec{S}$ is an oriented surface element. This measures how much of the field passes through the surface.

Applications:

- In fluid mechanics, the surface integral of velocity gives the volumetric flow rate.
- In electromagnetism, Gauss's law relates the surface integral of the electric field to the total charge enclosed.
- In engineering, heat transfer across a surface is expressed as the flux of the heat current density.

Figure 1.6 is a visual representation of line and surface integrals.

These integral formulations are connected by powerful theorems—Green's, Gauss's, and Stokes's—which relate line integrals, surface integrals, and volume integrals. These are encountered when solving problems in electromagnetism, QM, and fluid dynamics.

1.4.1 Fundamental Theorems of Vector Calculus

The line and surface integrals are connected by powerful theorems that are essential in physics and engineering:

Gauss's Divergence Theorem

$$\iiint_V (\nabla \cdot \vec{F}) d\tau = \oiint_S \vec{F} \cdot d\vec{S}$$

The flux through a closed surface equals the volume integral of the divergence inside.

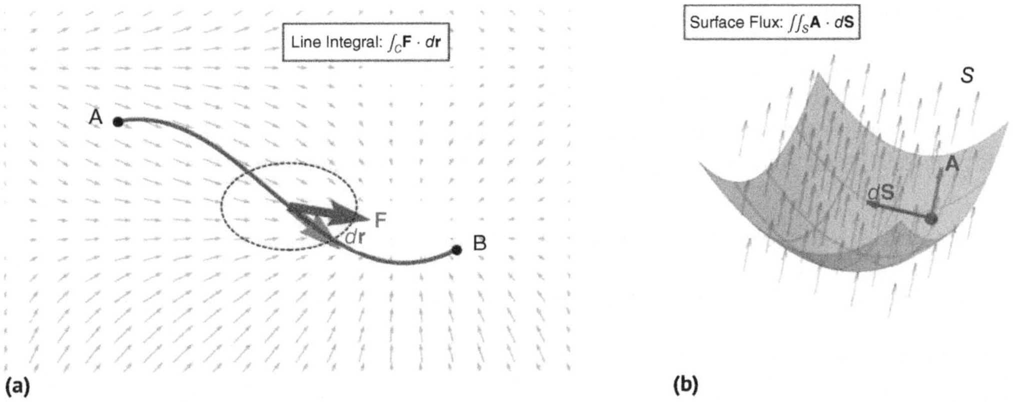

Figure 1.6 Line and surface integrals: (a) a line integral sums the tangential component of a field along a path; (b) a surface integral (flux) sums the normal component of a field through a surface.

Stokes's Theorem

$$\iint_S (\nabla \times \vec{F}) \cdot d\vec{S} = \oint_C \vec{F} \cdot d\vec{r}$$

The circulation around a closed loop equals the flux of curl through any surface bounded by the loop.

Applications:

- Gauss's theorem is the mathematical basis for Gauss's law in electromagnetism.
- Stokes's theorem underlies Faraday's law of induction.
- Both are crucial for deriving conservation laws (mass, energy, charge).

1.4.2 The Helmholtz Decomposition

A fundamental result in vector calculus is the **Helmholtz decomposition**, which states that any sufficiently smooth and rapidly decaying vector field $\vec{F}$ in three dimensions can be decomposed into the sum of an irrotational (curl-free) field and a solenoidal (divergence-free) field:

$$\vec{F} = -\nabla\varphi + \nabla \times \vec{A}$$

where:

- φ is a scalar potential
- $\vec{A}$ is a vector potential

This decomposition is unique given appropriate boundary conditions.

Fact Box 1.15: Helmholtz Decomposition

The Helmholtz theorem says that any vector field can be split into two parts: one that comes from a **potential** (like gravity) and one that is purely **rotational** (like a whirlpool). This is why physicists can often classify forces as "conservative" or "solenoidal." The theorem elegantly unifies gradients, divergences, and curls in one framework.

Physical Significance:

- In electromagnetism, this separates the electric field into conservative (electrostatic) and non-conservative (induced) components.
- In fluid dynamics, it separates flow into compressional and rotational parts.
- It provides the mathematical foundation for potential theory in physics.

Example 1.8: Helmholtz Decomposition of a Vector Field

Show that the field $\vec{F} = (2x + y, x - 3y, z)$ can be decomposed according to Helmholtz.

Step 1: Verify Helmholtz Theorem Conditions

First, let us check if $\vec{F}$ is sufficiently well behaved (smooth and decays rapidly). Since it is a polynomial vector field, it is smooth everywhere, so Helmholtz decomposition applies.

Let us calculate the divergence and curl:

Divergence:

$$\nabla \cdot \vec{F} = \frac{\partial}{\partial x}(2x + y) + \frac{\partial}{\partial y}(x - 3y) + \frac{\partial}{\partial z}(z) = 2 - 3 + 1 = 0.$$

Curl:

$$\nabla \times \vec{F} = \begin{vmatrix} \hat{i} & \hat{j} & \hat{k} \\ \dfrac{\partial}{\partial x} & \dfrac{\partial}{\partial y} & \dfrac{\partial}{\partial z} \\ 2x + y & x - 3y & z \end{vmatrix}$$

$$= \hat{i}\left(\frac{\partial z}{\partial y} - \frac{\partial(x - 3y)}{\partial z}\right) - \hat{j}\left(\frac{\partial z}{\partial x} - \frac{\partial(2x + y)}{\partial z}\right) + \hat{k}\left(\frac{\partial(x - 3y)}{\partial x} - \frac{\partial(2x + y)}{\partial y}\right)$$

$$= \hat{i}(0 - 0) - \hat{j}(0 - 0) + \hat{k}(1 - 1) = (0, 0, 0) = 0.$$

So, we have:

- $\nabla \cdot \vec{F} = 0$ (solenoidal/divergence-free)
- $\nabla \times \vec{F} = \vec{0}$ (irrotational/curl-free)

Step 2: Find the Scalar Potential φ

Since $\nabla \times \vec{F} = \vec{0}$, the field is conservative and we can write $\vec{F} = -\nabla\varphi$.

This gives us the system:

$$-\frac{\partial \varphi}{\partial x} = 2x + y, \tag{1}$$

$$-\frac{\partial \varphi}{\partial y} = x - 3y, \tag{2}$$

$$-\frac{\partial \varphi}{\partial z} = z. \tag{3}$$

From (1):

$$\varphi(x, y, z) = -x^2 - xy + f(y, z).$$

From (2):

$$\frac{\partial \varphi}{\partial y} = -x + \frac{\partial f}{\partial y} = -x + 3y$$

$$\Rightarrow \frac{\partial f}{\partial y} = 3y \Rightarrow f(y, z) = \frac{3}{2}y^2 + g(z).$$

So: $\varphi(x, y, z) = -x^2 - xy + \frac{3}{2}y^2 + g(z)$.

From (3):

$$\frac{\partial \varphi}{\partial z} = g'(z) = -z \Rightarrow g(z) = -\frac{1}{2}z^2 + C.$$

Final scalar potential:

$$\varphi(x, y, z) = -x^2 - xy + \frac{3}{2}y^2 - \frac{1}{2}z^2 + C.$$

Step 3: Verify the Decomposition

Let us verify that $\vec{F} = -\nabla\varphi$:

$$-\nabla\varphi = -\left(\frac{\partial \varphi}{\partial x}, \frac{\partial \varphi}{\partial y}, \frac{\partial \varphi}{\partial z}\right),$$

$$\frac{\partial \varphi}{\partial x} = -2x - y \Rightarrow -\frac{\partial \varphi}{\partial x} = 2x + y\checkmark,$$

$$\frac{\partial \varphi}{\partial y} = -x + 3y \Rightarrow -\frac{\partial \varphi}{\partial y} = x - 3y\checkmark,$$

$$\frac{\partial \varphi}{\partial z} = -z \Rightarrow -\frac{\partial \varphi}{\partial z} = z\checkmark.$$

Since $\nabla \cdot \vec{F} = 0$ and $\nabla \times \vec{F} = \vec{0}$, we can choose $\vec{A} = \vec{0}$ for the vector potential part.

Step 4: Final Helmholtz Decomposition

The Helmholtz decomposition is

$$\vec{F} = -\nabla\varphi + \nabla \times \vec{A},$$

where:

- $\varphi(x, y, z) = -x^2 - xy + \frac{3}{2}y^2 - \frac{1}{2}z^2$,
- $\vec{A} = \vec{0}$.

Or equivalently,

$$\vec{F} = (2x + y, x - 3y, z) = -\nabla\left(-x^2 - xy + \frac{3}{2}y^2 - \frac{1}{2}z^2\right) + \nabla \times \vec{0}.$$

Step 5: Physical Interpretation

This is a special case of Helmholtz decomposition where:

- The field is **both irrotational and solenoidal**.
- Therefore, it can be represented entirely as $-\nabla\varphi$ (the conservative/irrotational part).
- The solenoidal part $\nabla \times \vec{A}$ is zero.
- This is characteristic of a **Laplacian field** $(\nabla^2\varphi = 0)$.

Indeed, we can verify

$$\nabla^2\varphi = \frac{\partial^2\varphi}{\partial x^2} + \frac{\partial^2\varphi}{\partial y^2} + \frac{\partial^2\varphi}{\partial z^2} = -2 + 3 - 1 = 0.$$

✅ **The field $\vec{F}$ is successfully decomposed according to the Helmholtz theorem.**

Summary:

Final Answer: The field $\vec{F} = (2x + y, x - 3y, z)$ is both irrotational $(\nabla \times \vec{F} = \vec{0})$ and solenoidal $(\nabla \cdot \vec{F} = 0)$. Its Helmholtz decomposition is therefore particularly simple:

$$\vec{F} = -\nabla\varphi + \nabla \times \vec{A},$$

with scalar potential $\varphi(x, y, z) = -x^2 - xy + \frac{3}{2}y^2 - \frac{1}{2}z^2$ and vector potential $\vec{A} = \vec{0}$. This represents the special case of a Laplacian field where the entire field is conservative.

Figure 1.7 illustrates the Helmholtz decomposition of a vector field.

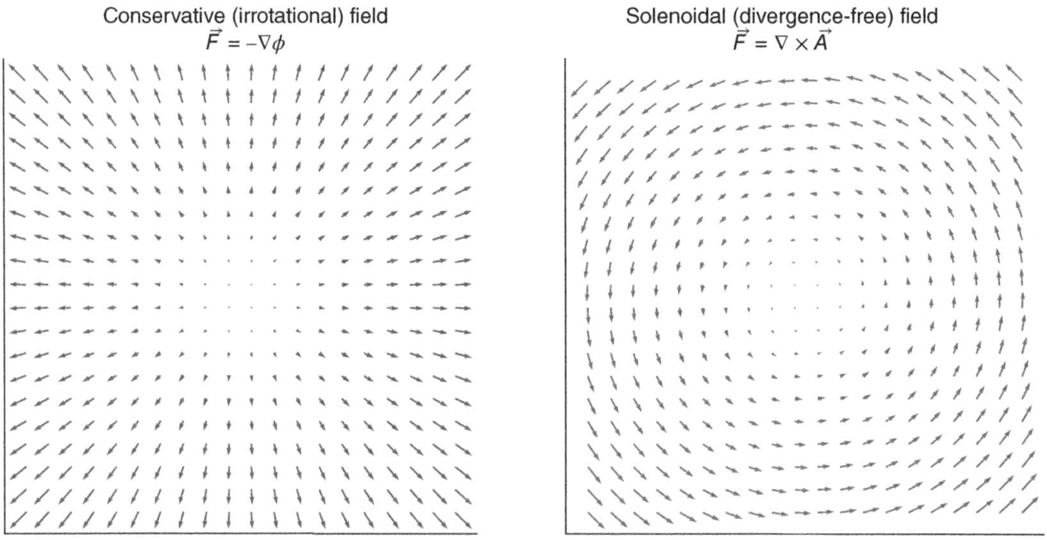

Figure 1.7 Helmholtz decomposition of a vector field.
It shows:

- **Left panel:** Conservative (irrotational) field $\vec{F} = -\nabla\varphi$, with arrows radiating outward.
- **Right panel:** Solenoidal (divergence-free) field $\vec{F} = \nabla \times \vec{A}$, with arrows circulating around the origin.

Helmholtz decomposition theorem

Any vector field F can be decomposed as:

$$\boldsymbol{F} = -\nabla\varphi + \nabla\times\boldsymbol{A} + \boldsymbol{H}$$

$-\nabla\varphi$	Irrotational (curl-free) part *Conservative field*	
$\nabla\times\boldsymbol{A}$	Solenoidal (div-free) part *Incompressible field*	
$\boldsymbol{H}$	Harmonic part *Both curl-free and div-free*	

Properties:

- $\nabla\times(-\nabla\varphi) = 0$
- $\nabla\cdot(\nabla\times A) = 0$
- $\nabla^2 H = 0$
- Decomposition is unique
- Applies to smooth, decaying fields

Applications:

Electromagnetism: $E = -\nabla V$ (electrostatic) + ... (induced)

Fluid dynamics: $u = \nabla\varphi$ (potential flow) + $\nabla\times\psi$ (vortical)

Quantum mechanics: vector potential formulations

1.5 Orthogonal Curvilinear Coordinates

Until now, we have worked with **Cartesian coordinates**, which are convenient when boundary conditions are rectangular. However, many problems in chemistry, physics, and engineering possess natural symmetries that are better expressed in other coordinate systems. For example, **spherical polar coordinates** are useful for atoms and elliptical coordinates for diatomic molecules, while **cylindrical coordinates** are appropriate for linear systems such as polymers or conducting wires.

When we move from Cartesian to a new set of variables, the differentials transform with a scaling factor given by the **Jacobian** determinant. This is why we cannot just write $dx\,dy\,dz = dr\,d\theta\,d\phi$, but instead need the Jacobian factor.

Let an arbitrary orthogonal coordinate system be represented by (u_1, u_2, u_3). A small displacement in this system is

$$d\vec{r} = h_1\,du_1\,\hat{e}_1 + h_2\,du_2\,\hat{e}_2 + h_3\,du_3\,\hat{e}_3,$$

where $\hat{e}_i$ are mutually perpendicular unit vectors, and the coefficients h_i are called **scale factors**. The squared distance is

$$ds^2 = h_1^2\,du_1^2 + h_2^2\,du_2^2 + h_3^2\,du_3^2,$$

and the **volume element** is

$$d\tau = h_1 h_2 h_3\,du_1 du_2 du_3.$$

The scale factors are related to the Jacobian of the **transformation** between Cartesian and curvilinear coordinates:

$$J(u_1, u_2, u_3) = \frac{\partial(x, y, z)}{\partial(u_1, u_2, u_3)}.$$

Thus, the volume element may also be expressed in terms of the Jacobian determinant.

Fact Box 1.16: The Jacobian in Coordinate Transformations

Suppose we change from Cartesian coordinates x, y, z to a new set of coordinates (u_1, u_2, u_3). The transformation is given by the functions

$$x = x(u_1, u_2, u_3), y = y(u_1, u_2, u_3), z = z(u_1, u_2, u_3).$$

The **Jacobian matrix** of this transformation is

$$J = \begin{bmatrix} \frac{\partial x}{\partial u_1} & \frac{\partial x}{\partial u_2} & \frac{\partial x}{\partial u_3} \\ \frac{\partial y}{\partial u_1} & \frac{\partial y}{\partial u_2} & \frac{\partial y}{\partial u_3} \\ \frac{\partial z}{\partial u_1} & \frac{\partial z}{\partial u_2} & \frac{\partial z}{\partial u_3} \end{bmatrix}.$$

The determinant of the Jacobian,

$$\left| \frac{\partial(x, y, z)}{\partial(u_1, u_2, u_3)} \right| = \det(J),$$

tells us how **volume elements** transform under the change of coordinates:

$$dx\, dy\, dz = \left| \frac{\partial(x, y, z)}{\partial(u_1, u_2, u_3)} \right| du_1\, du_2\, du_3.$$

This is why, in spherical coordinates, the volume element is $r^2 \sin\theta\, dr\, d\theta\, d\phi$, and in cylindrical coordinates, it is $\rho\, d\rho\, d\phi\, dz$.

Mini Simple Example (Polar Coordinates in Two Dimensions)

$$x = r\cos\theta, y = r\sin\theta,$$

Jacobian determinant:

$$\left| \frac{\partial(x, y)}{\partial(r, \theta)} \right| = \begin{vmatrix} \cos\theta & -r\sin\theta \\ \sin\theta & r\cos\theta \end{vmatrix} = r.$$

So:

$$dx\, dy = r\, dr\, d\theta.$$

Fact Box 1.17: Jacobian in Probability and Statistical Mechanics

Jacobians are not just for coordinate transformations—they also appear in probability distributions when changing variables.

! Red Flag 1.9: Jacobian Misuse

Mixing row/column order changes sign—always take the determinant properly.

Vector Operators in Orthogonal Coordinates

In an orthogonal system, the gradient, divergence, curl, and Laplacian take the general forms:

- **Gradient of a scalar** φ:

$$\nabla\varphi = \sum_{i=1}^{3} \hat{e}_i \frac{1}{h_i}\frac{\partial\varphi}{\partial u_i}.$$

- **Divergence of a vector field** $\vec{A} = \sum A_i\hat{e}_i$:

$$\nabla\cdot\vec{A} = \frac{1}{h_1h_2h_3}\sum_{i=1}^{3}\frac{\partial}{\partial u_i}\left(\frac{h_1h_2h_3}{h_i}A_i\right).$$

- **Curl of a vector field**:

$$\nabla\times\vec{A} = \frac{1}{h_1h_2h_3}\begin{vmatrix} h_1\hat{e}_1 & h_2\hat{e}_2 & h_3\hat{e}_3 \\ \partial/\partial u_1 & \partial/\partial u_2 & \partial/\partial u_3 \\ h_1A_1 & h_2A_2 & h_3A_3 \end{vmatrix}.$$

- **Laplacian of a scalar** φ:

$$\nabla^2\varphi = \frac{1}{h_1h_2h_3}\sum_{i=1}^{3}\frac{\partial}{\partial u_i}\left(\frac{h_1h_2h_3}{h_i^2}\frac{\partial\varphi}{\partial u_i}\right).$$

These formulas allow one to compute vector operations in any orthogonal curvilinear coordinate system, once the scale factors h_i are known.

Detailed derivations of these results are given in Appendix J.1.1.

1.5.1 The Metric Tensor and Scale Factors

Earlier, we introduced the Jacobian to describe how **volume elements** transform when moving from Cartesian to generalized coordinates. Closely related to this is the **metric tensor**, which tells us how to measure **lengths, areas**, and **angles** in the new coordinate system.

1.5.1.1 From Distance to Metric Tensor

In Cartesian coordinates, the square of the infinitesimal displacement is simply

$$ds^2 = dx^2 + dy^2 + dz^2.$$

If we now express x, y, z as functions of generalized coordinates (u_1, u_2, u_3), then

$$dx = \frac{\partial x}{\partial u_1}du_1 + \frac{\partial x}{\partial u_2}du_2 + \frac{\partial x}{\partial u_3}du_3,$$

and similarly, for dy, dz. Substituting and expanding, we obtain

$$ds^2 = \sum_{i=1}^{3} \sum_{j=1}^{3} g_{ij}\, du_i\, du_j,$$

with

$$g_{ij} = \frac{\partial x}{\partial u_i}\frac{\partial x}{\partial u_j} + \frac{\partial y}{\partial u_i}\frac{\partial y}{\partial u_j} + \frac{\partial z}{\partial u_i}\frac{\partial z}{\partial u_j}.$$

The coefficients g_{ij} form the **metric tensor** g, which is symmetric by construction.

In the familiar Cartesian system, the metric tensor is just the identity matrix, which is why we never think about it.

1.5.1.2 Orthogonal Coordinates and Scale Factors

If the coordinate system is **orthogonal**, the cross terms vanish ($g_{ij} = 0$ for $i \neq j$). Then,

$$ds^2 = h_1^2 du_1^2 + h_2^2 du_2^2 + h_3^2 du_3^2,$$

where the **scale factors** are

$$h_i = \sqrt{g_{ii}}.$$

For example:

- **Cylindrical coordinates** ρ, ϕ, z:

$$ds^2 = d\rho^2 + \rho^2 d\phi^2 + dz^2 (h_\rho = 1, h_\phi = \rho, h_z = 1).$$

- **Spherical coordinates** r, θ, ϕ:

$$ds^2 = dr^2 + r^2 d\theta^2 + r^2 \sin^2\theta\, d\phi^2$$
$$(h_r = 1,\ h_\theta = r,\ h_\phi = r \sin\theta).$$

1.5.1.3 Volume Element From Metric Tensor

In general, the **differential volume element** is given by

$$d\tau = \sqrt{\det \mathrm{g}}\, du_1 du_2 du_3,$$

where det g is the determinant of the metric tensor.

For orthogonal coordinates, this reduces to a simple product of scale factors:

$$d\tau = h_1 h_2 h_3\, du_1 du_2 du_3.$$

- Cylindrical: $d\tau = \rho\, d\rho\, d\phi\, dz$.
- Spherical: $d\tau = r^2 \sin\theta\, dr\, d\theta\, d\phi$.

Fact Box 1.18

- The **Jacobian** tells us how volumes transform.
- The **metric tensor** tells us how distances, areas, and angles transform.

- Together, they provide the foundation for defining vector calculus operators (grad, div, curl, Laplacian) in any coordinate system.

Spherical polar and cylindrical coordinates are frequently used in quantum chemistry. Spherical polar coordinates are useful for problems with spherical symmetry, helping simplify equations and clarify physical phenomena, especially for systems like the rigid rotator and hydrogen atom. Similarly, cylindrical coordinates are well suited for problems with rotational symmetry around an axis, making them especially useful in quantum chemistry when analyzing systems like linear molecules or particles constrained to move in a circular path.

Since we are frequently going to use spherical polar and cylindrical coordinates in quantum chemistry, it is important to describe these coordinates in some detail. Spherical polar coordinates provide a convenient framework for analyzing problems with spherical symmetry, which is common in quantum chemical systems. Understanding the specifics of this coordinate system allows for a more systematic approach to solving relevant equations and interpreting physical phenomena within these contexts.

1.6 Spherical Polar Coordinates

In spherical polar coordinates (Figure 1.8 [left]), a point P is represented by r, θ, ϕ, where r is the radial distance, θ the polar angle (from the z-axis), and ϕ the azimuthal angle (from the x-axis in the xy-plane).

The transformation to Cartesian coordinates is

$$x = r \sin\theta \cos\phi, y = r \sin\theta \sin\phi, z = r \cos\theta.$$

The scale factors are

$$h_r = 1, h_\theta = r, h_\phi = r \sin\theta.$$

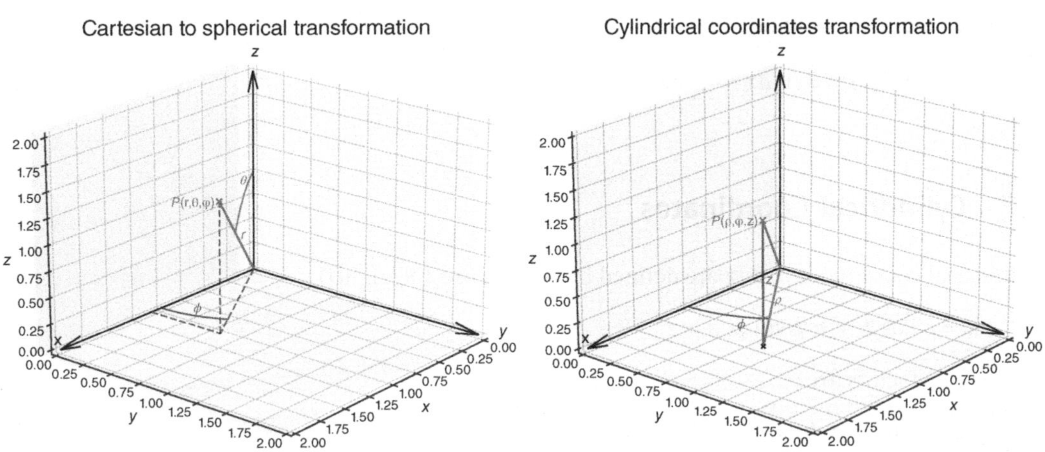

Figure 1.8 Spherical and cylindrical coordinate systems. *Left*: spherical polar coordinates. *Right*: cylindrical coordinates. Note how the unit vectors (except $\hat{k}/\hat{e}_z$) are position-dependent.

Hence the **volume element** is

$$d\tau = r^2 \sin\theta \, dr \, d\theta \, d\phi.$$

The del operator in spherical coordinates is

$$\nabla = \hat{e}_r \frac{\partial}{\partial r} + \hat{e}_\theta \frac{1}{r} \frac{\partial}{\partial \theta} + \hat{e}_\phi \frac{1}{r \sin\theta} \frac{\partial}{\partial \phi}.$$

The **Laplacian** is particularly important:

$$\nabla^2 = \frac{1}{r^2} \frac{\partial}{\partial r}\left(r^2 \frac{\partial}{\partial r}\right) + \frac{1}{r^2 \sin\theta} \frac{\partial}{\partial \theta}\left(\sin\theta \frac{\partial}{\partial \theta}\right) + \frac{1}{r^2 \sin^2\theta} \frac{\partial^2}{\partial \phi^2}.$$

Applications:

- Hydrogen atom wavefunctions are obtained by solving the **Schrödinger equation** in spherical coordinates.
- The rigid rotator model for diatomic molecules also uses spherical polar coordinates.

Application Box 1.6: Spherical Polar Coordinates

The hydrogen atom problem in quantum mechanics separates neatly only in spherical polar coordinates.

Red Flag 1.10: Angle Conventions

θ is polar (from z-axis); ϕ is azimuthal (in xy-plane).

Red Flag 1.11: Volume Element

Don't forget the $\sin\theta$ in $d\tau = r^2 \sin\theta dr \, d\theta \, d\phi$.

1.7 Cylindrical Coordinates

Cylindrical coordinates ρ, ϕ, z are defined by (see Figure 1.8 [right])

$$x = \rho \cos\phi, y = \rho \sin\phi, z = z.$$

The scale factors are

$$h_\rho = 1, h_\phi = \rho, h_z = 1,$$

so, the **volume element** is

$$d\tau = \rho \, d\rho \, d\phi \, dz.$$

The **Laplacian** in cylindrical coordinates is

$$\nabla^2 = \frac{1}{\rho}\frac{\partial}{\partial \rho}\left(\rho \frac{\partial}{\partial \rho}\right) + \frac{1}{\rho^2}\frac{\partial^2}{\partial \phi^2} + \frac{\partial^2}{\partial z^2}.$$

Applications:

- Cylindrical coordinates are natural for describing problems with **axial symmetry** such as vibrations in linear molecules, current in conducting wires, or diffusion in cylindrical pores.

Application Box 1.7: Cylindrical Coordinates

Used for modeling current flow in wires, diffusion in nanopores, and vibrations of long polymers.

Red Flag 1.12: Cylindrical Laplacian

The $\frac{1}{\rho}\frac{\partial}{\partial \rho}$ term is essential.

This completes the general treatment of orthogonal curvilinear coordinates.

1.8 Summary

In this chapter, we introduced vectors as the simplest tensors and reviewed their representation in Cartesian coordinates. After a brief reminder of vector algebra (dot, cross, and scalar triple products), the focus shifted to concepts central to advanced applications.

We discussed **vector spaces** and the idea of linear independence, which provides the mathematical basis for choosing sets of functions or orbitals in chemistry and physics. This leads naturally to the more abstract notion of Hilbert spaces, taken up in the next chapter.

We then developed the essential tools of **vector calculus**. The gradient, divergence, curl, and Laplacian describe how scalar and vector fields vary in space and time. These operators are not merely mathematical constructs; they correspond to physical realities: forces as gradients of potential energy surfaces, divergence in continuity equations, curl in electromagnetism and angular momentum, and the Laplacian in the Schrödinger equation and diffusion problems.

To solve problems with different symmetries, we generalized the Cartesian system to **orthogonal curvilinear coordinates**. The Jacobian of transformation was introduced, allowing us to express volume elements and vector operators in new coordinate systems. In particular, spherical polar coordinates were treated in detail because they are indispensable for atomic and molecular systems, while cylindrical coordinates were outlined for problems with axial symmetry.

These results establish a firm foundation for the study of **linear vector spaces** and **operators**, which form the core mathematical framework of QM. In the next chapter, we extend the notion of vectors from three-dimensional Euclidean space to abstract spaces of arbitrary dimension, and develop the formalism required for modern chemistry, physics, and engineering.

Fact Box 1.19: Looking Ahead

Vector calculus in 3D Euclidean space is the prototype for abstract Hilbert spaces to be studied in Chapter 2.

For chemists, the most important leap is to understand that **functions can be vectors**. The solutions to the Schrödinger equation, for instance, behave just like the orthogonal basis vectors $\hat{i}, \hat{j}, \hat{k}$. To see this, we need a way to define a "dot product" for functions.

While the formal machinery for this, called the **inner product**, will be developed in **Chapter 2**, we can get the main idea by defining the generalized dot product between two wavefunctions, ψ_a and ψ_b, as the integral $\int \psi_a(x)\psi_b(x)dx$.

Looking Ahead: An Example from QM

Consider the simple wavefunctions for a particle-in-a-box of length $L\left(\sqrt{\frac{2}{L}}\sin\left(\frac{n\pi x}{L}\right)\right)$, which are types of **special functions** we will explore in detail in **Chapter 9**. Let us take the first two states,

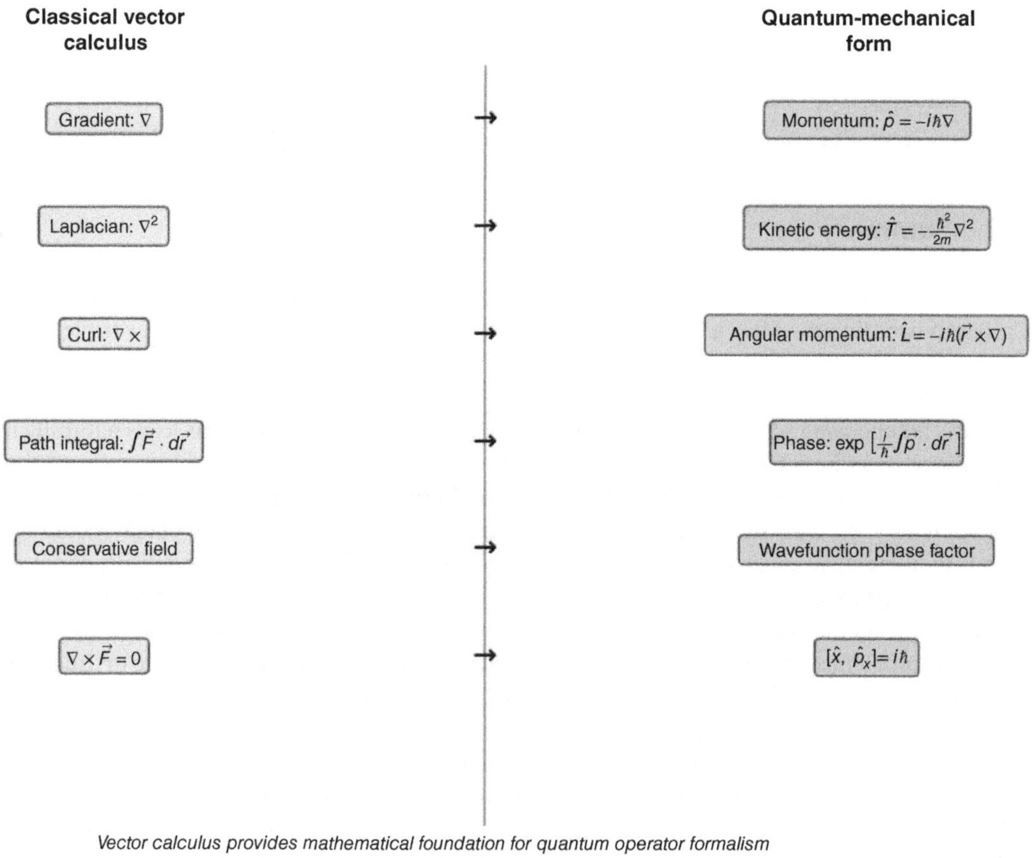

Figure 1.9 Operator correspondence—classical to quantum.

ψ_1 and ψ_2 (i.e., $n = 1$ and 2). If we calculate their generalized dot product, we find

$$\int_0^L \psi_1(x)\psi_2(x)dx = \frac{2}{L}\int_0^L \sin\left(\frac{\pi x}{L}\right)\sin\left(\frac{2\pi x}{L}\right)dx = 0.$$

The result is zero. This means these two wavefunctions are **orthogonal**—the function equivalent of being perpendicular. This single concept is the foundation of QM, basis sets, and spectroscopy. We have just shown that wavefunctions are not just mathematical formulas; they are vectors in an abstract space called a Hilbert space. We shall develop the idea further in the next chapter. For now, we will list how the classical operators we have studied in this chapter make their way into QM (Figure 1.9).

Further Reading

G. B. Arfken, H. J. Weber and F. E. Harris, *Mathematical Methods for Physicists*, Elsevier, 2011.

M. L. Boas, *Mathematical Methods in the Physical Sciences*, 3rd ed., Wiley, 2005.

D. J. Griffiths, *Introduction to Electrodynamics*, Pearson, 2015—for applications of divergence, curl, and Laplacian.

D. A. McQuarrie, *Mathematical Methods for Scientists and Engineers*, University Science Books, 2003.

D. A. McQuarrie and J. D. Simon, *Physical Chemistry: A Molecular Approach*, University Science Books, 2023—for chemical applications of vector calculus.

Exercises

A. Conceptual

1. Explain why the dipole moment convention differs between chemistry and physics.

2. What does the gradient of a scalar field physically represent? Give two chemical examples.

3. Why is the divergence of the velocity field zero for incompressible fluids?

4. Why are Jacobians essential in statistical mechanics and QM?

5. Let $\vec{a} = (1, 1, 0)$, $\vec{b} = (2, 2, 0)$, and $\vec{c} = (1, -1, 2)$ be three vectors.
   a. Show that $\vec{a}, \vec{b}$ are collinear.
   b. Show that $\vec{a} \times \vec{b}$ is perpendicular to both $\vec{a}$ and $\vec{b}$.
   c. Show that $\vec{a}, \vec{b}, \vec{c}$ are coplanar.

6. Show that three vectors are coplanar if their scalar triple product vanishes.

7. Identify the tensor rank of the following physical quantities: (a) Temperature (T), (b) Magnetic Field (B), (c) Molecular Polarizability (α), and (d) The stress on a solid (σ).

B. Vector Algebra and Geometry

1. Form the scalar and vector products of the given vectors:

$$\vec{a} = \hat{i}\cos\theta + \hat{j}\sin\theta,$$
$$\vec{b} = \hat{i}\cos\phi + \hat{j}\sin\phi,$$

and hence prove the identities:

$$\cos(\theta - \phi) = \cos\theta\cos\phi + \sin\theta\sin\phi,$$
$$\sin(\theta - \phi) = \sin\theta\cos\phi - \cos\theta\sin\phi.$$

2. Show that the volume V of the parallelepiped formed by $\vec{a}, \vec{b}, \vec{c}$ is given by

$$V = \vec{a} \cdot \left(\vec{b} \times \vec{c}\right),$$

and express this as a determinant:
   - Prove that the triple product is unchanged under cyclic permutations, i.e., $\vec{a} \cdot \vec{b} \times \vec{c} = \vec{b} \cdot \vec{c} \times \vec{a} = \vec{c} \cdot \vec{a} \times \vec{b}$.
   - Deduce that $\hat{i} \cdot \hat{j} \times \hat{k} = \hat{j} \cdot \hat{k} \times \hat{i} = \hat{k} \cdot \hat{i} \times \hat{j} = 1$ form a right-handed coordinate system.

3. Show that a given vector $\vec{b} = 2\hat{i} + \hat{j} - 6\hat{k}$ cannot be expressed as a linear combination of the vectors $\vec{a_1} = (1, 1, 1)$, $\vec{a_2} = (1, -1, 0)$, $\vec{a_3} = (2, 0, 1)$, while another vector $\vec{c} = -2\hat{j} - 3\hat{k}$ can be expressed in more than one way:
   - Provide both algebraic and geometric explanations.
   - Relate this to linear dependence and independence.

4. Form the scalar and vector products of vectors $\vec{p} = (1, 3, -2)$ and $\vec{q} = (4, -1, 2)$ and hence prove the identities:

$$|\vec{p} \times \vec{q}|^2 = |\vec{p}|^2|\vec{q}|^2 - (\vec{p} \cdot \vec{q})^2.$$

C. Problem-solving Questions

1. Let $\varphi(x, y) = x^2 - y^2$. Evaluate $\int_{0,0}^{2,8} (\nabla\varphi \cdot d\vec{r})$ along the curve $y = x^3$.

2. Show that the vector field $\vec{F} = \hat{i}\cos x - 2yz\hat{j} - y^2\hat{k}$ is conservative. Evaluate $\int_{0,-1,-1}^{7,3,2} \vec{F} \cdot d\vec{r}$.

3. Evaluate $\oint \vec{F} \cdot d\vec{r}$ for $\vec{F} = \dfrac{\hat{i}x + \hat{j}y}{\sqrt{x^2 + y^2}}$ around a square whose sides are $x = 0, x = a, y = 0, y = a$. Determine whether the field is conservative, and if so, find φ such that $\vec{F} = \nabla\varphi$.

4. The electrostatic field of a point charge is $\vec{E} = \dfrac{q}{4\pi\epsilon_0 r^2}\hat{r}$:
   - Calculate $\nabla \cdot \vec{E}$.
   - Show that the potential φ satisfies Laplace's equation $\left(\nabla^2\varphi = 0\right)$ away from the origin.

5. Verify directly that $\nabla \cdot \left(\nabla \times \vec{A}\right) = 0$ for any vector field $\vec{A}$.

6. Show that the work done in moving a particle under a conservative force field is path-independent.

7. Calculate the divergence of $\vec{F} = \left(x^2, y^2, z^2\right)$. Interpret the result.

8. Verify that $\nabla \times \nabla\varphi = 0$ for an arbitrary scalar field φ.

9. If $\varphi = 2x^3y^2z^4$, find the divergence of gradient of φ. What is the physical significance of the divergence of a vector field? Compare it with the curl.

10. Compute the Laplacian of r^n in spherical coordinates, where n is an integer.

11. Find the volume of a sphere of radius R using spherical coordinates and Jacobian formalism.

12. In cylindrical coordinates, evaluate the Laplacian of $\varphi(\rho, \phi, z) = \ln \rho$.

D. Orthogonal Coordinates: Spherical Polar

1. Verify the spherical polar transformations:

$$x = r \sin\theta \cos\phi, \; y = r \sin\theta \sin\phi, \; z = r \cos\theta.$$

Compute the scale factors h_r, h_θ, h_ϕ and verify them using $ds_i = h_i du_i$.

2. Derive the general expression for the gradient in spherical polar coordinates, and show that

$$\nabla\psi = \hat{e}_r \frac{\partial\psi}{\partial r} + \frac{1}{r}\hat{e}_\theta \frac{\partial\psi}{\partial\theta} + \frac{1}{r\sin\theta}\hat{e}_\phi \frac{\partial\psi}{\partial\phi}.$$

Show that the volume element in spherical polar coordinates is $d\tau = r^2 \sin\theta \, dr \, d\theta \, d\phi$.

3. Resolve the spherical polar unit vectors into Cartesian components and conversely resolve Cartesian unit vectors into spherical polar components.

E. Advanced/Applied/Challenge Problems

1. Show by direct calculation that the Jacobian of the spherical polar transformation equals $r^2 \sin\theta$.

2. Derive the Laplacian in spherical polar coordinates from first principles.

3. In a two-dimensional coordinate system with metric tensor

$$g_{ij} = \begin{bmatrix} 2 & 1 \\ 1 & 2 \end{bmatrix},$$

evaluate the dot product of the vectors

$\vec{a} = (1, 2)$ and $\vec{b} = (2, 1)$.

Explain how the metric modifies the standard Cartesian definition of the dot product.

4. Show that if ψ is independent of r, the Laplacian reduces to the angular momentum operator.

5. The angular momentum, $\vec{L}$, is given by $\vec{L} = \vec{r} \times \vec{p}$, where $\vec{p}$ is the linear momentum. Write expressions for the Cartesian components, L_x, L_y, and L_z, in terms of x, y, z and p_x, p_y, p_z. Quantum-mechanical operators can be obtained by replacing p_x by $-i\hbar\frac{\partial}{\partial x}$, p_y by $-i\hbar\frac{\partial}{\partial y}$, and p_z by $-i\hbar\frac{\partial}{\partial z}$. Show that

$$\widehat{L}_x = yp_z - zp_y = i\hbar\left(z\frac{\partial}{\partial y} - y\frac{\partial}{\partial z}\right).$$

Also show that these operators obey commutation rules of the form

$$[\widehat{L}_x, \widehat{L}_y] = \widehat{L}_x\widehat{L}_y - \widehat{L}_y\widehat{L}_x = i\hbar\widehat{L}_z,$$

and that

$$[\widehat{L}_x, \widehat{L}^2] = [\widehat{L}_y, \widehat{L}^2] = [\widehat{L}_z, \widehat{L}^2] = 0.$$

6. Show that the Schrödinger equation for the hydrogen atom separates in spherical polar coordinates.

7. Derive the expression for the kinetic energy operator in cylindrical coordinates.

8. Consider the diffusion of particles in a cylindrical pore. Write down the governing equation using the divergence and Laplacian in cylindrical coordinates.

Metric Tensor in Spherical Coordinates

9. Derive the metric tensor for spherical polar coordinates r, θ, ϕ:

$$g_{ij} = \begin{bmatrix} 1 & 0 & 0 \\ 0 & r^2 & 0 \\ 0 & 0 & r^2 \sin^2\theta \end{bmatrix},$$

and show that it gives the squared length element

$$ds^2 = dr^2 + r^2 d\theta^2 + r^2 \sin^2\theta \, d\phi^2.$$

Differential Forms and Vector Identities

10. Using the language of differential forms, show that

$$\nabla \cdot \left(\nabla \times \vec{A}\right) = 0$$

follows directly from the property $d^2 = 0$. Discuss why this unifies vector calculus identities.

2

Linear Vector Spaces

Linearity is the gateway to understanding.

—Anonymous (widely cited in teaching literature)

Learning Outcomes

After studying this chapter, students should be able to:

1. **Define** a linear vector space and **explain the rules** of vector addition and scalar multiplication.
2. **Distinguish** between finite- and infinite-dimensional vector spaces, with examples from chemistry and physics.
3. **Understand** the concept of inner product, orthogonality, and completeness, and **apply** them to function spaces (e.g., wavefunctions in quantum mechanics).
4. **Use** Dirac bra–ket notation and **connect** it to function spaces.

2.1 Introduction

The previous chapter prepared the ground for our study of **linear vector spaces**. The three-dimensional Euclidean space, with vectors expressed in terms of their Cartesian components, is one example of a **finite-dimensional** real vector space. The key idea is that a vector space need not consist only of "arrows" with magnitude and direction. Any collection of objects that obey certain rules of addition and scalar multiplication qualifies as a vector space. These objects can be conventional vectors, matrices, or even functions.

This generalization is essential in physics and chemistry. For example, the solutions of the Schrödinger equation—the wavefunctions—belong to an **infinite-dimensional** vector space, more precisely a **Hilbert space**, in which an **inner product** is defined. In such spaces, concepts like **orthogonality** and **completeness** acquire fundamental importance, since they determine whether a chosen set of functions (basis functions, such as atomic orbitals or sine functions in a box) can represent physical states fully.

In this chapter, we will first review the defining features of vector spaces and their basic properties. We then move to inner products, orthogonality, and completeness.

2.2 What Is a Linear Vector Space?

A set of objects (e.g., vectors, matrices, or functions) is said to form a **linear vector space** $\mathbb{V}$ if it obeys the following rules of **addition** and **scalar multiplication**.

Fact Box 2.1: VOCABULARY ALERT 1: "Vector"

In Mathematics: Any object (arrow, matrix, function, polynomial) that obeys the rules of a vector space.
In Chemistry/Physics: Often means a geometric arrow with magnitude and direction (e.g., a force or dipole moment vector).
In This Text: We use the mathematical definition. An atomic orbital is a vector. A wavefunction is a vector. This abstract view is the foundation of quantum mechanics.

Notation
In this chapter, vectors belonging to abstract linear vector spaces will be written in **boldface** (e.g., $\mathbf{v}, \mathbf{w}, \mathbf{x}$). This distinguishes them from the geometric vectors of the previous chapter, which were denoted with arrows above their symbols — that arrow notation is retained here whenever the vector is an explicit element of three-dimensional Euclidean space, with boldface reserved for vectors in general N-dimensional or abstract spaces. Matrices, though themselves elements of a vector space (Example 2.2), are conventionally written in italic capitals rather than boldface, reserving boldface for column-vector-like objects; this also anticipates their later role as operators acting on vectors (Chapter 3).

Addition

- **Closure and Commutativity:** For any two vectors $\mathbf{u}, \mathbf{v} \in \mathbb{V}$, their sum $\mathbf{u} + \mathbf{v}$ is also in $\mathbb{V}$, and $\mathbf{u} + \mathbf{v} = \mathbf{v} + \mathbf{u}$.
- **Associativity:** For any $\mathbf{u}, \mathbf{v}, \mathbf{w} \in \mathbb{V}$, $(\mathbf{u} + \mathbf{v}) + \mathbf{w} = \mathbf{u} + (\mathbf{v} + \mathbf{w})$.
- **Existence of the Zero (Null) Vector:** There exists a vector $\mathbf{0} \in \mathbb{V}$ such that $\mathbf{v} + \mathbf{0} = \mathbf{v}$ for all $\mathbf{v} \in \mathbb{V}$.
- **Existence of Additive Inverse:** For every vector $\mathbf{v} \in \mathbb{V}$, there exists a vector $-\mathbf{v} \in \mathbb{V}$ such that $\mathbf{v} + (-\mathbf{v}) = \mathbf{0}$.

Scalar Multiplication

- **Closure:** For any scalar $c \in \mathbb{C}$ (or $\mathbb{R}$) and vector $\mathbf{v} \in \mathbb{V}$, the product $c\mathbf{v}$ is also in $\mathbb{V}$.
- **Distributivity:**

$$a(\mathbf{u} + \mathbf{v}) = a\mathbf{u} + a\mathbf{v},$$

$$(a + b)\mathbf{v} = a\mathbf{v} + b\mathbf{v}, \text{ for all scalars } a, b.$$

- **Associativity:** $(ab)\mathbf{v} = a(b\mathbf{v})$, for all scalars a, b.
- **Multiplicative Identity:** There exists a scalar 1 such that $1 \cdot \mathbf{v} = \mathbf{v}$ for all $\mathbf{v} \in \mathbb{V}$.

A vector space defined over the complex numbers is called a **complex vector space**, while one defined over the real numbers is a **real vector space**. For example, the ordinary three-dimensional Euclidean Cartesian space is a real vector space.

The following example demonstrates that functions act as vectors in vector space.

Example 2.1: Wavefunctions as a Vector Space

Consider the set of wavefunctions for a particle confined in a one-dimensional box of length L. These are given by

$$\psi_n(x) = \sqrt{\frac{2}{L}}\sin\left(\frac{n\pi x}{L}\right), n = 1, 2, 3, \dots .$$

We ask: does this set form a vector space?

Solution:

1. **Closure under addition:** The sum of any two such functions, say $\psi_1(x) + \psi_2(x)$, is still a well-defined function on the interval $[0, L]$. Although it is not itself a single eigenfunction of the particle in a box, it still belongs to the function space under consideration.
2. **Closure under scalar multiplication:** Multiplying any function $\psi_n(x)$ by a scalar c gives another function $c\psi_n(x)$ in the same space.
3. **Zero element:** The null function, defined by $\psi_0(x) = 0$ for all x, is also part of this set.
4. **Additive inverse:** For every function $\psi_n(x)$, the function $-\psi_n(x)$ is also included.

Thus, the set of all linear combinations of the $\psi_n(x)$ constitutes a vector space.

Remark This example illustrates that "vectors" in a linear vector space need not be arrows in Euclidean space; they may be functions. Such functional vector spaces form the foundation of **quantum mechanics**, where the **Hilbert space** (see Section 2.4) of wavefunctions provides the stage for physical laws.

The next example (Example 2.2) shows that even matrices can be vectors in a linear vector space.

Example 2.2: Matrices as Vectors in a Vector Space

Consider the set of all 2×2 real matrices:

$$M = \begin{bmatrix} a & b \\ c & d \end{bmatrix}, a, b, c, d \in \mathbb{R}.$$

Show that this set forms a linear vector space.

Solution:

- **Closure under addition:** If

$$M_1 = \begin{bmatrix} a_1 & b_1 \\ c_1 & d_1 \end{bmatrix}, M_2 = \begin{bmatrix} a_2 & b_2 \\ c_2 & d_2 \end{bmatrix},$$

then

$$M_1 + M_2 = \begin{bmatrix} a_1 + a_2 & b_1 + b_2 \\ c_1 + c_2 & d_1 + d_2 \end{bmatrix},$$

which is again a 2×2 real matrix.
- **Additive identity:** The zero matrix

$$0 = \begin{bmatrix} 0 & 0 \\ 0 & 0 \end{bmatrix}$$

satisfies $M + 0 = M$.

- **Additive inverse:** For each M, its negative

$$-M = \begin{bmatrix} -a & -b \\ -c & -d \end{bmatrix}$$

satisfies $M + (-M) = 0$.
- **Closure under scalar multiplication:** For any scalar $\lambda \in \mathbb{R}$,

$$\lambda M = \begin{bmatrix} \lambda a & \lambda b \\ \lambda c & \lambda d \end{bmatrix},$$

which is again in the set.

Hence, the set of all 2×2 real matrices forms a real vector space of dimension 4, since each matrix can be uniquely expressed as a linear combination of the four basis matrices

$$E_{11} = \begin{bmatrix} 1 & 0 \\ 0 & 0 \end{bmatrix}, E_{12} = \begin{bmatrix} 0 & 1 \\ 0 & 0 \end{bmatrix}, E_{21} = \begin{bmatrix} 0 & 0 \\ 1 & 0 \end{bmatrix}, E_{22} = \begin{bmatrix} 0 & 0 \\ 0 & 1 \end{bmatrix}.$$

Thus, understanding matrices as elements of vector spaces prepares us to treat them systematically as vectors in a higher-dimensional space, a perspective that becomes crucial when we study linear operators and basis sets.

Example 2.3: Linear Combinations

Express the matrix

$$M = \begin{bmatrix} 3 & -1 \\ 2 & 4 \end{bmatrix}$$

as a linear combination of the basis matrices $E_{11}, E_{12}, E_{21}, E_{22}$.

Solution:

We recall that any 2×2 real matrix can be written as

$$M = aE_{11} + bE_{12} + cE_{21} + dE_{22}.$$

Solving,

$$a = 3, b = -1, c = 2, d = 4.$$

Thus,

$$M = 3\begin{bmatrix} 1 & 0 \\ 0 & 0 \end{bmatrix} + (-1)\begin{bmatrix} 0 & 1 \\ 0 & 0 \end{bmatrix} + 2\begin{bmatrix} 0 & 0 \\ 1 & 0 \end{bmatrix} + 4\begin{bmatrix} 0 & 0 \\ 0 & 1 \end{bmatrix}.$$

This decomposition demonstrates explicitly that the set $\{E_{11}, E_{12}, E_{21}, E_{22}\}$ spans the space of 2×2 matrices.

Examples of Vector Spaces

- $\mathbb{R}^3$: Real 3D space with arrow addition.
- $\mathbb{C}^n$: Complex n-tuples (quantum state spaces).
- M_{mn}: Space of $m \times n$ matrices.
- P_n: Polynomials of degree $\leq n$.
- Function spaces: Wavefunctions in quantum mechanics.

Quantum Chemistry Connection

In quantum chemistry, matrices are not just abstract objects—they represent physically meaningful quantities. For example:

- The **density matrix** encodes how electrons are distributed among orbitals.
- The **Fock matrix** represents the effective one-electron Hamiltonian in Hartree–Fock theory.
- Overlap, Hamiltonian, and coefficient matrices used in molecular orbital calculations all inhabit such vector spaces.

Thus, understanding matrices as elements of vector spaces prepares us to treat them systematically as vectors in a higher-dimensional space, a perspective that becomes crucial when we study linear operators and basis sets.

The power of this abstraction is that we can use the same geometric intuition for electron orbitals as we do for simple arrows in space.

2.2.1 Dimension of a Linear Vector Space

The **dimension** of a vector space is the smallest number of linearly independent vectors in that space, which **span** the entire space, and in terms of which all other vectors can be expressed. A **basis** for an N-dimensional space is any set of N linearly independent vectors (see Example 2.3). Any vector in the space can be uniquely expressed as a **linear combination** of basis vectors.

Fact Box 2.2: VOCABULARY ALERT 2: "Basis"

In Mathematics: A minimal set of linearly independent vectors that span the entire space.
In Computational Chemistry: A "basis set" is a set of atomic orbital functions (e.g., STO-3G, 6-31G*) used to construct molecular orbitals.
The Connection: The computational "basis set" is a direct application of the mathematical concept. Each atomic orbital function is a basis vector in the space of possible wavefunctions.

Recall from three-dimensional Euclidean space that three vectors are said to be linearly independent if none of them can be written as a linear combination of the others. More formally, the set $\vec{v}_1, \vec{v}_2, \vec{v}_3$ is linearly independent if no scalars c_1, c_2, c_3 exist such that

$$c_1\vec{v}_1 + c_2\vec{v}_2 + c_3\vec{v}_3 = \vec{0}, \tag{2.1}$$

unless all the coefficients are zero.

Any vector $\vec{a}$ in real 3D space can be expressed in terms of its components as

$$\vec{a} = a_x \hat{i} + a_y \hat{j} + a_z \hat{k}, \tag{2.2}$$

where $\hat{i}, \hat{j}, \hat{k}$ are the unit vectors along the Cartesian coordinate axes. In matrix notation, the same vector may be written as a column vector:

$$\vec{a} = \begin{bmatrix} a_x \\ a_y \\ a_z \end{bmatrix}. \tag{2.3}$$

The vectors $\hat{i}, \hat{j}, \hat{k}$ are known as the **canonical basis vectors** of the coordinate system, and can themselves be expressed as

$$\hat{i} = \begin{bmatrix} 1 \\ 0 \\ 0 \end{bmatrix}, \hat{j} = \begin{bmatrix} 0 \\ 1 \\ 0 \end{bmatrix}, \hat{k} = \begin{bmatrix} 0 \\ 0 \\ 1 \end{bmatrix}. \tag{2.4}$$

The same idea extends naturally to higher-dimensional spaces, denoted $\mathbb{V}^N$, where N is the dimension of the space. In such a space, a vector can be expressed as

$$\mathbf{u} = u_1\mathbf{e}_1 + u_2\mathbf{e}_2 + \cdots + u_N\mathbf{e}_N, \tag{2.5}$$

where $\mathbf{e}_1, \mathbf{e}_2, \ldots, \mathbf{e}_N$ form a basis. In column notation, this becomes

$$\mathbf{u} = \begin{bmatrix} u_1 \\ u_2 \\ \vdots \\ u_N \end{bmatrix}.$$

The u_i values are the **components** or **coordinates** of the vector in this basis.

Quantum Chemistry Connection

In molecular orbital (MO) theory, the atomic orbitals form a basis in which we expand molecular orbitals—the coefficients C_i in LCAO-MO (Linear Combination of Atomic Orbitals-MO) are precisely the components in this basis. The LCAO-MO method is essentially a change of basis in the vector space of electronic orbitals.

For finite-dimensional real vector spaces, linear independence of a set of vectors can be tested by arranging them as columns of a square matrix: if the determinant is nonzero, the vectors are linearly independent; if zero, they are linearly dependent (see Chapter 1).

Example 2.4: Linear Independence

Determine whether the following sets of vectors are linearly independent:

a. $\vec{v}_1 = (1, 0, 0), \vec{v}_2 = (0, 1, 0)$

b. $\mathbf{v}_1 = (1, 1, 0, 0), \mathbf{v}_2 = (0, 1, 1, 0), \mathbf{v}_3 = (0, 0, 1, 1), \mathbf{v}_4 = (1, 0, 0, 1)$

c. $\mathbf{v}_1 = (1, 1, 1, 1, 1), \mathbf{v}_2 = (1, 2, 3, 4, 5),\ \mathbf{v}_3 = (2, 3, 4, 5, 6),\ \mathbf{v}_4 = (3, 4, 5, 6, 7),$
$\mathbf{v}_5 = (4, 5, 6, 7, 8)$
d. $\vec{v}_1 = (3, 0, 0),\ \vec{v}_2 = (0, -2, 0),\ \vec{v}_3 = (0, 0, -1)$

Solution:

a. In 3D space, these two vectors lie along the x- and y-axes. They are not multiples of each other, so no nontrivial linear relation exists. Hence, they are **linearly independent,** though they cannot span all of $\mathbb{R}^3$.

b. Place the four vectors as columns of a 4×4 matrix. Its determinant is zero. Explicitly,

$$\mathbf{v}_1 - \mathbf{v}_2 + \mathbf{v}_3 - \mathbf{v}_4 = \mathbf{0}.$$

Thus, the vectors are **linearly dependent.** Any one vector (say $\mathbf{v}_4$) may be removed without reducing the span.

c. Write the vectors as rows of a 5×5 matrix. Inspection shows that the third row is the sum of the first two, so the determinant vanishes. Hence, the set is **linearly dependent.**

d. Arrange the vectors as columns:

$$\begin{vmatrix} 3 & 0 & 0 \\ 0 & -2 & 0 \\ 0 & 0 & -1 \end{vmatrix}.$$

The determinant is 6, which is nonzero. Therefore, the vectors are **linearly independent** and span 3D space.

Quantum Chemistry Connection

This same criterion applies in quantum chemistry. The concept generalizes directly to orbital basis sets in quantum chemistry, where the choice of a minimal independent set of functions ensures efficiency and avoids redundancy. Basis functions used in electronic structure calculations must be linearly independent. If two basis functions are too similar, the overlap matrix becomes nearly singular, leading to unstable or ill-defined molecular orbitals.

Key Takeaways

- Dimension = the minimum number of linearly independent vectors needed to span the space.
- In 3D Euclidean space, three mutually independent vectors form the basis.
- Basis vectors can be represented as canonical unit vectors.
- Linear independence is checked by ensuring the determinant of the matrix of vectors is nonzero.

2.2.2 Inner Product

To qualify as a **Hilbert space**, an N-dimensional vector space must be equipped with an **inner product**. We recall the definitions from the previous chapter. For real 3D Euclidean space, the inner product of two vectors $\vec{a}$ and $\vec{b}$ is their dot product,

$$\vec{a} \cdot \vec{b} = a_x b_x + a_y b_y + a_z b_z. \tag{2.6}$$

Two vectors are orthogonal (perpendicular) if their inner product is zero. The length (or **norm**) of a vector $\vec{a}$ is the square root of its inner product with itself,

$$|\vec{a}| = \sqrt{\vec{a} \cdot \vec{a}}.$$

In general, for vectors $\mathbf{u}, \mathbf{v} \in \mathbb{V}^N$, the inner product in N-dimensional inner product space is

$$\langle u, v \rangle = \sum_{i=1}^{N} u_i v_i. \tag{2.7}$$

In quantum mechanics, our vector spaces are complex, so the inner product definition (2.7) must include complex conjugation:

$$\langle u, v \rangle = \sum_{i=1}^{N} u_i^* v_i.$$

This definition assumes that the basis vectors are **orthonormal**. If $\langle u, v \rangle = 0$, then **u** and **v** are orthogonal. The unit vector in the direction of **u** is obtained by dividing by its norm,

$$\hat{\mathbf{u}} = \frac{\mathbf{u}}{|\mathbf{u}|}.$$

In this way, we can construct orthonormal bases from linearly independent sets of vectors, using orthogonalization procedures such as Gram–Schmidt (to be introduced later).

In **quantum mechanics**, we must often extend this definition of inner products beyond finite-dimensional spaces. The relevant objects are **function spaces**, denoted $\mathbb{F}^{\mathrm{N}}$, where the "vectors" are functions $\phi_i(x)$. For such spaces, the **inner product** is defined by an integral:

$$\langle \phi_i | \phi_j \rangle = \int \phi_i^*(x) \phi_j(x)\, dx,$$

where $\phi_i^*(x)$ is the complex conjugate of $\phi_i(x)$. The inner product $\langle \phi_i | \phi_j \rangle$ measures the "overlap" between quantum states—this becomes the probability amplitude in quantum mechanics. If the inner product is zero, the functions are orthogonal. This definition provides the mathematical foundation for **Hilbert spaces**, the natural setting for quantum mechanics.

Fact Box 2.3: VOCABULARY ALERT 3: "Orthogonality"

In Geometry: Two vectors are perpendicular (at a 90° angle).
In Mathematics: Two vectors are orthogonal if their inner product is zero: $\langle u, v \rangle = 0$.
|**In Quantum Chemistry:** Two wavefunctions are orthogonal if $\int \psi_i^* \psi_j \, d\tau = 0$. This has a profound physical meaning: if a system is in state ψ_i, there is zero probability of finding it in state ψ_j upon measurement.

Unified Notation

To express inner products uniformly for both vectors and functions, **Dirac's bra–ket notation** is employed. A vector **v** is written as a **ket** $|v\rangle$, while its dual (conjugate transpose) is the **bra** $\langle v|$. The inner product of **u** and **v** is then written as $\langle u|v\rangle$. This compact notation is invaluable in quantum mechanics, where it simplifies expressions and makes operator manipulations transparent. It will be discussed in detail in Section 2.3.

Example 2.5 demonstrates the calculation of inner products of functions in a vector space and the concept of orthogonality.

Example 2.5: Inner Product of Functions

Evaluate the inner product of the functions $\phi_1(x) = \sin(x)$ and $\phi_2(x) = \cos(x)$ in the interval $[0, \pi]$.

Solution:

By definition,

$$\langle \phi_1 | \phi_2 \rangle = \int_0^{\pi} \sin(x)\cos(x)\, dx.$$

Using the identity $\sin(x)\cos(x) = \frac{1}{2}\sin(2x)$,

$$\begin{aligned}
\langle \phi_1 | \phi_2 \rangle &= \frac{1}{2}\int_0^{\pi} \sin(2x)\, dx \\
&= \frac{1}{2}\left[-\frac{1}{2}\cos(2x)\right]_0^{\pi} = \frac{1}{2}\left(-\frac{1}{2}[\cos(2\pi) - \cos(0)]\right) \\
&= \frac{1}{2}\left(-\frac{1}{2}[1-1]\right) = 0.
\end{aligned}$$

Thus, $\sin(x)$ and $\cos(x)$ are **orthogonal functions** on $[0, \pi]$.

Quantum Mechanics Connection

This simple calculation illustrates the orthogonality of wavefunctions, a key property in quantum mechanics. For example, the eigenfunctions of a particle in a one-dimensional box are orthogonal under the inner product $\langle \psi_m | \psi_n \rangle = \int \psi_m^*(x)\psi_n(x)\, dx$. Orthogonality ensures that different quantum states are independent and can form a complete basis for expanding arbitrary wavefunctions.

In molecular orbital theory, basis functions (such as Gaussian-type orbitals) must be orthonormalized before constructing molecular orbitals. The overlap integral $\langle \phi_i | \phi_j \rangle$ plays the same role as the dot product in vector spaces. Linear dependence or poor orthogonality among basis functions can cause serious numerical problems in self-consistent field calculations.

Example 2.6 illustrates the concept of linear independence of functions in a vector space, followed by a demonstration of the linear independence of atomic orbitals.

Example 2.6: Linear Independence of Functions

Determine whether the following sets of functions are linearly independent.

a. $\{1, x, x^2\}$

b. $\{1, x, 5x - 1\}$

Solution:

a. If $a(1) + b(x) + c(x^2) = 0$ for all x, then necessarily $a = b = c = 0$. Hence, the three functions are linearly independent.

b. Consider $a(1) + b(x) + c(5x - 1) = 0$. Then, $a(1) + b(x) + c(5x - 1) = (a - c) + (b + 5c)x = 0$. This leads to the system: $a - c = 0$ and $b + 5c = 0$. Choosing $a = 1$, $b = -5$, $c = 1$ gives a nontrivial linear relation. Thus, the functions are linearly dependent.

We will now demonstrate that the 2s and 3s atomic orbitals of a hydrogen atom are linearly independent functions. To test for linear independence, we must show that the only solution to the equation $c_1\psi_{2s}(r) + c_2\psi_{3s}(r) = 0$ is $c_1 = 0$ and $c_2 = 0$. This equation must hold true for all possible values of the position r.

The key to the solution lies in the **radial nodes** of the orbitals. Atomic orbitals are functions with specific shapes and locations where their value is zero, called nodes. In the following plot (Figure 2.1):

- The ψ_{2s} orbital (solid curve) has **one** radial node at a specific distance from the nucleus, let us call it $r_2(2.0\ a_0)$. At this distance, $\psi_{2s}(r_2) = 0$.
- The ψ_{3s} orbital (dashed curve) has **two** radial nodes at different positions (1.9 and 7.1 a_0), neither of which is at r_2. Therefore, $\psi_{3s}(r_2)$ is not zero.

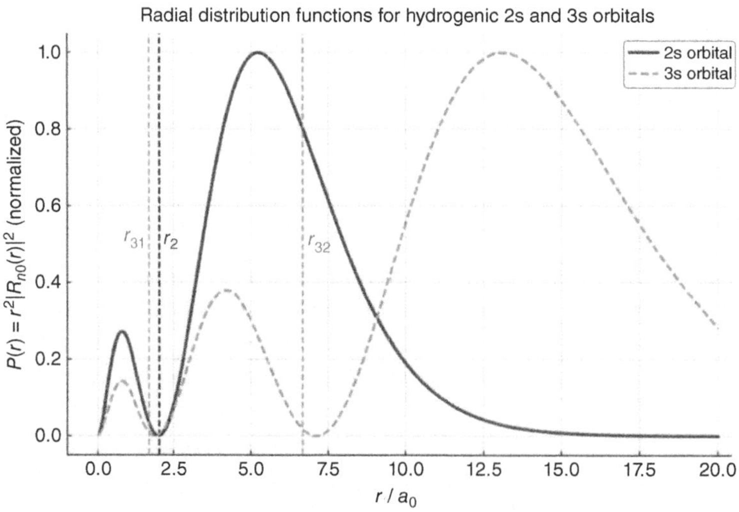

Figure 2.1 Radial distribution functions for hydrogenic 2s and 3s orbitals.

Let us evaluate our equation at the specific position of the 2s node, $r = r_2$:

$$c_1\psi_{2s}(r_2) + c_2\psi_{3s}(r_2) = 0.$$

Substituting what we know:

$$c_1 \times (0) + c_2 \times (\text{some nonzero value}) = 0$$

This simplifies to $c_2 \times$ (some nonzero value) $= 0$, which can only be true if $c_2 = 0$.

Now that we have proven c_2 must be zero, our original equation simplifies to $c_1\psi_{2s}(r) = 0$. Since the $\psi_{2s}(r)$ function is not zero everywhere, the only way for this equation to hold true for all r is if $c_1 = 0$.

Because the only solution is the trivial one ($c_1 = 0$, $c_2 = 0$), the 2s and 3s atomic orbitals are **linearly independent**.

Quantum Chemistry Connection

The node-counting argument just used generalizes: any two hydrogenic orbitals with different principal quantum number n are automatically linearly independent, since each has a distinct number of radial nodes. This is why STO and Gaussian basis sets built from different n shells (1s, 2s, 3s, ...) start from a linearly independent foundation — the real numerical risk in practice comes not from orbitals of different n, but from near-linear-dependence between diffuse functions added within the same shell, e.g., closely spaced polarization or diffuse functions in extended basis sets like aug-cc-pVTZ.

Normalization of Vectors and Functions

A vector $\mathbf{u}$ is said to be **normalized** if its norm is unity, i.e., $|\mathbf{u}| = 1$. If a vector does not already have unit length, we can always construct a normalized version by dividing by its norm $\hat{\mathbf{u}} = \frac{\mathbf{u}}{|\mathbf{u}|}$.

The same idea applies to functions in Hilbert spaces. A wavefunction $\psi(x)$ is normalized if

$$\langle\psi|\psi\rangle = \int |\psi(x)|^2 dx = 1.$$

Normalization ensures that vectors (or functions) can be compared on equal footing and is a prerequisite for constructing orthonormal bases, where each vector has a unit norm and is mutually orthogonal to the others. This will be essential later for quantum mechanics, where normalized wavefunctions correspond to physically realizable states.

Example 2.7: Normalization of a Function

Normalize the function $\phi(x) = \sin(x)$ over the interval $[0, \pi]$.

Solution:

We require that

$$\langle\phi|\phi\rangle = \int_0^{\pi} \sin^2(x)\, dx = 1.$$

Using the identity $\sin^2(x) = \frac{1}{2}(1 - \cos(2x))$,

$$\int_0^{\pi} \sin^2(x)dx = \frac{1}{2}\int_0^{\pi}(1 - \cos(2x))dx$$

$$= \frac{1}{2}\left[x - \frac{1}{2}\sin(2x)\right]_0^{\pi}$$

$$= \frac{1}{2}[\pi - 0] = \frac{\pi}{2}.$$

Thus,

$$\langle\phi|\phi\rangle = \frac{\pi}{2}.$$

To normalize, we divide by the square root of this value:

$$\phi_{\text{norm}}(x) = \frac{1}{\sqrt{\pi/2}}\sin(x) = \sqrt{\frac{2}{\pi}}\sin(x).$$

Quantum Mechanics Connection

This is the normalized form of the **first eigenfunction** of a particle in a one-dimensional box of length π. Normalization guarantees that the total probability of finding the particle somewhere inside the box is exactly 1.

Generalization

For the family of functions $\phi_n(x) = \sin(nx)$, defined on the interval $[0, \pi]$, the normalization condition requires

$$\langle\phi_n|\phi_n\rangle = \int_0^{\pi} \sin^2(nx)\,dx = 1.$$

Using the identity $\sin^2(nx) = \frac{1}{2}(1 - \cos(2nx))$,

$$\int_0^{\pi} \sin^2(nx)dx = \frac{1}{2}\left[x - \frac{1}{2n}\sin(2nx)\right]_0^{\pi}.$$

Since $\sin(2n\pi) = \sin(0) = 0$, this reduces to

$$\int_0^{\pi} \sin^2(nx)\,dx = \frac{\pi}{2}.$$

Thus, all the functions $\sin(nx)$ have the same squared norm, independent of n. The normalized functions are therefore

Quantum Mechanics Connection

These normalized functions are precisely the eigenfunctions of a **particle in a one-dimensional box of length** π. They form an **orthonormal basis set** in the Hilbert space of square-integrable functions on $[0, \pi]$. Any admissible wavefunction for the particle can be expressed as a linear combination of these functions.

$$\phi_n^{\text{norm}}(x) = \sqrt{\frac{2}{\pi}} \sin(nx), \; n = 1, 2, 3, \ldots .$$

Orthonormality of the Sine Basis

For integers $n, m \geq 1$, consider the inner product on $[0, \pi]$:

$$\langle \phi_n | \phi_m \rangle = \int_0^{\pi} \sin(nx) \sin(mx) \, dx.$$

Using trigonometric identities,

$$\sin A \sin B = \frac{1}{2}[\cos(A - B) - \cos(A + B)].$$

Substituting $A = nx$ and $B = mx$,

$$I = \frac{1}{2} \int_0^{\pi} [\cos((n - m)x) - \cos((n + m)x)] \, dx.$$

Integrating term by term,

$$I = \frac{1}{2}\left[\frac{\sin((n - m)x)}{n - m} - \frac{\sin((n + m)x)}{n + m}\right]_0^{\pi}.$$

Evaluating the limits:
At $x = 0$, both sine terms vanish.
At $x = \pi$,

$$\sin((n \pm m)\pi) = 0,$$

because n and m are integers.
Therefore, the expression becomes

$$I = 0$$

for $n \neq m$.
The special case is $n = m$.
If $n = m$,

$$I = \int_0^{\pi} \sin^2(nx) \, dx = \frac{\pi}{2}$$

(using $\sin^2 x = \frac{1}{2}(1 - \cos 2x)$).
Therefore,

$$\int_0^\pi \sin(nx)\sin(mx)\,dx = \begin{cases} 0, & n \neq m, \\ \frac{\pi}{2}, & n = m. \end{cases}$$

If $n = m$,

$$\langle \phi_n | \phi_n \rangle = \frac{\pi}{2}.$$

Normalization gives $\phi_n^{\text{norm}}(x) = \sqrt{\frac{2}{\pi}}\sin(nx)$.
If $n \neq m$,

$$\int_0^\pi \sin(nx)\sin(mx)\,dx = 0.$$

Thus, the functions are orthogonal.

This orthogonality property of sine functions underlies Fourier series (Chapter 9) and the orthonormal basis used in many areas of quantum mechanics and mathematical physics.

Hence, the family $\left\{\sqrt{\frac{2}{\pi}}\sin(nx)\right\}_{n=1}^{\infty}$ forms an **orthonormal set of functions** on $[0, \pi]$.

Quantum Mechanics Connection

This result underlies the **Fourier sine series** and provides the complete orthonormal basis set for the one-dimensional particle-in-a-box problem in quantum mechanics. Each $\phi_n(x)$ represents a quantum state with definite energy, and the orthogonality condition reflects the independence of different quantum states.

Fact Box 2.4: VOCABULARY ALERT 4: "State"

In Chemistry: Often refers to a physical condition (e.g., solid state, ground state).
In Quantum Mechanics: The "state" of a system (e.g., its energy, angular momentum) is completely described by a vector in a Hilbert space, written as a ket $|\psi\rangle$ (see Section 2.3).
Key Idea: The ket $|\psi\rangle$ is the state vector itself. Saying "the system is in the state $|\psi\rangle$" is analogous to saying "the vector is **v**."

In the earlier discussion, we briefly used Dirac's bra–ket notation, which provides a compact and elegant way to express inner products and norms. In the next section, we formally introduce this notation and explore its full structure and applications.

Key Takeaways

- The inner product generalizes the dot product to any N-dimensional space.
- Orthogonality $\Leftrightarrow$ inner product $= 0$.
- Norm $= \sqrt{(\text{inner product with itself})}$.

- Unit vectors are obtained by dividing by the norm.
- Inner products extend naturally to function spaces.

Quantum Mechanics Connection

Wavefunctions are treated as vectors in Hilbert space, with overlaps (integrals of product functions) as inner products. Orthogonality underpins selection rules in spectroscopy.

2.3 Dirac Notation

In quantum mechanics, vectors are most conveniently represented using **Dirac's bra–ket notation**, which is designed to work equally well for finite- and infinite-dimensional vector spaces.

Kets:

A ket is a column vector, written as $|v\rangle$. In this notation, there is no need to use an arrow above the symbol or boldface.

Bras:

For every ket $|v\rangle$, there exists a corresponding bra, written as $\langle v|$. This is the **Hermitian conjugate** (complex conjugate transpose) or adjoint of the ket. If $|v\rangle$ is an $N \times 1$ column vector, then $\langle v|$ is a $1 \times N$ row vector with conjugated entries. Thus, the bra lives in the **dual space** $\mathbb{V}^\dagger$.

Bra–ket (inner product):

The inner product of two vectors is written as a bra followed by a ket, $\langle u|v\rangle$. By definition,

$$\langle v|v\rangle = |v|^2 \geq 0,$$

which is always real and nonnegative. If $\langle v|v\rangle = 1$, the ket $|v\rangle$ is said to be **normalized**.

Note that order matters: $\langle u|v\rangle = (\langle v|u\rangle)^*$.

Concept	Traditional	Dirac Notation
Vector	$\mathbf{v}$	$\|v\rangle$
Dual vector	—	$\langle v\|$
Inner product	$\mathbf{v} \cdot \mathbf{w}$	$\langle v\|w\rangle$
Component	v_x	$\langle x\|v\rangle$

Example 2.8: Inner Products

Let

$$|v_1\rangle = \begin{bmatrix} 2+i \\ -1 \\ 3-i \end{bmatrix}, \quad |v_2\rangle = \begin{bmatrix} 1-i \\ 2 \\ 0 \end{bmatrix}.$$

i. Find the bra $\langle v_1|$.
ii. Compute the inner products $\langle v_1|v_2\rangle$ and $\langle v_2|v_1\rangle$.

Solution:

i. Take the complex conjugates and the transpose: $\langle v_1| = (2-i,\ -1,\ 3+i)$.
ii. $\langle v_1|v_2\rangle = (2-i)(1-i) + (-1)(2) + (3+i)(0) = 2 - 2i - i + i^2 - 2 = 2 - 3i + -1 - 2 = -1 - 3i$.
iii. The reverse inner product is the complex conjugate:
$\langle v_2|v_1\rangle = -1 + 3i$.

Historical Note 2.1: Paul A. M. Dirac (1902–1984)

Paul Dirac was a central figure in the development of quantum mechanics and is best known for placing the theory on a rigorous mathematical foundation. He introduced the abstract vector-space formulation of quantum mechanics and the bra–ket notation, which clarified the role of states as vectors and physical observables as operators. His relativistic wave equation for the electron naturally incorporated spin and led to the theoretical prediction of antimatter. Dirac's work established much of the operator formalism that underlies modern quantum chemistry and physics.

Red Flag 2.1

Order matters: $\langle u|v\rangle = (\langle v|u\rangle)^*$.

2.4 Hilbert Spaces

A finite vector space with an inner product is a Hilbert space, usually denoted $\mathcal{H}^N$ if it has dimension N. A Hilbert space is also a **complete metric space**. In finite dimensions, every inner product space is complete, but in the infinite-dimensional function spaces used for wavefunctions, completeness becomes essential and cannot be assumed.

For a canonical orthonormal basis $\{|i\rangle\}$, the completeness relation holds:

$$\langle i|j\rangle = \delta_{ij}, \quad \sum_i |i\rangle\langle i| = I. \tag{2.8}$$

This identity is extremely useful: it allows any ket $|v\rangle$ to be expanded in terms of basis kets, and the coefficients can be extracted via inner products.

In Equation (2.8), the expression $|i\rangle\langle i|$ is simply the standard Dirac notation used to represent the contribution of the basis ket $|i\rangle$ to the completeness relation. It represents an **outer product**. This is best illustrated with an example.

Example

Consider the orthonormal basis

$$|1\rangle = \begin{bmatrix} 1 \\ 0 \end{bmatrix}, \quad |2\rangle = \begin{bmatrix} 0 \\ 1 \end{bmatrix},$$

which satisfies

$$\langle 1|1\rangle = 1, \quad \langle 2|2\rangle = 1, \quad \langle 1|2\rangle = \langle 2|1\rangle = 0.$$

i.e., it is an orthonormal set, and the first condition in Equation (2.8) is satisfied.

Then

$$|1\rangle\langle 1| = \begin{bmatrix} 1 \\ 0 \end{bmatrix}\begin{bmatrix} 1 & 0 \end{bmatrix} = \begin{bmatrix} 1 & 0 \\ 0 & 0 \end{bmatrix}, \quad |2\rangle\langle 2| = \begin{bmatrix} 0 & 0 \\ 0 & 1 \end{bmatrix}.$$

Adding them gives

$$|1\rangle\langle 1| + |2\rangle\langle 2| = \begin{bmatrix} 1 & 0 \\ 0 & 1 \end{bmatrix} = I.$$

This satisfies the second condition of Equation (2.8).

Example 2.9 explores the orthogonality condition.

Example 2.9: Orthogonality Condition

Let

$$|u\rangle = |0\rangle + a|1\rangle, \quad |v\rangle = |0\rangle - 2|1\rangle,$$

where $|0\rangle$ and $|1\rangle$ are orthonormal basis vectors, and a is a constant.
Find the value of a such that $|u\rangle$ and $|v\rangle$ are orthogonal.

Solution:

We require

$$\langle v|u\rangle = 0.$$

Compute:

$$\langle v|u\rangle = (\langle 0| - 2\langle 1|)(|0\rangle + a|1\rangle) = 1 - 2a.$$

So, the orthogonality condition gives

$$1 - 2a = 0 \Rightarrow a = \frac{1}{2}.$$

Extension to Functions

This notation also applies to functions, which can be regarded as vectors in an infinite-dimensional Hilbert space. A function $\psi(x)$ is written as $|\psi\rangle$, and its inner product with another function $\phi(x)$ is defined as

$$\langle\phi|\psi\rangle = \int \phi^*(x)\,\psi(x)d\tau. \tag{2.9}$$

Here, the sum over discrete components has been replaced by an integral, corresponding to the limit of an infinitesimal step size.

The bra–ket notation can be applied to any representation of the wavefunction—coordinate space $\psi(x)$, momentum space $\phi(p)$, or expansion in terms of eigenfunctions.

Outer Product

For two kets $|u\rangle$ and $|v\rangle$, the **outer product** is

$$|u\rangle\langle v|,$$

which produces an operator (matrix) that maps kets to kets. This construction will be central when we discuss projection operators and observables in quantum mechanics, which will be the subject of the next chapter.

Fact Box 2.5: Dirac Bra–Ket Notation

- **Ket** $|v\rangle \rightarrow$ column vector in Hilbert space.
- **Bra** $\langle v| \rightarrow$ Hermitian conjugate (row vector with complex conjugates).
- **Inner product** (bra–ket):

$$\langle u|v\rangle = \sum_i u_i^* v_i = \int u^*(x)v(x)d\tau$$

  - Real, positive if $u = v$.
  - Zero if vectors/functions are orthogonal.
- **Norm:**

$$|v| = \sqrt{\langle v|v\rangle}.$$

- **Outer product** (ket–bra):

$$|u\rangle\langle v|$$

  $\rightarrow$ defines an operator (matrix) that acts on kets.
- Completeness relation:

$$\sum_i |i\rangle\langle i| = I.$$

Key Takeaways

- A ket $|\psi\rangle$ is a column vector; the bra $\langle\psi|$ is its Hermitian conjugate.
- Inner product: $\langle\phi|\psi\rangle$ is scalar; order matters due to conjugation.
- Basis states: $\langle i|j\rangle = \delta_{ij}$ ensures orthonormality.
- Outer product: $|\psi\rangle\langle\phi|$ produces operators (matrices).
- Dirac notation applies equally to discrete vectors and continuous functions.

Quantum Chemistry Connection

Schrödinger wavefunctions and Heisenberg matrix mechanics are unified in this notation, simplifying the mathematics of quantum chemistry.

Practical note for chemists: In quantum chemistry, we constantly move between explicit wavefunctions $\psi(x)$ and abstract kets $|\psi\rangle$. The coordinate representation is for calculation; the Dirac notation is for general reasoning.

2.5 Some Useful Inequalities

Two inequalities frequently encountered in vector spaces, with important applications in quantum mechanics, are the Schwarz inequality and the triangle inequality.

Schwarz Inequality

For any two vectors $|u\rangle$ and $|v\rangle$, we have

$$\boxed{|\langle u|v\rangle|^2 \leq \langle u|u\rangle\langle v|v\rangle.} \tag{2.10}$$

If $|u\rangle$ and $|v\rangle$ are linearly dependent, that is, if $|u\rangle = c\,|v\rangle$ for some scalar c, then the inequality becomes an equality.

The Schwarz inequality generalizes a key concept from Euclidean geometry:

$$|\vec{u}\cdot\vec{v}| \leq |\vec{u}||\vec{v}|.$$

This captures the idea that a straight line between two points is the shortest path. For abstract vectors, this means that the overlap of two vectors is never greater than the product of their lengths.

In quantum mechanics, the Schwarz inequality is a foundational principle that supports the derivation of the **Heisenberg Uncertainty Principle**. This inequality provides the mathematical basis for the uncertainty principle by confirming that the product $\Delta x \Delta p_x$ possesses a defined lower bound.

Triangle Inequality

For any two vectors $|u\rangle$ and $|v\rangle$,

$$\boxed{|\,|u\rangle + |v\rangle| \leq |\,|u\rangle| + |\,|v\rangle|.} \tag{2.11}$$

Equality holds when $|u\rangle$ and $|v\rangle$ are linearly dependent with a positive real proportionality constant.

The triangle inequality is a general version of the well-known Euclidean principle that states that the length of any side of a triangle cannot be greater than the sum of the lengths of the other two sides.

Example 2.10: Inequalities

Consider the kets

$$|u\rangle = 2|0\rangle - i|1\rangle, \ |v\rangle = |0\rangle + 3|1\rangle,$$

where $|0\rangle$ and $|1\rangle$ are orthonormal.

i. Write the bra forms of $|u\rangle$ and $|v\rangle$.
ii. Calculate the scalar products $\langle u|v\rangle$ and $\langle v|u\rangle$.
iii. Verify that $|u\rangle$ and $|v\rangle$ satisfy the Schwarz inequality.
iv. Verify the triangle inequality for these kets.

Solution:

i. Bras corresponding to the kets

$$\langle u| = 2\langle 0| + i\langle 1|, \langle v| = \langle 0| + 3\langle 1|.$$

ii. Scalar products

$$\langle u|v\rangle = (2\langle 0| + i\langle 1|)(|0\rangle + 3|1\rangle) = 2 + 3i,$$
$$\langle v|u\rangle = (\langle 0| + 3\langle 1|)(2|0\rangle - i|1\rangle) = 2 - 3i.$$

As expected, they are complex conjugates.

iii. Schwarz inequality

Compute norms:

$$\langle u|u\rangle = (2\langle 0| + i\langle 1|)(2|0\rangle - i|1\rangle) = 4 + 1 = 5,$$
$$\langle v|v\rangle = (\langle 0| + 3\langle 1|)(|0\rangle + 3|1\rangle) = 1 + 9 = 10.$$

Left-hand side:

$$|\langle u|v\rangle|^2 = |2 + 3i|^2 = 4 + 9 = 13.$$

Right-hand side:

$$\langle u|u\rangle\langle v|v\rangle = 5 \times 10 = 50.$$

Since $13 \leq 50$, the Schwarz inequality is satisfied.

iv. Triangle inequality

$$|u\rangle + |v\rangle = (2 + 1)|0\rangle + (-i + 3)|1\rangle = 3|0\rangle + (3 - i)|1\rangle.$$

Its norm:

$$|u + v| = \sqrt{9 + |3 - i|^2} = \sqrt{9 + (9 + 1)} = \sqrt{19}.$$

Right-hand side:

$$|u| + |v| = \sqrt{5} + \sqrt{10}.$$

Since $\sqrt{19} \leq \sqrt{5} + \sqrt{10}$, the triangle inequality holds.

Remark This concrete example illustrates the abstract inequalities in a simple two-dimensional Hilbert space. The same principles extend directly to higher-dimensional vector spaces and to function spaces in quantum mechanics. These inequalities hold for infinite-dimensional spaces too, which is crucial for quantum mechanics where our spaces are typically infinite-dimensional. In quantum chemistry, wavefunctions also form a vector space.

Fact Box 2.6: VOCABULARY ALERT 5: "Space"

In Everyday Language: The physical void in which the universe exists.
In Mathematics: A "vector space" or "Hilbert space" is an abstract set of elements with defined addition and multiplication rules.
Key Idea: Quantum states "live" in an abstract Hilbert space. This is not a physical location, but the mathematical arena where all quantum mechanical calculations take place. An electron's wavefunction is a point in this space.

The following example (Example 2.11) verifies the orthogonality of the wavefunctions of a particle in a one-dimensional box and demonstrates the Schwarz inequality within this functional space.

Example 2.11: Function Space—Orthogonality and Schwarz Inequality

Consider the functions (wavefunctions of a particle in a 1D box of length L):

$$\psi_1(x) = \sqrt{\frac{2}{L}} \sin\left(\frac{\pi x}{L}\right), \psi_2(x) = \sqrt{\frac{2}{L}} \sin\left(\frac{2\pi x}{L}\right).$$

i. Show that $\psi_1(x)$ and $\psi_2(x)$ are orthogonal.
ii. Verify the Schwarz inequality in this function space.

Solution:

i. The inner product in function space is defined as

$$\langle \psi_m | \psi_n \rangle = \int_0^L \psi_m^*(x)\psi_n(x)dx.$$

Thus,

$$\langle \psi_1 | \psi_2 \rangle = \frac{2}{L} \int_0^L \sin\left(\frac{\pi x}{L}\right) \sin\left(\frac{2\pi x}{L}\right) dx.$$

Using the identity $\sin A \sin B = \frac{1}{2}[\cos(A - B) - \cos(A + B)]$, the integral becomes

$$\langle \psi_1 | \psi_2 \rangle = \frac{1}{L} \int_0^L \left[\cos\left(\frac{\pi x}{L}\right) - \cos\left(\frac{3\pi x}{L}\right)\right] dx.$$

Both cosine terms integrate to zero over the interval $[0, L]$. (Since the states are distinct eigenfunctions, they are orthogonal, making the LHS zero.) Hence,

$$\langle \psi_1 | \psi_2 \rangle = 0,$$

showing that ψ_1 and ψ_2 are orthogonal.

ii. The norms are

$$\langle \psi_1 | \psi_1 \rangle = \int_0^L |\psi_1(x)|^2 dx = 1,$$

$$\langle \psi_2 | \psi_2 \rangle = \int_0^L |\psi_2(x)|^2 dx = 1.$$

The Schwarz inequality requires

$$|\langle \psi_1 | \psi_2 \rangle|^2 \leq \langle \psi_1 | \psi_1 \rangle \langle \psi_2 | \psi_2 \rangle.$$

Here, LHS $= 0^2 = 0$, and RHS $= 1 \times 1 = 1$.
Thus,

$$0 \leq 1,$$

so, the Schwarz inequality is satisfied.

Comment

This illustrates that wavefunctions form a Hilbert space: the orthogonality and normalization properties of $\psi_n(x)$ mirror those of finite-dimensional basis vectors.

Key Takeaways

- **Schwarz inequality:** $|\langle \phi | \psi \rangle|^2 \leq \langle \phi | \phi \rangle \langle \psi | \psi \rangle$. Equality holds if vectors are linearly dependent.
- **Triangle inequality:** $|\phi + \psi| \leq |\phi| + |\psi|$. Equality holds for collinear vectors.
- These generalize familiar Euclidean results.

Quantum Mechanics Connection

- The Schwarz inequality underlies the proof of the Heisenberg Uncertainty Principle.
- The triangle inequality appears in discussions of distances in state spaces.

2.6 Summary

In this chapter we extended the ideas of three-dimensional Euclidean vectors to the broader concept of **linear vector spaces**, which provide the mathematical foundation for quantum mechanics, spectroscopy, and many areas of theoretical chemistry and physics:

- **Vector Spaces**
  A vector space is a set of objects closed under vector addition and scalar multiplication. The constituents need not be geometric arrows; they can be matrices, functions, or wavefunctions. Real vector spaces involve real scalars, while complex vector spaces involve complex scalars.
- **Dimension and Basis**
  The **dimension** of a vector space is the number of linearly independent vectors needed to span it. Basis vectors provide the framework for expressing all other vectors as linear combinations. An **orthonormal basis** is often preferred for convenience.
- **Inner Product and Norm**
  The inner product generalizes the dot product and defines orthogonality, length (norm), and angles in any dimension. In function spaces, the inner product is expressed as an integral. **Normalization** ensures vectors (or wavefunctions) have unit length.
- **Dirac Notation**
  The bra–ket formalism unifies vectors and functions in a compact notation. Kets $|v\rangle$ represent column vectors, bras $\langle v|$ their Hermitian conjugates, and their product $\langle u|v\rangle$ defines the inner product. This notation is indispensable in quantum mechanics.
- **Useful Inequalities**
  The Schwarz inequality and triangle inequality set fundamental constraints on vectors and wavefunctions. They underpin important physical principles such as the **Heisenberg Uncertainty Principle**.
- **Key Connections**
  Linear vector spaces provide the mathematical scaffold for **Hilbert spaces**, where quantum states (wavefunctions) live.

In this chapter, we introduced the essential structure of linear vector spaces—vectors, linear combinations, bases, inner products, norms, and the inequalities that govern them. These ideas form the foundation on which more advanced concepts are built. In many applications, especially in quantum chemistry, physics, and engineering, we frequently need to not only describe vectors but also **transform** them, compare different bases, and construct mappings that act on vectors in systematic ways. These topics naturally lead us to the study of **linear transformations** and the operators associated with them, which we now develop in the next chapter.

Fact Box 2.7: Key Results From Chapter 2

- **Vector space:** A set closed under addition and scalar multiplication.
- **Dimension:** The number of linearly independent vectors in a space.
- Inner product:

$$\langle u, v\rangle = \sum_i u_i v_i$$

generalizes the dot product, defines orthogonality, and gives the norm: $|\mathbf{u}| = \sqrt{\langle u, u\rangle}$.

- Dirac notation:
  - $|v\rangle$: ket (column vector)
  - $\langle v|$: bra (Hermitian conjugate)
  - $\langle u|v\rangle$: inner product
  - $|u\rangle\langle v|$: outer product
- Inequalities:
  - Schwarz: $|\langle u|v\rangle|^2 \leq \langle u|u\rangle\langle v|v\rangle$
  - Triangle: $|\mathbf{u} + \mathbf{v}| \leq |\mathbf{u}| + |\mathbf{v}|$

Further Reading

P. W. Atkins and R. S. Friedman, *Molecular Quantum Mechanics*, Oxford University Press, 2010—chapters on symmetry and quantum states.

I. N. Levine, *Quantum Chemistry*, 7th ed., Pearson, 2014—sections on vector spaces and operators.

D. A. McQuarrie, *Mathematical Methods for Scientists and Engineers*, University Science Books, 2003—clear introduction to linear algebra in physics/chemistry contexts.

N. Zettili, *Quantum Mechanics: Concepts and Applications*, 2nd ed., Wiley, 2009—offers an in-depth treatment of the practical mathematical tools of quantum mechanics.

Exercises

A. Conceptual Understanding

1. **Orthogonality of Functions**
   Show that the functions

$$\sin\left(\frac{n\pi x}{L}\right) \text{ and } \sin\left(\frac{m\pi x}{L}\right)(n \neq m)$$

   are orthogonal on the interval $[0, L]$. What does this imply for Hilbert space?

2. **Inner Product with Complex Conjugation**
   Explain why the inner product for quantum mechanical wavefunctions must be defined with complex conjugation. What would go wrong physically if we used $\langle\psi|\phi\rangle = \int \psi\phi dx$ instead of $\int \psi^*\phi dx$?

3. **Linear Dependence in Function Spaces**
   Are the functions $1, x$, and x^2 linearly independent on the interval $[0, 1]$? Justify your answer.

4. **Zero Vector Uniqueness**
   Explain why no nonzero vector can be orthogonal to itself.

B. Problems

1. Checking Orthonormality

Given the vectors in $\mathbb{R}^3$:

$$\vec{u} = (1, 1, 0), \quad \vec{v} = (-1, 1, 0), \quad \vec{w} = (0, 0, 1),$$

normalize them and show that the set $\vec{u}$, $\vec{v}$, $\vec{w}$ becomes orthonormal after normalization.

2. Completeness in a Finite Basis

Let

$$|1\rangle = \begin{bmatrix} 1 \\ 0 \end{bmatrix}, \ |2\rangle = \begin{bmatrix} 0 \\ 1 \end{bmatrix}, \ |v\rangle = \begin{bmatrix} a \\ b \end{bmatrix}.$$

Use the completeness relation

$$I = |1\rangle\langle 1| + |2\rangle\langle 2|$$

to show that

$$|v\rangle = a\,|1\rangle + b\,|2\rangle.$$

3. Linear Independence

Show that the vectors

$$(1, 2, 3), (2, 4, 6), (1, 0, 1)$$

are linearly dependent. Find a nontrivial linear relation between them.

4. Norms and the Schwarz Inequality

Verify the Schwarz inequality for the vectors

$$\vec{u} = (1, -1, 2), \quad \vec{v} = (2, 0, 1)$$

in $\mathbb{R}^3$.

C. Advanced Problems

1. Orthogonality of Hydrogenic s-Orbitals

Show that the hydrogenic 2s and 3s orbitals are orthogonal by evaluating

$$\int_0^\infty R_{2s}(r)R_{3s}(r)r^2 dr = 0,$$

or by explaining why hydrogenic radial functions with different n are orthogonal.

2. Infinite-dimensional Vector Space

Show that the set of functions x^n for $n = 0, 1, 2 \ldots$ are linearly independent, implying an infinite dimension.

3. Dimension Argument

Show that no set of more than N vectors in an N-dimensional vector space can be linearly independent.

D. Chemistry/Physics/Engineering Applications

1. Quantum States as Vectors

If

$$|\psi\rangle = a\,|1\rangle + b\,|2\rangle + c\,|3\rangle,$$

and the basis is orthonormal, show that normalization requires

$$|a|^2 + |b|^2 + |c|^2 = 1.$$

2. Signal Processing Analogy

In electrical engineering, three signals

$$\sin(\omega t),\ \cos(\omega t),\ \sin(2\omega t)$$

form vectors in a function space. Are the first two orthogonal? Why does this matter for circuit analysis?

3

Linear Transformations and Operators

The book of nature you can only understand, if you have previously learnt the language and the letters. It is written in mathematical language and the letters on geometrical figures, and without these means it is impossible for human beings to understand even a word of it.

—Galileo Galilei, 17th Century

Learning Outcomes

After completing this chapter, the student will be able to:

1. **Define** linear transformations and linear operators and **verify** linearity using the principle of superposition.
2. **Construct** matrix representations of linear transformations and **interpret** matrix multiplication as operator action on vectors.
3. **Relate** operators to physical observables in quantum mechanics, including the role of Hermitian operators and basis dependence of matrix representations.
4. **Use** outer products to construct projection operators and **interpret** their action on vectors and quantum states.
5. **Represent** geometrical operations (rotations, reflections, inversions, improper rotations) as matrices and **understand** their role in molecular symmetry and group theory.
6. **Distinguish** between linear and affine transformations and **apply** affine mappings to molecular coordinates, neural networks, and computational models.
7. **Recognize** linear and affine mappings as the mathematical core of neural networks, including the roles of weight matrices, bias vectors, activation functions, and gradient-based optimization.

3.1 Introduction

Linear transformations play a central role in mathematics and in the physical sciences. They allow us to map vectors from one representation to another, express changes of basis, and formulate operators that describe measurable quantities, symmetries, and physical processes. In chemistry, physics, and engineering, matrices and operators arise whenever we describe rotations, projections, reflections, translations, or changes of coordinate frames.

In this chapter, we formalize the idea of a linear transformation, relate it to its **matrix representation**, and examine how vectors and operators behave under changes of basis. Matrices will naturally appear as finite-dimensional representations of such operators. In quantum mechanics, the Hamiltonian itself is a **linear operator**, and its mathematical properties are central to understanding physical observables. These mathematical structures are not abstract curiosities—they form the very language in which molecular quantum mechanics is written, from atomic orbitals (AOs) to molecular vibrations.

We then introduce outer products, **projection operators**, and **affine transformations**—key tools for working with quantum states, molecular orbitals (MOs), vector decompositions, geometric manipulations, and computational models. These ideas extend the framework developed in the previous chapter and prepare the ground for eigenvalue problems, diagonalization, and operator-based formulations that appear throughout the physical sciences.

3.2 Linear Transformations

A transformation T that takes an N-dimensional vector and transforms it into another M-dimensional vector is said to be **linear** if it satisfies the following condition:

$$T(a\mathbf{u} + b\mathbf{v}) = aT(\mathbf{u}) + bT(\mathbf{v}), \tag{3.1}$$

for all N-dimensional vectors $\mathbf{u}$, $\mathbf{v}$ and for all scalars a, b.

3.2.1 Properties of a Linear Transformation

- A linear transformation always maps the zero vector to the zero vector:

  $$T(\mathbf{0}) = \mathbf{0}.$$

  Proof: $T(\mathbf{0}) = T(0 \cdot \mathbf{v}) = 0 \cdot T(\mathbf{v}) = \mathbf{0}$.
- For any set of vectors $\mathbf{v}_1, \mathbf{v}_2, \dots, \mathbf{v}_k$ in N-dimensional space and scalars $c_1, c_2, \dots, c_k$, we have the **principle of superposition**:

  $$T(c_1\mathbf{v}_1 + c_2\mathbf{v}_2 + \cdots + c_k\mathbf{v}_k) = c_1T(\mathbf{v}_1) + c_2T(\mathbf{v}_2) + \cdots + c_kT(\mathbf{v}_k). \tag{3.2}$$

3.2.2 Matrix Representation of a Linear Transformation

Let us now construct an $M \times N$ **matrix** A whose columns are the vectors $T(\mathbf{e}_1), T(\mathbf{e}_2), \dots, T(\mathbf{e}_N)$, where $\mathbf{e}_j$ denotes the jth standard **basis** vector in N-space. Then, for any vector $\mathbf{x}$,

$$T(\mathbf{x}) = A\mathbf{x}.$$

Equivalently, multiplying a matrix A by the coordinate vector $\mathbf{e}_j$ simply selects its jth column. For example, if the matrix

$$A = \begin{bmatrix} 1 & 2 \\ 3 & 4 \\ 5 & 6 \end{bmatrix}$$

is multiplied by

$$\mathbf{e}_2 = \begin{bmatrix} 0 \\ 1 \end{bmatrix},$$

the result is simply the second column of A:

$$A\mathbf{e}_2 = \begin{bmatrix} 2 \\ 4 \\ 6 \end{bmatrix}.$$

Thus, **every linear transformation can be represented as a matrix multiplication** (Examples 3.1 and 3.2).

Example 3.1: Linear Transformation

Find the linear transformation associated with the matrix

$$A = \begin{bmatrix} 1 & -2 \\ 0 & 3 \\ 4 & 5 \end{bmatrix}.$$

Solution:
Because the matrix is 3×2, it represents a linear transformation from a 2D vector space into a 3D vector space. Explicitly,

$$T\begin{bmatrix} x \\ y \end{bmatrix} = \begin{bmatrix} 1 & -2 \\ 0 & 3 \\ 4 & 5 \end{bmatrix}\begin{bmatrix} x \\ y \end{bmatrix} = \begin{bmatrix} x - 2y \\ 3y \\ 4x + 5y \end{bmatrix}$$

Example 3.2: Matrices as Linear Transformations

Find the standard matrix for the 3D to 4D linear transformation defined by

$$T(x, y, z) = (x + y, y + z, x + z, x + y + z).$$

Solution:
The matrix associated with this transformation will be 4×3:
- for $T(1, 0, 0) : (1, 0, 1, 1)$
- for $T(0, 1, 0) : (1, 1, 0, 1)$
- for $T(0, 0, 1) : (0, 1, 1, 1)$

Thus, the standard matrix is

$$A = \begin{bmatrix} 1 & 1 & 0 \\ 0 & 1 & 1 \\ 1 & 0 & 1 \\ 1 & 1 & 1 \end{bmatrix}$$

You may verify that for any vector (x, y, z),

$$T(x, y, z) = A \begin{bmatrix} x \\ y \\ z \end{bmatrix}.$$

3.3 Applications to Quantum Chemistry

Every linear transformation is associated with a unique matrix, and, conversely, every matrix defines a linear transformation.

Quantum Chemistry Connection: Linear Transformations in Quantum Chemistry

One of the most important applications of linear transformations in chemistry is the **basis transformation** between AOs and MOs.

- In the **LCAO-MO method**, each MO $|\psi_i\rangle$ is expressed as a linear combination of AOs $|\chi_\mu\rangle$:

  $$|\psi_i\rangle = \sum_\mu C_{\mu i} |\chi_\mu\rangle.$$

  Here, the coefficients $C_{\mu i}$ form the **transformation matrix** C.
- In matrix notation, the relationship can be written compactly as

  $$\Psi = XC,$$

  where X is the matrix of AOs, C is the transformation matrix, and Ψ contains the MOs.
- The transformation matrix C in LCAO-MO is typically unitary $(C^\dagger C = I)$, which preserves orthonormality of the basis—this is why MOs remain orthogonal if AOs are orthogonal.

As we shall see in the next chapter, finding the optimal C matrix for MOs involves solving an eigenvalue problem.

Fact Box 3.1: VOCABULARY ALERT 1: "Transformation"

In Mathematics: A function that maps vectors to vectors.

In Organic Chemistry: A series of reactions that convert one molecule into another.

In This Context: A "basis transformation" is a change of perspective. It is the mathematical operation of re-expressing the same quantum state (an MO) in a different set of basis functions (from AOs to MOs). The physical system does not change; only our description of it does.

Thus, the abstract mathematical idea of a linear transformation corresponds directly to the physical process of **constructing MOs from a chosen basis of AOs**. This is a cornerstone of computational quantum chemistry.

Mini-example: Bonding and Antibonding Orbitals of H_2

Consider two hydrogen 1s AOs, $|\chi_A\rangle$ and $|\chi_B\rangle$. We can form MOs as linear combinations of these AOs:

$$|\psi_+\rangle = \frac{1}{\sqrt{2}}(|\chi_A\rangle + |\chi_B\rangle)\text{(bonding MO)},$$
$$|\psi_-\rangle = \frac{1}{\sqrt{2}}(|\chi_A\rangle - |\chi_B\rangle)\text{(antibonding MO)}.$$

This transformation can be written in matrix form:

$$\begin{bmatrix} |\psi_+\rangle \\ |\psi_-\rangle \end{bmatrix} = \frac{1}{\sqrt{2}}\begin{bmatrix} 1 & 1 \\ 1 & -1 \end{bmatrix}\begin{bmatrix} |\chi_A\rangle \\ |\chi_B\rangle \end{bmatrix}.$$

Here, the 2×2 matrix performs the **linear transformation** from the AO basis $\{|\chi_A\rangle, |\chi_B\rangle\}$ to the MO basis $\{|\psi_+\rangle, |\psi_-\rangle\}$.

Takeaway: The concept of linear transformations provides the mathematical backbone for understanding orbital mixing and MO theory.

Formation of Molecular Orbitals from Atomic Orbitals

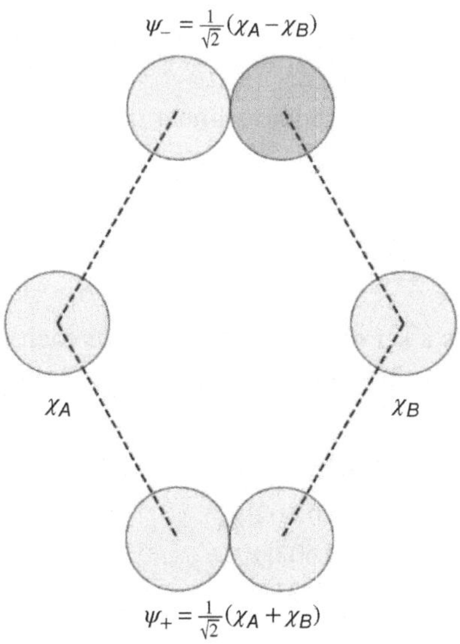

Key Takeaways

- A transformation is linear if $T(a\phi + b\psi) = aT\phi + bT\psi$.
- Every linear transformation can be represented by a matrix.
- Standard matrix is built by evaluating T on basis vectors.
- The superposition principle is a direct consequence.

Quantum Mechanics Connection

- Linear operators (like the Hamiltonian) describe physical observables.
- Transformation properties explain how orbitals and wavefunctions respond to rotations, reflections, and so on.

Let us expand this idea further for a slightly advanced case.

Worked Example: LCAO-MO Theory as a Vector Space Problem

Let us use the π-system of 1,3-butadiene to illustrate how AOs form a basis for the resulting MOs.

1,3-Butadiene

Problem Setup

1. **The System:** The 1,3-butadiene molecule has a π-system formed by four carbon atoms, each contributing one p-orbital.
2. **The Basis Set:** Our basis consists of the four atomic p-orbitals, which we can label as kets: $\{|\chi_1\rangle, |\chi_2\rangle, |\chi_3\rangle, |\chi_4\rangle\}$. This is a four-dimensional vector space.
3. **The Goal:** Find the four π MOs ($|\psi_1\rangle$ to $|\psi_4\rangle$) by expressing them as linear combinations of this p-orbital basis.

The MOs as Vectors

Solving the Hückel theory equations for butadiene yields a set of four coefficients for each of the four MOs. Each MO is a vector in our p-orbital space, and the coefficients are its components in that basis:

- **First MO (ψ_1):** $|\psi_1\rangle = 0.372|\chi_1\rangle + 0.602|\chi_2\rangle + 0.602|\chi_3\rangle + 0.372|\chi_4\rangle$.
- **Second MO (ψ_2):** $|\psi_2\rangle = 0.602|\chi_1\rangle + 0.372|\chi_2\rangle - 0.372|\chi_3\rangle - 0.602|\chi_4\rangle$.
- **Third MO (ψ_3):** $|\psi_3\rangle = 0.602|\chi_1\rangle - 0.372|\chi_2\rangle - 0.372|\chi_3\rangle + 0.602|\chi_4\rangle$.
- **Fourth MO (ψ_4):** $|\psi_4\rangle = 0.372|\chi_1\rangle - 0.602|\chi_2\rangle + 0.602|\chi_3\rangle - 0.372|\chi_4\rangle$.

As Column Vectors:

$\|\psi_1\rangle$	$\|\psi_2\rangle$	$\|\psi_3\rangle$	$\|\psi_4\rangle$
0.372	0.602	0.602	0.372
0.602	0.372	−0.372	−0.602
0.602	−0.372	−0.372	0.602
0.372	−0.602	0.602	−0.372

Conceptual Connection

This example perfectly illustrates the key concepts from this and the previous chapter:

- **Vector Space:** The set of all possible π-orbitals for butadiene forms a vector space.
- **Basis:** The four atomic p-orbitals serve as a convenient, physically intuitive **basis** for this space. The dimension of the space is 4.
- **Vectors:** We started with four basis vectors and obtained the same number of linearly independent MOs. Each resulting MO ($|\psi_1\rangle$, $|\psi_2\rangle$, etc.) is a **vector** within this space.
- **Components:** The familiar LCAO-MO coefficients are simply the **components** (or coordinates) of the MO vector in the chosen AO basis.

Changing our description from the set of four p-orbitals to the set of four MOs is a **change of basis**. Both sets are complete and can describe any π-state of the molecule, but the MO basis is more useful because in that basis, the energy operator (the Hamiltonian) is diagonal.

Transforming from an AO basis to an MO basis simplifies our description of the molecule. This idea of changing to a "better" basis is central to quantum mechanics. The ultimate goal, which we will explore in the next chapter, is to find the one special basis—the **eigenbasis**—where the matrix of an operator becomes a simple diagonal line. The values on that diagonal are the **eigenvalues**, which correspond to the only possible results of a physical measurement.

3.4 Matrices as Linear Operators

We can also consider a matrix as the operator for a linear transformation that converts one vector into another. An operator is simply a symbol for a specific mathematical procedure that transforms one function (or vector) into another:

$$A|v\rangle = |w\rangle. \tag{3.3}$$

In this equation, we say that the operator A acts on $|v\rangle$ to give a new vector $|w\rangle$. We have already encountered operators such as the gradient operator and the Laplacian; the latter will appear frequently in quantum mechanics.

Fact Box 3.2: VOCABULARY ALERT 2: "Operator"

In Mathematics: A rule that transforms one vector into another.

In Everyday Language: A person who operates equipment.

In Quantum Mechanics: An operator represents a physical observable (e.g., the Hamiltonian $\hat{H}$ is the energy operator). Acting on a state vector, $\hat{H}\,|\psi\rangle$ gives a new vector related to the energy measurement.

All the quantum mechanical operators are **linear** and obey the properties already stated in Equations (3.1) and (3.2). In addition, linear operators satisfy the following rules:

$$(A + B)|v\rangle = A|v\rangle + B|v\rangle, \tag{3.4}$$

$$(AB)|v\rangle = A(B|v\rangle). \tag{3.5}$$

Multiplication of operators is **associative**,

$$A(BC) = (AB)C,$$

but not necessarily **commutative**, i.e., in general,

$$AB \neq BA. \tag{3.6}$$

Two operators are said to be equal if their action is identical on all vectors of the space. We also define a **null operator** 0 such that $0|v\rangle = 0$, and the **identity operator** I, which leaves a vector unchanged, $I|v\rangle = |v\rangle$.

Now, consider an **orthonormal** basis $\{|e_i\rangle\}$ for the vector space. Any ket can be written as

$$|v\rangle = \sum_j v_j |e_j\rangle. \tag{3.7}$$

Then, the action of an operator A is

$$A|v\rangle = \sum_i \left(\sum_j a_{ij} v_j \right) |e_i\rangle, \tag{3.8}$$

so that a_{ij} is the ith component of the transformed vector in this basis. Substituting into Equation (3.3),

$$|w\rangle = A|v\rangle, \tag{3.9}$$

with $|w\rangle$ expanded as

$$|w\rangle = \sum_i w_i |e_i\rangle, \tag{3.10}$$

we obtain

$$w_i = \sum_j a_{ij} v_j, \tag{3.11}$$

which may also be written in matrix form as

$$\begin{bmatrix} w_1 \\ w_2 \\ \vdots \\ w_N \end{bmatrix} = \begin{bmatrix} a_{11} & a_{12} & \cdots & a_{1N} \\ a_{21} & a_{22} & \cdots & a_{2N} \\ \vdots & \vdots & \ddots & \vdots \\ a_{N1} & a_{N2} & \cdots & a_{NN} \end{bmatrix} \begin{bmatrix} v_1 \\ v_2 \\ \vdots \\ v_N \end{bmatrix}. \tag{3.12}$$

Thus, the operator A that converts one ket into another can be represented by a square matrix of order N. More generally, an operator transforming vectors from an N-dimensional space into an M-dimensional one corresponds to an $M \times N$ matrix. The set of all $M \times N$ matrices also forms a vector space of dimension MN.

Importantly, matrices inherit the properties of the operators they represent: addition, scalar multiplication, and linearity. In real Euclidean space, a matrix describes the transformation of one geometrical entity into another, which justifies its interpretation as a **linear operator**.

In quantum mechanics, the act of **measurement** is represented by an operator acting on a state vector. The postulate that observables correspond to Hermitian operators is the cornerstone that connects the entire mathematical formalism to experimental reality. Physical observables (like energy, momentum, and position) are represented by **Hermitian operators** because their eigenvalues are always real, corresponding to measurable quantities. This concept is fundamental. An operator is Hermitian if $\langle u|\hat{H}v\rangle = \langle \hat{H}u|v\rangle$ for any two vectors $|u\rangle$ and $|v\rangle$ in the space. More about these operators in the next chapter.

Crucially, the matrix representation of an operator depends on the chosen basis. The same quantum mechanical operator (like the Hamiltonian) has different matrix representations in different basis sets. While the individual elements of a matrix representation depend on the basis chosen, the underlying operator has intrinsic properties that are the same in any basis. In the next chapter, we will see that quantities like the **trace**—the sum of the diagonal elements—remain constant regardless of the basis, reflecting a fundamental characteristic of the operator itself.

Equation (3.12) shows that the action of a linear operator can always be expressed in terms of matrix multiplication. To see this more concretely, let us consider a simple geometric example.

Mini-example: A Simple Operator as a Matrix

Consider the operator that rotates vectors in the plane counterclockwise by 90°. In matrix form this operator is

$$R = \begin{bmatrix} 0 & -1 \\ 1 & 0 \end{bmatrix}.$$

Acting on the vector

$$|v\rangle = \begin{bmatrix} 1 \\ 0 \end{bmatrix},$$

we obtain

$$R|v\rangle = \begin{bmatrix} 0 & -1 \\ 1 & 0 \end{bmatrix}\begin{bmatrix} 1 \\ 0 \end{bmatrix} = \begin{bmatrix} 0 \\ 1 \end{bmatrix}.$$

Thus, the vector originally pointing along the x-axis is rotated to a point along the y-axis. This simple case illustrates the essence of operator action: a matrix encodes the transformation, while the vector records its outcome. With this in mind, we can now proceed to study more general classes of operators that arise naturally in physics and chemistry.

Key Takeaways

- Operators act on vectors to produce new vectors.
- Quantum mechanical operators (Hamiltonian, Laplacian, momentum, etc.) are all linear.
- Matrix elements a_{ij} represent operator action in a chosen basis.
- Null operator = 0, identity operator = I, with standard algebraic rules.

Quantum Chemistry Connection

- Observables correspond to Hermitian operators.
- Matrix representation allows computation of energies, overlaps, and transformations.

Fact Box 3.3: Computational Chemistry

In computational chemistry, we do not work with operators directly—we work with their matrix representations in some chosen AO basis.

3.5 Outer Products

In Dirac notation, we can combine a ket $|u\rangle$ and a bra $\langle v|$ to form a new object

$$|u\rangle\langle v|,$$

called the **outer product** of $|u\rangle$ and $|v\rangle$. Unlike inner products, which produce a **number**, the outer product produces an **operator**—a linear transformation that acts on vectors.

To see what it does, let it act on an arbitrary ket $|w\rangle$:

$$(|u\rangle\langle v|)|w\rangle = |u\rangle\langle v|w\rangle.$$

Thus, the bra $\langle v|$ takes an inner product with $|w\rangle$ (producing a scalar), and the result multiplies the ket $|u\rangle$. The operator therefore:

- **extracts the component of** $|w\rangle$ along $|v\rangle$; and
- **replaces the direction of that component** by $|u\rangle$.

If $|u\rangle = |v\rangle$, we obtain

$$|u\rangle\langle u|,$$

which maps any vector to its component **along** $|u\rangle$. This is the simplest form of a **projection operator**.

Earlier, in the previous chapter, we used the identity

$$\sum_i |i\rangle\langle i| = I$$

for an orthonormal basis $\{|i\rangle\}$. We can now understand this expression: each term $|i\rangle\langle i|$ projects any vector onto the direction $|i\rangle$. Summing over all basis directions reconstructs the entire vector—hence the identity operator.

3.5.1 Matrix Form

If

$$|u\rangle = \begin{bmatrix} u_1 \\ u_2 \end{bmatrix}, \qquad \langle v| = [v_1^*, v_2^*],$$

then

$$|u\rangle\langle v| = \begin{bmatrix} u_1 v_1^* & u_1 v_2^* \\ u_2 v_1^* & u_2 v_2^* \end{bmatrix}.$$

Thus, outer products directly produce matrices.

3.5.2 Projection Operator

The **outer product** can be used to construct projection operators. If $|u\rangle$ is a **normalized** state ($\langle u \mid u\rangle = 1$), then

$$P_u = |u\rangle\langle u|$$

is a projection operator onto the direction of $|u\rangle$.

3.5.3 Action on Any Vector

$$P_u|v\rangle = |u\rangle\langle u|v\rangle.$$

The scalar $\langle u \mid v\rangle$ is the "component" of $|v\rangle$ along $|u\rangle$:

- The operator P_u extracts this component and multiplies it by the vector $|u\rangle$.

Worked Example (2D Space)

Let

$$|u\rangle = \begin{bmatrix} 1 \\ 0 \end{bmatrix}, \quad |v\rangle = \begin{bmatrix} 2 \\ 3 \end{bmatrix}.$$

Then

$$P_u = |u\rangle\langle u| = \begin{bmatrix} 1 \\ 0 \end{bmatrix}\begin{bmatrix} 1 & 0 \end{bmatrix} = \begin{bmatrix} 1 & 0 \\ 0 & 0 \end{bmatrix}.$$

Acting on $|v\rangle$:

$$P_u|v\rangle = \begin{bmatrix} 1 & 0 \\ 0 & 0 \end{bmatrix}\begin{bmatrix} 2 \\ 3 \end{bmatrix} = \begin{bmatrix} 2 \\ 0 \end{bmatrix}.$$

This is precisely the component of $|v\rangle$ along the direction of $|u\rangle$.

Quantum Chemistry Connection

In quantum mechanics, if $|\phi\rangle$ is a normalized state, the operator $P = |\phi\rangle\langle\phi|$ is the probability of finding the system in state $|\phi\rangle$. Acting on a state $|\psi\rangle$, $P|\psi\rangle = |\phi\rangle\langle\phi|\psi\rangle$ gives the component of $|\psi\rangle$ along $|\phi\rangle$, which is used to calculate probabilities and expectation values.

3.6 Geometrical Operations

To illustrate the action of matrices as operators, it is convenient to examine geometrical operations. These are easy to visualize and, importantly, they form the basis for the **group theoretical treatment of molecular symmetry**. Group theory is useful in chemistry for many reasons: it predicts whether a molecule is chiral or polar, determines which AOs contribute to bond formation in MO theory, identifies infrared (IR) and Raman activity of molecular vibrations, and explains spectroscopic selection rules, among other applications.

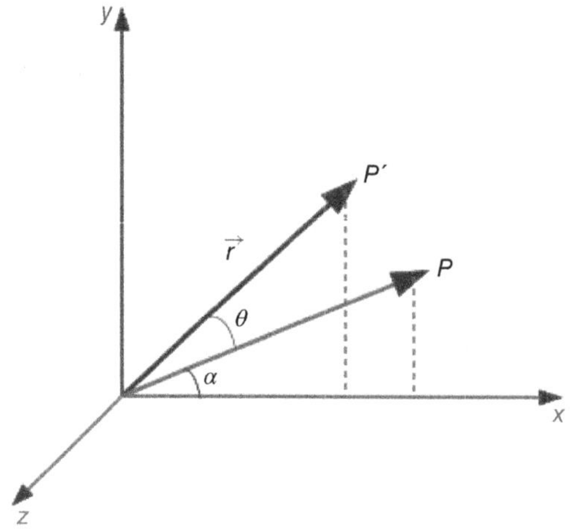

Figure 3.1 Counterclockwise rotation of a vector about the z-axis. A vector $\vec{r}$ undergoes a counterclockwise rotation by angle θ about the z-axis. The transformation from $\vec{r} = (x, y, z)$ to (x', y', z') is governed by the rotation matrix $R_z(\theta)$. Such rotations are fundamental symmetry operations in molecules; for example, a C_4 rotation axis corresponds to $\theta = 90°$. This demonstrates how a geometrical operation is encoded in a linear transformation. Note that for a clockwise rotation, θ is replaced by $-\theta$.

3.6.1 Rotation, *R*

Consider the counterclockwise rotation in the xy-plane about the z-axis of vector $\vec{r}$. The vector lies in the xy-plane and makes an angle α with the x-axis. The rotation is through an angle θ, as shown in Figure 3.1.

In Cartesian components, the coordinates of point P are

$$x = r\cos\alpha,\ y = r\sin\alpha. \tag{3.13}$$

The new coordinates after rotation are

$$x' = r\cos(\theta + \alpha) = r\cos(\theta)\cos(\alpha) - r\sin(\theta)\sin(\alpha) = x\cos\theta - y\sin\theta,$$

$$y' = r\sin(\theta)\cos(\alpha) + r\cos(\theta)\sin(\alpha) = x\sin\theta + y\cos\theta, \tag{3.14}$$

while $z' = z$, since the z-coordinate remains unchanged.

This transformation can be expressed in matrix form as

$$\begin{bmatrix} x' \\ y' \\ z' \end{bmatrix} = \begin{bmatrix} \cos\theta & -\sin\theta & 0 \\ \sin\theta & \cos\theta & 0 \\ 0 & 0 & 1 \end{bmatrix} \begin{bmatrix} x \\ y \\ z \end{bmatrix}.$$

Thus, the operator for counterclockwise rotation about the z-axis is the **rotation matrix**

$$R_z(\theta) = \begin{bmatrix} \cos\theta & -\sin\theta & 0 \\ \sin\theta & \cos\theta & 0 \\ 0 & 0 & 1 \end{bmatrix}. \tag{3.15}$$

For clockwise rotation, we replace θ by $-\theta$. The diagonal elements remain the same, but the off-diagonal terms change sign.

Rotation about the x- or y-axis is obtained by cyclic permutation of coordinates. Such rotations are called **active rotations** (the vector rotates and axes fixed). A clockwise rotation of the coordinate axes has the same effect as a counterclockwise rotation of the vector. This is called a **passive rotation**. The two are sometimes referred to as **alibi** and **alias** descriptions, respectively.

In group theory, rotations are represented as C_n, where C stands for cyclic rotation and $\theta = 2\pi/n$, i.e., $C_n \equiv R_z(\theta)$. For $\theta = 0°$, the rotation reduces to the identity matrix I_3, and the operation is denoted as E.

Fact Box 3.4: VOCABULARY ALERT 3: "Group"

In Mathematics: A set of elements (like matrices) with a binary operation that satisfies certain axioms (closure, associativity, identity, inverses).

In Chemistry: A "functional group" is a specific cluster of atoms in a molecule (e.g., a hydroxyl group).

In This Context: The "point group" of a molecule is the set of all its symmetry operations (rotations, reflections, etc.). These operations mathematically form a group, and their matrix representations do too.

As an example, consider a water molecule (Figure 3.2). Rotation about the z-axis by 180° (i.e., π) restores its appearance. Water, therefore, has a C_2 axis of rotation.

- The $\boldsymbol{C_2}$ **axis** - shown as a dashed line (green online) corresponds to the rotation matrix $R_z(\pi)$ with $\theta = 180°$.
- The $\boldsymbol{\sigma_v(yz)}$ **mirror plane** - shown in the lighter shaded plane (pink online) corresponds to the reflection matrix $R(\sigma_{yz})$, which negates the x-coordinate.
- The $\boldsymbol{\sigma'_v(xz)}$ **mirror plane** - shown in the darker shaded plane (blue online) corresponds to $R(\sigma_{xz})$, which negates the y-coordinate.
- This set of symmetry operations forms the point group C_{2v}.

3.6.2 Reflection

Suppose a mirror lies in the xz-plane through the oxygen atom (Figure 3.2). Reflection exchanges the two hydrogens while leaving the oxygen unchanged. The effect is to transform

$$(x, y, z) \longrightarrow (x, -y, z).$$

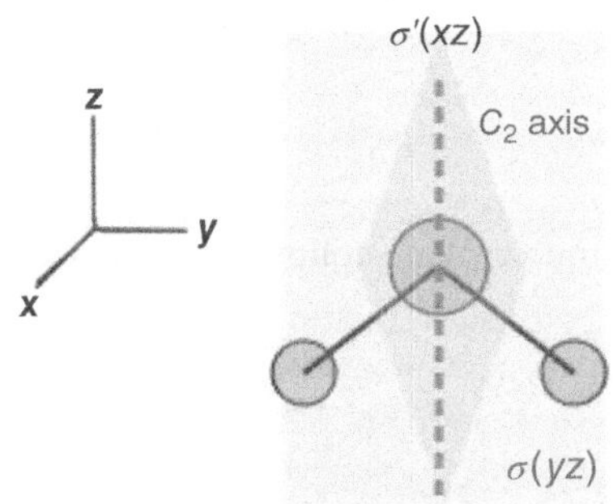

Symmetry elements of H_2O showing the C_2 axis (green) and two σ_v planes—$\sigma_v(yz)$ (pink) and $\sigma'_v(xz)$ (blue).

Figure 3.2 Symmetry elements of water and their associated linear operators. The water molecule possesses a C_2 rotation axis and two mirror planes, providing a concrete example of the geometrical operations.

The matrix form is

$$R(\sigma_{xz}) = \begin{bmatrix} 1 & 0 & 0 \\ 0 & -1 & 0 \\ 0 & 0 & 1 \end{bmatrix}. \tag{3.16}$$

This reflection is denoted σ_{xz}.

Another reflection plane is the yz-plane. Its action leaves y and z unchanged but reverses x:

$$R(\sigma_{yz}) = \begin{bmatrix} -1 & 0 & 0 \\ 0 & 1 & 0 \\ 0 & 0 & 1 \end{bmatrix}. \tag{3.17}$$

3.6.3 Inversion

This operation reverses the sign of each coordinate:

$$(x, y, z) \longrightarrow (-x, -y, -z).$$

Common Symmetry Operations

Operation	Symbol	Matrix (3D)
Identity	E	$\begin{bmatrix} 1 & 0 & 0 \\ 0 & 1 & 0 \\ 0 & 0 & 1 \end{bmatrix}$
Reflection	σ_{xy}	$\begin{bmatrix} 1 & 0 & 0 \\ 0 & 1 & 0 \\ 0 & 0 & -1 \end{bmatrix}$
Inversion	i	$\begin{bmatrix} -1 & 0 & 0 \\ 0 & -1 & 0 \\ 0 & 0 & -1 \end{bmatrix}$

The noncommutativity of matrices directly explains why the order of symmetry operations matters in molecules—rotations and reflections generally do not commute.

Example 3.3 illustrates the combined effect of two operations.

Example 3.3: Combination of Operations

Calculate the image of a point (x, y, z) after reflection in the xz-plane, followed by rotation about the z-axis by 180°.

Solution:

Reflection in the xz-plane gives

$$(x, y, z) \longrightarrow (x, -y, z).$$

Rotation by π about the z-axis is given by

$$R_z(\pi) \equiv R(C_2) = \begin{bmatrix} -1 & 0 & 0 \\ 0 & -1 & 0 \\ 0 & 0 & 1 \end{bmatrix}.$$

Therefore,

$$(x, -y, z) \longrightarrow (-x, y, z).$$

The combined operator is the product

$$R = R_z(\pi)R(\sigma_{xz}),$$

illustrating how the combined effect of successive operations corresponds to matrix multiplication. Since matrix multiplication is not commutative, the order of operations is important.

Examples 3.4 and 3.5 examine matrices for **compound operations** (S_n), with Example 3.5 focusing on the S_4 operation in methane.

Example 3.4: Compound Operations

Write down the matrices representing the following transformations applied to a point (x, y, z):

i. Reflection in the xy-plane.

ii. Rotation through 90° counterclockwise about the z-axis.

iii. First rotation through 90° counterclockwise about the z-axis, then reflection in the xy-plane.

Solution:

i.

$$R(\sigma_{xy}) = \begin{bmatrix} 1 & 0 & 0 \\ 0 & 1 & 0 \\ 0 & 0 & -1 \end{bmatrix}.$$

ii. This is the C_4 operation

$$R_z(90°) = \begin{bmatrix} 0 & -1 & 0 \\ 1 & 0 & 0 \\ 0 & 0 & 1 \end{bmatrix}.$$

iii. The compound operator is

$$R(\sigma_{xy})R(C_4) = \begin{bmatrix} 0 & -1 & 0 \\ 1 & 0 & 0 \\ 0 & 0 & -1 \end{bmatrix}.$$

This operation, rotation followed by reflection, is an **improper rotation**, denoted S_n. For $\theta = 2\pi/n$, the above is an S_4 operation.

Inversion

Another operation is inversion, denoted by i, which reverses all coordinates:

$$(x, y, z) \longrightarrow (-x, -y, -z).$$

The matrix is

$$R(i) = \begin{bmatrix} -1 & 0 & 0 \\ 0 & -1 & 0 \\ 0 & 0 & -1 \end{bmatrix}.$$

This is a special case of an improper rotation: rotation by π followed by reflection,

$$R(\sigma_{xy})R(C_2) = \begin{bmatrix} 1 & 0 & 0 \\ 0 & 1 & 0 \\ 0 & 0 & -1 \end{bmatrix}\begin{bmatrix} -1 & 0 & 0 \\ 0 & -1 & 0 \\ 0 & 0 & 1 \end{bmatrix} = \begin{bmatrix} -1 & 0 & 0 \\ 0 & -1 & 0 \\ 0 & 0 & -1 \end{bmatrix}$$

Example 3.5: S_4 Operation in Methane Molecule

Methane (CH_4) can be represented by placing a carbon atom at the center of a cube at coordinate (0, 0, 0) and the four hydrogen atoms at alternating corners.

a. Provide a set of coordinates for the four hydrogen atoms that correctly describes a regular tetrahedron.

b. One of the characteristic symmetry operations for a tetrahedral molecule like methane is the improper rotation, S_4. An S_4 operation consists of a 90° rotation about an axis (C_4), followed by a reflection through a plane perpendicular to that axis (σ_h). Show that an S_4 operation performed along the z-axis is a symmetry operation for methane by demonstrating that it leaves the set of atomic positions unchanged.

Solution:

a. Coordinates for Hydrogen Atoms

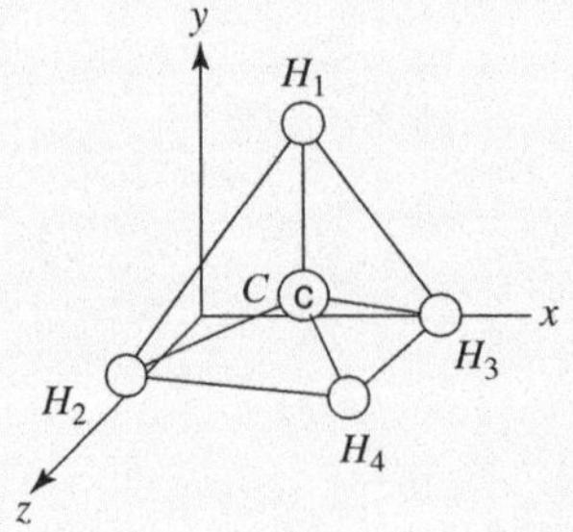

To form a regular tetrahedron inside a unit cube, we place the hydrogen atoms at alternating corners. A correct set of coordinates is:

- H_1: (1, 1, 1)
- H_2: (1, −1, −1)
- H_3: (−1, 1, −1)
- H_4: (−1, −1, 1)

With these coordinates, the distance from the central carbon at (0, 0, 0) to each hydrogen is identical ($\sqrt{3}$), and the distance between any two hydrogen atoms is also identical ($2\sqrt{2}$). This correctly represents the geometry of methane.

b. Demonstrating the S_4 Symmetry Operations

We will now perform the two steps of the $\boldsymbol{S_4}$ **operation** about the z-axis:
Step 1: Rotate by 90° about the z-axis (C_4 operation).
Step 2: Reflect through the xy-plane (σ_h operation).

From Example 3.4, the compound operator is

$$R(S_4) = R(\sigma_{xy})R(C_4) = \begin{bmatrix} 0 & -1 & 0 \\ 1 & 0 & 0 \\ 0 & 0 & -1 \end{bmatrix}.$$

Applying this to the positions in (a):

- H_1 (1, 1, 1) → (−1, 1, −1) ... This is the original position of $\boldsymbol{H_3}$.
- H_2 (1, −1, −1) → (1, 1, 1) ... This is the original position of $\boldsymbol{H_1}$.
- H_3 (−1, 1, −1) → (−1, −1, 1) ... This is the original position of $\boldsymbol{H_4}$.
- H_4 (−1, −1, 1) → (1, −1, −1) ... This is the original position of $\boldsymbol{H_2}$.

Conclusion: The S_4 operation results in the permutation $H_1 \longrightarrow H_3$, $H_2 \longrightarrow H_1$, $H_3 \longrightarrow H_4$, and $H_4 \longrightarrow H_2$. Although the individual hydrogen atoms have moved, the new set of coordinates is identical to the original set. Therefore, the molecule is indistinguishable after the operation, proving that S_4 is a valid symmetry operation for methane.

Key Takeaways: Geometrical Operations and Matrices

- **Matrices as Operators:** Rotations, reflections, inversions, and improper rotations can all be expressed as **matrix transformations** acting on vectors.
- **Active Versus Passive Rotations:** Rotating a vector or rotating the coordinate system in the reverse direction is mathematically equivalent but conceptually distinct.
- **Group Theory Connection:** These symmetry operations form mathematical **groups**. In chemistry, this underpins the analysis of **molecular symmetry**—explaining chirality, orbital mixing, vibrational selection rules, and spectroscopy.
- **Matrix Multiplication as Sequential Operations:** Performing two or more operations corresponds to multiplying their matrices. The order of multiplication matters (matrix multiplication is **not commutative**). In successive symmetry or matrix operations, the operation written first (on the right) is the one that is performed first, followed by those to its left, in accordance with the standard convention of operator multiplication.

Red Flag 3.1

The operation with the right-hand operator is performed first.

Connection to Chemistry

- Symmetry operations explain molecular properties: chirality, polarity, orbital mixing, vibrational activity, and selection rules.
- Group theory tools (characters, irreducible representations) all stem from these basic operations.

In the previously given examples, we changed from one set of coordinates to another using a linear transformation, which preserved the origin and scaled or mixed the coordinates through matrix operations. However, not all useful transformations in chemistry are purely linear. Many structural changes, shifts of reference frames, and coordinate redefinitions involve a **translation** in addition to a linear part. Transformations of this more general type are called **affine transformations**. They extend the idea of a linear transformation by allowing us to move the origin, making them especially useful in molecular geometry, conformational analysis, and coordinate shifts. We introduce them here because they arise naturally whenever a system is translated as well as stretched, rotated, or reflected.

3.7 Affine Transformations

An **affine transformation** combines a **linear transformation** and a **translation**, expressed as

$$\mathbf{y} = A\mathbf{x} + \mathbf{b}, \tag{3.18}$$

where:

- $\mathbf{x}$ = column vector of input coordinates,
- A = linear transformation matrix,
- $\mathbf{b}$ = translation vector.

3.7.1 Interpretation

- The term $A\mathbf{x}$ refers to a **linear transformation**, such as rotation, reflection, scaling, or shear.
- The inclusion of $\mathbf{b}$ results in a shift by means of a **translation**.

Examples

1. **2D Rotation and Translation**

$$\begin{bmatrix} x' \\ y' \end{bmatrix} = \begin{bmatrix} \cos\theta & -\sin\theta \\ \sin\theta & \cos\theta \end{bmatrix} \begin{bmatrix} x \\ y \end{bmatrix} + \begin{bmatrix} a \\ b \end{bmatrix}.$$

Here, the matrix rotates the point by angle θ, and the vector (a, b) shifts it.

2. **Scaling and Translation**

$$x' = 3x + 2, \qquad y' = 0.5y - 1.$$

This scales x by 3, y by 0.5, then shifts the coordinates.

3. **In Chemistry**
   - Transformations between **molecular coordinate frames** (e.g., local to global coordinates) are affine.
   - In **computer graphics or visualization of MOs**, affine matrices position objects in 3D scenes.
   - In **machine learning** (ML), each neuron performs an affine transformation before applying a nonlinear activation $f(\mathbf{z}) = f(W\mathbf{x} + \mathbf{b})$ (see next section).

Properties

1. The composition of two affine transformations is also affine.
2. The inverse of a nonsingular affine transformation is also affine.
3. Straight lines map to straight lines, and parallel lines remain parallel.
4. The origin is not necessarily fixed (unlike pure linear transformations).

Special Cases

Transformation	Equation	Description
Linear	$\mathbf{y} = A\mathbf{x}$	Origin fixed, no translation
Translation	$\mathbf{y} = \mathbf{x} + \mathbf{b}$	Identity matrix for A
Affine	$\mathbf{y} = A\mathbf{x} + \mathbf{b}$	Combination of both

3.7.2 Geometrical Significance

Affine transformations preserve

- **straightness** of lines,
- **parallelism** of lines,
- **ratios of segment lengths along a line**,

but do **not** preserve

- **angles**
- **lengths**
- **areas or volumes** (unless $|\det(A)| = 1$)

3.7.3 Connection to Vector Spaces

Affine transformations operate on **affine spaces**, which are like vector spaces but **without a fixed origin**. You can think of an affine space as a "vector space that allows translation."

Worked Example: A Linear Transformation followed by a Translation

Consider a point in two-dimensional space with coordinates

$$\mathbf{x} = \begin{bmatrix} 1 \\ 2 \end{bmatrix}.$$

We first apply the linear transformation

$$A = \begin{bmatrix} 1 & 1 \\ 0 & 1 \end{bmatrix},$$

which performs a shear, and then translate by

$$\mathbf{b} = \begin{bmatrix} 2 \\ -1 \end{bmatrix}.$$

The affine transformation is

$$\mathbf{x}' = A\mathbf{x} + \mathbf{b}.$$

Compute

$$A\mathbf{x} = \begin{bmatrix} 1 & 1 \\ 0 & 1 \end{bmatrix}\begin{bmatrix} 1 \\ 2 \end{bmatrix} = \begin{bmatrix} 3 \\ 2 \end{bmatrix}.$$

Adding the translation

$$\mathbf{x}' = \begin{bmatrix} 3 \\ 2 \end{bmatrix} + \begin{bmatrix} 2 \\ -1 \end{bmatrix} = \begin{bmatrix} 5 \\ 1 \end{bmatrix}.$$

Thus, the original point (1, 2) is mapped to (5, 1) by this affine transformation.

Fact Box 3.5: Affine Transformations in Chemistry and Machine Learning

In Chemistry

- **Molecular coordinates** are often transformed from local to global frames using affine transformations.
  Example: Aligning a molecular fragment to a reference structure involves
  $\mathbf{r}' = R\mathbf{r} + \mathbf{t}$,
  where R is a rotation (orthogonal matrix) and $\mathbf{t}$ is the translation vector.
- Used in **molecular graphics** and **geometry optimization** to visualize or superimpose molecules.

In Machine Learning

Each neuron in a feed-forward neural network applies an affine transformation before the activation function:

$$\mathbf{z} = W\mathbf{x} + \mathbf{b}, \qquad \mathbf{y} = f(\mathbf{z}).$$

Here W represents the weights (linear mapping) and $\mathbf{b}$ the bias (translation).

The affine form allows flexible shifting of decision boundaries in high-dimensional feature spaces.

Geometrical Illustration

Transformation type	Equation	Origin fixed?	Preserves lines?
Linear	$\mathbf{y} = A\mathbf{x}$	Yes	Yes
Translation	$\mathbf{y} = \mathbf{x} + \mathbf{b}$	No	Yes
Affine	$\mathbf{y} = A\mathbf{x} + \mathbf{b}$	No	Yes

Summary

Affine transformations form the bridge between **purely linear mappings** and **real-world coordinate shifts.**

They appear naturally in both **molecular geometry** and **neural network computation**, making them a unifying concept across chemistry and applied mathematics.

We now look at the application of the concepts studied in this chapter to discuss their application in **neural network (NN)** computation. A neural network is an **ML** model that uses interconnected artificial neurons to learn patterns from data, inspired by the human brain.

3.8 Affine Transformations as the Building Blocks of Learning: Neural Networks

Neural networks may seem far removed from quantum chemistry or spectroscopy, yet their inner workings rely almost entirely on the mathematical ideas you have just learned—**vectors, matrices, affine maps**, and **composition of transformations**. What makes them fascinating is that they take the familiar operations of linear algebra and repeatedly combine them in ways that allow a computer to "learn" complex patterns.

For chemistry students, neural networks are increasingly unavoidable. They now appear in:

- predicting molecular properties,
- analyzing IR, Raman, UV–Vis, and NMR spectra,
- classifying electronic structures,
- learning potential energy surfaces (PESs),
- accelerating DFT calculations, and
- molecular design and data-driven reaction prediction.

At their heart lies an unexpectedly simple mathematical structure. Before examining the mathematical structure of neural networks, it is useful to visualize a simple feed-forward architecture.

3.8.1 Neural Networks as Applied Linear Algebra

As a chemist, you are trained to find relationships. You might run an experiment to see how changing a reactant's concentration (a variable, x) affects the reaction rate (a result, y). You then try to find a function, $y = f(x)$, that models this relationship.

A **neural network** is a powerful mathematical tool for "learning" this function, f, even when the relationship is incredibly complex (e.g., predicting a molecule's toxicity from 2,000 different structural features).

You might think of NNs as complex artificial intelligence (AI), but, at their core, they are built almost entirely from the concepts in this and the previous chapter: **vector spaces, linear transformations, and vector addition.**

3.8.1.1 The Core Idea: A Layer Is Just $f(W\mathbf{x} + \mathbf{b})$

Forget the "brain" analogy. Think of a single layer in a neural network as one compact mathematical operation that transforms an input vector **x** into a new output vector **y**.

Each layer in a network performs two steps: a **linear transformation** ($\mathbf{z} = W\mathbf{x} + \mathbf{b}$), followed by a **nonlinear activation** ($\mathbf{y} = f(\mathbf{z})$), making it a practical extension of the concepts discussed in this chapter.

Let us break down the math.

- **x**: This is your **input vector.**
- W: This is a **weight matrix**. This is the **linear transformation**. It rotates, scales, and projects your input vector, effectively changing its basis.
- **b**: This is a **bias vector**. This is the **vector addition**. It shifts the entire vector space, just like an intercept. (The combination $W\mathbf{x} + \mathbf{b}$ is technically an *affine transformation* [3.18].) The bias vector $\boldsymbol{b}$ plays a role analogous to the translation vector in an affine transformation.
- f: This is a simple, nonlinear **activation function** (like ReLU, $f(\mathbf{z}) = \max(0, \mathbf{z})$) that is applied elementwise (see Fact Box 3.6). This is the "secret sauce." Without f, stacking 100 layers would just be one big linear transformation. This nonlinearity is what allows the network to model complex, curved relationships—like a PES or a nonlinear rate law.

Fact Box 3.6: ReLU Activation Function

ReLU, or Rectified Linear Unit, is a popular activation function in artificial neural networks defined by the formula

$$f(\mathbf{x}) = \max(0, \mathbf{x})$$

It outputs **x** for any positive input and 0 for any negative input, which introduces a simple nonlinearity to the model. This simplicity makes it computationally efficient and helps neural networks learn complex patterns, often used as the default activation function in hidden layers.

Other nonlinear activation functions are sigmoid and tanh.

Figure 3.3 illustrates the data for a single neural network layer where a three-dimensional input vector **x** gives a two-dimensional output vector **y**.

The operation $W\mathbf{x} + \mathbf{b}$ is precisely the **affine transformation** discussed in Section 3.7 and introduced in Equation (3.18). The function f then introduces controlled nonlinearity, which makes the overall operation $\mathbf{y} = f(W\mathbf{x} + \mathbf{b})$ nonlinear.

3.8.1.2 Vector–Matrix Interpretation

- W defines a change of basis or projection in a vector space.
- **b** shifts the origin.
- f warps the space, allowing nonlinear surfaces.

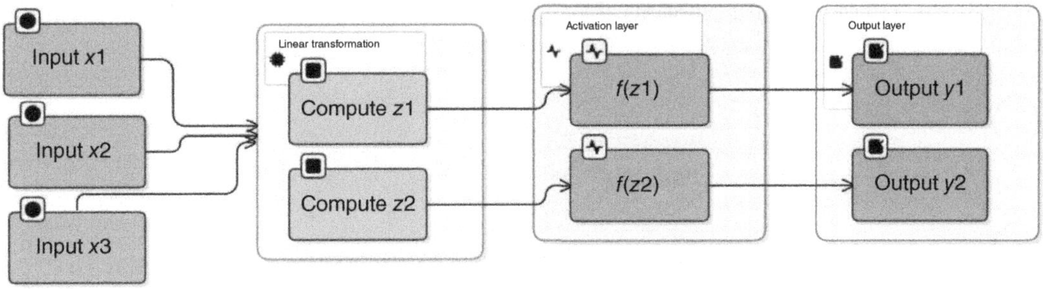

Figure 3.3 Data flow in a single neural network layer. An input vector $\mathbf{x} \in \mathbb{R}^3$ is transformed by the weight matrix W and bias vector $\mathbf{b}$, resulting in the pre-activation vector $\mathbf{z}$. The nonlinear activation function f is then applied elementwise to produce the layer's output $\mathbf{y} \in \mathbb{R}^2$.

Successive layers apply successive transformations—effectively **compositions of matrices and nonlinear maps**.

3.8.1.2.1 What Is a "Vector" in Chemistry?

This is the key to appreciating the relevance. In this context, a vector isn't just an abstract arrow; it is a **chemical object**.

The "vector space" the network operates in is a **"chemical property space."**

Imagine you want to predict a molecule's solubility (a single number, a scalar). Your input vector $\mathbf{x}$ could be:

$$\mathbf{x} = \begin{bmatrix} \text{Molar Mass} \\ \log P \\ \text{Number of H-bond donors} \\ \text{Polar surface area} \end{bmatrix}$$

The neural network's job is to learn the transformation matrix W and bias vector $\mathbf{b}$ that map this high-dimensional vector $\mathbf{x}$ to a final output $\mathbf{y}$ (solubility). The matrix W literally "learns" the *importance* and *interactions* of these features.

To visualize how the affine transformations discussed earlier are applied in practice, we now look at a minimal feed-forward neural network architecture. This diagram (Figure 3.4) shows how input descriptors propagate through a **hidden layer** to produce the final prediction (**output layer**).

3.8.1.3 Concrete Mini-example: 2D Input → Hidden Layer → 1D Output

Worked Example: A Tiny Network (2D → 2 → 1)

Let the input be

$$\mathbf{x} = \begin{bmatrix} x_1 \\ x_2 \end{bmatrix}.$$

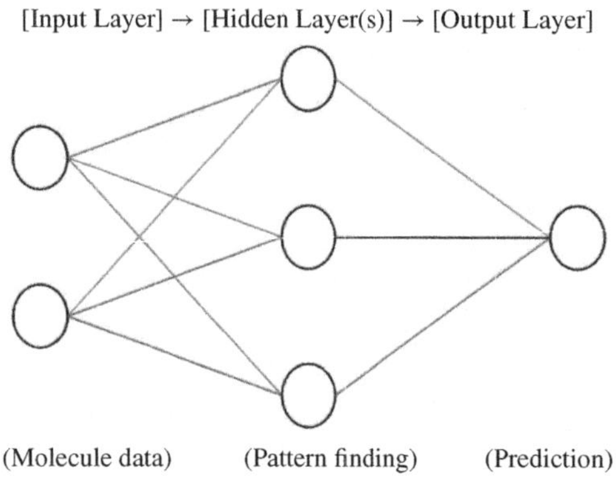

Figure 3.4 A simple feed-forward neural network. A two-input, three-neuron hidden-layer network connected to a single-output neuron, illustrating weighted connections of different strengths. Each circle represents a neuron. Data is transmitted from left to right, undergoing processing at every stage.

Layer 1: Affine Transformation

$$W_1 = \begin{bmatrix} 2 & -1 \\ 1 & 3 \end{bmatrix}, \quad \mathbf{b}_1 = \begin{bmatrix} 1 \\ -2 \end{bmatrix}.$$

Compute

$$\mathbf{z}_1 = W_1\mathbf{x} + \mathbf{b}_1 = \begin{bmatrix} 2x_1 - x_2 + 1 \\ x_1 + 3x_2 - 2 \end{bmatrix}.$$

Apply a nonlinearity (ReLU):

$$\mathbf{a}_1 = [\max(0, z_{1,1}), \max(0, z_{1,2})]^T.$$

Layer 2: Affine Transformation to Output

$$W_2 = \begin{bmatrix} 4 & 1 \end{bmatrix}, \quad \mathbf{b}_2 = -3.$$

Output:

$$y = W_2\mathbf{a}_1 + \mathbf{b}_2 = 4a_{1,1} + a_{1,2} - 3.$$

Each step is nothing more than the affine transformations we discussed—now chained together.

We now apply these concepts to the real example of predicting the solubility (Example 3.6). Instead of four parameters, we take only two for simplicity.

Example 3.6: Predicting Solubility

Predict a molecule's solubility ($\log S$, a scalar) from two molecular descriptors: x_1 ($\log P$, hydrophobicity) and x_2 (molar mass).

Solution:

Network Architecture:
Input: $\mathbf{x} = [x_1, x_2]^T$.
Hidden Layer: two neurons with ReLU activation: $f(\mathbf{z}) = \max(0, \mathbf{z})$.
Output Layer: one neuron with linear activation (for regression).
Step-by-step Forward Pass:
Input to Hidden Layer:

$$\mathbf{z}^{(\text{hidden})} = W^{(\text{hidden})}\mathbf{x} + \mathbf{b}^{(\text{hidden})}.$$

Let us assume specific values:

$$W^{(\text{hidden})} = \begin{bmatrix} 0.5 & -0.2 \\ -1.0 & 0.8 \end{bmatrix}, \quad \mathbf{b}^{(\text{hidden})} = \begin{bmatrix} 1.0 \\ -0.5 \end{bmatrix}, \quad \mathbf{x} = \begin{bmatrix} 2.0 \\ 150.0 \end{bmatrix},$$

$$\mathbf{z}^{(\text{hidden})} = \begin{bmatrix} 0.5 & -0.2 \\ -1.0 & 0.8 \end{bmatrix}\begin{bmatrix} 2.0 \\ 150.0 \end{bmatrix} + \begin{bmatrix} 1.0 \\ -0.5 \end{bmatrix} = \begin{bmatrix} (1.0 - 30.0) + 1.0 \\ (-2.0 + 120.0) - 0.5 \end{bmatrix} = \begin{bmatrix} -28.0 \\ 117.5 \end{bmatrix}.$$

Apply Activation (ReLU):

$$\mathbf{a}^{(\text{hidden})} = f\left(\mathbf{z}^{(\text{hidden})}\right) = \begin{bmatrix} \max(0, -28.0) \\ \max(0, 117.5) \end{bmatrix} = \begin{bmatrix} 0.0 \\ 117.5 \end{bmatrix}.$$

This demonstrates how nonlinearity can "turn off" neurons.

Hidden to Output Layer:

$$\mathbf{z}^{(\text{output})} = W^{(\text{output})}\mathbf{a}^{(\text{hidden})} + \mathbf{b}^{(\text{output})}.$$

Let us assume:

$$W^{(\text{output})} = \begin{bmatrix} 0.1 & -0.05 \end{bmatrix}, \qquad \mathbf{b}^{(\text{output})} = [0.2],$$

$$\mathbf{z}^{(\text{output})} = \begin{bmatrix} 0.1 & -0.05 \end{bmatrix}\begin{bmatrix} 0.0 \\ 117.5 \end{bmatrix} + [0.2] = (0.0 - 5.875) + 0.2 = -5.675.$$

The final predicted solubility ($\log S$) is $\boldsymbol{y} = \mathbf{-5.675}$.

In Example 3.6, using a raw value of 150.0 for mass will dominate the calculation compared to $\log P$ (~2.0). This is a crucial practical point in ML.

ML Pro-tip box

In practice, input features are usually standardized (scaled to have zero mean and unit variance) to ensure no single feature dominates the learning process due to its scale.

3.8.2 How Networks Learn: Gradient Descent and Backpropagation

3.8.2.1 The Learning Goal: Minimizing Error

Once a neural network makes a prediction, we need to measure how good it is. We define a **loss function** $L(W, \mathbf{b})$ (also called a cost function) that quantifies the error between the network's predictions and the true values.

Crucially, $L(W, \mathbf{b})$ is a function of all the weights and biases in the network. Think of it as defining a surface in a very high-dimensional space:

The Coordinates: Every possible combination of all weights and biases $(W, \mathbf{b})$ defines a point in this space. If the network has millions of parameters, this is a million-dimensional space.

The Height: The value of the loss function L at each point is the "height," representing the total error for that particular network configuration.

Our goal is to find the point on this surface with the lowest error—the global minimum.

3.8.2.2 The Chemist's Analogy: Optimization on a Potential Error Surface

This process is exactly analogous to a computational chemistry task you already know: **geometry optimization on a PES**.

Neural network training	Molecular geometry optimization
Loss function $L(W, \mathbf{b})$	Potential energy $E(\mathbf{r})$
Network parameters $(W, \mathbf{b})$	Atomic coordinates $(\mathbf{r})$
Goal: Find $(W, \mathbf{b})$ that minimize L	Goal: Find $\mathbf{r}$ that minimize E
High loss = bad prediction	High energy = unstable structure

In both cases, we start from an initial guess (random weights/initial molecular geometry) and iteratively "walk downhill" to find a minimum.

3.8.2.3 The Algorithm: Gradient Descent

How do we find the downhill direction? We use calculus. The gradient of the loss function, ∇L, is a vector that points in the direction of the steepest ascent.

Therefore, to minimize the loss, we take a small step in the opposite direction:

$$W_{\text{new}} = W_{\text{old}} - \eta \frac{\partial L}{\partial W}, \qquad \mathbf{b}_{\text{new}} = \mathbf{b}_{\text{old}} - \eta \frac{\partial L}{\partial \mathbf{b}}. \tag{3.19}$$

Here, η (eta) is the **learning rate**, which controls the size of our step (analogous to the step size in a geometry optimization algorithm). This process is called **gradient descent**.

3.8.2.4 The Learning Algorithm: Backpropagation and Gradient Descent

We now understand that training a network means finding the parameters $(W, \mathbf{b})$ that minimize the loss function L. This is done through gradient descent, where we iteratively take small steps "downhill" on the loss surface. But this raises a critical question: How do we actually calculate the gradient $\partial L/\partial W$ for every weight matrix in a network that might be millions of parameters deep?

The answer is the **backpropagation algorithm**. It is the efficient, clever application of the chain rule from calculus that allows us to compute all the necessary gradients in one forward and one backward pass through the network.

3.8.2.5 The Core Idea of Backpropagation: Distributing "Blame"

Imagine the network makes a bad prediction. The output layer is "wrong," but it is not entirely its fault—it received poor inputs from the hidden layers. Those hidden layers, in turn, received poor inputs from the layers before them.

Backpropagation is the process of distributing the "blame" for the final error backward through the network, from the output layer all the way back to the input layer. It answers the question: "How much did each neuron, in each layer, contribute to the final error?"

3.8.2.6 The Two-pass Procedure

3.8.2.6.1 Step 1: The Forward Pass

We feed an input vector $\mathbf{x}$ through the network, layer by layer, to compute a prediction $\mathbf{y}$. This is exactly what we did in Example 3.6 and the worked example. At the end of this pass, we calculate the loss L based on how wrong the prediction was.

3.8.2.6.2 Step 2: The Backward Pass (Backpropagation)

This is where the learning happens. We work backward from the output layer to the input layer:

Output Layer: We calculate how much a tiny change in each output neuron's pre-activation ($\mathbf{z}$) would affect the final loss L. This gives us an initial **error signal**, $\delta^{(\text{output})}$.

Hidden Layers: For each hidden layer, we take the error signal from the layer above it and ask: "How much of this error was caused by the pre-activations in *this* layer?" We calculate a new, localized error signal $\delta^{(\text{hidden})}$ for the current layer. This step uses the derivative of the activation function, telling us how sensitive the output is to changes in its input.

Calculate Gradients: Once we have the error signal $\delta^{(l)}$ for a layer l, calculating the gradient for its weights is straightforward. The gradient for a weight connecting neuron j (in the previous layer) to neuron i (in the current layer) is proportional to

$$\frac{\partial L}{\partial w_{ij}} \propto \delta_i \times a_j,$$

where:

- δ_i is the error signal of neuron i (how "guilty" it is).
- a_j is the activation of neuron j (how "active" it was during the forward pass).

A weight is adjusted more if it is connected to a very active neuron and a neuron that was very responsible for the error.

3.8.2.7 The Chain Rule in Action

Backpropagation is, at its heart, a repeated application of the **chain rule**. The loss L is a function of the outputs, which are functions of the pre-activations, which are functions of the weights, and so on. The chain rule allows us to break down the complex derivative $\frac{\partial L}{\partial W}$ into a product of simpler, local derivatives at each layer,

$$\frac{\partial L}{\partial W_{\text{layer}}}.$$

Each term in this product is something we can easily compute locally at a specific layer. Backpropagation is an efficient algorithm that organizes this calculation for the entire network.

3.8.2.8 Conceptual Summary

Forward Pass: "What is the current output and error?" (Calculation).
Backward Pass: "How is the error shared across all neurons?" (Learning).
Weight Update: "How should I adjust each weight to reduce the error?" (Optimization).

This cycle—forward pass, backward pass, parameter update—is repeated for thousands of examples until the network's performance converges. The result is a set of tuned weights and biases (W, $\mathbf{b}$) that have learned to map input features (molecular descriptors) to the desired output (solubility, energy, etc.) with high accuracy.

3.8.2.9 Real-world Case Studies in Chemistry

Neural networks are transforming chemical research in ways that go far beyond abstract mathematics. The following examples show how the linear-algebra machinery you have just learned—matrix multiplications, affine maps, and nonlinear activations—drives real breakthroughs across the chemical sciences.

Case Study 1: Predicting Reaction Conditions in Organic Synthesis

Modern synthetic chemistry today is assisted by deep neural networks that recommend **reagents, solvents, catalysts, and temperatures**, learned from thousands of recorded reactions.

These models do not simply memorize recipes—they learn **patterns** linking structure, reactivity, and reaction environment.

Impact:

- Correctly matched solvents/reagents in the **top 10 predictions 73%** of the time.
- Predicted reaction temperatures within **±20 °C in 89%** of cases.
- Suggested alternative solvents and conditions **not present** in the training data.
- Helps chemists **explore new synthetic pathways** instead of relying purely on heuristics.

This is linear algebra assisting creativity.

Case Study 2: Neural Networks for PESs

Accurate PES calculations from quantum chemistry (CCSD(T), MP2, DFT) are computationally expensive. Neural networks learn the PES from high-level data and then predict energies at new geometries **orders of magnitude faster**.

Impact:

- Enables long-time-scale molecular dynamics.
- Allows exploration of reaction pathways otherwise too expensive to compute.
- Makes accurate PES modeling accessible for **larger molecules**.

This is one of the clearest examples of where chemistry students immediately see the power of a learned affine map.

Case Study 3: Drug Discovery With Graph Neural Networks

Graph neural networks (GNNs) treat a molecule as a **graph**—atoms are nodes; bonds are edges. This representation naturally captures:

- connectivity
- aromatic rings
- heteroatom environments
- coordination patterns

Pharmaceutical researchers use GNNs to predict activity, selectivity, toxicity, ADMET properties, and binding affinity.

Impact:

- Accelerates hit identification.
- Reduces costly wet-lab screening.
- Discovers **nonintuitive scaffolds** missed by traditional QSAR methods.

This is vector spaces applied directly to molecules.

Case Study 4: Machine Learning for Spectroscopy

Neural networks trained on NMR, IR, Raman, or UV–Vis spectra can identify functional groups, classify molecular families, and even predict connectivity.

Impact:

- Automated peak assignments.
- Faster identification of unknown compounds.
- Improved analysis of complex mixtures.
- Reduction of human interpretation bias.

This is exactly the kind of task where affine transformations distill high-dimensional signals into chemically meaningful patterns.

3.8.2.10 Why Neural Networks Matter in Chemistry

The landscape of chemical data continues to grow: spectra, simulations, reaction databases, molecular libraries, and materials datasets. Traditional statistical or rule-based models plateau after a point.

Neural networks, however, continue to improve as data grows.

3.8.2.10.1 Performance Versus Data Availability

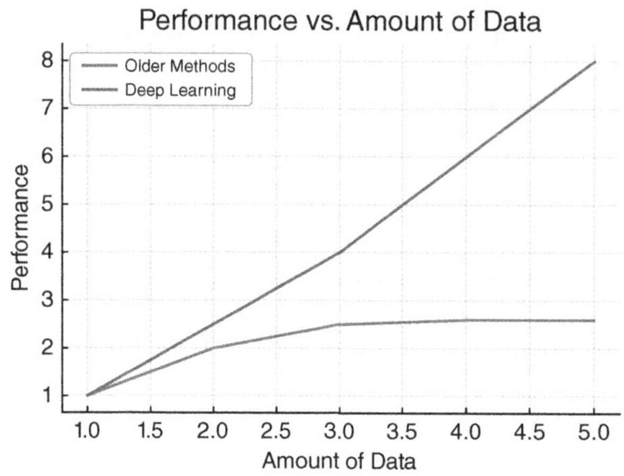

As datasets expand:

- **Older methods** plateau as data increases.
- **Deep learning** continues to improve markedly.

3.8.2.11 Summary: The Math–Chemistry Dictionary

To tie it all together, here is a "translation" of the mathematical concepts from the chapter into their roles inside a neural network and their familiar analogs in chemistry.

Mathematical concept	Neural network role	Chemistry analogy
Vector space	The set of all possible input vectors	**Chemical property space** (e.g., all possible combinations of log P, molar mass, etc.)
Vector ($\mathbf{x}$)	A single data point (e.g., one molecule)	A **molecular descriptor vector** or **fingerprint**
Linear transformation ($W\mathbf{x}$)	The **weight matrix** W mixes and scales features	**Basis change**. Very similar to an LCAO transformation, which mixes atomic orbitals (old basis) into MOs (new basis)—the butadiene example
Vector addition ($+\mathbf{b}$)	The **bias vector b** shifts the origin	Adjusting a **baseline** or **reference energy**
Function composition ($f_2(W_2(f_1(W_1\mathbf{x})))$)	Stacking multiple layers ("deep learning")	Creating a **hierarchy of features** (e.g., layer 1 finds "bonds," layer 2 finds "functional groups," layer 3 finds "reactivity")
Scalar field ($L(W, \mathbf{b})$)	The **loss function** (measures error)	A **potential energy surface (PES)**
Gradient vector (∇L)	The derivative of the error	The **force vector** on a PES (points "uphill")
Optimization (gradient descent)	Taking small steps opposite the gradient to find the minimum L	**Geometry optimization** (energy minimization)

Key Takeaways

- Neural networks build powerful models using **only** linear algebra and simple nonlinear functions.
- Each layer performs an **affine transformation** followed by a scalar nonlinearity.
- The entire network is a **composition of linear maps**, exactly the topic of this chapter.
- Backpropagation is just the **chain rule** applied repeatedly.
- Modern chemical applications (property prediction, spectroscopy, potentials) rely heavily on these ideas.

3.9 Summary

In this chapter, we extended the ideas of three-dimensional Euclidean vectors to the broader concept of linear transformations and operators, which provide the mathematical foundation for quantum mechanics, spectroscopy, and many areas of theoretical chemistry and physics:

- **Linear Transformations and Operators:** A linear transformation maps one vector into another while preserving linearity. Such transformations can always be represented by a matrix, which acts as the operator. Operators in quantum mechanics (Hamiltonian, Laplacian, etc.) are linear.
- **Matrices as Operators:** Every operator can be represented by a matrix relative to a chosen basis. The action of the operator is independent of the basis, but the representation (the matrix elements) depends on it.
- **Geometrical Operations and Symmetry:** Rotations, reflections, and inversions can be represented as matrices. These operations are central to group theory, which classifies molecular symmetry and predicts physical properties such as chirality, vibrational activity, and selection rules in spectroscopy. Compound operations correspond to matrix multiplication, showing why order matters in symmetry operations.

3.10 Key Connections

- Linear vector spaces provide the mathematical scaffold for Hilbert spaces, where quantum states (wavefunctions) live.
- Symmetry operations, expressed as matrices, are crucial in MO theory and spectroscopy.
- The interplay between linear operators and matrices sets the stage for the study of eigenvalues and eigenvectors, which link directly to observables in quantum mechanics.

The concepts introduced earlier—linear transformations, basis changes, and affine mappings—extend far beyond classical vector analysis. In many contemporary areas of science, including computational chemistry and data-driven modeling, these same operations form the mathematical core of predictive algorithms. Every layer of a modern neural network, for instance, performs a linear mapping of an input vector followed by a controlled nonlinear modification. Thus, the ideas developed in this chapter provide the algebraic framework on which such models are built, connecting abstract vector-space theory to one of the most powerful analytical tools in current research.

In the next chapter, we will explore matrices in greater depth—their properties, types, and operations. This will prepare us for the crucial concept of eigenvalues and eigenvectors, which connect directly to measurable quantities in quantum chemistry: orbital energies, vibrational frequencies, and spectroscopic transitions.

Further Reading

P. W. Atkins and R. S. Friedman, *Molecular Quantum Mechanics*, Oxford University Press, 2010—chapters on symmetry and quantum states.

K. T. Butler, D. W. Davies, H. Cartwright, O. Isayev and A. Walsh, *Machine learning for molecular and materials science*, Nature, 559, 547–555 (2018).

I. Goodfellow, Y. Bengio, and A. Courville, *Deep Learning*, MIT Press, 2016.

I. N. Levine, *Quantum Chemistry*, Pearson, 2014—sections on vector spaces and operators.

D. A. McQuarrie, *Mathematical Methods for Scientists and Engineers*, University Science Books, 2003—clear introduction to linear algebra in physics/chemistry contexts.

M. Rupp, *Machine learning for quantum mechanics in a nutshell*, Int. J. Quantum Chem., 115(16), 1058–1073 (2015).

R. Shankar, *Principles of Quantum Mechanics*, 2nd ed., Plenum Press, 1994—very clear operator treatment.

M. Tinkham, *Group Theory and Quantum Mechanics*, Courier Corporation, 2004—standard reference for molecular symmetry.

O. A. von Lilienfeld, *Quantum machine learning in chemical compound space*, Angew. Chem. Int. Ed., 57(16), 4164–4169 (2018)—a standard modern reference.

H. Weyl, *The Theory of Groups and Quantum Mechanics*, Courier Corporation, 2014—classical treatment of group theory in physics.

Exercises

A. Conceptual Understanding

1. **Properties of Rotation Matrices**
   Matrix A represents the rotation of axes in a plane. Show that

   $$AA^T = A^T A = I,$$

   and provide a geometrical interpretation of this result.

2. **Symmetry Operations**
   For the reflection in the xz-plane and rotation C_2 about the z-axis, describe their physical/geometrical meaning without computing matrices. Why are these considered group operations?

3. Consider the π-orbitals of *cis*-butadiene. Explain how the coefficients in the Hückel MOs can be understood as vector components in a basis set of p_x orbitals on each carbon atom.

4. If a vector is written in two different bases as

$$\mathbf{v} = 3\mathbf{e}_1 - \mathbf{e}_2 = \alpha\mathbf{f}_1 + \beta\mathbf{f}_2,$$

what does this tell you about the relationship between the two bases?

5. What is the mathematical difference between a linear and an affine transformation?

6. In a neural network, why is the bias vector $\mathbf{b}$ necessary if an activation function f follows the linear transformation?

7. Explain how the gradient vector guides the process of optimization in neural network learning.

8. Describe how the chain rule of calculus applies in the backpropagation algorithm.

9. Why is a nonlinear activation function essential for a neural network to model complex relationships?

B. Problem-solving

1. **3D Rotation Operator**
   Show that the matrix

$$A = \begin{bmatrix} \cos\theta & -\sin\theta & 0 \\ \sin\theta & \cos\theta & 0 \\ 0 & 0 & 1 \end{bmatrix}$$

represents a rotation of all vectors.

- Identify the axis, angle, and direction.
- Show that the same matrix also represents passive rotation of the coordinate axes.

2. **Compound Transformation**
   A vector (x, y, z) is rotated by $\pi/6$ about the z-axis. Then, it is rotated by $\pi/4$ about the y-axis (both counterclockwise as viewed along the negative axis).
- Write the two rotation matrices A and B.
- Show that the overall transformation is given by the matrix BA.
- Determine the new coordinates of the vector $(1, 1, 1)$.

3. **Determinants and Rotations**
   a. Write the triple scalar product of three vectors in determinantal form. Prove that if a 3×3 matrix is orthogonal, then its determinant can only be ± 1.
   b. For the matrices

$$M_1 = \begin{bmatrix} 0 & -1 & 0 \\ 1 & 0 & 0 \\ 0 & 0 & 1 \end{bmatrix}, M_2 = \begin{bmatrix} -1 & 0 & 0 \\ 0 & -1 & 0 \\ 0 & 0 & -1 \end{bmatrix},$$

determine whether they represent proper or improper rotations and describe their geometrical action.

4. **Orthonormal Systems**
   Consider the vectors
   $\vec{u}_1 = (1, 0, 0), \vec{u}_2 = (0, 1, 0), \vec{u}_3 = (0, 0, 1)$.
   a. Verify that they are unit, mutually orthogonal, and form a right-handed system.
   b. Construct the rotation matrix that transforms an old set of coordinates into this basis.
   c. Apply the matrix to $(0, 3, 2)$, $(1, 4, 3)$, and $(2, -2, -2)$. Interpret the peculiar behavior of the last vector.

5. **ReLU Computation**
   Let

$$W = \begin{bmatrix} 1 & -2 \\ 0 & 3 \end{bmatrix}, \mathbf{b} = \begin{bmatrix} 1 \\ -1 \end{bmatrix}, \mathbf{x} = \begin{bmatrix} 2 \\ 1 \end{bmatrix}.$$

   Compute $\mathbf{z} = W\mathbf{x} + \mathbf{b}$ and the activated output $\mathbf{y} = f(\mathbf{z})$ for $f(\mathbf{z}) = \max(0, \mathbf{z})$ (ReLU activation).

6. **Partial Derivatives (tanh activation)**
   For a single neuron defined as $\mathbf{y} = f(w_1\mathbf{x}_1 + w_2\mathbf{x}_2 + \mathbf{b})$,
   compute the partial derivatives $\frac{\partial y}{\partial w_1}$ and $\frac{\partial y}{\partial w_2}$ when $f(z) = \tanh(z)$.

7. **Chain Rule for Cost Function**
   A cost function is given by $L = \left(\mathbf{y} - \mathbf{y}_{\text{target}}\right)^2$.
   Derive $\frac{\partial L}{\partial w_i}$ for a neuron $\mathbf{y} = f(w_1\mathbf{x}_1 + w_2\mathbf{x}_2 + \mathbf{b})$ using the chain rule.

8. **Physical Interpretation of Neural Network for Dipole Moment**
   Consider a neural network layer designed to predict a molecule's dipole moment from its structure. If the input vector is $\mathbf{x} =$[partial charge on O, partial charge on H, O–H bond length]^T, interpret what the elements of the weight matrix W and bias vector $\mathbf{b}$ might represent physically.

9. **ReLU for Bond Dissociation Energy**
   Explain why the ReLU activation function $f(\mathbf{z}) = \max(0, \mathbf{z})$ might be useful for modeling a property like "bond dissociation energy," which is always positive.

C. Advanced/Challenge

1. **Rotation vs. Improper Rotation**
   Consider the matrices

$$A = \begin{bmatrix} 0 & -1 & 0 \\ 1 & 0 & 0 \\ 0 & 0 & 1 \end{bmatrix}, B = \begin{bmatrix} -1 & 0 & 0 \\ 0 & -1 & 0 \\ 0 & 0 & -1 \end{bmatrix}.$$

   Determine which one(s) represent rotation and explain the geometrical meaning of the other(s).

2. **Potential Energy Surface**
   In computational chemistry, the PES $E(\mathbf{r})$ of a molecule can be modeled using a neural network where $\mathbf{r}$ represents atomic coordinates. Explain which part of the network corresponds to the linear transformation and which to the nonlinear mapping of physical behavior.

3. **Affine Transformations for Boiling Points**
   Consider a dataset of molecular descriptors (dipole moment, polarizability, ionization energy).
   Describe qualitatively how a neural network would use affine transformations and activation functions to predict boiling points.

4. **Energy Minimization vs. Gradient Descent**
   Compare the process of energy minimization in molecular geometry optimization with gradient descent in neural network training. In what sense are both methods searching for minima on a surface?

4

Matrix Characteristics and Quantum-mechanical Operators

The mathematics of quantum mechanics is the mathematics of linear operators on Hilbert space.

— Paul A. M. Dirac

Learning Outcomes

After studying this chapter, the student should be able to:

1. **Explain** the properties and significance of Hermitian, unitary, orthogonal, and normal matrices in the context of quantum mechanics.
2. **Distinguish** between different classes of matrices and **identify** their physical meanings as linear operators.
3. **Recognize** the implications of commuting and non-commuting operators for the simultaneous measurement of observables.
4. **Relate** matrix invariants such as trace and determinant to physical quantities.
5. **Interpret** the role of unitary transformations in preserving norms, probabilities, and quantum states.

4.1 Introduction

In the previous two chapters, we introduced **linear vector spaces** and showed that their basis elements could be vectors, matrices, functions, or any other set of objects that obey specific algebraic rules. We also adopted a unified framework—the **Dirac bra–ket notation**—which provides a concise and versatile way to express relationships across different kinds of vector spaces.

Matrices naturally emerged as **linear operators** that transform one vector into another, and we illustrated their use through examples of **symmetry operations**. We briefly mentioned **Hermitian operators** and **unitary matrices**, noting their importance but deferring a detailed discussion.

In quantum mechanics, every observable—such as energy, momentum, or angular momentum—is represented by a **Hermitian operator**. This is because Hermitian operators possess **real eigenvalues**, a necessary condition for measurable physical quantities.

In this chapter, we develop these ideas further by defining the key characteristics of matrices that qualify them to serve as **quantum-mechanical operators**, exploring their properties, eigenvalue equations, and physical interpretations.

We begin by defining key **matrix characteristics**, including the **trace**, **determinant**, and **rank**, and discuss their significance in linear transformations. We then move on to **Hermitian**, **unitary**, and **orthogonal matrices**, emphasizing their physical relevance in quantum mechanics.

This leads naturally to the study of **commuting operators**, **simultaneous eigenfunctions**, and the **spectral properties** of Hermitian operators. Finally, we explore how these mathematical structures underpin quantum-mechanical formulations such as **operator algebra**, **expectation values**, and the **representation of observables** in Hilbert space.

Quantum Chemistry Connection: The Mathematical Toolkit of QM

To build the theory of quantum mechanics, we need:

1. **Hermitian Operators** to represent measurable quantities (observables).
2. **Unitary Operators** to represent changes of state that conserve probability (e.g., time evolution).
3. **The Eigenvalue Problem** to find the possible states and values of these observables.
4. **Commutation Rules** to understand what can be measured simultaneously.

This chapter develops this entire mathematical toolkit.

4.2 Some Definitions

4.2.1 Matrix Invariants: Trace, Determinant, and Rank

In Chapter 3, we briefly mentioned that while the operator matrix is sensitive to a change of basis, its **trace** remains invariant. In this section, we explain these terms in greater detail.

4.2.1.1 Trace of a Matrix

The trace of a matrix A is represented as $\mathrm{Tr}(A)$ and is defined as the sum of its diagonal elements, that is,

$$\mathrm{Tr}(A) = \sum_i A_{ii}. \tag{4.1}$$

Some important properties of the trace are as follows:

- **Linearity:**

  $$\mathrm{Tr}(A \pm B) = \mathrm{Tr}(A) \pm \mathrm{Tr}(B).$$

- **Cyclic property:**

  $$\mathrm{Tr}(AB) = \mathrm{Tr}(BA).$$

  Hence, the trace of the product of two matrices is independent of the order of multiplication, irrespective of whether the matrices commute.
- **Basis invariance:**
  The trace of a matrix is independent of the basis chosen for its representation.
- **Characteristic property:**
  The trace is a characteristic of the linear operator that the matrix represents. For example, for the geometric operation C_n corresponding to proper rotation by θ discussed in the previous chapter,

  $$C_n = C_{2\pi/\theta} : \mathrm{Tr}(C_n) = 1 + 2\cos\theta.$$

For the special case $\theta = 0°$, $\mathrm{Tr}(E) = 3$.

Similarly, for improper rotation,

$$S_n = S_{2\pi/\theta} : \mathrm{Tr}(S_n) = -1 + 2\cos\theta,$$

and for $\theta = 0°$, $\mathrm{Tr}(\sigma) = 1$.

In 3D space, any reflection reverses the sign of one coordinate and leaves the other two unchanged; hence, its trace is always unity.

For the inversion matrix,

$$\mathrm{Tr}(i) = -1 + 2\cos\pi = -3.$$

All three diagonal elements (in a Cartesian basis) are -1, so the trace is straightforwardly -3.

- **Transpose property:**

  For real matrices, the transpose is denoted by A^T, obtained by interchanging rows and columns. Since the diagonal elements remain unchanged,

  $$\mathrm{Tr}(A^T) = \mathrm{Tr}(A). \tag{4.2}$$

- **Hermitian conjugate:**

  For matrices with complex elements, the Hermitian conjugate $A^\dagger$ is defined as the transpose of the complex conjugate:

  $$[A^\dagger]_{ij} = ([A]_{ji})^*.$$

  Hence,

  $$\mathrm{Tr}(A^\dagger) = [\mathrm{Tr}(A)]^*. \tag{4.3}$$

- **Product of matrices:**

  The transpose of a product of two matrices is given by

  $$(AB)^T = B^T A^T. \tag{4.4}$$

  This can be extended to the product of several matrices:

  $$(AB \ldots G)^T = G^T \ldots B^T A^T.$$

  For matrices containing complex elements, the analogous relations are

  $$(AB)^\dagger = B^\dagger A^\dagger, \qquad (AB \ldots G)^\dagger = G^\dagger \ldots B^\dagger A^\dagger.$$

  Note that the order of multiplication is reversed.

4.2.1.2 Rank of a Matrix

Let A be an $M \times N$ matrix. Then A consists of N column vectors, each of which is an M-dimensional vector. The **rank** of A is the maximal number of linearly independent column vectors in A, that is, the maximal number of linearly independent vectors among $\{\mathbf{a}_1, \mathbf{a}_2, \ldots, \mathbf{a}_N\}$.

The rank of a matrix has several important applications. If A represents a set of N simultaneous nonhomogeneous equations in its elements, its rank r is the number of independent equations:

- If $r = N$, there are as many independent equations as unknowns; A has an inverse, and a unique solution exists.
- If $r < N$, the number of independent equations is less than the number of unknowns; A has no inverse, and no unique solution exists.

Note that we may compute the rank of any matrix—square or not. To find the rank, we determine the largest square submatrix whose determinant is nonzero. Thus, the rank of a matrix is r if there exists an $r \times r$ submatrix S whose determinant is nonzero, but no $(r+1) \times (r+1)$ submatrix with a nonzero determinant.

Example 4.1: Rank of the Stoichiometric Matrix

The rank of the $N \times R$ stoichiometric matrix S, where N is the number of substances and R is the number of reactions, gives the number of linearly independent reactions in a reaction system. Determine the rank of the stoichiometric matrix for the Michaelis–Menten mechanism of enzyme catalysis:

$$[1]\quad \mathrm{E} + \mathrm{S} \rightarrow \mathrm{ES},$$

$$[2]\quad \mathrm{ES} \rightarrow \mathrm{E} + \mathrm{S},$$

$$[3]\quad \mathrm{ES} \rightarrow \mathrm{P} + \mathrm{E}.$$

We number the species as (1) E, (2) S, (3) ES, and (4) P.

The stoichiometric matrix is obtained by writing the coefficients of the reactants as negative and those of the products as positive:

	[1]	[2]	[3]
E	−1	1	1
S	−1	1	0
ES	1	−1	−1
P	0	0	1

Hence,

$$S = \begin{bmatrix} -1 & 1 & 1 \\ -1 & 1 & 0 \\ 1 & -1 & -1 \\ 0 & 0 & 1 \end{bmatrix}.$$

The two 3 × 3 submatrices, S_{123} and S_{124}, have zero determinants (123 and 124 are row numbers).

Among the 2 × 2 submatrices formed from the first two rows and columns 12 and 13, S_{12} and S_{13}, the first has a zero determinant, while the second has a determinant of 1,

$$\det(S_{13}) = 1.$$

Therefore, the **rank** of the matrix is **two**. That is, among the three reactions, only two are linearly independent. It is evident that the columns corresponding to reactions [1] and [2] differ only in sign, confirming their linear dependence,

$$\text{Rank of } S = 2.$$

To avoid confusion, the symbol for the substrate is written as upright S and the matrix in italics S.

Chemistry Connection

Rank and Reaction Networks
In chemical kinetics, the **rank** of the stoichiometric matrix provides direct insight into the **number of independent reactions** and **conservation relationships** in a system:

- Each row of the matrix corresponds to a **chemical species**, and each column represents a **reaction step**.
- The rank, r, equals the number of **linearly independent reaction vectors**.
- The difference $(N - r)$ indicates the number of **independent conservation laws** (such as conservation of total enzyme, total mass, or charge).

For the Michaelis–Menten mechanism, $r = 2$ implies that while three reaction steps are written, only **two are independent**, and one **conservation constraint** (total enzyme) limits the system's degrees of freedom.

4.3 Special Types of Matrices and Their Physical Roles

4.3.1 Diagonal Matrices

We have already encountered a diagonal matrix—the unit matrix—which has ones along its leading diagonal and zeros elsewhere. Any other matrix with nonzero elements only on the diagonal is also a diagonal matrix.

The Laplacian expansion of a diagonal matrix yields, for the determinant,

$$\det(A) = a_{11}a_{22}a_{33} \ldots a_{NN}.$$

The inverse is easily obtained by replacing each diagonal element with its reciprocal:

$$A^{-1} = \begin{bmatrix} 1/a_{11} & 0 & \cdots & 0 \\ 0 & 1/a_{22} & \cdots & 0 \\ \vdots & \vdots & \ddots & \vdots \\ 0 & 0 & \cdots & 1/a_{NN} \end{bmatrix}.$$

It may also be noted that if two matrices A and B are both diagonal, then $AB = BA$; that is, they **commute** with each other.

4.3.2 Symmetric and Antisymmetric Matrices

For a real matrix A, if $A^T = A$, the matrix is said to be **symmetric**, for example,

$$A = \begin{bmatrix} 1 & 2 & 3 \\ 2 & 4 & 5 \\ 3 & 5 & 6 \end{bmatrix}.$$

If A is symmetric, its inverse A^{-1} is also symmetric.

If, on the other hand, $A^T = -A$, the matrix is **antisymmetric** or **skew-symmetric**; for example,

$$A = \begin{bmatrix} 0 & 2 & -3 \\ -2 & 0 & 5 \\ 3 & -5 & 0 \end{bmatrix}.$$

This condition can be satisfied only if all the diagonal elements are zero.

Any square $N \times N$ matrix can be decomposed into symmetric and antisymmetric parts as

$$A = \frac{1}{2}(A + A^T) + \frac{1}{2}(A - A^T) = B + C, \tag{4.5}$$

where $B = {}^1\!/\!_2(A + A^T)$ is the symmetric part $B^T = B$ and $C = {}^1\!/\!_2(A - A^T)$ is the antisymmetric part $C^T = -C$.

4.3.3 Hermitian and Anti-Hermitian Matrices

The complex analogs of symmetric and antisymmetric matrices are **Hermitian** and **anti-Hermitian** matrices.

A matrix A is **Hermitian** if $A^\dagger = A$ and **anti-Hermitian** if $A^\dagger = -A$.

As before, any complex matrix can be decomposed into Hermitian and anti-Hermitian components:

$$A = \frac{1}{2}(A + A^\dagger) + \frac{1}{2}(A - A^\dagger) = B + C, \tag{4.6}$$

where $B^\dagger = B$ (Hermitian) and $C^\dagger = -C$ (anti-Hermitian).

If A is Hermitian or anti-Hermitian, its inverse retains the same property since

$$\left(A^{-1}\right)^\dagger = \left(A^\dagger\right)^{-1} = \pm A^{-1}.$$

To appreciate the distinction between symmetric and Hermitian matrices, consider

$$A_1 = \begin{bmatrix} 1 & i \\ i & 1 \end{bmatrix}, \quad A_2 = \begin{bmatrix} 1 & i \\ -i & 1 \end{bmatrix}.$$

The first matrix is symmetric but not Hermitian, whereas the second is Hermitian but not symmetric.

Quantum Chemistry Connection

Anti-Hermitian matrices play a crucial role in quantum mechanics as the generators of unitary time evolution operators; specifically, the time evolution operator $U(t) = e^{-iHt/\hbar}$ is unitary because the Hamiltonian H is Hermitian, making $-iH$ anti-Hermitian, which ensures the preservation of probability and norm in quantum states during time evolution.

4.3.4 Orthogonal Matrices

An **orthogonal matrix** is one for which $A^T = A^{-1}$; that is, its inverse equals its transpose. Hence,

$$A^T A = I \Rightarrow \det(A^T A) = \det(I) = 1,$$

and, therefore,

$$[\det(A)]^2 = 1 \Rightarrow \det(A) = \pm 1. \tag{4.7}$$

This shows that the determinant of an orthogonal matrix must be either +1 (for a proper rotation) or −1 (for an improper rotation, such as reflection or inversion). For orthogonal matrices, all entries are real, and the determinant has unit modulus, meaning it is either 1 or −1.

It can also be shown that the inverse of an orthogonal matrix is itself orthogonal, since

$$\left(A^{-1}\right)^T = \left(A^T\right)^{-1} = \left(A^{-1}\right)^{-1} = A.$$

Let the orthogonal matrix A represent a linear operator acting on a vector $|v\rangle$.

If the transformed vector is $|v'\rangle = A|v\rangle$, the **inner product** (or norm) satisfies

$$\langle v' | v' \rangle = \langle v | A^T A | v \rangle = \langle v | v \rangle, \tag{4.8}$$

which shows that the transformation preserves the **length** (norm) of the vector.

The same reasoning applies to the inner product between any two vectors $|u\rangle$ and $|v\rangle$:

$$\langle u' | v' \rangle = \langle u | A^T A | v \rangle = \langle u | v \rangle. \tag{4.9}$$

Hence, orthogonal transformations preserve both the lengths of vectors and the **angles** between them.

4.3.5 Unitary Matrices

For complex matrices, the corresponding property is **unitarity**. A matrix A is **unitary** if $A^\dagger = A^{-1}$, that is, its inverse equals its Hermitian conjugate. The inverse of a unitary matrix is also unitary:

$$\left(A^{-1}\right)^\dagger = \left(A^\dagger\right)^{-1} = \left(A^{-1}\right)^{-1}.$$

As with orthogonal matrices, the determinant of a unitary matrix has unit modulus ($|\det A| = 1$).

All results derived above remain valid if A^T is replaced by $A^\dagger$, the Hermitian conjugate.

Thus, for unitary operators, the transformations

$$\langle v' | v' \rangle = \langle v | A^\dagger A | v \rangle = \langle v | v \rangle,$$
$$\langle u' | v' \rangle = \langle u | A^\dagger A | v \rangle = \langle u | v \rangle$$

confirm that a unitary transformation preserves both norms and inner products in complex vector spaces.

For functions, the corresponding equation shows that unitary transformations preserve the norms (lengths) of complex vectors:

$$\langle A\psi \mid A\psi\rangle = \langle \psi \mid \psi\rangle.$$

Such transformations play a central role in quantum mechanics, where they correspond to operations that conserve total probability.

Example 4.2: Unitary Matrices

Show that the matrix

$$A = \begin{bmatrix} \frac{3}{5} & \frac{4i}{5} \\ \frac{4i}{5} & \frac{3}{5} \end{bmatrix}$$

is unitary.

Solution:

To show that the matrix is unitary, we must verify that

$$A^{\dagger}A = I,$$

where $A^{\dagger}$ denotes the Hermitian conjugate of A.

Taking the complex conjugate and transpose, we obtain

$$A^{\dagger} = \begin{bmatrix} \frac{3}{5} & -\frac{4i}{5} \\ -\frac{4i}{5} & \frac{3}{5} \end{bmatrix}.$$

Multiplying,

$$A^{\dagger}A = \begin{bmatrix} \frac{3}{5} & -\frac{4i}{5} \\ -\frac{4i}{5} & \frac{3}{5} \end{bmatrix}\begin{bmatrix} \frac{3}{5} & \frac{4i}{5} \\ \frac{4i}{5} & \frac{3}{5} \end{bmatrix} = \begin{bmatrix} 1 & 0 \\ 0 & 1 \end{bmatrix} = I.$$

Hence, the condition $A^{\dagger}A = I$ is satisfied, and **A is a unitary matrix.**

Concept Note 4.1

The matrix A preserves both the inner product and the norm of any complex vector on which it operates. Such transformations correspond to *rotations in complex space*, conserving probability amplitudes—a fundamental requirement for operators in quantum mechanics.

Fact Box 4.1: The Pauli Matrices

The **Pauli matrices**, denoted $\sigma_x, \sigma_y, \sigma_z$, form a fundamental set of 2×2 Hermitian and unitary matrices that appear throughout quantum mechanics,

$$\sigma_x = \begin{bmatrix} 0 & 1 \\ 1 & 0 \end{bmatrix}, \sigma_y = \begin{bmatrix} 0 & -i \\ i & 0 \end{bmatrix}, \sigma_z = \begin{bmatrix} 1 & 0 \\ 0 & -1 \end{bmatrix}.$$

Key Properties

1. Each σ_i is **Hermitian** $(\sigma_i^\dagger = \sigma_i)$ and **unitary** $(\sigma_i^\dagger \sigma_i = I)$.
2. Their eigenvalues are ± 1; the corresponding eigenvectors form an orthonormal basis.
3. They obey the commutation relations

$$[\sigma_i, \sigma_j] = 2i\, \epsilon_{ijk} \sigma_k,$$

ϵ_{ijk} is the **Levi-Civita** symbol (or **permutation symbol).** In the Pauli commutation relation, it compactly encodes all three relations at once:

- $[\sigma_x, \sigma_y] = 2i\, \sigma_z$
- $[\sigma_y, \sigma_z] = 2i\, \sigma_x$
- $[\sigma_z, \sigma_x] = 2i\, \sigma_y$

and **anticommutation relations**

$$\{\sigma_i, \sigma_j\} = 2\delta_{ij} I.$$

Here, δ_{ij} is the Kronecker delta.
The set $\{I, \sigma_x, \sigma_y, \sigma_z\}$ forms a complete basis for all 2×2 matrices.

Quantum Significance

- Represent **spin-½ operators**: $\widehat{S}_i = \frac{\hbar}{2}\sigma_i$.
- Serve as the building blocks for two-level systems (qubits, magnetic resonance, etc.).
- Provide simple, concrete examples of **Hermitian operators with real eigenvalues** that also generate **unitary rotations** $U = e^{i\theta\sigma_i/2}$.

4.3.6 Normal Matrices

A **normal matrix** is defined as one that **commutes with its Hermitian conjugate**, that is,

$$A^\dagger A = AA^\dagger.$$

Examples of normal matrices include **Hermitian (symmetric)** and **unitary (orthogonal)** matrices.

For a **Hermitian matrix**,

$$A^\dagger = A \Rightarrow AA^\dagger = AA = A^\dagger A.$$

For a **unitary matrix**,

$$A^{-1} = A^\dagger \Rightarrow AA^\dagger = I = A^{-1}A = A^\dagger A.$$

For **symmetric** and **orthogonal** matrices, the **transpose** replaces the Hermitian conjugate in the above relations.

It can also be shown that **if A is a normal matrix, its inverse A^{-1} is also normal**, since

$$(A^{-1})^{\dagger}(A^{-1}) = (A^{\dagger}A)^{-1} = (AA^{\dagger})^{-1} = (A^{-1})(A^{-1})^{\dagger}.$$

Concept Note 4.2

The most *physically significant* classes of operators in quantum mechanics—Hermitian and unitary operators—are both subclasses of **normal matrices**. Conversely, a matrix is normal if and only if it is unitarily diagonalizable. This is the feature that makes them so physically significant.

The key property of normal matrices is that they are **diagonalizable by a unitary transformation**, which is why we can always find a nice orthonormal basis of eigenstates for our observables:

- **Hermitian operators** correspond to *observables* (with real eigenvalues).
- **Unitary operators** represent *reversible transformations*, such as rotations and time evolution.

Both classes preserve the inner product structure of the Hilbert space, ensuring that physical probabilities remain invariant.

Example 4.3: Normal Matrices

Show that the matrix

$$A = \begin{bmatrix} 1 & i \\ i & 1 \end{bmatrix}$$

is **normal** but **not Hermitian.**

Solution:

1. **Test for Hermiticity:** A matrix is Hermitian if $A^{\dagger} = A$. Let us find the Hermitian conjugate (adjoint) of A by taking the complex conjugate of each element and transposing the matrix:

$$A^{\dagger} = \begin{bmatrix} 1 & i \\ i & 1 \end{bmatrix}^{*T} = \begin{bmatrix} 1 & -i \\ -i & 1 \end{bmatrix}^{T} = \begin{bmatrix} 1 & -i \\ -i & 1 \end{bmatrix}.$$

Comparing this to the original matrix:

$$\begin{bmatrix} 1 & -i \\ -i & 1 \end{bmatrix} \neq \begin{bmatrix} 1 & i \\ i & 1 \end{bmatrix}.$$

Since $A^{\dagger} \neq A$, the matrix is **not Hermitian**.

2. **Test for Normality:** A matrix is normal if it commutes with its adjoint, meaning $A^\dagger A = AA^\dagger$.
   Calculate $A^\dagger A$:

$$A^\dagger A = \begin{bmatrix} 1 & -i \\ -i & 1 \end{bmatrix}\begin{bmatrix} 1 & i \\ i & 1 \end{bmatrix} = \begin{bmatrix} (1)(1)+(-i)(i) & 1(i)+(-i)(1) \\ -i(1)+(1)(i) & (-i)(i)+(1)(1) \end{bmatrix}.$$

   Recalling that $i^2 = -1$, so $(-i)(i) = -i^2 = -(-1) = 1$:

$$= \begin{bmatrix} 1+1 & i-i \\ -i+i & 1+1 \end{bmatrix} = \begin{bmatrix} 2 & 0 \\ 0 & 2 \end{bmatrix} = 2I.$$

   Now calculate $AA^\dagger$:

$$AA^\dagger = \begin{bmatrix} 1 & i \\ i & 1 \end{bmatrix}\begin{bmatrix} 1 & -i \\ -i & 1 \end{bmatrix} = \begin{bmatrix} (1)(1)+(i)(-i) & 1(i)+(-i)(1) \\ i(1)+(1)(-i) & (i)(-i)+(1)(1) \end{bmatrix}$$

$$= \begin{bmatrix} 1+1 & -i+i \\ i-i & 1+1 \end{bmatrix} = \begin{bmatrix} 2 & 0 \\ 0 & 2 \end{bmatrix} = 2I.$$

Conclusion

Since $A^\dagger A = AA^\dagger$ (both equal $2I$), matrix A is normal. This demonstrates that while all Hermitian matrices are normal, not all normal matrices are Hermitian.

Concept Note 4.3

This example shows that **normality** is a broader property than **Hermiticity**.

All Hermitian and unitary matrices are normal, but not all normal matrices are Hermitian.

Normal matrices can always be **diagonalized by a unitary transformation**, making them especially useful in quantum mechanics and spectroscopy for analyzing coupled systems.

Matrix type	Defining property	Key feature	Eigenvalue property	Physical role in QM
Hermitian	$A^\dagger = A$	Real eigenvalues	Real $\lambda \in R$	**Observables** (energy, momentum)
Unitary	$A^\dagger = A^{-1}$	Preserves inner products	Complex, modulus 1 ($\lvert\lambda\rvert = 1$)	**Transformations** (time evolution, rotation)
Orthogonal	$A^T = A^{-1}$	Preserves norm in real space	Complex, modulus 1, or real ± 1 ($\lvert\lambda\rvert = 1$)	Rotations in 3D space
Normal	$A^\dagger A = AA^\dagger$	Orthogonal eigenvectors	No universal constraint (can be real or complex)	Guarantees a "nice" eigenbasis

Now that we are equipped with the tools, here is the fascinating geometric question we are going to investigate. The question we need to ask ourselves is: "Why do we need to find eigenvalues and eigenvectors?" We will try to answer this question through this geometric example.

4.4 The Core Idea: Finding the "Axes" of a Transformation

Imagine you have a linear transformation. It is a function that takes a vector and maps it to a new vector (see Chapter 3). This transformation, represented by a matrix A, might stretch, compress, rotate, or shear the entire vector space. When you apply it, most vectors get knocked off their original direction.

Eigenvectors are the special, exceptional vectors that do *not* change their direction under the transformation. When you apply the transformation A to an **eigenvector** $|v\rangle$, the resulting vector $A|v\rangle$ points in the exact same (or exact opposite) direction as the original vector $|v\rangle$. The transformation only scales the eigenvector by a certain amount.

This is what the defining equation $A|v\rangle = \lambda|v\rangle$ means geometrically:

$A|v\rangle$ is the vector after the transformation. $\lambda|v\rangle$ is the original vector $|v\rangle$ scaled by a factor λ.

So, the equation says the transformed vector is just a scaled version of the original. The line spanned by the eigenvector is **invariant**.

4.4.1 The Role of the Eigenvalue λ: The Scaling Factor

The eigenvalue λ tells you *how* the eigenvector $|v\rangle$ is scaled.

- **If** $\lambda > 1$: The eigenvector is **stretched**. The transformation points it in the same direction but makes it longer.
- **If** $0 < \lambda < 1$: The eigenvector is **compressed**. It points in the same direction but becomes shorter.
- **If** $\lambda < 0$: The eigenvector is **flipped and scaled**. It points in the exact opposite direction. If $\lambda = -1$, it is a perfect reflection across the origin.
- **If** $\lambda = 1$: The eigenvector is **unchanged**.
- **If** $\lambda = 0$: The eigenvector is **squashed into the zero vector**. These eigenvectors form the null space (or **kernel**) of the transformation.

The Key Takeaway—Eigenvectors reveal the fundamental "action" of a transformation. They are the directions along which the transformation is purely a stretch, compression, or reflection.

4.4.2 A Powerful Visual Example: The Shear Transformation

A horizontal shear is a fantastic way to see the importance of eigenvectors. This transformation slants everything to the side.

Consider the matrix $A = \begin{bmatrix} 1 & 1 \\ 0 & 1 \end{bmatrix}$. Let us see what it does to a few vectors:

- The vector $|u\rangle = \begin{bmatrix} 0 \\ 1 \end{bmatrix}$ (y-axis pointing up) gets transformed to $A|u\rangle = \begin{bmatrix} 1 \\ 1 \end{bmatrix}$. Its direction completely changes.
- The vector $|w\rangle = \begin{bmatrix} 0.5 \\ 0.5 \end{bmatrix}$ gets transformed to $A|w\rangle = \begin{bmatrix} 1 \\ 0.5 \end{bmatrix}$. Its direction also changes.

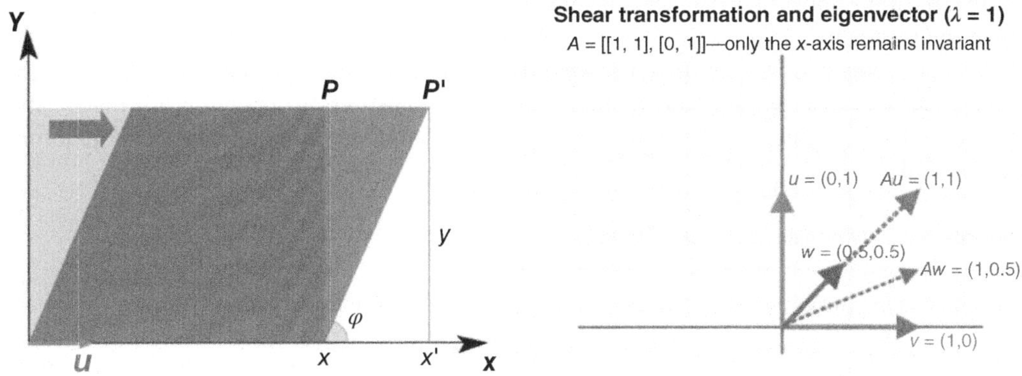

Figure 4.1 Shear transformation and eigenvector $\lambda = 1$.

4.4.3 But What About the Eigenvector?

For this matrix, the only eigenvector is any vector along the x-axis, like $|\,v\rangle = \begin{bmatrix} 1 \\ 0 \end{bmatrix}$. Let us test it:

$$A|\,v\rangle = \begin{bmatrix} 1 & 1 \\ 0 & 1 \end{bmatrix}\begin{bmatrix} 1 \\ 0 \end{bmatrix} = \begin{bmatrix} 1 \\ 0 \end{bmatrix}.$$

The vector comes out completely unchanged (Figure 4.1)! Here, the eigenvalue is $\lambda = 1$.

4.4.4 The Eigenvalue Problem in Quantum Mechanics

Perhaps the most important application of matrices in quantum mechanics is in determining the **eigenvectors** and **eigenvalues** of linear operators. Quantum-mechanical systems can exist only in the *eigenstates* of the Hamiltonian operator, and the eigenvalues are the only measurable values for the corresponding observable, such as energy.

An eigenvalue equation for a linear operator A can be written as

$$A\,|v\,\rangle = \lambda\,|v\,\rangle. \tag{4.10}$$

When the matrix A acts on a vector $|v\,\rangle$, it returns the same vector multiplied by a constant λ. The vector $|v\,\rangle$ is then the **eigenvector** of the operator, and λ is its **eigenvalue**. The term “eigen” originates from the German word *eigen*, meaning “peculiar to” or “characteristic.”

We now turn to the general properties of eigenvalues and eigenvectors, which form the bridge between linear algebra and quantum mechanics.

4.5 Properties of Hermitian Matrices

There are some important theorems concerning the eigenvectors and eigenvalues of **Hermitian matrices** that must be understood before proceeding further.

In the following, we shall use the general **Dirac notation** to describe scalar products so that the theorems apply both to Hermitian (self-adjoint) matrices that act on ordinary vectors, and to

Hermitian linear operators that act in a Hilbert space—for example, the space of wavefunctions in quantum mechanics.

In the latter context, one often discusses **eigenfunctions** rather than eigenvectors. The two are mathematically equivalent; only the context changes. By substituting "operator" for matrix and "function" for vector, the same definitions and theorems apply to the Hilbert space of quantum wavefunctions.

Theorem 4.1 *The eigenvalues of a Hermitian matrix are real.*

Proof: The eigenvalue equation for an operator represented by the matrix A is

$$A\,|v\rangle = \lambda\,|v\rangle. \tag{4.11}$$

Taking the inner product with $\langle v\,|$ gives

$$\langle v|A|v\rangle = \lambda\langle v|v\rangle. \tag{4.12}$$

Now take the Hermitian conjugate of this equation. Since $A^\dagger = A$ for a Hermitian matrix, we have

$$\langle v|A|v\rangle = \lambda^*\langle v|v\rangle.$$

Subtracting the two equations gives

$$(\lambda - \lambda^*)\langle v|v\rangle = 0. \tag{4.13}$$

Since $\langle v|v\rangle \neq 0$, it follows that $\lambda^* = \lambda$; that is, each eigenvalue is equal to its own complex conjugate and must therefore be **real**.

Hence proved. ■

Theorem 4.2 *The eigenvectors of a Hermitian matrix corresponding to different eigenvalues are orthogonal.*

Proof: The eigenvalue equation for the ith eigenvalue λ_i is

$$A\,|v_i\rangle = \lambda_i\,|v_i\rangle. \tag{4.14}$$

Taking the Hermitian conjugate gives

$$\langle v_i\,|A = \lambda_i^*\,\langle v_i\,|.$$

Multiplying by $|v_j\rangle$ on the right,

$$\langle v_i|A|v_j\rangle = \lambda_i^*\langle v_i|v_j\rangle. \tag{4.15}$$

Using Equation (4.14) again, we can write

$$\langle v_i|A|v_j\rangle = \lambda_j\langle v_i|v_j\rangle. \tag{4.16}$$

Subtracting (4.15) from (4.16) gives

$$(\lambda_j - \lambda_i^*)\langle v_i|v_j\rangle = 0. \tag{4.17}$$

For $i = j$, this simply shows that $\lambda_i^* = \lambda_i$, confirming that the eigenvalues are real.

For $i \neq j$, since the eigenvalues are different $(\lambda_i \neq \lambda_j)$, we have

$$\langle v_i | v_j \rangle = 0.$$

Thus, **eigenvectors corresponding to different eigenvalues of a Hermitian matrix are orthogonal.** ■

Real symmetric matrices are special cases of Hermitian matrices and hence also have real, mutually orthogonal eigenvectors.

4.5.1 Significance

These two properties—*real eigenvalues* and *orthogonal eigenvectors*—are of prime importance in quantum mechanics. Observables such as energy, linear momentum, and angular momentum are represented by **Hermitian operators**, ensuring that measurable quantities are real.

The orthogonality condition does not hold when $\lambda_i = \lambda_j$; in that case, the eigenvalues are said to be **degenerate**. The corresponding eigenvectors can be made orthogonal through the **Gram–Schmidt orthogonalization** procedure, which will be discussed in the next chapter. They can also be normalized to form an **orthonormal set**.

If an $N \times N$ Hermitian matrix has N distinct eigenvalues, then all its eigenvectors are mutually orthogonal and form a **complete set**. This means any vector in the space can be expressed as a linear combination of these eigenvectors—the mathematical statement of the **superposition principle**.

To understand this, consider a three-dimensional real space. Two orthogonal vectors define only a plane, and only vectors in that plane can be expressed as their linear combinations. With three mutually orthogonal vectors, however, we can represent any vector in the entire 3D space. The same principle holds in higher-dimensional Hilbert spaces in quantum mechanics.

4.6 Eigenvectors and Eigenvalues of Unitary Matrices

We learned that a unitary matrix is one for which $A^\dagger = A^{-1}$, that is, $A^\dagger A = I$.

Let $|v_i\rangle$ and $|v_j\rangle$ be two eigenvectors of the unitary matrix A. Then,

$$A|v_i\rangle = \lambda_i |v_i\rangle$$

and

$$A|v_j\rangle = \lambda_j |v_j\rangle.$$

We take the Hermitian conjugate of the first equation and apply it to the second:

$$\langle v_i | A^\dagger A | v_j \rangle = \lambda_i^* \lambda_j \langle v_i | v_j \rangle.$$

Since $A^\dagger A = I$ for unitary matrices, we have

$$\langle v_i | v_j \rangle = \lambda_i^* \lambda_j \langle v_i | v_j \rangle.$$

Set $i = j$ in the above equation. Because $|v_i\rangle$ is a nonzero vector, the inner product $\langle v_i | v_i \rangle \neq 0$. Hence,

$$\lambda_i^* \lambda_i = |\lambda_i|^2 = 1.$$

Therefore, **the eigenvalues of unitary (and orthogonal) matrices have unit modulus**, i.e., they lie on the unit circle in the complex plane.

We have now seen that Hermitian matrices have **real eigenvalues** and **orthogonal eigenvectors**, while unitary matrices possess **complex eigenvalues of unit modulus** and **orthonormal eigenvectors**.

An important question arises when two or more such operators act on the same vector space: *Can they share a common set of eigenvectors?*

This leads naturally to the concept of **simultaneous diagonalization**, which is central to quantum mechanics—particularly in defining *commuting observables* such as energy and angular momentum.

4.7 Commuting Operators and Simultaneous Measurement

Theorem 4.3 *If two matrices A and B have the same set of eigenvectors, then they commute.*

Proof: Let A and B have common eigenvectors. We must prove that, in this case, A and B commute. Then,

$$A|v_i\rangle = \lambda_i|v_i\rangle, \qquad B|v_i\rangle = \mu_i|v_i\rangle.$$

Let

$$|\psi\rangle = \sum_i c_i|v_i\rangle$$

be the linear combination of all eigenvectors. Then,

$$AB|\psi\rangle = A\sum_i c_i B|v_i\rangle = A\sum_i c_i\mu_i|v_i\rangle = \sum_i c_i\mu_i\lambda_i|v_i\rangle. \tag{4.18}$$

Similarly,

$$BA|\psi\rangle = B\sum_i c_i A|v_i\rangle = B\sum_i c_i\lambda_i|v_i\rangle = \sum_i c_i\lambda_i\mu_i|v_i\rangle. \tag{4.19}$$

Since λ_i and μ_i are scalars, the order of multiplication does not matter. Subtracting Equation (4.19) from (4.18) gives

$$(AB - BA)|\psi\rangle = 0$$

because $\lambda_i\mu_i = \mu_i\lambda_i$, they being scalars.

Hence, $AB - BA = 0$; that is, A commutes with B. ■

4.7.1 Example Table: Commuting Matrices and Their Shared Eigenvectors

Consider the following two commuting matrices:

$$A = \begin{bmatrix} 1 & 0 \\ 0 & 2 \end{bmatrix}, \qquad B = \begin{bmatrix} 3 & 0 \\ 0 & 4 \end{bmatrix}.$$

Both matrices are diagonal and share the same set of eigenvectors:

Matrix	Eigenvalue	Eigenvector
A	1	$\begin{bmatrix}1\\0\end{bmatrix}$
A	2	$\begin{bmatrix}0\\1\end{bmatrix}$
B	3	$\begin{bmatrix}1\\0\end{bmatrix}$
B	4	$\begin{bmatrix}0\\1\end{bmatrix}$

Interpretation

Both matrices are diagonalized by the same set of eigenvectors. Since $AB = BA$, they commute and can be simultaneously diagonalized. The shared eigenbasis here is $\begin{bmatrix}1\\0\end{bmatrix}$ and $\begin{bmatrix}0\\1\end{bmatrix}$.

Theorem 4.4 *If two operators commute, then there exists a complete set of common eigenvectors.*
This means we can choose the eigenvectors within any degenerate subspaces to also be eigenvectors of the other operator.

Proof: Let $|v\rangle$ be an eigenvector of A with eigenvalue λ, that is,

$$A|v\rangle = \lambda|v\rangle.$$

Since A and B commute,

$$AB|v\rangle = BA|v\rangle = B(\lambda|v\rangle) = \lambda B|v\rangle.$$

This is only possible if $B|v\rangle$ is a scalar multiple of $|v\rangle$; that is,

$$B|v\rangle = \mu|v\rangle.$$

Therefore, every eigenvector of A is also an eigenvector of B. ∎

This simple proof given applies to the nondegenerate case.

Theorems 4.3 and 4.4 together establish that two matrices commute **if and only if** they share a common set of eigenvectors.

4.7.2 Discussion and Significance

Theorems 4.3 and 4.4 together express two complementary aspects:

- If two operators share the same set of eigenvectors, they **must commute**.
- Conversely, if they **commute**, they necessarily **share a common eigenbasis**.

In quantum mechanics, this relationship has profound implications. **Commuting operators** represent **compatible observables**—quantities that can be measured simultaneously with arbitrary precision (e.g., energy and angular momentum components along the same axis).

Non-commuting operators, on the other hand, do not have common eigenvectors; their corresponding observables cannot be simultaneously determined. This is the mathematical foundation of the **Heisenberg Uncertainty Principle**.

4.8 Orthogonality of Eigenvectors of Normal Matrices

Lemma. *If A is normal and $A\,|v\rangle = \lambda\,|v\rangle$, then $A^\dagger\,|v\rangle = \lambda^*\,|v\rangle$.*
Proof: Let $B = A - \lambda I$. Since A is normal,

$$\begin{aligned} B^\dagger B &= (A^\dagger - \lambda I)(A - \lambda I) = A^\dagger A - \lambda A^\dagger - \lambda A + |\lambda|^2 I \\ &= AA^\dagger - \lambda A^\dagger - \lambda A + |\lambda|^2 I = (A - \lambda I)(A^\dagger - \lambda I) = BB^\dagger, \end{aligned}$$

so B is also normal. For any normal operator, $\langle v|B^\dagger B|v\rangle = \langle v|BB^\dagger|v\rangle$, i.e. $\|B|v\rangle\|^2 = \left\|B^\dagger|v\rangle\right\|^2$. Since $|v\rangle$ is an eigenvector of A with eigenvalue λ, $B\,|v\rangle = 0$, so $\left\|B^\dagger|\,v\rangle\right\| = 0$ too, giving $B^\dagger\,|v\rangle = 0$ — that is, $A^\dagger\,|v\rangle = \lambda^*\,|v\rangle$. ■

An important property of normal matrices is that their **eigenvectors corresponding to distinct eigenvalues are orthogonal**.

This result generalizes the familiar case of Hermitian matrices.

Let A be a normal matrix, so that

$$A^\dagger A = AA^\dagger.$$

Let $|u\rangle$ and $|v\rangle$ be two eigenvectors of A, corresponding to distinct eigenvalues λ_u and λ_v, respectively:

$$A\,|u\rangle = \lambda_u\,|u\rangle, \qquad A\,|v\rangle = \lambda_v\,|v\rangle.$$

Taking the inner product of the first equation with $|v\rangle$, we obtain

$$\langle v\mid A\mid u\rangle = \lambda_u\langle v\mid u\rangle.$$

Applying the Lemma to $|v\rangle$ with eigenvalue λ_v gives $A^\dagger\,|v\rangle = \lambda_v^*\,|v\rangle$, so

$$\langle v|A = (A^\dagger|v\rangle)^\dagger = \lambda_v\langle v|.$$

Now evaluate $\langle v|A|u\rangle$ two ways. Acting on the right with $A|\,u\rangle = \lambda_u|u\rangle$:

$$\langle v|A|u\rangle = \lambda_u\langle v|u\rangle.$$

Acting on the left with $\langle v\,|A = \lambda\,\langle v\,|$:

$$\langle v|A|u\rangle = \lambda_v\langle v|u\rangle.$$

Subtracting,

$$(\lambda_u - \lambda_v^*)\langle v\mid u\rangle = 0.$$

For distinct eigenvalues $\lambda_u \neq \lambda_v$, the only possible solution is

$$\boxed{\langle v\mid u\rangle = 0.}$$

Thus, **eigenvectors of a normal matrix corresponding to distinct eigenvalues are orthogonal**.

Concept Note 4.4

Orthogonality of eigenvectors ensures that normal matrices—whether Hermitian or unitary—can be **diagonalized by a unitary transformation**. This forms the mathematical foundation for expressing operators in **orthonormal bases** in quantum mechanics.

4.9 Transition to Eigenvalue Problems

The orthogonality of eigenvectors of normal matrices has deep implications in quantum mechanics. Every observable quantity—such as energy, momentum, or angular momentum—is represented by a Hermitian operator, whose eigenvalues are real and eigenvectors form a complete orthonormal basis for the system's state space. The problem of finding these eigenvalues and eigenvectors, therefore, lies at the heart of quantum theory. In matrix form, it appears as a purely algebraic relation, while in wave mechanics it takes the form of a differential equation—the **Schrödinger equation**.

The results discussed so far establish the fundamental properties of eigenvalues and eigenvectors of Hermitian and unitary matrices, as well as the conditions under which different operators may possess common eigenvectors. In practical applications, however, we frequently need to determine the eigenvalues and eigenvectors of a given operator or matrix explicitly. This process not only provides the energy levels and stationary states in quantum mechanics but also forms the foundation for the diagonalization of matrices and the solution of coupled linear equations. We therefore turn to methods for determining eigenvalues and eigenvectors of finite-dimensional matrices in the next chapter.

The concepts of Hermitian, unitary, and normal matrices have revealed how quantum-mechanical operators act and what their eigenvalues signify. In the next chapter, we turn from these general properties to the **explicit determination of eigenvalues and eigenvectors**, learning how matrices are diagonalized and how orthogonality and completeness emerge naturally from this process.

Key Takeaways

- **Hermitian Operators:**
  All quantum-mechanical observables are represented by Hermitian operators because their eigenvalues are *real* and correspond to *measurable physical quantities.*
- **Matrix–Operator Correspondence:**
  Any linear transformation can be represented as a matrix operation. The matrix elements depend on the chosen basis, but *characteristic quantities* such as the **trace, determinant,** and **eigenvalues** remain invariant under basis transformation.
- **Special Matrices:**
  - Symmetric/Hermitian: $A^T = A$ or $A^\dagger = A$.
  - Orthogonal/Unitary: $A^T = A^{-1}$ or $A^\dagger = A^{-1}$.

  These preserve the inner product and hence the *norm* of vectors.

- **Normal Matrices:**
  Matrices that commute with their Hermitian conjugate ($A^\dagger A = AA^\dagger$) include all Hermitian and unitary matrices. Their eigenvectors form a complete orthogonal set.
- **Commuting Operators:**
  Operators that commute share the same eigenvectors and can be *simultaneously diagonalized*. Their corresponding observables can be *measured simultaneously*.
- **Unitary Transformations:**
  These preserve norms and inner products—hence, they represent *physically allowable transformations* in quantum mechanics (e.g., rotations, time evolution).

4.10 Summary

In this chapter, we explored the properties of matrices as linear operators in quantum mechanics:

- The concepts of trace and rank were defined, along with their mathematical significance and invariance properties.
- Various special types of matrices—diagonal, symmetric, antisymmetric, Hermitian, anti-Hermitian, orthogonal, unitary, and normal—were introduced and discussed in relation to physical operators.
- The concept of eigenvalues and eigenvectors was developed, showing that Hermitian matrices possess real eigenvalues and orthogonal eigenvectors.
- The physical significance of these properties in quantum mechanics was emphasized, where observables correspond to Hermitian operators.
- The conditions for commutation of operators and the resulting simultaneous eigenvectors were discussed.

Further Reading

P. W. Atkins and R. S. Friedman, *Molecular Quantum Mechanics*, 5th ed., Oxford University Press, 2010.

D. J. Griffiths and D. F. Schroeter, *Introduction to Quantum Mechanics*, 3rd ed., Cambridge University Press, 2018.

I. N. Levine, *Quantum Chemistry*, 7th ed., Pearson, 2013.

H. Margenau and G. M. Murphy, *The Mathematics of Physics and Chemistry*, 2nd ed., Van Nostrand, 1956.

D. A. McQuarrie, *Quantum Chemistry*, University Science Books, 2008.

Exercises

A. Conceptual Questions

1. Define the following: (a) Hermitian matrix, (b) unitary matrix, and (c) normal matrix.

2. Discuss why Hermitian operators are used to represent observables in quantum mechanics.

3. Explain why the eigenvalues of a Hermitian operator must be real and why its eigenvectors corresponding to distinct eigenvalues are orthogonal.

4. Show that the trace of a matrix is invariant under a change of basis. Give one physical example where this invariance is significant.

5. If two operators commute, what does this imply about their simultaneous eigenvectors? Explain its quantum-mechanical significance.

6. Distinguish among Hermitian, unitary, and normal matrices. Give one quantum-mechanical example of each.

7. Why must an observable in quantum mechanics be represented by a Hermitian operator?

8. Show mathematically that for a unitary operator, probability is conserved.

9. Explain the difference between symmetry under rotation and invariance under unitary transformation.

10. If two operators commute, why can their eigenvectors form a common basis? Discuss the measurement consequences.

11. Why are anti-Hermitian operators useful in time evolution or generator formulations?

12. What is the physical significance of the determinant and trace of a matrix operator?

B. Problem-solving Exercises

1. For each of the following matrices, determine whether it is Hermitian, unitary, or normal:

$$A = \begin{bmatrix} 0 & i \\ -i & 0 \end{bmatrix}, B = \frac{1}{\sqrt{2}}\begin{bmatrix} 1 & 1 \\ -1 & 1 \end{bmatrix}, C = \begin{bmatrix} 2 & 1 \\ 0 & 2 \end{bmatrix}.$$

Interpret each physically.

2. Show that the Pauli matrices are both Hermitian and unitary up to a constant factor. Explain their role as spin operators.

3. Verify explicitly that for any 2×2 orthogonal matrix A, $|\det A| = 1$. Give one representing a reflection and one a rotation.

4. Let

$$H = \begin{bmatrix} 3 & 2i \\ -2i & 3 \end{bmatrix}.$$

   a. Show that H is Hermitian.

b. Determine its eigenvalues and normalized eigenvectors.
c. Interpret the physical meaning if H represents a two-level system Hamiltonian.

5. Prove that the product of two commuting Hermitian matrices is Hermitian if and only if they commute.

6. Consider a matrix $U = e^{i\theta\sigma_z}$. Show that U is unitary for any real θ and interpret this transformation physically.

7. Show that the eigenvalues of a normal matrix are invariant under a unitary similarity transformation.

C. Advanced/Analytical Exercises

1. Write the triple scalar product of three vectors in determinantal form. Hence, prove that if a 3×3 matrix is orthogonal (or unitary), its determinant can only have two possible values: +1 or −1.

2. Consider the matrices

$$A = \begin{bmatrix} 0 & -1 & 0 \\ 1 & 0 & 0 \\ 0 & 0 & 1 \end{bmatrix}, \quad B = \begin{bmatrix} 1 & 0 & 0 \\ 0 & -1 & 0 \\ 0 & 0 & 1 \end{bmatrix}.$$

Indicate which one represents a proper rotation and which an improper rotation.

3. The stoichiometric matrix of the Michaelis–Menten mechanism was shown to have rank 2:
a. Explain what this means physically.
b. Give another chemical example where the reaction matrix has a rank lower than the number of reactions.

4. Show that the commutator $[A, B] = 0$ is both necessary and sufficient for simultaneous diagonalization of A and B.

5. Given two operators A and B such that $AB = -BA$, show that if ψ is an eigenvector of A with eigenvalue a, then $B\psi$ is an eigenvector of A with eigenvalue $-a$. Discuss examples from angular momentum theory.

6. Demonstrate that any Hermitian operator can be expressed as $A = \sum_i \lambda_i |\psi_i\rangle\langle\psi_i|$, where λ_i and $|\psi_i\rangle$ are its eigenvalues and eigenvectors, respectively.

7. In a two-state quantum system, a Hamiltonian operator is represented by

$$H = \begin{bmatrix} E_0 & V \\ V & E_0 \end{bmatrix}.$$

Find its eigenvalues and eigenvectors and relate them to bonding/antibonding combinations in molecular orbital theory.

8. For a 3×3 rotation matrix about the z-axis by angle θ, show that its trace $= 1 + 2\cos\theta$ and discuss its physical interpretation.

D. Research/Open-ended

1. Discuss how the diagonalization of Hermitian matrices underlies electronic-structure methods such as Hartree–Fock and DFT.

2. Explore how unitary transformations are implemented numerically to ensure orthonormality preservation in quantum-chemical computations.

5

Eigenvalue Problems, Orthogonality, and Completeness

The fundamental laws necessary for the mathematical treatment of a large part of physics and the whole of chemistry are thus completely known, and the difficulty lies only in the fact that the exact application of these laws leads to equations much too complicated to be soluble.

—Paul A. M. Dirac

This quotation perfectly captures the motivation behind this chapter. While the fundamental equations of quantum mechanics are known, solving them in their exact form is often intractable. Mathematical tools such as eigenvalue problems, orthogonality, and completeness make these equations solvable in approximate or reduced form, revealing the essential physical content while maintaining mathematical rigor.

Learning Outcomes

After completing this chapter, students will be able to:

1. **Define** eigenvalues and eigenvectors and **explain** their geometric and physical significance.
2. **Formulate and solve** the characteristic (eigenvalue) equation of a matrix or operator.
3. **Demonstrate** the orthogonality of eigenvectors of Hermitian and normal matrices.
4. **Explain** the concept of degeneracy and its implications in quantum systems.
5. **Apply** the principles of completeness and spectral decomposition to represent operators.
6. **Relate** matrix diagonalization to the interpretation of quantum-mechanical observables.

5.1 Introduction

The study of eigenvalues and eigenvectors lies at the heart of both mathematics and quantum mechanics. Every linear transformation, when expressed in its most revealing form, acts by stretching or compressing vectors along characteristic directions—its eigenvectors—by definite factors called eigenvalues. These quantities not only describe the intrinsic behavior of a matrix or operator but also determine its diagonal form, thereby simplifying both calculation and interpretation.

In quantum mechanics, the same concept acquires profound physical meaning: the measurable values of an observable correspond to the eigenvalues of the associated Hermitian operator, while the eigenfunctions (or eigenvectors) represent the system's possible states after measurement. Hence, solving an eigenvalue problem is equivalent to predicting the possible outcomes of experiments.

This chapter develops the mathematical framework for determining eigenvalues and eigenvectors of matrices and linear operators, analyzing their orthogonality and completeness, and understanding how these results lead to diagonalization and spectral decomposition.

Special emphasis is placed on **Hermitian, unitary**, and **normal matrices**—the fundamental operator classes encountered in quantum chemistry and spectroscopy. Through these ideas, the chapter builds a bridge between **linear algebra** and the **operator formalism** of physical theory.

Because this chapter (and the next one) moves between **matrix algebra**, **geometric vector analysis**, and **quantum-mechanical operator language**, the mix of notations (boldface, arrows, and Dirac's $|v\rangle$ notation) can be confusing. The notations are listed in Appendix K.1.

5.2 Determination of Eigenvalues and Eigenvectors of a Matrix

We need to solve the eigenvalue equation

$$A|v\rangle = \lambda|v\rangle. \tag{5.1}$$

This can be equivalently written as

$$(A - \lambda I)|v\rangle = 0, \tag{5.2}$$

where I is the identity matrix. Since the eigenvectors $|v\rangle$ are nonzero, this implies that

$$\det(A - \lambda I) = 0. \tag{5.3}$$

This determinant, on expansion, provides the **characteristic equation** of the matrix—a polynomial of degree N that yields N roots, the required eigenvalues.

Step-by-step Logical Argument

1. We want to find a scalar λ and a **nonzero** vector such that $A|v\rangle = \lambda|v\rangle$.
2. Rearrange the equation to $(A - \lambda I)|v\rangle = 0$.
3. This is a homogeneous system. For a **nonzero** solution $|v\rangle$ to exist, the coefficient matrix $(A - \lambda I)$ must be singular.
4. A square matrix is singular if and only if its determinant is zero.
5. Therefore, the values of λ we seek are the solutions to the characteristic equation: $\det(A - \lambda I) = 0$.

For a diagonal matrix, the eigenvalues are simply the diagonal elements, and the characteristic equation is

$$\det(A - \lambda I) = (a_{11} - \lambda)(a_{22} - \lambda) \ldots (a_{NN} - \lambda), \tag{5.4}$$

which implies $\lambda_i = a_{ii}$. The sum of the eigenvalues is therefore

$$\sum_i \lambda_i = \mathrm{Tr}(A). \tag{5.5}$$

In general, this relation holds for all matrices—the sum of the eigenvalues equals the trace of the matrix. This provides a convenient verification on the correctness of any calculation.

Relation Between Determinant, Trace, and Eigenvalues

For any $N \times N$ matrix A, the characteristic polynomial is

$$\det(A - \lambda I) = (-1)^N \left(\lambda^N - a_1\lambda^{N-1} + a_2\lambda^{N-2} - \cdots \pm a_N\right),$$

where:

- the sum of the eigenvalues satisfies $\sum_i \lambda_i = \mathrm{Tr}(A)$, where $\mathrm{Tr}(A)$ is the sum of the diagonal elements,
- $\prod_i \lambda_i = \det(A)$, and
- the coefficients a_k are elementary symmetric functions of the eigenvalues λ_i.

Thus,

$$\sum_i \lambda_i = \mathrm{Tr}(A), \quad \prod_i \lambda_i = \det(A).$$

These relationships provide convenient checks on computed eigenvalues:

- The **trace** equals the sum of eigenvalues.
- The **determinant** equals their product.

In Example 5.1, we make use of these concepts to determine the eigenvalues of two symmetric matrices without solving the eigenvalue equation.

Example 5.1: Determination of Eigenvalues

Calculate the eigenvalues λ_k of the matrices

$$A = \begin{bmatrix} 0 & 1 & 0 \\ 1 & 0 & 0 \\ 0 & 0 & 1 \end{bmatrix}, \quad B = \begin{bmatrix} 0 & 1 & 0 & 0 \\ 1 & 0 & 0 & 0 \\ 0 & 0 & 0 & 1 \\ 0 & 0 & 1 & 0 \end{bmatrix}.$$

Solution:

Both matrices are symmetric; therefore, they have real eigenvalues. They are also orthogonal, since $A^T A = I$ and $B^T B = I$. Hence, their eigenvalues are ± 1. Moreover, the sum of their eigenvalues equals their traces.

For matrix A, $\mathrm{Tr}(A) = 1$, so the combination of eigenvalues that yields a sum of $+1$ is $(-1, 1, 1)$.

For matrix B, $\mathrm{Tr}(B) = 0$, hence the eigenvalues are $(-1, -1, 1, 1)$.

Note that this is a **degenerate** spectrum.

In Examples 5.2 and 5.3, the eigenvalues of real symmetric matrices are obtained by explicitly solving the eigenvalue equation. In Example 5.3, the eigenvectors are also evaluated.

Example 5.2: Eigenvalues of a Real Symmetric Matrix

Find the eigenvalues of the real symmetric matrix

$$A = \begin{bmatrix} 3 & 2 & 0 \\ 2 & 4 & 0 \\ 0 & 0 & 5 \end{bmatrix}.$$

Solution:

The determinant to be solved is

$$\begin{vmatrix} 3-\lambda & 2 & 0 \\ 2 & 4-\lambda & 0 \\ 0 & 0 & 5-\lambda \end{vmatrix} = 0.$$

Expanding along the third row (choose the row with the most zeroes for simplicity) gives the characteristic equation

$$(5-\lambda)[(3-\lambda)(4-\lambda)-4] = 0,$$

This gives one eigenvalue immediately: $\lambda_1 = 5$. The remaining two eigenvalues are found by solving the quadratic:

$$\lambda^2 - 7\lambda + 8 = 0 \Rightarrow \lambda = \frac{7 \pm \sqrt{(49-32)}}{2} = \frac{7 \pm \sqrt{17}}{2},$$

giving $\lambda_2 = \left(7+\sqrt{17}\right)/2$ and $\lambda_3 = \left(7-\sqrt{17}\right)/2$.

Thus, $\lambda_1 = 5, \lambda_2 = \frac{7+\sqrt{17}}{2}, \lambda_3 = \frac{7-\sqrt{17}}{2}$.

Leaving in surd form for λ_2 and λ_3 is preferred for exactness.

Example 5.3: Eigenvalues and Eigenvectors of a Real Symmetric Matrix

Determine the eigenvalues and eigenvectors of the matrix

$$A = \begin{bmatrix} 2 & 0 & 2 \\ 0 & -2 & 0 \\ 2 & 0 & 2 \end{bmatrix}.$$

Solution:

Step 1: Characteristic Equation

We solve the determinantal equation

$$\det(A - \lambda I) = \begin{vmatrix} 2-\lambda & 0 & 2 \\ 0 & -2-\lambda & 0 \\ 2 & 0 & 2-\lambda \end{vmatrix} = 0.$$

Expand along the second row (simplest):

$$(-2-\lambda)\begin{vmatrix} 2-\lambda & 2 \\ 2 & 2-\lambda \end{vmatrix} = 0.$$

Compute the determinant inside:

$$(2-\lambda)^2 - 4 = \lambda^2 - 4\lambda.$$

Hence:

$$(-2-\lambda)(\lambda^2 - 4\lambda) = 0.$$

Step 2: Eigenvalues

Setting each factor to zero:

$$-2-\lambda = 0 \Rightarrow \lambda_1 = -2, \lambda^2 - 4\lambda = 0 \Rightarrow \lambda(\lambda - 4) = 0 \Rightarrow \lambda_2 = 0, \lambda_3 = 4,$$

$$\boxed{\lambda_1 = -2, \lambda_2 = 0, \lambda_3 = 4.}$$

Step 3: Eigenvectors

a. **For $\lambda_1 = -2$,**

$$A + 2I = \begin{bmatrix} 4 & 0 & 2 \\ 0 & 0 & 0 \\ 2 & 0 & 4 \end{bmatrix}.$$

Equations:

$$4x_1 + 2x_3 = 0, \quad 2x_1 + 4x_3 = 0.$$

From the first equation, $x_3 = -2x_1$. Substituting into the second gives $2x_1 + 4(-2x_1) = -6x_1 = 0$, which forces $x_1 = 0$ and hence $x_3 = 0$. The second row of $A + 2I$ gives $0 \times x_2 = 0$, so x_2 is a free variable; setting $x_2 = 1$ gives the eigenvector:

$$\mathbf{v}_1 = (0, 1, 0)^T.$$

b. For $\lambda_2 = 0$,

$$A = \begin{bmatrix} 2 & 0 & 2 \\ 0 & -2 & 0 \\ 2 & 0 & 2 \end{bmatrix}.$$

Equations:

$$2x_1 + 2x_3 = 0, \quad -2x_2 = 0 \Rightarrow x_2 = 0.$$

From the first equation, $x_3 = -x_1$,

$$\mathbf{v}_2 = (1, 0, -1)^T.$$

c. For $\lambda_3 = 4$,

$$A - 4I = \begin{bmatrix} -2 & 0 & 2 \\ 0 & -6 & 0 \\ 2 & 0 & -2 \end{bmatrix}.$$

Equations:

$$-2x_1 + 2x_3 = 0 \Rightarrow x_1 = x_3, -6x_2 = 0 \Rightarrow x_2 = 0,$$

$$\mathbf{v}_3 = (1, 0, 1)^T.$$

Step 4: Normalize and Check Orthogonality

$$|\mathbf{v}_1| = \sqrt{0 + 1 + 0} = 1, \quad |\mathbf{v}_2| = \sqrt{2}, \quad |\mathbf{v}_3| = \sqrt{2}.$$

Normalized:

$$\mathbf{v}_1 = (0, 1, 0), \mathbf{v}_2 = \frac{1}{\sqrt{2}}(1, 0, -1), \mathbf{v}_3 = \frac{1}{\sqrt{2}}(1, 0, 1).$$

All dot products vanish:

$$\mathbf{v}_i \cdot \mathbf{v}_j = \delta_{ij}.$$

Hence, the eigenvectors form an **orthonormal basis** for the three-dimensional space.

✅ **Final Results**

$$\begin{aligned} &\lambda_1 = -2, \quad \mathbf{v}_1 = (0, 1, 0)^T, \\ &\lambda_2 = 0, \quad \mathbf{v}_2 = \frac{1}{\sqrt{2}}(1, 0, -1)^T, \\ &\lambda_3 = 4, \quad \mathbf{v}_3 = \frac{1}{\sqrt{2}}(1, 0, 1)^T. \end{aligned}$$

All eigenvectors are mutually orthogonal, as expected for a **real symmetric matrix**.

Summary Note

The eigenvalue problem reduces to solving the *characteristic equation* $\det(A - \lambda I) = 0$ for λ, followed by finding the corresponding normalized eigenvectors from $(A - \lambda I)|v\rangle = 0$. These eigenvectors form an **orthonormal basis**, forming the foundation for the **diagonalization** of Hermitian and unitary matrices in quantum mechanics.

Example 5.2 is an example of a matrix in block diagonal form, where the characteristic equations reduce to two blocks: a 2×2 block and a 1×1 block. Example 5.3 is also in block diagonal form in disguise. If you exchange rows and columns 2 and 3, you get the same form as Example 5.2.

5.3 Algebraic Versus Geometric Multiplicity: The Key to Diagonalization

After finding the eigenvalues of a matrix, a crucial question arises: for a given eigenvalue, how many linearly independent eigenvectors can we find? The answer to this question is the single most important factor in determining if a matrix can be diagonalized. To formalize this concept, we introduce two types of multiplicity: algebraic multiplicity (AM) and geometric multiplicity (GM).

Let us first take a simple example. Consider a 2×2 matrix

$$B = \begin{bmatrix} 1 & 1 \\ 0 & 1 \end{bmatrix}.$$

Its characteristic equation is found by solving the determinant

$$\det(B - \lambda I) = \begin{vmatrix} 1-\lambda & 1 \\ 0 & 1-\lambda \end{vmatrix} = 0.$$

Thus, the characteristic polynomial is $(\lambda - 1)^2 = 0$.

1. **Eigenvalues:** $\lambda = 1$: Multiplicity = **2**.
2. **Eigenvectors:** We find them by solving $(B - 1I)\mathbf{v} = \mathbf{0}$,

$$\begin{bmatrix} 0 & 1 \\ 0 & 0 \end{bmatrix}\begin{bmatrix} x \\ y \end{bmatrix} = \begin{bmatrix} 0 \\ 0 \end{bmatrix} \Longrightarrow y = 0.$$

The second row yields $0x + 0y = 0$, which provides no constraint. The first row yields $0x + 1y = 0 \to y = 0$. This clarifies why there is only one free variable (x).

Hence, the solution vectors are of the form $\begin{bmatrix} x \\ 0 \end{bmatrix} = x\begin{bmatrix} 1 \\ 0 \end{bmatrix}$. The eigenspace is spanned by only one vector. The multiplicity of eigenvectors is 1.

The multiplicity of the eigenvalue as a root of the characteristic polynomial is called the **AM**. In this case $(\lambda - 1)^2 = 0$, so for $\lambda = 1$ we have AM = 2, but the **GM**, which is the multiplicity of the eigenvectors, is 1 (GM = 1).

5.3.1 Algebraic Multiplicity

The **AM** of an eigenvalue λ is its multiplicity as a root of the characteristic polynomial. In simpler terms it is the number of times $(\lambda - \lambda_i)$ appears as a factor in the polynomial.

For example, if the characteristic equation is $(\lambda - 5)^2(\lambda - 2)$, then:

- The eigenvalue $\lambda = 5$ has **AM 2**.
- The eigenvalue $\lambda = 2$ has **AM 1**.

5.3.2 Geometric Multiplicity

The **GM** of an eigenvalue λ is the dimension of its corresponding eigenspace, $E_\lambda = \ker(A - \lambda I)$:

- "Ker" is short for the **kernel** of a linear transformation or a matrix. It is the same thing as the **null space**.
- The kernel is the set of all input vectors that get mapped to the zero vector.

In practical terms, it is the number of linearly independent eigenvectors we can find for that eigenvalue.

The GM is always at least 1, because an eigenvalue must have at least one eigenvector. The fundamental relationship between the two multiplicities is given by the following property.

Property 5.1 *For any eigenvalue λ of a matrix A, its GM is always less than or equal to its AM,*

$$1 \leq \mathrm{GM}(\lambda) \leq \mathrm{AM}(\lambda).$$

The comparison between AM and GM is the **diagnostic test** for diagonalizability.

The lower bound holds because every eigenvalue must have at least one eigenvector. The upper bound is a deeper result whose proof relies on the theory of Jordan normal forms, discussed in advanced linear algebra texts (see Further Reading).

5.3.3 Examples: Testing for a Full Set of Eigenvectors

Let us see this in action with two contrasting examples.

Example 1: A Matrix With a Full Set of Eigenvectors

Consider the matrix $A = \begin{bmatrix} 2 & -1 & 1 \\ 0 & 1 & 1 \\ 0 & 0 & 2 \end{bmatrix}$.

1. **Eigenvalues (AM):** Since A is a triangular matrix, its eigenvalues are its diagonal entries. The characteristic equation is $(\lambda - 2)^2(\lambda - 1) = 0$:
   - $\lambda_1 = 2$: **AM = 2**
   - $\lambda_2 = 1$: **AM = 1**
2. **Eigenspaces (GM):**
   - For $\lambda_1 = 2$: We find the null space of $(A - 2I)|v\rangle = 0$:

$$\begin{bmatrix} 0 & -1 & 1 \\ 0 & -1 & 1 \\ 0 & 0 & 0 \end{bmatrix}\begin{bmatrix} x \\ y \\ z \end{bmatrix} = \begin{bmatrix} 0 \\ 0 \\ 0 \end{bmatrix} \Longrightarrow -y + z = 0 \Longrightarrow y = z,$$

$0 \times x = 0 \Longrightarrow x$ is indeterminate.

Hence, the solution vectors are of the form $\begin{bmatrix} x \\ y \\ z \end{bmatrix} = x\begin{bmatrix} 1 \\ 0 \\ 0 \end{bmatrix} + y\begin{bmatrix} 0 \\ 1 \\ 1 \end{bmatrix}$. The eigenspace is spanned by two linearly independent vectors. Thus, for $\lambda_1 = 2$, $\mathbf{GM = 2}$.

- For $\lambda_2 = 1$: When an eigenvalue has AM $= 1$, its GM is guaranteed to be 1 as well ($1 \leq \text{GM} \leq 1$). Thus, for $\lambda_2 = 1$, $\mathbf{GM = 1}$.

Diagnosis: For every eigenvalue, the GM equals the AM (AM = GM). This matrix has a full set of three linearly independent eigenvectors and is therefore diagonalizable.

Example 2: A Matrix With a Deficient Set of Eigenvectors

An example of a matrix with a deficient set of eigenvectors is the matrix B, discussed at the beginning of this section: $B = \begin{bmatrix} 1 & 1 \\ 0 & 1 \end{bmatrix}$.

This matrix has two eigenvalues: $\lambda = 1$: $\mathbf{AM = 2}$, but the eigenspace is spanned by only one vector. Thus, for $\lambda = 1$, $\mathbf{GM = 1}$.

Diagnosis: Here, GM (1) < AM (2). Even though the matrix is 2×2, we can only find one linearly independent eigenvector. This matrix is **not diagonalizable**.

5.3.4 The Condition for Diagonalizability

The examples above lead to the central theorem that connects this chapter to the next.

Theorem 5.1 *An $N \times N$ matrix A is **diagonalizable** if and only if the sum of the geometric multiplicities of its eigenvalues is equal to N. This condition is met if and only if, for every eigenvalue λ of A, its GM is equal to its AM.*

Justification: The two examples in Section 5.3.3 illustrate both cases:

- In Example 1, GM = AM for every eigenvalue, the sum of the GMs is $3 = N$, and the matrix is diagonalizable.
- In Example 2, GM < AM for $\lambda = 1$, the sum of GMs is less than $N = 2$, and the matrix is not diagonalizable.

The general proof of this theorem relies on the theory of Jordan normal forms and is beyond the scope of this text (see Further Reading).

Practical rule: A matrix is diagonalizable if and only if you can find N linearly independent eigenvectors (sum of GMs = N).

5.4 Degenerate Eigenvalues

We now consider the case in which two or more eigenvalues of a matrix are **degenerate. Degeneracy** means AM ≥ 2 for some eigenvalue, and GM can range from 1 up to AM. Before proceeding, let us recall some key properties of the eigenvectors corresponding to degenerate eigenvalues.

Suppose there are k eigenvectors of a matrix associated with the same eigenvalue λ_i. Then any linear combination of these eigenvectors is also an eigenvector with the same eigenvalue.

Consider a linear combination

$$|v\rangle = \sum_{j=1}^{k} c_j |v_j\rangle,$$

where the vectors $|v_1\rangle, |v_2\rangle, \dots, |v_k\rangle$ are degenerate eigenvectors of an operator represented by matrix A with the same eigenvalue λ_i, i.e.,

$$A|v_j\rangle = \lambda_i |v_j\rangle.$$

Operating on the linear combination, we obtain

$$A|v\rangle = A\left(\sum_{j=1}^{k} c_j |v_j\rangle\right) = \sum_{j=1}^{k} c_j (A|v_j\rangle) = \lambda_i \sum_{j=1}^{k} c_j |v_j\rangle = \lambda_i |v\rangle.$$

Hence, for degenerate eigenvalues, **any linear combination of degenerate eigenvectors is also an eigenvector** of A with the same eigenvalue λ_i.

Example 5.4: Eigenvectors for Degenerate Eigenvalues

Find the eigenvalues and eigenvectors of the real symmetric matrix

$$A = \begin{bmatrix} 1 & 2 & 2 \\ 2 & 1 & 2 \\ 2 & 2 & 1 \end{bmatrix}.$$

Solution:

Step 1: Characteristic Equation

$$\det(A - \lambda I) = \begin{vmatrix} 1-\lambda & 2 & 2 \\ 2 & 1-\lambda & 2 \\ 2 & 2 & 1-\lambda \end{vmatrix} = 0.$$

By symmetry, the vector (1, 1, 1) is an eigenvector, since each row sums to 5, and therefore $A(1, 1, 1)^T = 5(1, 1, 1)^T$.

Hence, one eigenvalue is $\lambda_1 = 5$.

For the remaining subspace orthogonal to (1, 1, 1), we can choose vectors satisfying $x + y + z = 0$, e.g., $(1, -1, 0)$ and $(1, 0, -1)$.

Operating with A on either gives

$$A(1, -1, 0)^T = (-1, 1, 0)^T, \quad A(1, 0, -1)^T = (-1, 0, 1)^T,$$

so, the corresponding eigenvalue is $\lambda_2 = \lambda_3 = -1$.

Step 2: Eigenvalues and Eigenvectors

$$\boxed{\lambda_1 = 5, \lambda_2 = -1, \lambda_3 = -1,}$$

$$\mathbf{v}_1 = (1, 1, 1)^T, \mathbf{v}_2 = (1, -1, 0)^T, \mathbf{v}_3 = (1, 0, -1)^T.$$

Note: $\mathbf{v}_2$ and $\mathbf{v}_3$ are not yet orthogonal. We must apply Gram–Schmidt to these two vectors to obtain a truly orthogonal basis for the subspace. Alternatively, a vector orthogonal to both $\mathbf{v}_1$ and $\mathbf{v}_2$ can be obtained directly by taking their cross product, as demonstrated in Section 5.4.1.

Normalized vectors:

$$\mathbf{v}_1 = \frac{1}{\sqrt{3}}(1, 1, 1), \mathbf{v}_2 = \frac{1}{\sqrt{2}}(1, -1, 0), \mathbf{v}_3 = \frac{1}{\sqrt{6}}(1, 1, -2).$$

5.4.1 Example: Finding a Vector Orthogonal to Two Orthogonal Vectors in 3D-space

Before describing the Gram–Schmidt process, we describe a simple method for finding the third vector of an orthonormal set, provided two are known. In Example 5.4, consider the two orthogonal vectors $\mathbf{v}_1$ and $\mathbf{v}_2$. To find the third vector orthogonal to both of them, we take their cross product $\mathbf{v}_1 \times \mathbf{v}_2$:

$$\mathbf{v}_3 = \begin{vmatrix} \hat{i} & \hat{j} & \hat{k} \\ 1 & 1 & 1 \\ 1 & -1 & 0 \end{vmatrix} = \hat{i}(0+1) - \hat{j}(0-1) + \hat{k}(-1-1) = 1\hat{i} + 1\hat{j} - 2\hat{k}.$$

Therefore, $\mathbf{v}_3 = (1, 1, -2)$ and $|\mathbf{v}_3| = \sqrt{1^2 + 1^2 + (-2)^2} = \sqrt{6}$.

Hence, for degenerate eigenvalues, we can always construct a set of orthogonal eigenvectors by appropriate linear combinations (here we used symmetry)—or equivalently, by the Gram–Schmidt orthogonalization procedure.

5.4.2 Discussion

The presence of degenerate eigenvalues introduces **freedom** in choosing the corresponding eigenvectors. When eigenvalues are degenerate, the associated eigenvectors need not be orthogonal, but they can always be made orthogonal through the **Gram–Schmidt orthogonalization procedure** (discussed in the next section). Such orthogonalized eigenvectors are particularly valuable in quantum mechanics, where eigenfunction orthonormality simplifies both interpretation and computation.

5.5 Gram–Schmidt Orthogonalization Process

To obtain an **orthogonal** or **orthonormal set of eigenvectors** from a linearly independent but nonorthogonal set, we use the **Gram–Schmidt orthogonalization** procedure.

We define a **projection operator** as

$$P = \frac{|\psi_1\rangle\langle\psi_1|}{\langle\psi_1 \mid \psi_1\rangle}, \tag{5.6}$$

which orthogonally projects any vector $|\psi\rangle$ onto the line spanned by $|\psi_1\rangle$.

Let $\mathbb{V}$ be an N-dimensional vector space with an inner product.

Suppose $\{|v_1\rangle, |v_2\rangle, \ldots, |v_N\rangle\}$ is a basis for $\mathbb{V}$.
To construct an orthogonal basis $\{|u_1\rangle, |u_2\rangle, \ldots, |u_N\rangle\}$, we proceed as follows:

$$\begin{aligned}
|u_1\rangle &= |v_1\rangle,\\
|u_2\rangle &= |v_2\rangle - \frac{\langle u_1 \mid v_2\rangle}{\langle u_1 \mid u_1\rangle}|u_1\rangle,\\
|u_3\rangle &= |v_3\rangle - \frac{\langle u_1 \mid v_3\rangle}{\langle u_1 \mid u_1\rangle}|u_1\rangle - \frac{\langle u_2 \mid v_3\rangle}{\langle u_2 \mid u_2\rangle}|u_2\rangle,\\
&\vdots\\
|u_k| &= |v_k| - \sum_{j=1}^{k-1} \frac{\langle u_j \mid v_k\rangle}{\langle u_j \mid u_j\rangle}|u_j\rangle.
\end{aligned} \tag{5.7}$$

If each $|u_i\rangle$ is normalized, the resulting set forms an **orthonormal basis,** $\{|\hat{u}_1\rangle, |\hat{u}_2\rangle, \ldots, |\hat{u}_N\rangle\}$ for the vector space $\mathbb{V}$.

From this point onward, column vectors in square brackets are used for clarity in finite-dimensional examples, but the same formulas apply in Dirac notation. In N-dimensional space, Equation (5.7) can be written as

$$\mathbb{R}^N: u_k = v_k - \sum((u_j \cdot v_k)/(u_j \cdot u_j))u_j.$$

Example 5.5: Orthonormal Bases

Let a plane in three-dimensional space be spanned by the vectors

$$|a_1\rangle = \begin{bmatrix}1\\1\\0\end{bmatrix}, \quad |a_2\rangle = \begin{bmatrix}1\\2\\3\end{bmatrix}.$$

i. Find an orthonormal basis for the plane.
ii. Extend it to an orthonormal basis in three-dimensional space.

Solution:

i. The two vectors are not orthogonal because

$$\langle a_1|a_2\rangle = 3.$$

We first construct $|u_1\rangle$ by normalizing $|a_1\rangle$:

$$|u_1\rangle = \frac{1}{\sqrt{2}}\begin{bmatrix}1\\1\\0\end{bmatrix}.$$

Next, we remove from $|a_2\rangle$ its projection on $|u_1\rangle$:

$$|u_2\rangle = |a_2\rangle - \frac{\langle u_1 \mid a_2\rangle}{\langle u_1 \mid u_1\rangle}|u_1\rangle = \begin{bmatrix}1\\2\\3\end{bmatrix} - \frac{3}{2}\begin{bmatrix}1\\1\\0\end{bmatrix} = \begin{bmatrix}-0.5\\0.5\\3\end{bmatrix}.$$

Normalizing $|u_2\rangle$:

$$|\hat{u}_2\rangle = \frac{1}{\sqrt{9.5}}\begin{bmatrix}-0.5\\0.5\\3\end{bmatrix}.$$

Hence, the orthonormal basis for the plane is

$$|\hat{u}_1\rangle = \frac{1}{\sqrt{2}}\begin{bmatrix}1\\1\\0\end{bmatrix}, \quad |\hat{u}_2\rangle = \frac{1}{\sqrt{9.5}}\begin{bmatrix}-0.5\\0.5\\3\end{bmatrix}.$$

ii. To extend this to three dimensions, we construct a vector perpendicular to both $|\hat{u}_1\rangle$ and $|\hat{u}_2\rangle$ using their **cross product**:

$$|\hat{u}_3\rangle = |\hat{u}_1\rangle \times |\hat{u}_2\rangle = \begin{bmatrix}3\\-3\\1\end{bmatrix}.$$

After normalization,

$$|\hat{u}_3\rangle = \frac{1}{\sqrt{19}}\begin{bmatrix}3\\-3\\1\end{bmatrix}.$$

Thus, the complete orthonormal basis in three dimensions is

$$\{|\hat{u}_1\rangle, |\hat{u}_2\rangle, |\hat{u}_3\rangle\}.$$

Remark The scalar factors in Equation (5.7) are ratios of scalar products; hence each new vector is a **linear combination** of the original vectors.

Key Insight

For degenerate eigenvalues, any linear combination of the corresponding eigenvectors is also an eigenvector with the same eigenvalue. These vectors can always be made orthogonal—and normalized—to form an orthonormal set, using the Gram–Schmidt procedure.

If the original vectors were eigenvectors corresponding to the same eigenvalue λ_i, each newly constructed vector remains an eigenvector of A with the same eigenvalue.

Using **functions instead of coordinate vectors** shows the real power of the Gram–Schmidt process. Here is a simple, instructive example often used in quantum mechanics and mathematical methods courses.

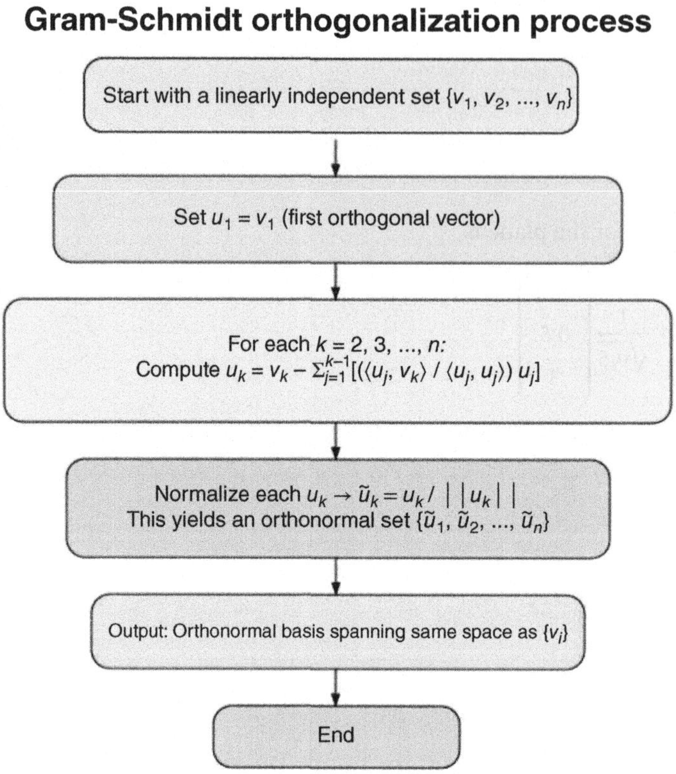

Example 5.6: Orthogonalization of 1 and *x* on [0, 1]

Orthonormalize the two simple functions: $f_1(x) = 1,\ f_2(x) = x$ on the interval [0, 1] with the standard inner product

$$\langle f, g\rangle = \int_0^1 f(x)g(x)dx.$$

Solution:

Step 1: Normalize $f_1(x)$

$$|f_1| = \sqrt{\langle f_1, f_1\rangle} = \sqrt{\int_0^1 1^2\, dx} = 1.$$

Hence,

$$\phi_1(x) = f_1(x) = 1.$$

Step 2: Orthogonalize $f_2(x)$ against $\phi_1(x)$

Compute the projection of f_2 onto ϕ_1:

$$\text{proj}_{\phi_1}(f_2) = \langle \phi_1, f_2 \rangle \phi_1 = \left(\int_0^1 x\,dx \right)(1) = \frac{1}{2}.$$

Subtract this component:

$$u_2(x) = f_2(x) - \text{proj}_{\phi_1}(f_2) = x - \frac{1}{2}.$$

Step 3: Normalize $u_2(x)$

$$|u_2| = \sqrt{\int_0^1 \left(x - \frac{1}{2}\right)^2 dx} = \sqrt{\frac{1}{12}} = \frac{1}{2\sqrt{3}}.$$

Hence, the normalized second function is

$$\phi_2(x) = \frac{u_2(x)}{|u_2|} = 2\sqrt{3}\left(x - \frac{1}{2}\right).$$

Final Orthonormal Set

$$\boxed{\phi_1(x) = 1, \quad \phi_2(x) = 2\sqrt{3}\left(x - \frac{1}{2}\right).}$$

Verification

$$\langle \phi_1, \phi_2 \rangle = \int_0^1 1 \cdot 2\sqrt{3}\left(x - \frac{1}{2}\right) dx = 0.$$

Thus, $\{\phi_1, \phi_2\}$ is an **orthonormal set** on $[0, 1]$.

Let us now extend this example to **three functions**—say, 1, x, x^2—to show how **Legendre polynomials** naturally emerge from the Gram–Schmidt procedure. It is a beautiful connection to quantum mechanics and spectroscopy. This result is one of the most elegant demonstrations of how **Gram–Schmidt orthogonalization** connects to **Legendre polynomials**, which in turn appear throughout **quantum mechanics** (e.g., angular momentum and hydrogen atom wavefunctions).

Example 5.7: Orthogonalization of 1, x, and x^2 on $[-1, 1]$

Take the set of linearly independent functions

$$f_1(x) = 1, \quad f_2(x) = x, \quad f_3(x) = x^2$$

defined on the interval $[-1, 1]$ with the inner product

$$\langle f, g \rangle = \int_{-1}^1 f(x)g(x)dx$$

and orthogonalize them.

Solution:

Step 1: Normalize $f_1(x)$

$$|f_1| = \sqrt{\int_{-1}^{1} 1^2\, dx} = \sqrt{2}.$$

Hence,

$$\phi_1(x) = \frac{f_1(x)}{|f_1|} = \frac{1}{\sqrt{2}}.$$

Step 2: Orthogonalize $f_2(x)$ With Respect to $\phi_1(x)$

$$\text{proj}_{\phi_1}(f_2) = \langle \phi_1, f_2 \rangle \phi_1 = \left(\frac{1}{\sqrt{2}} \int_{-1}^{1} x\, dx \right) \frac{1}{\sqrt{2}} = 0$$

(since the integrand is odd).
Therefore, $u_2(x) = f_2(x)$ and

$$|u_2| = \sqrt{\int_{-1}^{1} x^2\, dx} = \sqrt{\frac{2}{3}}.$$

So,

$$\phi_2(x) = \frac{u_2(x)}{|u_2|} = \sqrt{\frac{3}{2}}x.$$

Step 3: Orthogonalize $f_3(x)$

Subtract the projections of f_3 on both ϕ_1 and ϕ_2:

$$u_3(x) = f_3(x) - \langle \phi_1, f_3 \rangle \phi_1 - \langle \phi_2, f_3 \rangle \phi_2.$$

Compute each term:

$$\langle \phi_1, f_3 \rangle = \frac{1}{\sqrt{2}} \int_{-1}^{1} x^2\, dx = \frac{1}{\sqrt{2}} \cdot \frac{2}{3} = \frac{\sqrt{2}}{3},$$

$$\langle \phi_2, f_3 \rangle = \sqrt{\frac{3}{2}} \int_{-1}^{1} x^3 dx = 0.$$

Hence,

$$u_3(x) = x^2 - \frac{\sqrt{2}}{3} \cdot \frac{1}{\sqrt{2}} = x^2 - \frac{1}{3}.$$

Normalize it:

$$|u_3| = \sqrt{\int_{-1}^{1} \left(x^2 - \frac{1}{3} \right)^2 dx} = \sqrt{\frac{8}{45}} = \frac{2\sqrt{10}}{15}.$$

Therefore,

$$\phi_3(x) = \frac{u_3(x)}{\lceil u_3 \rceil} = \sqrt{\frac{5}{8}}(3x^2 - 1).$$

Final Orthonormal Set

$$\boxed{\begin{aligned} \phi_1(x) &= \frac{1}{\sqrt{2}}, \\ \phi_2(x) &= \sqrt{\frac{3}{2}}x, \\ \phi_3(x) &= \sqrt{\frac{5}{8}}\,(3x^2 - 1). \end{aligned}}$$

This elegant result shows how the normalized **Legendre polynomials** naturally emerge from the Gram–Schmidt process, illustrating the deep connection between **linear algebra** and **quantum mechanics**, where orthogonal functions such as $P_0(x), P_1(x)$, and $P_2(x)$ appear in angular momentum theory and the hydrogen atom wavefunctions (discussed in detail in Chapter 13).

These look different from the tabulated Legendre polynomials, which are not normalized, but are standardized so that $P_n(1) = 1$.

Concept Note 5.1

Gram–Schmidt generates **normalized** Legendre polynomials.

Standard Legendre polynomials (found in tables) are defined by $P_n(1) = 1$ and are *orthogonal* but not *orthonormal* (norm is $\frac{2}{2n+1}$).

Chemistry Connection: Orthogonality and Completeness of Basis Functions

In quantum chemistry and molecular spectroscopy, the mathematical concept of **orthonormal functions** plays a central role.

When the Gram–Schmidt process is applied to simple functions such as $1, x, x^2, \ldots$ over the interval $[-1, 1]$, the resulting orthonormal set corresponds to the **normalized Legendre polynomials**,

$$\phi_1(x) = \frac{1}{\sqrt{2}}, \quad \phi_2(x) = \sqrt{\frac{3}{2}}x, \quad \phi_3(x) = \sqrt{\frac{5}{8}}(3x^2 - 1).$$

The standard Legendre polynomials are:

$$P_0(x) = 1, P_1(x) = x, P_2(x) = \frac{1}{2}(3x^2 - 1), \ldots.$$

These functions are not only mutually orthogonal,

$$\int_{-1}^{1} P_m(x)P_n(x)\, dx = 0\ (m \neq n),$$

but also **complete**—meaning any reasonable function defined on $[-1, 1]$ can be expressed as a linear combination of them (see next section).

This property directly parallels the use of orthogonal basis functions in chemistry (*e.g.*, molecular orbitals, vibrational modes, or electronic wavefunctions), where each eigenfunction represents a distinct, independent mode of the system.

5.6 Completeness

In the previous sections, we established that the eigenvectors of a Hermitian operator are orthogonal and can be normalized to form an orthonormal set. When these eigenvectors span the entire vector space, they form a **complete set**. Completeness means that **any vector** in the space can be expressed as a linear combination of these eigenvectors.

If a Hermitian operator A has N orthonormal eigenvectors

$$\{|a_1\rangle, |a_2\rangle, \dots, |a_N\rangle\},$$

with corresponding real eigenvalues $\lambda_1, \lambda_2, \dots, \lambda_N$, then any vector $|v\rangle$ in the space can be represented as

$$|v\rangle = \sum_{i=1}^{N} c_i |a_i\rangle, \quad \text{where } c_i = \langle a_i | v\rangle.$$

The completeness of the eigenvectors can be expressed as the **completeness relation**

$$\sum_{i=1}^{N} | a_i\rangle\langle a_i | = I. \tag{5.8}$$

This relation states that the operator obtained by summing the outer products of all normalized eigenvectors equals the **identity operator** (see also Chapter 4).

When this operator acts on a state $|v\rangle$, it leaves it unchanged,

$$I \mid v\rangle = \left(\sum | a_i\rangle\langle a_i|\right) \mid v\rangle = \sum | a_i\rangle\langle a_i|v\rangle = \sum c_i \mid a_i\rangle = \mid v\rangle.$$

Thus, it is clear that the relation (5.8) is equivalent to the statement of completeness.

For a **continuous spectrum** (as in the case of free particles or unbound states), the basis states are labeled by a continuous variable, such as the position x, and the completeness relation takes the integral form

$$\int_{-\infty}^{\infty} |x\rangle\langle x| \, dx = \hat{I}$$

which ensures that any wavefunction can be expanded in terms of a complete set of eigenfunctions.

When this operator acts on a state $|v\rangle$, it yields the standard expansion of the state vector in terms of the position wavefunction $\psi_v(x) = \langle x|v\rangle$.

Significance in Quantum Mechanics

Completeness lies at the heart of the quantum-mechanical formalism. It guarantees that any physical state (wavefunction) can be represented as a superposition of eigenstates of a Hermitian operator, such as the Hamiltonian. In this framework, the coefficients c_i (or $c(x)$ for continuous spectra) describe the **probability amplitudes** for finding the system in each eigenstate.

Thus, completeness not only provides a mathematically rigorous basis for representing states but also offers a **physical interpretation**—it is the foundation of quantum superposition and measurement theory.

Example 5.8: Eigenvalues and Orthogonal Eigenvectors of a Hermitian Matrix

Consider the Hermitian matrix

$$A = \begin{bmatrix} 2 & i \\ -i & 2 \end{bmatrix}.$$

i. Find the eigenvalues.
ii. Find the corresponding eigenvectors.
iii. Normalize the eigenvectors.
iv. Verify orthogonality.

Solution:

i. The eigenvalue equation is $\det(A - \lambda I) = 0$, that is,

$$\begin{vmatrix} 2-\lambda & i \\ -i & 2-\lambda \end{vmatrix} = 0.$$

Expanding the determinant,

$$(2-\lambda)^2 - (-i)(i) = (2-\lambda)^2 - (1) = 0 \Rightarrow (2-\lambda)^2 = 1,$$
$$\therefore \lambda_1 = 3,\ \lambda_2 = 1.$$

Hence, the eigenvalues are **real**, as expected for a Hermitian matrix.

ii. For $\lambda_1 = 3$:

$$(A - 3I)|v_1\rangle = 0$$
$$\Rightarrow \begin{bmatrix} -1 & i \\ -i & -1 \end{bmatrix}\begin{bmatrix} x_1 \\ x_2 \end{bmatrix} = 0.$$

This gives the relation $ix_2 = x_1 \Rightarrow x_2 = -ix_1$.
Choosing $x_1 = 1$, the unnormalized eigenvector is

$$|\,v_1\rangle = \begin{bmatrix} 1 \\ -i \end{bmatrix}.$$

Similarly, for $\lambda_2 = 1$:

$$(A - I) \mid v_2\rangle = 0$$

$$\Rightarrow \begin{bmatrix} 1 & i \\ -i & 1 \end{bmatrix} \begin{bmatrix} x_1 \\ x_2 \end{bmatrix} = 0.$$

This gives $x_2 = ix_1$, so that

$$\mid v_2\rangle = \begin{bmatrix} 1 \\ i \end{bmatrix}.$$

iii. The norm of $\mid v_1\rangle$ is

$$\sqrt{\begin{bmatrix} 1 & i \end{bmatrix} \begin{bmatrix} 1 \\ -i \end{bmatrix}} = \sqrt{1^2 - i^2} = \sqrt{2},$$

since $i^2 = -1$.
So, the normalized eigenvectors are

$$\mid v_1\rangle = \frac{1}{\sqrt{2}} \begin{bmatrix} 1 \\ -i \end{bmatrix}, \quad \mid v_2\rangle = \frac{1}{\sqrt{2}} \begin{bmatrix} 1 \\ i \end{bmatrix}.$$

iv.

$$|v_1\rangle\langle v_1| + |v_2\rangle\langle v_2| = \frac{1}{2}\left(\begin{bmatrix} 1 & i \\ -i & 1 \end{bmatrix} + \begin{bmatrix} 1 & -i \\ i & 1 \end{bmatrix} \right) = \begin{bmatrix} 1 & 0 \\ 0 & 1 \end{bmatrix} = I.$$

Hence, the completeness relationship is verified. The eigenvectors are **orthogonal** and form an **orthonormal set**.

Significance:

This simple example illustrates the defining properties of **Hermitian matrices**:

1. Their eigenvalues are **real**.
2. Their eigenvectors corresponding to distinct eigenvalues are **orthogonal**.

These properties ensure that Hermitian operators representing observables in quantum mechanics yield real measurement outcomes and orthogonal state functions.

Key Takeaways: Orthogonality, Degeneracy, and Completeness

- **Gram–Schmidt Orthogonalization:**
This process systematically converts a set of linearly independent vectors $\{|v_i\rangle\}$ into an orthonormal set $\{|\hat{u}_i\rangle\}$:

$$|u_k\rangle = |v_k\rangle - \sum_{j=1}^{k-1} \frac{\langle u_j|v_k\rangle}{\langle u_j|u_j\rangle} |u_j\rangle.$$

After normalization, the resulting vectors satisfy $\langle \hat{u}_i | \hat{u}_j \rangle = \delta_{ij}$.

- **Degenerate Eigenvalues:**

When two or more eigenvectors correspond to the same eigenvalue, any linear combination of them is also an eigenvector:

$$A|v_j\rangle = \lambda_i \mid v_j\rangle \Rightarrow A(\sum c_j \mid v_j\rangle) = \lambda_i \sum c_j |v_j\rangle.$$

These degenerate vectors can be orthogonalized using the Gram–Schmidt process.

- **Completeness:**

A complete set of orthonormal eigenvectors spans the entire vector space:

$$\sum_i |a_i\rangle\langle a_i| = I.$$

This means any vector $|v\rangle$ can be expressed as

$$|v\rangle = \sum_i \langle a_i|v\rangle |a_i\rangle.$$

- Expressing operators in their eigenbasis simplifies analysis and paves the way to **diagonalization** and **spectral decomposition**.

Transition to Chapter 6

In this chapter, we learned how to find **eigenvalues** (from the characteristic polynomial) and **eigenvectors** (by determining the eigenspace).

We also saw that the equality of **algebraic and geometric multiplicities** guarantees a sufficient number of linearly independent eigenvectors to form a complete basis.

The ideas of eigenvalues, orthogonality, and completeness naturally lead to **matrix diagonalization**, the focus of the next chapter. There, you will see how every Hermitian or normal matrix can be represented in diagonal form using its eigenvectors—a principle that underlies the mathematical treatment of **quantum-mechanical operators, molecular orbitals, and vibrational modes**.

We will learn that if a matrix has a complete set of eigenvectors (i.e., it is diagonalizable), we can write it as $A = PDP^{-1}$ or, for Hermitian matrices, $A = UDU^{\dagger}$, where D is the diagonal eigenvalue matrix. This "spectral decomposition" is one of the most useful results in linear algebra.

5.7 Summary

In this chapter, we examined how the eigenvalue equation $A\mathbf{v} = \lambda\mathbf{v}$ (or $A|v\rangle = \lambda|v\rangle$) reveals the intrinsic structure of a matrix or operator. We learned that:

- **Eigenvalues** determine invariant scaling factors, and **eigenvectors** define characteristic directions.
- **Hermitian operators** yield **real eigenvalues** and **orthogonal eigenvectors**, forming the mathematical basis of **quantum observables**.
- The **Gram–Schmidt process** converts nonorthogonal eigenvectors into an orthonormal basis, ensuring a consistent and physically meaningful representation.

- The **completeness** of an orthonormal set guarantees that any function (or wavefunction) can be expanded in this basis—the foundation of the **superposition principle**.

Together, these results establish the groundwork for **diagonalization and spectral decomposition**, discussed in the next chapter, where we show how operators can be represented in their simplest diagonal form.

Further Reading

P. W. Atkins and R. S. Friedman, *Molecular Quantum Mechanics*, 5th ed., Oxford University Press, 2010.

A. R. Edmonds, *Angular Momentum in Quantum Mechanics*, Princeton University Press, 1996.

D. J. Griffiths and D. F. Schroeter, *Introduction to Quantum Mechanics*, 3rd ed., Cambridge University Press, 2018.

I. N. Levine, *Quantum Chemistry*, 7th ed., Pearson, 2013.

H. Margenau and G. M. Murphy, *The Mathematics of Physics and Chemistry*, 2nd ed., Van Nostrand, 1956.

D. A. McQuarrie, *Quantum Chemistry*, University Science Books, 2008.

Exercises

A. Conceptual Questions

1. Explain why Hermitian matrices always have real eigenvalues.

2. Show how orthogonality of eigenvectors leads to energy quantization in a molecular Hamiltonian.

3. Distinguish between completeness and linear independence with an example.

B. Problem-solving Exercises

1. Consider the matrix

$$A = \begin{bmatrix} 3 & 0 & 4 \\ 0 & 5 & 0 \\ 4 & 0 & 3 \end{bmatrix}.$$

Determine its eigenvalues and normalized eigenvectors. Comment on the symmetry of A and the orthogonality of its eigenvectors.

2. Verify that the matrix

$$A = \frac{1}{5}\begin{bmatrix} 3 & 4i \\ 4i & 3 \end{bmatrix}$$

is unitary, and determine its eigenvalues.

3. The transformation

$$A = \begin{bmatrix} 1 & 1 \\ 0 & 1 \end{bmatrix}$$

performs a **shear**. Find its eigenvalues and eigenvectors, and comment on whether it is **normal**.

4. For

$$A = \begin{bmatrix} 3 & 0 & 2 \\ 0 & 1 & 0 \\ 2 & 0 & 3 \end{bmatrix},$$

find the eigenvalues and normalized eigenvectors. Verify **orthogonality**.

5. Find a **unitary matrix** U such that $U^\dagger AU$ is diagonal for

$$A = \begin{bmatrix} 0 & 1 \\ 1 & 0 \end{bmatrix}.$$

6. Construct an **orthonormal set** from

$$\mathbf{a}_1 = (1, 1, 0), \mathbf{a}_2 = (1, 0, 1), \text{ and } \mathbf{a}_3 = (0, 1, 1)$$

using the Gram–Schmidt process.

7. **Gram–Schmidt Process**
   Apply Gram–Schmidt to the set

$$(1, 0, 1), (1, 1, 0)$$

   to obtain an orthonormal set.

C. Advanced/Analytical Exercises

1. Construct an orthonormal basis from the following nonorthogonal vectors using the **Gram–Schmidt procedure**:

$$|v_1\rangle = \begin{bmatrix} 1 \\ 1 \\ 0 \end{bmatrix}, \ |v_2\rangle = \begin{bmatrix} 1 \\ 0 \\ 1 \end{bmatrix}, \ |v_3\rangle = \begin{bmatrix} 0 \\ 1 \\ 1 \end{bmatrix}.$$

2. Prove that the **eigenvectors of a degenerate eigenvalue** can be made orthogonal by the **Gram–Schmidt process**.
   Illustrate with a 3×3 symmetric matrix of your choice.

3. **Investigating Diagonalizability**
For each of the following matrices, perform a complete analysis of its eigenvalues and eigenspaces to determine if the matrix is diagonalizable.

Part A: A Symmetric Matrix

$$\text{Consider the matrix } A = \begin{bmatrix} 1 & -1 & -1 \\ -1 & 1 & -1 \\ -1 & -1 & 1 \end{bmatrix}.$$

a. **Find the Characteristic Polynomial:** Calculate $\det(A - \lambda I)$ and find the eigenvalues.
b. **State the Algebraic Multiplicities (AMs):** For each distinct eigenvalue, state its AM.
c. **Find the Eigenspaces:** For each distinct eigenvalue, find a basis for its corresponding eigenspace, $E_\lambda = \ker(A - \lambda I)$.
d. **State the Geometric Multiplicities (GMs):** For each distinct eigenvalue, state its GM (which is the dimension of its eigenspace) and compare it to the AM.
e. Conclude whether A is diagonalizable.
f. **Conclude:** Compare the AM and GM for each eigenvalue. Is the matrix A diagonalizable? Justify your answer.

Part B: An Upper Triangular Matrix

$$\text{Consider the matrix } B = \begin{bmatrix} 2 & 1 & 0 \\ 0 & 2 & 1 \\ 0 & 0 & 3 \end{bmatrix}.$$

a. **Find the Eigenvalues and Algebraic Multiplicities (AM)s:** By inspection, find the eigenvalues and state their AMs.
b. **Find the Eigenspaces:** For each distinct eigenvalue, find a basis for its corresponding eigenspace.
c. **State the Geometric Multiplicities (GMs):** For each distinct eigenvalue, state its GM.
d. **Conclusion:** Compare the AM and GM for each eigenvalue. Is the matrix B diagonalizable? Justify your answer.

4. The matrix

$$H = \begin{bmatrix} 5 & 1 & 0 \\ 1 & 5 & 1 \\ 0 & 1 & 5 \end{bmatrix}$$

represents a simple **three-atom vibrational coupling**:
a. Find its eigenvalues and normalized eigenvectors.
b. Comment on the physical meaning of **degeneracy** and **mode symmetry**.

5. Prove that for any **Hermitian operator A**,

$$\langle v_i \mid A \mid v_j \rangle = \lambda_i \delta_{ij}.$$

Explain its significance in the context of quantum measurement.

6

Diagonalization of Matrices

Quadratic forms occur frequently in applications of linear algebra to engineering (in design criteria and optimization) and signal processing (as output noise power). They also arise, for example, in physics (as potential and kinetic energy), differential geometry (as normal curvature of surfaces), economics (as utility functions), and statistics (in confidence ellipsoids).

—David C. Lay, Linear Algebra and Its Applications, 3rd Edition

Learning Outcomes

After studying this chapter, the student should be able to:

1. **Explain** the concept of a change of basis and **derive** the corresponding similarity transformation between two matrix representations of the same linear operator.
2. **Demonstrate** the process of diagonalization of real symmetric and Hermitian matrices using eigenvalues and eigenvectors.
3. **Transform** a quadratic form into canonical (diagonal) form through an orthogonal transformation and **interpret** the resulting coefficients geometrically as principal axes and radii of an ellipse or ellipsoid.
4. **Classify** quadratic surfaces (ellipsoid, hyperboloid, cylinder, sphere) using the signs and magnitudes of eigenvalues.
5. **Apply** diagonalization techniques to physical systems, such as the moment of inertia tensor and polarizability ellipsoid, and **interpret** eigenvalues as principal moments or principal polarizabilities.
6. **Explain** the concept of simultaneous diagonalization and the condition of positive definiteness required for it.
7. **Relate** simultaneous diagonalization to normal-mode analysis and **describe** how mass-weighted coordinates convert the kinetic energy matrix to an identity matrix.
8. **Compute** characteristic frequencies and normal modes for simple molecular models (e.g., diatomic and linear triatomic systems) using the generalized eigenvalue approach.
9. **Appreciate** the geometric interpretation of diagonalization, linking coordinate transformations, rotation of axes, and invariant quantities across disciplines such as chemistry, physics, and engineering.

6.1 Introduction

The quotation from David Lay's book says it all. We frequently need to diagonalize matrices for various applications and to simplify the algebra. For example, the determinant and inverse of a diagonal matrix are significantly easier to calculate. In this chapter, we also explore the geometric interpretation of a diagonal matrix. We shall see that diagonalization of a matrix requires a **change of basis** to the **eigenvectors** of the matrix. We have already studied a procedure for determining the eigenvalues and eigenvectors of a matrix in the previous chapter. We also considered the conditions under which a matrix is non-diagonalizable, i.e., if the **geometric multiplicity is less than the algebraic multiplicity** for any eigenvalue.

6.2 Change of Basis and Similarity Transformation

6.2.1 Change of Basis

Let $|v\rangle$ be a vector in an N-dimensional space. With respect to an initial basis $\{|a_i\rangle\}$, it can be expanded as

$$|v\rangle = \sum_i c_i |a_i\rangle,$$

where the coefficients c_i are the components of the vector in this basis; that is, the column vector of its components in this basis is $\mathbf{c}$.

Now, suppose we introduce another basis $\{|b_i\rangle\}$, which is related to the first by a linear transformation

$$|b_i\rangle = \sum_j S_{ji} |a_j\rangle. \tag{6.1}$$

Equation (6.1) states that each **new** basis vector $|b_i\rangle$ is a linear combination of the **old** basis vectors $|a_j\rangle$.

Here, S_{ji} is the jth component of the new basis vector $|b_i\rangle$ in the old basis $\{|a_j\rangle\}$. In matrix form, this can be written as

$$[|b_1\rangle|b_2\rangle \ldots |b_N\rangle] = [|a_1\rangle|a_2\rangle \ldots |a_N\rangle]\, S. \tag{6.2}$$

So, the **columns** of S are the coordinates of the new basis vectors expressed in the old basis.

To express the same vector $|v\rangle$ in the new basis, we require its new components c'_j such that

$$|v\rangle = \sum_j c'_j |b_j\rangle. \tag{6.3}$$

Using Equation (6.1) and the fact that S is invertible, we obtain the relation between the old and new component column vectors $\mathbf{c}$ and $\mathbf{c}'$:

$$\mathbf{c}' = S^{-1}\mathbf{c}. \tag{6.4}$$

Thus, the **basis vectors** transform as S, whereas the **components** transform as S^{-1}, so that the physical vector itself remains unchanged:

$$|v\rangle = \sum_i c_i |a_i\rangle = \sum_j c'_j |b_j\rangle.$$

This relationship forms the foundation for the **similarity transformation** of operators, which we now derive. But, before that, a brief numerical example to illustrate change of basis.

6.2.2 Numerical Example: Change of Basis in Two Dimensions

Consider a vector in $\mathbb{R}^2$ with coordinates in the standard basis $\{|e_1\rangle,\ |e_2\rangle\}$:

$$\mathbf{c} = \begin{bmatrix} 3 \\ 1 \end{bmatrix}.$$

Define a new (oblique) basis given by vectors $|b_1\rangle = \begin{bmatrix} 1 \\ 0 \end{bmatrix}$, $|b_2\rangle = \begin{bmatrix} 1 \\ 1 \end{bmatrix}$.

Form the change of basis matrix S whose columns are the new basis vectors expressed in the old basis:

$$S = \begin{bmatrix} 1 & 1 \\ 0 & 1 \end{bmatrix}.$$

To find the coordinates $\mathbf{c}'$ of the vector in the new basis, solve

$$\mathbf{c} = S\mathbf{c}' \Longrightarrow \mathbf{c}' = S^{-1}\mathbf{c}.$$

Calculate S^{-1}:

$$S^{-1} = \begin{bmatrix} 1 & -1 \\ 0 & 1 \end{bmatrix}.$$

Then,

$$\mathbf{c}' = S^{-1}\mathbf{c} = \begin{bmatrix} 1 & -1 \\ 0 & 1 \end{bmatrix} \begin{bmatrix} 3 \\ 1 \end{bmatrix} = \begin{bmatrix} 2 \\ 1 \end{bmatrix}.$$

So, in the new basis, the vector has components

$$\mathbf{c}' = [2, \ 1]^{\mathrm{T}}.$$

Now consider a linear operator given in the old basis by

$$A = \begin{bmatrix} 2 & 1 \\ 0 & 3 \end{bmatrix}.$$

The matrix representation of A in the new basis is

$$A' = S^{-1}AS.$$

Calculate

$$AS = \begin{bmatrix} 2 & 1 \\ 0 & 3 \end{bmatrix} \begin{bmatrix} 1 & 1 \\ 0 & 1 \end{bmatrix} = \begin{bmatrix} 2 & 3 \\ 0 & 3 \end{bmatrix},$$

and then

$$A' = S^{-1}(AS) = \begin{bmatrix} 1 & -1 \\ 0 & 1 \end{bmatrix} \begin{bmatrix} 2 & 3 \\ 0 & 3 \end{bmatrix} = \begin{bmatrix} 2 & 0 \\ 0 & 3 \end{bmatrix}.$$

Thus, in the new basis, the operator A is diagonal.

6.2.3 Similarity Transformation of Operators

Let A be a **linear operator** that acts on vectors in the original basis $\{|a_i\rangle\}$. By definition, when A acts on the vector $|v\rangle$, it produces another vector $|v'\rangle$:

$$|v'\rangle = A|v\rangle. \tag{6.5}$$

If the components of $|v\rangle$ in this basis are collected into a column vector $\mathbf{c}$, then the components of the transformed vector $|v'\rangle$ are given by

$$\mathbf{c}' = A\mathbf{c}.$$

Now, let us represent this same operation in the **new basis** $\{|b_i\rangle\}$, which is related to the old basis by the transformation S defined earlier:

$$|b_i\rangle = \sum_j S_{ji}|a_j\rangle.$$

As shown before in Equation (6.4), the coordinates of the vector transform as

$$\mathbf{c}' = S^{-1}\mathbf{c}.$$

We would now like to find how the operator A itself transforms so that the relation between the operator and the vector remains valid in the new basis.

In the new basis, the same geometric transformation is expressed as

$$|v'\rangle = A'|v\rangle,$$

where A' is the **matrix representation of the same operator** in the new basis. Because $|v\rangle$ is the same vector in both representations, we can write

$$S|v\rangle_{\text{new}} = |v\rangle_{\text{old}},$$

and, therefore,

$$|v'\rangle_{\text{old}} = A|v\rangle_{\text{old}} = A\big(S|v\rangle_{\text{new}}\big).$$

Multiplying on the left by S^{-1}, we obtain

$$S^{-1}|v'\rangle_{\text{old}} = S^{-1}AS|v\rangle_{\text{new}}.$$

But the left-hand side is simply $|v'\rangle_{\text{new}}$, since the coordinates of $|v'\rangle$ in the new basis are related to the old ones by the same transformation S^{-1}. Hence,

$$|v'\rangle_{\text{new}} = \big(S^{-1}AS\big)|v\rangle_{\text{new}}.$$

Therefore, the operator in the new basis is given by the **similarity transformation**

$$A' = S^{-1}AS. \tag{6.6}$$

This is called a **similarity transformation**, and the matrices A and A' are said to be **similar**. We can now state explicitly:

If matrix A represents a linear operator in some basis $\{|a_i\rangle\}$, then $A' = S^{-1}AS$ represents the same operator in another basis $\{|b_i\rangle\}$.

Remarks

1. The two matrices A and A' represent the *same linear operator*, but with respect to different bases.
2. Since the geometric vector $|v\rangle$ remains unchanged, the transformation only affects the numerical representation of both the vector components and the operator.
3. The relationship $A' = S^{-1}AS$ ensures that the **form of the operator equation** $A|v\rangle = |v'\rangle$ is preserved in all bases.

6.2.4 Properties of Similar Matrices

From the preceding discussion, we have shown that if A represents a linear operator in some basis $\{|a_i\rangle\}$, then in another basis $\{|b_i\rangle\}$ related to it by the transformation S, $A' = S^{-1}AS$. The two matrices A and A' are called similar matrices. They represent the *same* linear operator in two different bases.

Several important properties follow directly from this definition:

1. **The identity operator is invariant under any change of basis**
   If $A = I$, then

   $$A' = S^{-1}IS = S^{-1}S = I.$$

   Thus, the identity matrix remains unchanged under a similarity transformation because the identity operator commutes with all matrices.
2. **Similar matrices have equal determinants**
   The determinant of A' is obtained as

   $$\det(A') = \det\left(S^{-1}AS\right) = \det\left(S^{-1}\right)\det(A)\det(S) = \det(A),$$

   since $\det\left(S^{-1}\right)\det(S) = 1$.
   Therefore, $\det(A') = \det(A)$.
   This implies that similar matrices have the same determinant.
3. **Similar matrices have the same eigenvalues**
   Consider the characteristic equation for A':

   $$\det(A' - \lambda I) = \det\left(S^{-1}AS - \lambda I\right).$$

   Multiplying by S on the left and S^{-1} on the right does not change the determinant:

   $$\det\left(S^{-1}AS - \lambda I\right) = \det\left(S^{-1}(A - \lambda I)S\right) = \det(A - \lambda I).$$

   Hence, the eigenvalues of A' are identical to those of A.
4. **The traces of similar matrices are equal**
   The trace is invariant under a similarity transformation:

   $$\mathrm{Tr}(A') = \mathrm{Tr}\left(S^{-1}AS\right) = \mathrm{Tr}\left(AS^{-1}S\right) = \mathrm{Tr}(A),$$

   because the trace of a product is invariant under cyclic permutation of the factors, $\mathrm{Tr}(XY) = \mathrm{Tr}(YX)$.
5. **Unitary and congruent transformations**
   If the transformation matrix S is **unitary**, i.e.,

   $$S^{-1} = S^{\dagger},$$

   then the similarity transformation takes the special form (congruent transformation)

   $$A' = S^{\dagger}AS. \tag{6.7}$$

   This is still a similarity transformation—it preserves eigenvalues, trace, and determinant—because $S^{\dagger} = S^{-1}$. For real orthogonal matrices, $S^{\dagger} = S^{\mathrm{T}}$, so the transformation reduces to $A' = S^{\mathrm{T}}AS$.

   Note that a **congruent transformation**, by contrast, has the form $A \to S^{\mathrm{T}}AS$ for a general (not necessarily orthogonal) matrix S, and preserves quadratic forms but not eigenvalues. Congruent and unitary (or orthogonal) transformations coincide only when S is real orthogonal.

Unitary transformations preserve the **inner product** between vectors, and hence all physical quantities derived from it (such as probability amplitudes in quantum mechanics). If both the old and new bases are orthonormal, S is unitary by construction.

6. **Additional properties of unitary transformations**
   Unitary transformations preserve key matrix characteristics:
   i. If A is Hermitian (or anti-Hermitian), then A' is also Hermitian (or anti-Hermitian):
   Since
   $$A^\dagger = \pm A,$$
   we have
   $$(A')^\dagger = \left(S^\dagger AS\right)^\dagger = S^\dagger A^\dagger S = \pm S^\dagger AS = \pm A'.$$
   ii. If A is unitary ($A^{-1} = A^\dagger$), then A' is also unitary:
   $$(A')^\dagger A' = \left(S^\dagger AS\right)^\dagger \left(S^\dagger AS\right) = S^\dagger A^\dagger SS^\dagger AS = S^\dagger A^\dagger AS = S^\dagger IS = I.$$

Thus, the property of being unitary is invariant under a unitary transformation.

The similarity transformation $A' = S^{-1}AS$, thus, expresses the same operator in a new basis, preserving eigenvalues, determinant, and trace. It forms the mathematical foundation for matrix diagonalization and the geometric interpretation of linear operators.

This completes the discussion of the change of basis and similarity transformations. We are now ready to use these concepts to diagonalize matrices and interpret their geometrical and physical meanings.

6.3 Matrix Diagonalization

Let A be a matrix representing a linear operator $\widehat{A}$ in a certain basis $\{|a_i\rangle\}$. We know that in a new basis, say $\{|b_i\rangle\}$, the corresponding matrix A' is obtained by a similarity transform (6.6): $A' = S^{-1}AS$, where S is the transformation matrix associated with the change of basis.

Now, let the columns of S contain the eigenvectors of A:

$$A\,|a_i\rangle = \lambda_i\,|a_i\rangle. \tag{6.8}$$

Then, the similarity transformation gives

$$S^{-1}AS = \begin{bmatrix} \lambda_1 & 0 & \cdots & 0 \\ 0 & \lambda_2 & \cdots & 0 \\ \vdots & \vdots & \ddots & \vdots \\ 0 & 0 & \cdots & \lambda_N \end{bmatrix} = D. \tag{6.9}$$

Thus, when S is chosen so that its ith column is the eigenvector $|a_i\rangle$ of A, we have $AS = S\Lambda$, and therefore the similarity transformation $S^{-1}AS = \Lambda$ yields the diagonal matrix of eigenvalues.

That is, the matrix A' (denoted D or Λ) is **diagonal**, with the eigenvalues of A as its diagonal elements:

$$A' = \begin{bmatrix} \lambda_1 & 0 & 0 \\ 0 & \lambda_2 & 0 \\ 0 & 0 & \lambda_3 \end{bmatrix}. \tag{6.10}$$

The process of diagonalization is therefore a specific application of the similarity transformation, where we choose a new basis composed of the eigenvectors of the operator. In this special "eigenbasis," the operator's matrix representation is diagonal.

Therefore, to diagonalize A, we find its eigenvectors $|a_i\rangle$, place them as the columns of S, and compute $A' = S^{-1}AS$. Because we require S^{-1}, the eigenvectors must be **linearly independent**.

Geometrically, the eigenvectors define the new coordinate axes (principal axes), and the eigenvalues represent the scaling factors along these axes when the transformation A is applied.

All quantum-mechanical operators are **Hermitian**; hence, their eigenvectors are linearly independent when the eigenvalues are distinct. However, even in the case of **degenerate eigenvalues**, the eigenvectors can be chosen to be **orthogonal** (as shown in the previous chapter).

Every diagonalizable matrix has a similarity transformation $S^{-1}AS$. However, only normal matrices (e.g., Hermitian, real symmetric) can be diagonalized with a unitary transformation $U^\dagger AU$ where U is unitary $(U^\dagger = U^{-1})$. This is the matrix version of the **spectral theorem**: *A matrix can be unitarily diagonalized iff it is normal* $(AA^\dagger = A^\dagger A)$. This is why the eigenvectors of Hermitian matrices are chosen to be orthonormal.

Example 6.1: Matrix Diagonalization

Diagonalize the matrix

$$A = \begin{bmatrix} 2 & 0 & 2 \\ 0 & -2 & 0 \\ 2 & 0 & 2 \end{bmatrix}.$$

Solution:

This is a symmetric (real Hermitian) matrix, and we have solved it previously in Example 5.3.

From Example 5.3, the normalized eigenvectors are $[0,\ 1,\ 0]^T$, $\left[1/\sqrt{2},\ 0, -1/\sqrt{2}\right]^T$, and $\left[1/\sqrt{2},\ 0,\ 1/\sqrt{2}\right]^T$ corresponding to the eigenvalues $-2, 0$, and 4.

Placing the normalized eigenvectors $|a_i\rangle$ as the columns of S,

$$S = \frac{1}{\sqrt{2}} \begin{bmatrix} 0 & 1 & 1 \\ \sqrt{2} & 0 & 0 \\ 0 & -1 & 1 \end{bmatrix}.$$

Since A is an orthogonal symmetric matrix, $S^{-1} = S^\dagger$, we perform the congruent transformation

$$S^\dagger AS = \begin{bmatrix} -2 & 0 & 0 \\ 0 & 0 & 0 \\ 0 & 0 & 4 \end{bmatrix},$$

where the diagonal elements are the eigenvalues corresponding to the eigenvectors' order in S.

If the eigenvectors are also normalized, they form an orthonormal set, and

$$(S^{\dagger}S)_{ij} = \langle a_i | a_j \rangle = \delta_{ij}. \tag{6.11}$$

Thus, S is **unitary** ($S^{-1} = S^{\dagger}$), and the matrix A can be diagonalized by the **unitary transformation**

$$A' = S^{-1}AS = S^{\dagger}AS. \tag{6.12}$$

If A is diagonalized by a similarity transformation $S^{-1}AS = D$, then

$$\det(A - \lambda I) = \det(D - \lambda I), \tag{6.13}$$

because the eigenvalues remain invariant.

If a matrix is of the form

$$A = \begin{bmatrix} A_1 & 0 \\ 0 & A_2 \end{bmatrix},$$

where A_1 and A_2 are themselves matrices, it is said to be **block diagonal**, and each block can be diagonalized separately.

Example 6.2: Diagonalization of a Symmetric Matrix

Find the eigenvalues and eigenvectors of

$$A = \begin{bmatrix} -2 & 0 & 0 \\ 0 & 1 & 2 \\ 0 & 2 & 1 \end{bmatrix}$$

and the similarity transformation that diagonalizes it.

Solution:

The matrix is in block-diagonal form—a 1×1 block and a 2×2 block.

The first block directly yields $\lambda_1 = -2$ with eigenvector $|a_1\rangle = (1, 0, 0)^T$.

For the 2×2 block,

$$\begin{bmatrix} 1 & 2 \\ 2 & 1 \end{bmatrix} \Rightarrow \det \begin{bmatrix} 1-\lambda & 2 \\ 2 & 1-\lambda \end{bmatrix} = 0 \Rightarrow (\lambda - 3)(\lambda + 1) = 0.$$

Hence, $\lambda_2 = 3$ and $\lambda_3 = -1$.

For $\lambda_2 = 3$:

$$(A - 3I)|a_2\rangle = 0 \Rightarrow |a_2\rangle = \frac{1}{\sqrt{2}}(0, 1, 1)^T.$$

For $\lambda_3 = -1$:

$$(A + I)|a_3\rangle = 0 \Rightarrow |a_3\rangle = \frac{1}{\sqrt{2}}(0, -1, 1)^T.$$

The transformation matrix $S = (|a_1\rangle, |a_2\rangle, |a_3\rangle) = \begin{bmatrix} 1 & 0 & 0 \\ 0 & \frac{1}{\sqrt{2}} & -\frac{1}{\sqrt{2}} \\ 0 & \frac{1}{\sqrt{2}} & \frac{1}{\sqrt{2}} \end{bmatrix}$ is orthogonal and has been chosen with sign conventions ensuring $\det S = +1$.

The diagonal form is

$$S^{-1}AS = \begin{bmatrix} -2 & 0 & 0 \\ 0 & 3 & 0 \\ 0 & 0 & -1 \end{bmatrix}.$$

Here is a Hermitian equivalent.

Example 6.3: Diagonalization of a Hermitian Matrix

Find the normalized eigenvectors of the Hermitian matrix

$$A = \begin{bmatrix} 0 & -i & 0 \\ i & 0 & 0 \\ 0 & 0 & 2 \end{bmatrix}$$

and verify that they are orthogonal.

Solution:

This is again block diagonal. The upper 2×2 block gives

$$\det\begin{bmatrix} -\lambda & -i \\ i & -\lambda \end{bmatrix} = \lambda^2 - 1 = 0 \Rightarrow \lambda = \pm 1.$$

For $\lambda_1 = 1$:

$$A|a_1\rangle = \lambda_1|a_1\rangle \Rightarrow |a_1\rangle = \frac{1}{\sqrt{2}}\begin{bmatrix} 1 \\ i \\ 0 \end{bmatrix}.$$

For $\lambda_2 = -1$:

$$|a_2\rangle = \frac{1}{\sqrt{2}}\begin{bmatrix} 1 \\ -i \\ 0 \end{bmatrix}.$$

The third eigenvector, corresponding to $\lambda_3 = 2$, is simply

$$|a_3\rangle = \begin{bmatrix} 0 \\ 0 \\ 1 \end{bmatrix}.$$

The matrix of eigenvectors is therefore (after multiplying the second column by -1 to make S unitary)

$$S = \frac{1}{\sqrt{2}} \begin{bmatrix} 1 & -1 & 0 \\ i & i & 0 \\ 0 & 0 & 1 \end{bmatrix}.$$

Since

$$S^\dagger S = I,$$

the eigenvectors are indeed **orthogonal** and **normalized**, as expected for a Hermitian matrix.

Procedure for Diagonalizing a Matrix A:

1. Find the eigenvalues λ_i of A by solving $\det(A - \lambda I) = 0$.
2. For each eigenvalue λ_i, find a basis for the eigenspace $(A - \lambda_i\, I)$.
3. If the total number of linearly independent eigenvectors is less than N, then A is not diagonalizable.
4. If you have N independent eigenvectors, form the matrix S whose columns are these eigenvectors.
5. The diagonal matrix $D = S^{-1}AS$ will have the eigenvalues on the diagonal, in the same order as their eigenvectors appear in S.

6.4 Quadratic and Hermitian Forms

6.4.1 Quadratic Forms

A general **quadratic form** in variables x_i is an expression of the type

$$Q = \sum_i \sum_j a_{ij} x_i x_j,$$

that is, each term has a total degree of 2.

In **matrix notation**, the same expression can be written compactly as $Q = \mathbf{x}^{\mathrm{T}} A\mathbf{x}$ or

$$Q = [x_1 \ x_2 \ \dots \ x_n]A\begin{bmatrix} x_1 \\ x_2 \\ \vdots \\ x_n \end{bmatrix},$$

or, equivalently, using the **Dirac notation** introduced earlier,

$$\boxed{Q = \langle x \mid A \mid x \rangle.} \tag{6.14}$$

Here, $|x\rangle$ represents a **column vector** of components x_i; $\langle x \mid$ is its **row-vector conjugate (transpose)**; and A is a **real symmetric** or **Hermitian** matrix representing the quadratic operator.

For real variables, $\langle x| = |x\rangle^T$, while for complex variables $\langle x| = |x\rangle^\dagger$.

For example, a quadratic function

$$f(x_1, x_2, x_3) = 3x_1^2 + 2x_1x_2 + 4x_2^2 + 2x_2x_3 + 5x_3^2$$

in three-dimensional space can be written as

$$f = \begin{bmatrix} x_1 & x_2 & x_3 \end{bmatrix}\begin{bmatrix} 3 & 1 & 0 \\ 1 & 4 & 1 \\ 0 & 1 & 5 \end{bmatrix}\begin{bmatrix} x_1 \\ x_2 \\ x_3 \end{bmatrix},$$

where A is a real symmetric matrix.

If we now perform a **change of basis** to $|x'\rangle$, where

$$|x'\rangle = S^{-1}|x\rangle, \tag{6.15}$$

the new matrix A' in this basis is obtained by a **similarity transformation,**

$$A' = S^{-1}AS. \tag{6.16}$$

If S is an **orthogonal transformation**, then $S^{-1} = S^T$. Furthermore, as shown earlier, if S is the **matrix of eigenvectors** of A, then A' is **diagonal**, that is,

$$A' = S^T AS = \Lambda, \tag{6.17}$$

where Λ is the diagonal matrix of eigenvalues.

Example 6.4: Quadratic Forms

Find an orthogonal transformation that converts the quadratic form

$$Q = 5x_1^2 + 4x_1x_2 + x_2^2$$

into the form

$$Q = \lambda_1 y_1^2 + \lambda_2 y_2^2.$$

Solution:

The corresponding symmetric matrix A is obtained by halving the coefficient of the cross-term x_1x_2 and placing the coefficients of x_1^2 and x_2^2 on the diagonal:

$$A = \begin{bmatrix} 5 & 2 \\ 2 & 1 \end{bmatrix}.$$

The eigenvalues of A are found from

$$\det(A - \lambda I) = \begin{vmatrix} 5-\lambda & 2 \\ 2 & 1-\lambda \end{vmatrix} = 0,$$

which gives

$$\lambda^2 - 6\lambda + 1 = 0 \Rightarrow \lambda_{1,2} = 3 \pm 2\sqrt{2}.$$

For $\lambda_1 = 3 + 2\sqrt{2}$,

$$(A - \lambda_1 I)|a_1\rangle = 0 \Rightarrow \begin{bmatrix} 5-\lambda_1 & 2 \\ 2 & 1-\lambda_1 \end{bmatrix}\begin{bmatrix} c_{11} \\ c_{21} \end{bmatrix} = 0,$$

which yields $c_{21}/c_{11} = (\lambda_1 - 5)/2 = \sqrt{2} - 1$.

After normalization,

$$|a_1\rangle = \frac{1}{\sqrt{1 + \left(\sqrt{2}-1\right)^2}}\begin{bmatrix} 1 \\ \sqrt{2}-1 \end{bmatrix} = \frac{1}{\sqrt{4 - 2\sqrt{2}}}\begin{bmatrix} 1 \\ \sqrt{2}-1 \end{bmatrix}.$$

For $\lambda_2 = 3 - 2\sqrt{2}$,
$c_{21}/c_{11} = (\lambda_2 - 5)/2 = -\left(\sqrt{2}+1\right)$, giving

$$|a_2| = \frac{1}{\sqrt{1 + \left(\sqrt{2}+1\right)^2}}\begin{bmatrix} 1 \\ -\left(\sqrt{2}+1\right) \end{bmatrix} = \frac{1}{\sqrt{4 + 2\sqrt{2}}}\begin{bmatrix} 1 \\ -\left(\sqrt{2}+1\right) \end{bmatrix}.$$

Hence, the **orthogonal matrix** S composed of these normalized eigenvectors is

$$S = \begin{bmatrix} \dfrac{1}{\sqrt{4-2\sqrt{2}}} & \dfrac{1}{\sqrt{4+2\sqrt{2}}} \\ \dfrac{\sqrt{2}-1}{\sqrt{4-2\sqrt{2}}} & \dfrac{-\sqrt{2}+1}{\sqrt{4+2\sqrt{2}}} \end{bmatrix}.$$

The similarity transformation is

$$S^T A S = \begin{bmatrix} \lambda_1 & 0 \\ 0 & \lambda_2 \end{bmatrix} = \begin{bmatrix} 3+2\sqrt{2} & 0 \\ 0 & 3-2\sqrt{2} \end{bmatrix}.$$

Thus, in the new coordinates (y_1, y_2), the quadratic form becomes

$$Q = \left(3 + 2\sqrt{2}\right) y_1^2 + \left(3 - 2\sqrt{2}\right) y_2^2,$$

and the cross-term is eliminated by the orthogonal transformation.

6.4.2 Hermitian Forms

In complex vector spaces, we encounter quadratic forms of the type

$$Q = \langle x \mid A \mid x \rangle, \tag{6.18}$$

where A is now a **Hermitian matrix** (that is, $A^\dagger = A$) and $|x\rangle$ is a complex column vector.

Because

$$Q^* = (\langle x|A|x\rangle)^* = \langle x|A^\dagger|x\rangle = \langle x|A|x\rangle,$$

the value of Q is **real**, even though both the matrix A and the vector $|x\rangle$ may have complex elements.

Hence, Hermitian forms generalize real quadratic forms to complex spaces while preserving a real scalar output. They arise throughout quantum mechanics, where expectation values such as

$$\langle \psi \mid \hat{H} \mid \psi \rangle$$

represent observable quantities (energy, momentum, angular momentum, etc.).

If $|x'\rangle$ is related to $|x\rangle$ through a **unitary transformation**

$$|x'\rangle = S^{-1}|x\rangle = S^\dagger|x\rangle,$$

then the transformed matrix A' becomes

$$A' = S^{-1}AS = S^\dagger AS. \tag{6.19}$$

Since A and A' are **unitarily similar**, they have identical eigenvalues. If A is Hermitian, A' is also Hermitian, because

$$(A')^\dagger = \left(S^\dagger AS\right)^\dagger = S^\dagger A^\dagger S = S^\dagger AS = A'. \tag{6.20}$$

Therefore, **Hermitian forms remain invariant under unitary transformations**, just as real quadratic forms remain invariant under orthogonal transformations.

Example 6.5: Hermitian Form Is Real

Verify that the Hermitian form

$$Q = \langle x|A|x\rangle, \quad A = \begin{bmatrix} 1 & i \\ -i & 2 \end{bmatrix}, \quad |x\rangle = \begin{bmatrix} x_1 \\ x_2 \end{bmatrix},$$

is real.

Solution:

$$Q = \langle x \mid A \mid x\rangle = [x_1^*, \ x_2^*]\begin{bmatrix} 1 & i \\ -i & 2 \end{bmatrix}\begin{bmatrix} x_1 \\ x_2 \end{bmatrix} = x_1^* x_1 + x_2^* x_2 + i(x_1^* x_2 - x_2^* x_1).$$

The factor $(x_1^* x_2 - x_2^* x_1)$ is purely imaginary, so $i(x_1^* x_2 - x_2^* x_1)$ is real, and thus Q is real.
Hence, Q is real for all complex x_1, x_2, confirming that Hermitian forms always yield real scalar values.

Key Insights: Real Versus Hermitian Quadratic Forms

Feature	Real quadratic form	Hermitian quadratic form
Definition	$Q = [x_1\, x_2\, \ldots\, x_n]A[x_1\, x_2\, \ldots\, x_n]^T$	$Q = \langle x \mid A \mid x\rangle$
Matrix property	A is **real symmetric** $(A^T = A)$	A is **Hermitian** $(A^\dagger = A)$
Nature of variables	Real-valued x_i	Complex-valued components of $\lvert x\rangle$
Value of Q	Always **real**	Always **real**, even though A and $\lvert x\rangle$ may be complex
Invariant under	Orthogonal transformations $(S^{-1} = S^T)$	Unitary transformations $(S^{-1} = S^\dagger)$
Physical context	Geometry, mechanics, statistics	Quantum mechanics (expectation values)

Interpretation:

- A *real quadratic form* expresses geometric quantities such as distance, energy, or curvature.
- A *Hermitian quadratic form* extends this idea to complex vector spaces—every expectation value $\langle\psi|\hat{A}|\psi\rangle$ in quantum mechanics is such a form.
- Both remain **invariant** under their respective transformations, preserving fundamental scalar quantities.

6.5 Quadratic Surfaces

The equation of a three-dimensional quadratic surface can be represented in the general form

$$\mathbf{x}^T A\mathbf{x} + B^T\mathbf{x} + C = 0, \tag{6.21}$$

or, equivalently,

$$\sum_{i=1}^{3}\sum_{j=1}^{3} a_{ij}x_i x_j + \sum_{i=1}^{3} b_i x_i + c = 0. \tag{6.22}$$

In three dimensions, the quadratic form $\mathbf{x}^T A\mathbf{x}$ defines a surface whose **principal axes** are aligned with the three mutually perpendicular **eigenvectors** of A.

If S is an orthogonal transformation whose columns are the eigenvectors of A, then, as derived earlier,

$$A' = S^T A S, \tag{6.23}$$

and since $S^{-1} = S^T$,

$$\mathbf{x} = S\mathbf{x}', \ \mathbf{x}' = S^T\mathbf{x}. \tag{6.24}$$

Here, A' is a diagonal matrix containing the **eigenvalues** of A as its diagonal elements. Thus, in the new coordinate system,

$$\mathbf{x}^{\mathrm{T}} A\,\mathbf{x} = \mathbf{x}'^{T} A'\mathbf{x}' = \lambda_1 x_1'^2 + \lambda_2 x_2'^2 + \lambda_3 x_3'^2. \tag{6.25}$$

In three-dimensional Euclidean space, the general quadratic surface therefore has the form

$$\lambda_1 {x_1'}^2 + \lambda_2 {x_2'}^2 + \lambda_3 {x_3'}^2 = 1. \tag{6.26}$$

This is the standard form of a quadratic surface with its principal axes aligned with the coordinate system.

Comparison with the standard equation of an ellipsoid,

$$\frac{{x_1'}^2}{a^2} + \frac{{x_2'}^2}{b^2} + \frac{{x_3'}^2}{c^2} = 1,$$

gives

$$\lambda_1 = \frac{1}{a^2},\ \lambda_2 = \frac{1}{b^2},\ \lambda_3 = \frac{1}{c^2}.$$

Hence, $a = 1/\sqrt{\lambda_1}$, $b = 1/\sqrt{\lambda_2}$, and $c = 1/\sqrt{\lambda_3}$.

When all $\lambda_i > 0$, Equation (6.26) represents an **ellipsoid** with principal radii a, b, c, ordered so that $a > b > c$.

If all three eigenvalues are equal, $a = b = c$, the surface is a **sphere**. If one eigenvalue vanishes ($\lambda_i = 0$), the corresponding axis becomes infinite, giving a **cylinder**. In two dimensions, Equation (6.26) reduces to an **ellipse**, and for $a = b$, to a **circle**.

Example 6.6: Quadratic Curves

Sketch the quadratic curve $5x_1^2 + 4x_1x_2 + 5x_2^2 = 21$.

Solution:

Rewriting the equation for the surface,

$$\frac{1}{21}(5x_1^2 + 4x_1x_2 + 5x_2^2) = 1,$$

the symmetric matrix A is

$$A = \begin{bmatrix} 5 & 2 \\ 2 & 5 \end{bmatrix}.$$

Diagonalization proceeds via the eigenvalue problem $|A - \lambda I| = 0$, giving

$$\lambda_1 = 7, \lambda_2 = 3.$$

For $\lambda_1 = 7$,

$$(A - \lambda_1 I)\, c_1 = 0 \Rightarrow \begin{bmatrix} -2 & 2 \\ 2 & -2 \end{bmatrix} \begin{bmatrix} c_{11} \\ c_{21} \end{bmatrix} = 0,$$

yielding $c_{21} = c_{11}$.

After normalization,

$$|v_1\rangle = \frac{1}{\sqrt{(1)^2 + (1)^2}} \begin{bmatrix} 1 \\ 1 \end{bmatrix} = \frac{1}{\sqrt{2}} \begin{bmatrix} 1 \\ 1 \end{bmatrix}.$$

Similarly, for $\lambda_2 = 3$,

$$|v_2\rangle = \frac{1}{\sqrt{2}} \begin{bmatrix} 1 \\ -1 \end{bmatrix}.$$

Thus, the transformation matrix is

$$S = \begin{bmatrix} \frac{1}{\sqrt{2}} & -\frac{1}{\sqrt{2}} \\ \frac{1}{\sqrt{2}} & \frac{1}{\sqrt{2}} \end{bmatrix},$$

which is orthogonal with $\det S = 1$. (We have reversed the signs for eigenvector 2 in order to make the determinant equal $+1$.)

Comparing the original equation with the general form $ax^2 + 2hxy + by^2 = 1$, we have $a = 5$, $b = 5$, $2h = 4 \implies h = 2$.

The cross-term xy is eliminated by rotation through an angle θ given by

$$\tan(2\theta) = \frac{2h}{a - b} = \frac{4}{0} = \infty,$$

$$2\theta = \tan^{-1}(\infty) = 90^\circ,$$

$$\boxed{\theta = 45^\circ.}$$

That is the **true geometrical rotation** that diagonalizes the quadratic form—that is, aligns the ellipse with the new coordinate axes.

This is the angle through which either (a) the ellipse must be rotated **clockwise**, or, equivalently, (b) the coordinate axes rotated **counterclockwise**, so that the principal axes of the ellipse coincide with the new coordinates x_1', x_2'.

The diagonal form of the equation is

$$7{x_1'}^2 + 3{x_2'}^2 = 21,$$

with semi-axes proportional to $1/\sqrt{7}$ and $1/\sqrt{3}$, respectively.

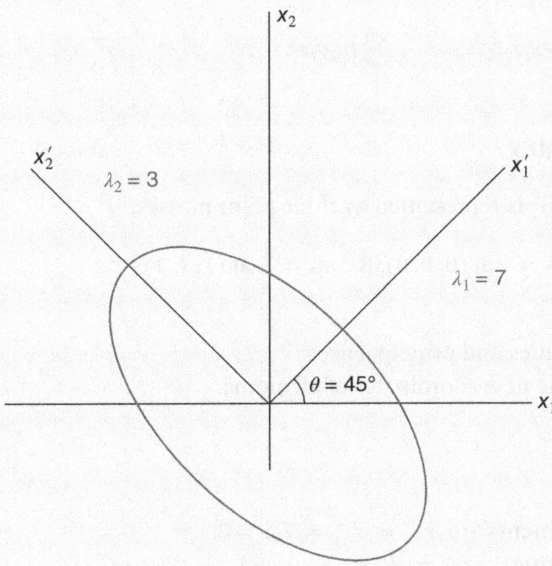

Mixed term $4x_1x_2$ removed; ellipse aligned with principal axes
Rotation of Coordinate Axes by 45°

6.5.1 Physical Significance

The diagonalization of quadratic surfaces provides a direct geometric interpretation: each eigenvalue represents the **inverse squared length** of a principal radius, and the corresponding eigenvector specifies the **orientation** of that principal axis.

This approach is particularly useful in **rotational spectroscopy**, where the *moment of inertia tensor* defines an analogous ellipsoid.

6.5.2 Moment of Inertia Tensor

For a rigid body, the **moment of inertia tensor** $\mathcal{I}$ (a second-rank tensor) is represented by a real symmetric 3×3 matrix.

It therefore possesses three mutually orthogonal eigenvectors associated with three real eigenvalues $\mathcal{I}_a, \mathcal{I}_b, \mathcal{I}_c$, such that

$$\mathcal{I}\omega_i = I_i\,\omega_i,\ \ i = 1, 2, 3. \tag{6.27}$$

The unit vectors ω_i define the **principal axes of rotation**, while the eigenvalues $\mathcal{I}_i$ are the **principal moments of inertia**.

The tensor itself is written as

$$\mathcal{I} = \begin{bmatrix} \mathcal{I}_{xx} & -\mathcal{I}_{xy} & -\mathcal{I}_{xz} \\ -\mathcal{I}_{xy} & \mathcal{I}_{yy} & -\mathcal{I}_{yz} \\ -\mathcal{I}_{xz} & -\mathcal{I}_{yz} & \mathcal{I}_{zz} \end{bmatrix}. \tag{6.28}$$

The diagonal elements are computed from

$$\mathcal{I}_{xx} = \sum_i m_i(y_i^2 + z_i^2),\ \mathcal{I}_{yy} = \sum_i m_i(x_i^2 + z_i^2),\ \mathcal{I}_{zz} = \sum_i m_i(x_i^2 + y_i^2), \tag{6.29}$$

and the off-diagonal terms as

$$\mathcal{I}_{xy} = -\sum_i m_i x_i y_i,\ \mathcal{I}_{xz} = -\sum_i m_i x_i z_i,\ \mathcal{I}_{yz} = -\sum_i m_i y_i z_i. \tag{6.30}$$

Example 6.7: Moment of Inertia Matrix

A hypothetical linear rigid molecule AB_2 is represented by three point masses:

$$B_1 : m_1 = \frac{1}{2} \text{ at } (-1, -1, -1),\ A : m_2 = 1 \text{ at } (0, 0, 0),\ B_2 : m_3 = \frac{1}{2} \text{ at } (1, 1, 1).$$

i. Compute the inertia matrix.
ii. Diagonalize it to find the eigenvalues and principal axes.
iii. Obtain the rotation matrix and the new coordinates of all atoms.
iv. Interpret the $\lambda = 0$ eigenvalue.

Solution:

From Equation (6.29), the diagonal elements are $\mathcal{I}_{xx} = \mathcal{I}_{yy} = \mathcal{I}_{zz} = 2$.

The off-diagonal terms (6.30) are identical: $\mathcal{I}_{xy} = \mathcal{I}_{xz} = \mathcal{I}_{yz} = -1$.

Hence,

$$\mathcal{I} = \begin{bmatrix} 2 & -1 & -1 \\ -1 & 2 & -1 \\ -1 & -1 & 2 \end{bmatrix}.$$

The characteristic equation gives the eigenvalues

$$\lambda_1 = 0, \lambda_2 = 3, \lambda_3 = 3.$$

The normalized eigenvectors are

$$|v_1\rangle = \frac{1}{\sqrt{3}}(1, 1, 1),\quad |v_2\rangle = \frac{1}{\sqrt{6}}(1, -2, 1),\quad |v_3\rangle = \frac{1}{\sqrt{2}}(1, 0, -1).$$

Thus, the rotation matrix is

$$S = \begin{bmatrix} \frac{1}{\sqrt{3}} & \frac{1}{\sqrt{6}} & \frac{1}{\sqrt{2}} \\ \frac{1}{\sqrt{3}} & \frac{-2}{\sqrt{6}} & 0 \\ \frac{1}{\sqrt{3}} & \frac{1}{\sqrt{6}} & \frac{-1}{\sqrt{2}} \end{bmatrix}, S^T \mathcal{I} S = \begin{bmatrix} 0 & 0 & 0 \\ 0 & 3 & 0 \\ 0 & 0 & 3 \end{bmatrix}.$$

After transformation, the molecule's coordinates become

$$B_1 : \left(-\sqrt{3}, 0, 0\right),\ A : (0, 0, 0),\ B_2 : \left(\sqrt{3}, 0, 0\right),$$

showing that all atoms lie on a single line—the x-axis.

The zero eigenvalue corresponds to **zero moment of inertia** for rotation about the molecular axis itself—a feature of any linear arrangement of point masses, and the reason linear molecules have only two rotational degrees of freedom instead of three.

6.5.3 Extension: Polarizability Ellipsoid

An analogous situation arises for the **polarizability tensor** in **Raman spectroscopy**.

When an electric field **E** is applied, the induced dipole moment $\boldsymbol{\mu}$ is given by

$$\boldsymbol{\mu} = \boldsymbol{\alpha}\,\mathbf{E}, \tag{6.31}$$

where the polarizability tensor $\boldsymbol{\alpha}$ is a second-rank symmetric tensor. It can likewise be diagonalized to yield its principal polarizabilities $\alpha_1, \alpha_2, \alpha_3$, defining a **polarizability ellipsoid** with semi-axes proportional to $1/\sqrt{\alpha_i}$.

6.6 Simultaneous Diagonalization

Frequently, two matrices must be diagonalized simultaneously using the same linear transformation. This is straightforward if the two matrices commute, for then they possess the same set of eigenvectors and the same rotation matrix. The difficulty arises when the two matrices do not commute.

To illustrate, suppose we have two quadratic surfaces S_1 and S_2 given by

$$S_1 : x^T A x = x^2 + 9y^2,\ S_2 : x^T B x = 5x^2 - 8xy + 5y^2, \tag{6.32}$$

and we wish to diagonalize them simultaneously. These surfaces may be represented by matrices A and B as follows:

$$A = \begin{bmatrix} 1 & 0 \\ 0 & 9 \end{bmatrix},\ B = \begin{bmatrix} 5 & -4 \\ -4 & 5 \end{bmatrix}.$$

These two matrices do not commute with each other.

Matrix A is already in diagonal form, so we diagonalize B in the usual manner. This gives

$$\begin{vmatrix} 5-\lambda & -4 \\ -4 & 5-\lambda \end{vmatrix} = 0 \Rightarrow (5-\lambda)^2 - 16 = 0 \Rightarrow \lambda_1 = 1, \lambda_2 = 9.$$

The normalized eigenvectors are

$$b_1 = \begin{bmatrix} \frac{1}{\sqrt{2}} \\ \frac{1}{\sqrt{2}} \end{bmatrix},\ b_2 = \begin{bmatrix} \frac{1}{\sqrt{2}} \\ -\frac{1}{\sqrt{2}} \end{bmatrix},$$

and the transformation matrix is

$$S = \frac{1}{\sqrt{2}} \begin{bmatrix} 1 & 1 \\ 1 & -1 \end{bmatrix}.$$

Applying this transformation to both A and B gives

$$A' = S^T A S = \begin{bmatrix} 5 & -4 \\ -4 & 5 \end{bmatrix},\ B' = S^T B S = \begin{bmatrix} 1 & 0 \\ 0 & 9 \end{bmatrix}.$$

In attempting to diagonalize one matrix (B), we have disrupted the diagonal form of the other (A). We therefore require another transformation that diagonalizes both matrices simultaneously—this procedure is called **simultaneous diagonalization**.

Theorem 6.1 *Two real symmetric matrices A and B can be simultaneously diagonalized by the transformations*

$$A' = S^T AS, \ B' = S^T BS,$$

*provided that one of them, say A, is **positive definite**.*

In that case, the generalized eigenvalues can be obtained by solving the **generalized eigenvalue equation**

$$(B - \lambda A)\,\mathbf{x} = 0.$$

By *positive definite* we mean a Hermitian (or real symmetric) matrix satisfying

$$\langle v|A|v\rangle > 0 \text{ for all nonzero } |v\rangle,$$

If $|v\rangle = 0$, the inner product is identically zero.

Physically, a positive definite matrix often represents a quantity that must always be positive, such as kinetic energy (where masses and velocity-squared are positive) or the potential energy of a system in a stable equilibrium.

An example of a positive definite matrix is the identity matrix.

The test for positive definiteness is that all **principal minors** of the matrix have positive determinants.

For a symmetric 3×3 matrix

$$A = \begin{bmatrix} a_{11} & a_{12} & a_{13} \\ a_{12} & a_{22} & a_{23} \\ a_{13} & a_{23} & a_{33} \end{bmatrix},$$

the principal minors are

$$M_1 = a_{11}, M_2 = \begin{vmatrix} a_{11} & a_{12} \\ a_{12} & a_{22} \end{vmatrix}, M_3 = \det(A),$$

and all must be positive.

For example, the real symmetric matrix

$$A = \begin{bmatrix} 2 & 1 \\ 1 & 3 \end{bmatrix}$$

is positive definite because $M_1 = 2$, $M_2 = 5$, and $\det(A) = 6 - 1 = 5 > 0$, whereas

$$A = \begin{bmatrix} 1 & 2 \\ 2 & 1 \end{bmatrix}$$

has $\det(A) = -3 < 0$ and is not positive definite, even though its elements are all positive.

In contrast, the real symmetric matrix,

$$A = \begin{bmatrix} 2 & -1 & 0 \\ -1 & 2 & -1 \\ 0 & -1 & 2 \end{bmatrix}$$

is positive definite because its minors $M_1 = a_{11} = 2$, $M_2 = 3$ and its determinant $= 4$, all of which are positive, though there are some negative elements in the matrix.

Fact Box 6.1: Derivation of the Transformation for Simultaneous Diagonalization

Two real symmetric (or Hermitian) matrices A and B can be simultaneously diagonalized provided that one of them, say A, is positive definite.

We begin with the **generalized eigenvalue equation**

$$B\,\mathbf{x} = \lambda\, A\,\mathbf{x}, \tag{1}$$

where λ are the generalized eigenvalues and $\mathbf{x}$ the corresponding generalized eigenvectors.

Because A is positive definite, it possesses a symmetric square root $A^{1/2}$ and an inverse square root $A^{-1/2}$ such that

$$A = A^{1/2}A^{1/2}. \tag{2}$$

Multiplying Equation (1) by $A^{-1/2}$ from the left gives

$$A^{-1/2}B\,\mathbf{x} = \lambda\, A^{1/2}\mathbf{x}. \tag{3}$$

Let $\mathbf{y} = A^{1/2}\mathbf{x}$, then $\mathbf{x} = A^{-1/2}\mathbf{y}$, and substitution into Equation (3) yields

$$A^{-1/2}BA^{-1/2}\,\mathbf{y} = \lambda\,\mathbf{y}. \tag{4}$$

This is now an ordinary eigenvalue problem for the **symmetric matrix** $A^{-1/2}BA^{-1/2}$.

Let X be the **orthogonal (or unitary) matrix** whose columns are the normalized eigenvectors of $A^{-1/2}BA^{-1/2}$.

Then

$$X^{-1}(A^{-1/2}BA^{-1/2})X = \Lambda, \tag{5}$$

where Λ is diagonal with elements λ_i.

Finally, define

$$S = A^{-1/2}X. \tag{6}$$

It follows that

$$S^T AS = I, \quad S^T BS = \Lambda. \tag{7}$$

Hence, both A and B are **simultaneously diagonalized** by the same transformation S.

This shows that with the transformation $S = A^{-1/2}X$, we simultaneously achieve $S^T AS = I$ and $S^T BS = \Lambda$.

Geometrical Interpretation

The operator $A^{-1/2}$ first scales the coordinate system so that the quadratic form defined by A becomes a sphere (or a circle in two dimensions).

The scaling by $A^{-1/2}$ corresponds exactly to going into mass-weighted coordinates in the later normal-mode section.

The subsequent rotation X then diagonalizes B in this normalized coordinate space.

Thus, $S = A^{-1/2}X$ performs both the scaling and the rotation necessary for simultaneous diagonalization.

6.6.1 Geometrical Interpretation

This procedure can be understood with reference to geometrical operations. Returning to the surfaces given in Equation (6.32), we aim to find a rotation that can bring the axes of both ellipses along the coordinate axes, which is not possible if their axes are not parallel; that is, if they are inclined at different angles to the coordinate axes. However, if we can perform a coordinate transformation to convert one of these ellipses (the one represented by the positive definite matrix) to a circle and also rotate the axes of the other ellipsoid to coincide with the coordinate axes, both matrices will be simultaneously diagonalized because a circle has no preferred axis.

We first scale the coordinates to turn the quadratic form of the positive definite matrix A into a circle (which is invariant under rotation), and then we rotate the coordinates to diagonalize the other matrix.

We illustrate this procedure for the two matrices as follows.

$$S_1 : x^T A x = x^2 + 9y^2, \ S_2 : x^T B x = 5x^2 - 8xy + 5y^2.$$

1. Scale x and y such that S_1 becomes a unit circle; that is, $x' = x$, $y' = 3y$. Hence $y = y'/3$ and

$$S_1' = x'^2 + y'^2, S_2' = 5x'^2 - \left(\frac{8}{3}\right)x'y' + \left(\frac{5}{9}\right)y'^2.$$

The new A' and B' matrices are then

$$A' = \begin{bmatrix} 1 & 0 \\ 0 & 1 \end{bmatrix}; B' = \begin{bmatrix} 5 & -\frac{4}{3} \\ -\frac{4}{3} & \frac{5}{9} \end{bmatrix}.$$

2. Rotate x' and y' such that S_2' is aligned with the coordinate axes. We do this by solving the following:

$$\begin{vmatrix} 5-\lambda & -\frac{4}{3} \\ -\frac{4}{3} & \frac{5}{9}-\lambda \end{vmatrix} = 0 \Rightarrow \lambda = 0.1862, 5.3693.$$

This yields the normalized transformation matrix

$$S = \begin{bmatrix} 0.2669 & -0.9637 \\ 0.9637 & 0.2669 \end{bmatrix}$$

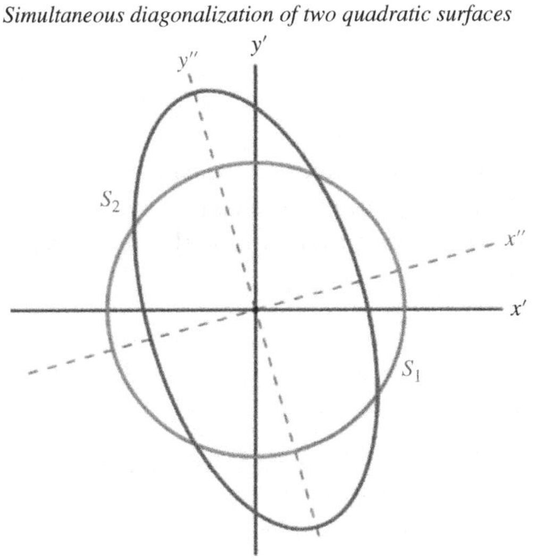

Figure 6.1 Simultaneous diagonalization of the two surfaces.

and the transformation $S^T B' S$ gives the diagonalized matrix B'' as

$$B'' = \begin{bmatrix} 0.1862 & 0 \\ 0 & 5.3693 \end{bmatrix}.$$

The A' matrix remains diagonal because it is the identity matrix, and any similarity transformation with an orthogonal matrix $(S^T = S^{-1})$ leaves it unchanged because $S^T I S = I S^{-1} S = I \times I = I$, since the identity matrix commutes with all matrices. Thus, both matrices are diagonalized simultaneously.

The equations of the two surfaces are therefore

$$S_1'' = x'^2 + y'^2 = 1, \ S_2'' = 0.1862x'^2 + 5.3693y'^2 = 1.$$

Figure 6.1 shows the two new surfaces.

6.6.2 Application in Normal-mode Analysis

This strategy of simultaneous diagonalization is also adopted in **normal-mode analysis**, where the goal is to convert the kinetic-energy matrix into an identity matrix (a circle in two-dimensional analogy) so that both kinetic and potential-energy forms become diagonal.

Fact Box 6.2: Lagrange's Equations and Normal Coordinates

For a system described by generalized coordinates q_i, the equations of motion follow from the Lagrangian

$$L = T - V,$$

where T and V denote the kinetic and potential energies.

Lagrange's equations,

$$\frac{d}{dt}\left(\frac{\partial L}{\partial \dot{q}_i}\right) - \frac{\partial L}{\partial q_i} = 0,$$

lead, under the harmonic approximation, to a set of coupled linear differential equations.

When T and V are quadratic forms in $\dot{q}_i$ and q_i, respectively, the vibrational problem reduces to the **simultaneous diagonalization of the kinetic-energy matrix and the potential-energy matrix**, yielding independent normal modes.

Historical Note 6.1

The use of Lagrange's equations in molecular vibration theory provides the classical foundation for the Wilson–Decius–Cross GF method, which systematized normal coordinate analysis for polyatomic molecules.

6.6.3 Example: The Diatomic Molecule

A diatomic molecule can be modeled as two atoms joined by a spring.

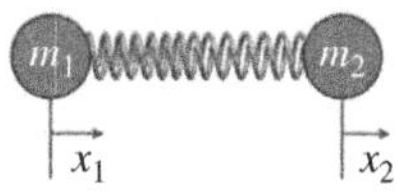

The **kinetic energy** is

$$T = \frac{1}{2}\left(m_1\dot{x}_1^2 + m_2\dot{x}_2^2\right), \tag{6.33}$$

where m_1 and m_2 are the atomic masses, and the dot denotes the time derivative (velocity).
The potential energy is

$$V = \frac{1}{2}\,k(x_2 - x_1)^2, \tag{6.34}$$

where k is the bond force constant (stiffness of the bond).
Both T and V are **quadratic forms**, which can be written compactly as

$$T = \frac{1}{2}\,\dot{\mathbf{x}}^T T\,\dot{\mathbf{x}}, \; V = \frac{1}{2}\,\mathbf{x}^T F\,\mathbf{x}, \tag{6.35}$$

where

$$T = \begin{bmatrix} m_1 & 0 \\ 0 & m_2 \end{bmatrix}, \; F = \begin{bmatrix} k & -k \\ -k & k \end{bmatrix}.$$

The kinetic-energy matrix T is already diagonal, whereas the force-constant matrix F must be diagonalized to yield independent equations. If we attempt to diagonalize F alone by a linear transformation, the same transformation disrupts T. Both matrices are positive definite because $m_i > 0$ and $k > 0$.

This problem can also be viewed as diagonalizing the Lagrangian (see Fact Box 6.2).

6.6.4 Generalized Eigenvalue Problem

We solve the **generalized eigenvalue equation**

$$(F - \lambda T)\,\mathbf{x} = 0, \tag{6.36}$$

which gives the eigenvalues

$$\lambda = 0,\ \lambda = \frac{k}{\mu},\ \mu = \frac{m_1 m_2}{m_1 + m_2},$$

where μ is the **reduced mass.**

The corresponding eigenvectors are

$$\lambda = 0\colon [1, 1]^T,\ \lambda = \frac{k}{\mu}\colon [1, -m_1/m_2]^T.$$

6.6.5 Simultaneous Diagonalization

The similarity transformation satisfying

$$S^T T S = I \tag{6.37}$$

is

$$S = \begin{bmatrix} \dfrac{1}{\sqrt{(m_1 + m_2)}} & \sqrt{\dfrac{m_2}{m_1(m_1 + m_2)}} \\ \dfrac{1}{\sqrt{(m_1 + m_2)}} & -\sqrt{\dfrac{m_1}{m_2(m_1 + m_2)}} \end{bmatrix},$$

which simultaneously diagonalizes F:

$$S^T F S = \begin{bmatrix} 0 & 0 \\ 0 & \dfrac{k}{\mu} \end{bmatrix}. \tag{6.38}$$

The first eigenvalue corresponds to a **translation mode** (no restoring force), and the second to a **vibrational mode**. The zero eigenvalue corresponds to motion along a flat direction of the potential, and the nonzero eigenvalues give ω^2.

Fact Box 6.3: Normalization Using Scaling Factors s and t

To ensure that the kinetic energy matrix becomes an identity matrix after transformation, we introduce scaling factors s and t for the two eigenvectors.

Let the transformation matrix be

$$S = \begin{bmatrix} s & t \\ s & -t\dfrac{m_1}{m_2} \end{bmatrix}.$$

The transformed kinetic energy matrix is

$$T' = S^T T S,\ \text{where } T = \begin{bmatrix} m_1 & 0 \\ 0 & m_2 \end{bmatrix}.$$

Explicitly,

$$T' = \begin{bmatrix} s^2(m_1 + m_2) & 0 \\ 0 & t^2\left(m_1 + \dfrac{m_1^2}{m_2}\right) \end{bmatrix}.$$

To make $T' = I$, we require that

$$s^2(m_1 + m_2) = 1,\ t^2\left(m_1 + \frac{m_1^2}{m_2}\right) = 1.$$

Hence,

$$s = \frac{1}{\sqrt{(m_1 + m_2)}},\ t = \sqrt{\frac{m_2}{m_1(m_1 + m_2)}}.$$

With these scaling factors, the kinetic energy matrix becomes the identity matrix, and the potential-energy matrix is simultaneously diagonalized using the same transformation.

6.6.6 Mass-weighted Coordinates

In normal-mode analysis, we often first transform to **mass-weighted coordinates**, where the kinetic energy automatically becomes an identity form.

Let

$$q_i = \sqrt{m_i}\, x_i,\quad \dot{q}_i = \sqrt{m_i}\, \dot{x}_i, \tag{6.39}$$

so that

$$T = \frac{1}{2}\dot{\mathbf{q}}^T\dot{\mathbf{q}},\ V = \frac{1}{2}\mathbf{q}^T F'\mathbf{q}, \tag{6.40}$$

with

$$F' = \begin{bmatrix} \dfrac{k}{m_1} & -\dfrac{k}{\sqrt{m_1 m_2}} \\ -\dfrac{k}{\sqrt{m_1 m_2}} & \dfrac{k}{m_2} \end{bmatrix}. \tag{6.41}$$

We then diagonalize F' by an orthogonal transformation

$$S = \frac{1}{\sqrt{(m_1 + m_2)}}\begin{bmatrix} \sqrt{m_1} & \sqrt{m_2} \\ \sqrt{m_2} & -\sqrt{m_1} \end{bmatrix} \tag{6.42}$$

giving

$$S^T F' S = \begin{bmatrix} 0 & 0 \\ 0 & \dfrac{k}{\mu} \end{bmatrix}. \tag{6.43}$$

The kinetic-energy matrix remains the identity matrix, since $S^T I S = I$.

To express the eigenvectors in displacement coordinates, divide the ith row of S by $\sqrt{m_i}$:

$$\mathbf{x}' = S^T \mathbf{x}, \qquad (6.44)$$

yielding displacements corresponding to the two normal modes.

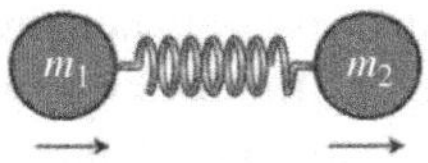

Translational

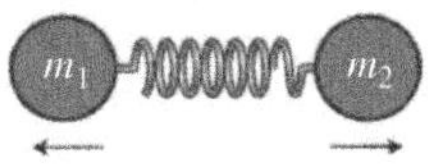

Stretching

6.6.7 Physical Interpretation

- The zero eigenvalue ($\lambda = 0$) corresponds to **translation**, where both atoms move together.
- The nonzero eigenvalue ($\lambda = k/\mu$) corresponds to **vibration**, where the atoms move in opposite directions.
- Geometrically, the transformation makes one of two ellipses (the kinetic energy) a circle and rotates the coordinate system so that the principal axes of both conic sections coincide.

Note: This two-atom case illustrates how simultaneous diagonalization of F and T leads naturally to independent normal coordinates. The same procedure extends to polyatomic molecules, where the transformation matrices are larger but follow the same principle.

A worked example for a linear triatomic molecule is provided in Appendix J.2.

In Fact Box 6.2, we introduced generalized coordinates in relation to Lagrange's equations of motion (see also Appendix J.3). When applied to molecular vibrations, generalized coordinates are often chosen as **internal coordinates**—bond lengths, bond angles, and dihedral angles—rather than Cartesian coordinates. This choice automatically eliminates the three translational and three rotational degrees of freedom, removing the zero eigenvalues that would otherwise appear in a Cartesian-coordinate treatment.

For an N-atomic nonlinear molecule:

- $3N$ Cartesian coordinates describe all atomic displacements.
- Internal coordinates consist of $N - 1$ bond lengths, $N - 2$ bond angles, and $N - 3$ dihedral angles, giving exactly $3N - 6$ vibrational degrees of freedom.
- These $3N - 6$ coordinates then form the basis for constructing the normal modes of vibration.

Marginal Note (Why Generalized Coordinates Are Used)

Why Generalized Coordinates?

Generalized (internal) coordinates automatically eliminate overall translation and rotation, allowing the vibrational problem to be formulated directly in terms of physically meaningful internal motions.

6.7 Summary

A brief summary of this chapter is provided below:

- Quadratic surfaces can be represented by **symmetric matrices**, which describe ellipses in two-dimensional space and ellipsoids in three-dimensional space.
- Two matrices can be **simultaneously diagonalized**, provided that one of them is **positive definite**. Geometrically, this involves converting the surface represented by the positive definite matrix into a **circle** (or a **sphere** in three-dimensional space) by a suitable change of coordinates and diagonalizing the other surface in the new set of coordinates.
- In **normal-mode analysis**, this procedure is implemented using **mass-weighted coordinates**.

Further Reading

P. W. Atkins and R. S. Friedman, *Molecular Quantum Mechanics*, 5th ed., Oxford University Press, 2010.
C. Cohen-Tannoudji, B. Diu and F. Laloë, *Quantum Mechanics*, Vol. 1 Wiley-VCH, 2005.
H. Goldstein, C. Poole and J. Safko, *Classical Mechanics*, 3rd ed., Pearson Education, 2002.
E. Kreyszig, *Advanced Engineering Mathematics*, 10th ed., Wiley, 2011.
P. Lancaster and M. Tismenetsky, *The Theory of Matrices*, 2nd ed., Academic Press, 1985.
D. C. Lay, *Linear Algebra and Its Applications*, 3rd ed., Pearson Education, 2003.
R. McWeeny, *Methods of Molecular Quantum Mechanics*, 2nd ed., Academic Press, 1992.
G. Strang, *Linear Algebra and Its Applications*, 4th ed., Cengage Learning, 2007.
E. B. Wilson Jr., J. C. Decius and P. C. Cross, *Molecular Vibrations: The Theory of Infrared and Raman Vibrational Spectra*, Dover Publications, 1980.

Exercises

A. Conceptual Questions

1. Define a *similarity transformation*. What does it mean geometrically when two matrices are similar?

2. Explain why the eigenvalues of two similar matrices are the same, but their eigenvectors are generally different.

3. What is the significance of diagonalizing a matrix? Discuss how diagonalization simplifies physical or mathematical problems.

4. State the conditions under which a real symmetric or Hermitian matrix can be diagonalized by an orthogonal or unitary transformation, respectively.

5. Why must the transformation matrix S be nonsingular in the diagonalization $A' = S^{-1}AS$?

6. Explain how the *trace* and *determinant* of a matrix relate to its eigenvalues.

7. Define a *positive definite* matrix. What is the physical meaning of this property in the context of kinetic or potential energy?

8. What is meant by *simultaneous diagonalization*? Under what conditions can two matrices be simultaneously diagonalized?

9. Why is it impossible, in general, to simultaneously diagonalize two matrices that do not commute?

10. Discuss the role of eigenvectors in changing the basis of representation of a linear operator.

11. Describe geometrically how the diagonalization of a quadratic form corresponds to a rotation of coordinate axes.

12. What is the physical significance of a zero eigenvalue in a normal-mode analysis?

13. Explain why, in normal-mode analysis, we often convert to *mass-weighted coordinates*. What advantage does this provide?

14. Show how the diagonalization of the moment of inertia tensor corresponds to finding the principal axes of a molecule.

15. Why can a circle (or sphere) be considered a special case of an ellipse (or ellipsoid) in the context of quadratic forms?

B. Problems

1. **Rotation matrix eigenvalues**
   a. Show that the sum of the three eigenvalues of a rotation matrix is $1 + 2\cos\theta$, where θ is the net angle of rotation about a single fixed axis.
   b. Given that one eigenvalue is 1, show that the other two eigenvalues must be $e^{i\theta}$ and $e^{-i\theta}$. *(Hint: An orthogonal rotation matrix with real elements has complex eigenvalues.)*

2. **Diagonalization of an ellipse**
   The equation $2x^2 + 4xy + 5y^2 = 1$ represents an ellipse in the xy-plane.
   Show that it can be transformed to either of the diagonal forms
   a. $x'^2 + 6y'^2 = 1$,
   b. $6x'^2 + y'^2 = 1$
   by rotating the coordinate axes. Display the matrix of rotation for both cases.

3. Ellipsoidal surface and eigenvectors

Consider the surface

$$5x^2 + 3y^2 + 3z^2 + 2xy - 2xz - 2yz = 1.$$

a. Display the matrix associated with the quadratic form.
b. By finding its eigenvalues, show that the surface is an ellipsoid.
c. Verify that the sum of the eigenvalues equals the trace of the matrix.
d. Find the corresponding eigenvectors.

4. Eigenstructure of a real 3 × 3 matrix

Let

$$A = \begin{bmatrix} 1 & 2 & 3 \\ 0 & 1 & 4 \\ 0 & 0 & 1 \end{bmatrix}.$$

a. Does A^{-1} exist? If so, find it.
b. Find the eigenvalues of A. For the repeated eigenvalue, find a basis for its eigenspace and state its geometric multiplicity.
c. Compare the algebraic and geometric multiplicities. Is A diagonalizable? Justify your answer.
d. Evaluate $B = A^3 - 6A^2 + 11A - 6I$.
e. Does A^{-1} exist? If so, find it.

5. Transformation of a linear operator

The matrix of a certain linear operator with respect to basis $\vec{v}_1, \vec{v}_2, \vec{v}_3$ is

$$A = \begin{bmatrix} 0 & 1 & 0 \\ 0 & 0 & 1 \\ 1 & 0 & 0 \end{bmatrix}.$$

Show that the matrix of the same operator with respect to the orthogonal basis

$$\frac{1}{\sqrt{3}}(1, 1, 1), \frac{1}{\sqrt{2}}(1, 0, -1), \frac{1}{\sqrt{6}}(1, -2, 1)$$

is diagonal. *(Hint: Use the transformation $A' = S^{-1}AS$.)*

6. Spin-1 matrices and commutation relations

The spin-1 matrices (in units of $\hbar$) are given by

$$S_x = \frac{1}{\sqrt{2}}\begin{bmatrix} 0 & 1 & 0 \\ 1 & 0 & 1 \\ 0 & 1 & 0 \end{bmatrix}, S_y = \frac{1}{\sqrt{2}}\begin{bmatrix} 0 & -i & 0 \\ i & 0 & -i \\ 0 & i & 0 \end{bmatrix}, S_z = \begin{bmatrix} 1 & 0 & 0 \\ 0 & 0 & 0 \\ 0 & 0 & -1 \end{bmatrix}.$$

(a) Show that $[S_x, S_y] = iS_z$, and similarly for cyclic permutations.
(b) Show that $S_x^2 + S_y^2 + S_z^2 = 2I$.
(c) Show that $S_+ = S_x + iS_y$ and $S_- = S_x - iS_y$ act as ladder operators.

7. Simultaneous diagonalization of commuting Hermitian operators

Verify that the spin-1 matrices are Hermitian. Since $[S^2, S_z] = 0$, the two operators may be simultaneously diagonalized.

(a) Find the eigenvectors of S_z and show that they are also eigenvectors of S^2.
(b) Determine the eigenvalues and discuss their physical significance.
(c) Show that

$$S_+|1, m\rangle = \hbar\sqrt{2 - m(m+1)}\,|1, m+1\rangle, S_-|1, m\rangle = \hbar\sqrt{2 - m(m-1)}\,|1, m-1\rangle.$$

8. Commuting matrices

You are given the following matrices:

$$A = \begin{bmatrix} 1 & 0 & 0 \\ 0 & 2 & 0 \\ 0 & 0 & 3 \end{bmatrix}, B = \begin{bmatrix} 2 & 0 & 0 \\ 0 & 3 & 0 \\ 0 & 0 & 1 \end{bmatrix}, C = \begin{bmatrix} 3 & 0 & 0 \\ 0 & 1 & 0 \\ 0 & 0 & 2 \end{bmatrix}.$$

(a) Prove that A, B, and C commute with each other.
(b) Determine their eigenvalues.
(c) Construct a common orthonormal set of eigenvectors.
(d) Construct a unitary matrix U that simultaneously diagonalizes them.
(e) Write the diagonal forms of A, B, and C.

9. Diagonalization of a Hermitian matrix

(a) Find the eigenvalues and eigenvectors of

$$A = \begin{bmatrix} 0 & i \\ -i & 0 \end{bmatrix}.$$

(b) Find the unitary transformation that diagonalizes A and write its diagonal form.

10. Positive definite quadratic forms

Let

$$A = \begin{bmatrix} 2 & 1 \\ 1 & 2 \end{bmatrix}$$

represent a quadratic form.
Write the orthogonal transformation S that diagonalizes A to $A' = S^T AS$.
(b) Does such a transformation exist for

$$A_1 = \begin{bmatrix} 1 & 2 \\ 2 & 4 \end{bmatrix}, A_2 = \begin{bmatrix} 3 & -4 \\ -4 & 5 \end{bmatrix}?$$

(Hint: Check whether each is positive definite.)

11. Matrix diagonalization check

Given the matrix

$$A = \begin{bmatrix} 2 & 1 \\ 1 & 2 \end{bmatrix},$$

show that $U^{-1}AU$ is diagonal, where U is the matrix of normalized eigenvectors of A.

12. Vibrational Modes as Vectors

In a three-atom molecule (e.g., H—O—H), internal vibrational displacements can be viewed as vectors in a coordinate space. Explain why the symmetric and antisymmetric stretching modes are orthogonal.

7

Group Theory and Symmetry

Symmetry, as wide or as narrow as you may define its meaning, is one idea by which man through the ages has tried to comprehend and create order, beauty, and perfection.

—Hermann Weyl, Symmetry (1952)

Learning Outcomes

After studying this chapter, students will be able to:

1. **Define** symmetry operations and elements and **identify** them in molecules.
2. **Explain** the mathematical meaning of a group and **verify** that symmetry operations satisfy group postulates.
3. **Classify** molecules into point groups using decision flowcharts.
4. **Represent** symmetry operations by matrices and **determine** their characters.
5. **Distinguish** between reducible and irreducible representations and **interpret** character tables.
6. **Use** Mulliken symbols to label molecular orbitals and vibrational modes.
7. **Relate molecular symmetry to observable properties** such as dipole moment, IR/Raman activity, and orbital degeneracy.

7.1 Motivation and Overview

This chapter introduces a fundamental shift in perspective, meriting a dedicated overview. It is all about a mathematical theory that simplifies science.

Mathematics is the language of patterns, and **group theory** is the study of patterns that *remain unchanged* when a system is transformed. Every physical law—whether describing a vibrating molecule, a rotating crystal, or the motion of an electron—contains an element of symmetry. Recognizing these symmetries helps us understand why certain quantities stay constant, why energy levels are degenerate, and why some transitions are allowed while others are forbidden.

In the earlier chapters, we developed the tools needed to describe change: differentiation, integration, and the solution of differential equations. We also examined matrices and transformations—operations that altered coordinates or variables while preserving relationships among them. In this chapter, we take a natural next step: we ask what happens when a system is transformed but its essential nature remains the same.

This is the central idea of *symmetry*. When a molecule is rotated, reflected, or inverted, its atoms may move in space, but the molecule as a whole can look identical. When a coordinate system is rotated, the physical law describing motion or energy must not change. Such invariance under transformation is the hallmark of all fundamental theories—classical mechanics, electromagnetism, and quantum mechanics alike.

Group theory gives a mathematical framework to describe these invariances. It provides rules for combining transformations, understanding their order and structure, and representing them through matrices. These representations allow us to treat symmetry operations just as we treated linear transformations—but now with a deeper physical meaning.

Symmetry lies at the heart of all natural structures—from the hexagonal perfection of snowflakes to the periodic repetition of crystalline lattices. Chemistry, in particular, finds in symmetry both an aesthetic principle and a predictive tool. The shapes of molecules, the degeneracies of energy levels, the rules of spectroscopic transitions—all emerge from symmetry considerations.

The year **2025** has brought renewed attention to this idea with the Nobel Prize in Chemistry awarded to Susumu Kitagawa, Richard Robson, and Omar Yaghi for the design and discovery of metal–organic frameworks (MOFs)—crystalline materials built from metal nodes and organic linkers into exquisitely symmetrical, porous networks. These frameworks illustrate how *group theory* translates the elegance of geometry into the utility of chemistry, guiding the understanding of connectivity, topology, and even gas adsorption properties. Indeed, MOFs are a striking example of how mathematical symmetry leads to tangible chemical beauty.

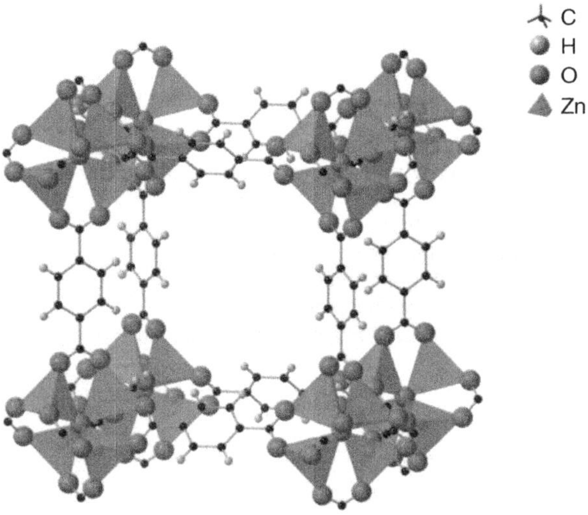

Crystal structure of MOF-5 or IRMOF-1—a cubic MOF compound with the formula $Zn_4O(BDC)_3$, where BDC^{2-} = 1,4-benzodicarboxylate (MOF-5). It was first synthesized in the lab of Omar M. Yaghi. Yaghi's work on MOF-5 is partially responsible for his 2025 Nobel Prize in Chemistry.

Group theory provides the formal language to express these patterns. It tells us not only *how* a molecule looks but also *why* it behaves the way it does—why CO_2 is linear and nonpolar, why benzene exhibits delocalized π-orbitals, and why MOFs adopt repeating, space-group symmetries extending to three dimensions.

The goal of this chapter is to develop this language systematically, beginning from the simple observation that some molecules "look the same" after certain transformations.

We begin by defining what constitutes a *group* and exploring simple examples. Then we introduce **matrix representations** of symmetry operations, leading to the classification of molecules into **point groups**. Finally, we link these abstract ideas to physical observables—how symmetry governs degeneracy, vibrational modes, and conservation laws—in the next chapter.

By the end of this chapter, the reader should see symmetry not as an aesthetic property but as a **powerful analytical tool**, one that simplifies complex problems by revealing what must remain unchanged.

7.2 Introduction

Group theory is the mathematical study of *symmetry*. Formally, it deals with algebraic structures called **groups**, but its importance lies far beyond pure mathematics. In chemistry and physics, group theory provides a unifying framework for understanding how the *form* of a system constrains its behavior. Spectroscopy, crystallography, molecular orbital theory, and optical activity—all draw their organizing principles from symmetry, and symmetry is expressed most precisely through group theory.

7.2.1 From Mathematics to Chemistry

Originally developed by mathematicians such as Évariste Galois to solve algebraic equations, group theory was later recognized as a powerful tool in the natural sciences. Chemists adopted it to classify molecules and interpret physical phenomena. What began as a set of abstract theorems became a practical method for predicting degeneracies, selection rules, and the splitting of energy levels. The theorems your mathematics teachers prove in abstraction—about closure, identity, and inverses—reappear here as the logic behind rotational symmetry, mirror planes, and inversion centers.

Historical Note 7.1: Évariste Galois (1811–1832)

Évariste Galois was a French mathematician whose ideas laid the foundations of modern group theory. He introduced the concept of studying mathematical objects through their *symmetries*, rather than through explicit formulas.

Galois showed that the solvability of polynomial equations depends on the structure of the group formed by permutations of their roots. This insight shifted mathematics from computation to structure and abstraction. Although his work was initially overlooked and published only after his death, it profoundly influenced algebra.

Group theory, abstracted from Galois's ideas, later became essential in physics and chemistry. In molecular symmetry, rotations, reflections, and inversions are classified using the same group-theoretic principles, making Galois's work central to spectroscopy, molecular orbital theory, and crystallography.

7.3 Symmetry Operations and Point Groups

Every molecule possesses certain **symmetry operations**—transformations that leave it apparently unchanged. These include rotation about an axis, reflection in a plane, inversion through a point, and the trivial operation of doing nothing. The collection of all such operations for a given molecule forms a **point group**, because all operations leave at least one point (usually the center of mass) unmoved.

For example, both H_2O and CH_2Cl_2 belong to the C_{2v} point group, characterized by four operations: E, C_2, σ_v, and σ'_v. The square-planar molecule XeF_4 is more symmetric, possessing:

- a principal axis (C_4),
- four perpendicular C_2 axes,
- one horizontal plane perpendicular to the main axis (σ_h), and
- several vertical and dihedral planes (σ_v, σ_d).

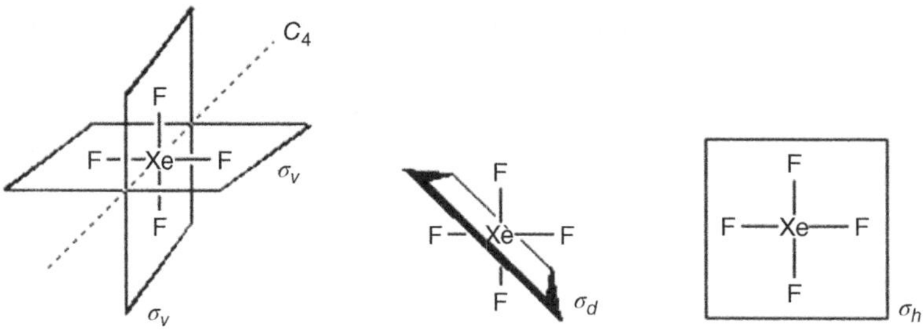

Because only a limited number of **symmetry elements** can exist in finite molecules, there are only a limited number of point groups. Hence, many seemingly different molecules share identical symmetry classifications.

When translational symmetry is also present—as in crystals—the relevant description extends to **space groups**, which no longer have a single fixed point (Section 7.14). Space groups govern the repeating patterns in lattices and explain the geometric order observed in crystalline solids.

7.3.1 Symmetry and Physical Properties

The symmetry of a molecule determines much more than its aesthetic appearance. It dictates whether the dipole moment is zero, whether certain vibrational modes are infrared (IR) or Raman active, and whether a particular electronic transition is allowed. Group theory provides the language to express these restrictions quantitatively through irreducible representations of point groups. In the next chapter, we shall see how these representations correspond to sets of functions—atomic orbitals, vibrational displacements, or transition moments—that transform together under the symmetry operations of the molecule.

7.3.2 Mathematical Continuity

The ideas developed in earlier chapters now appear in a new light. Matrices and linear transformations, previously introduced as abstract operations, become the working tools of symmetry. A symmetry operation is simply a transformation represented by a matrix that preserves certain physical quantities—length, energy, or potential. Thus, group theory connects directly with coordinate transformations and the invariance of physical laws: when a molecule or a coordinate system is rotated, the underlying equations must retain the same form.

7.3.3 Symmetry Operations

Symmetry operations are transformations that move all points of a molecule in space in such a way that the final configuration is indistinguishable from the original one. Each operation corresponds to a definite geometric element—an axis, a plane, or a point—about or through which the operation is performed.

All symmetry operations leave at least one point of the molecule unchanged (usually the **center of mass**), and together they form the **point group** to which the molecule belongs.

7.3.3.1 Axis of Rotation (C_n)

A **rotation axis** is denoted by C_n, where C stands for *cyclic* and n is an integer (1, 2, 3, ...). A rotation by an angle $\theta = 2\pi/n$ about this axis brings the molecule into an orientation indistinguishable from the original one. If this is true, then the molecule possesses an n-fold rotation axis.

A. Molecules with C_1 axes

The simplest case is C_1, present in all molecules, since a rotation through 360° (2π radians) always restores the original configuration. Examples of molecules having only a C_1 axis are bromochlorofluoromethane (CHBrClF) and 2-butanol.

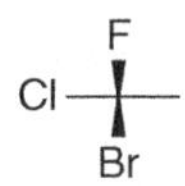

Bromochlorofluoromethane (CHBrClF)

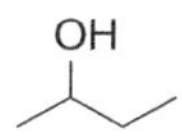

2-Butanol

B. Molecules with C_2 axes

The next higher case is C_2. The water molecule (H_2O) has a C_2 axis passing through the oxygen atom, bisecting the H—O—H angle in the plane of the molecule (Figure 7.1).

For example, we'll perform a **rotation** (a 180° turn) on **dichloromethane** (CH_2Cl_2). The axis passes through the carbon atom and bisects both the H—C—H and Cl—C—Cl angles.

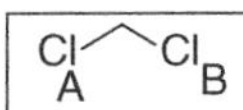	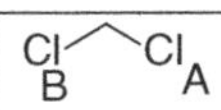
Before C_2 rotation	**After C_2 rotation**

The Cl atoms, though indistinguishable, are labeled to show their positions before and after rotation. The molecule remains unchanged in appearance.

Examples:

- Water, *s-cis-* and *s-trans*-1, 3-butadiene, and naphthalene (Figure 7.1) also possess C_2 axes.

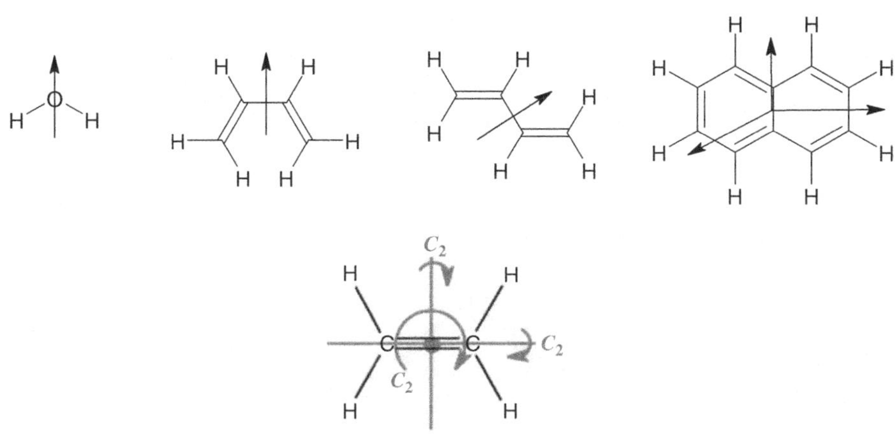

Figure 7.1 Molecules possessing a C_2 axis of symmetry.

In water, the C_2 axis lies in the molecular plane and bisects the single H—O—H bond angle. In *s-cis*-1,3-butadiene, the C_2 axis lies in the molecular plane and bisects the single C—C bond, whereas in the *trans*-isomer the axis is perpendicular to the plane. Naphthalene and ethylene each possess three mutually perpendicular C_2 axes (Figure 7.1).

C. Molecules with C_3 axes

- Ammonia (NH_3) and boron trifluoride (BF_3) both possess a C_3 axis.
  In NH_3 the molecule is pyramidal, while in BF_3 it is planar; in BF_3, the C_3 axis is perpendicular to the molecular plane (Figure 7.2). BF_3 also has three C_2 axes lying in the molecular plane, each passing through a B—F bond.

D. Molecules with C_4 axes

- Xenon tetrafluoride (XeF_4), $[PtCl_4]^{2-}$, and cyclobutadiene are square-planar molecules with a C_4 axis perpendicular to the molecular plane. Each also possesses several C_2 axes (Figure 7.3).

E. Molecules with C_5 and C_6 axes

- Cyclopentadienyl anion $(C_5H_5^-)$ and benzene (C_6H_6) are planar, with C_5 and C_6 axes, respectively, perpendicular to the molecular plane. Rotation by 72° or 60°, respectively, leaves their appearance unchanged (Figure 7.4).

Molecule	(a)	(b)
Ammonia	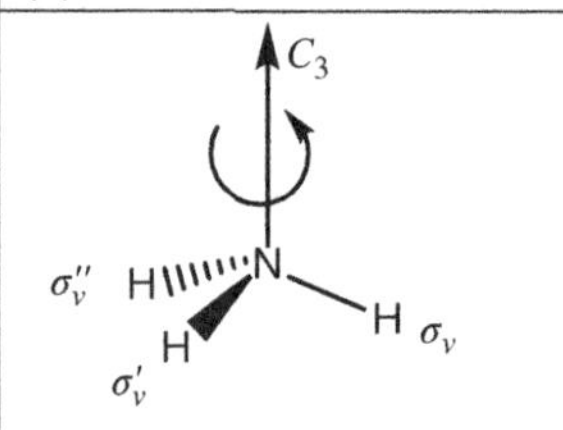	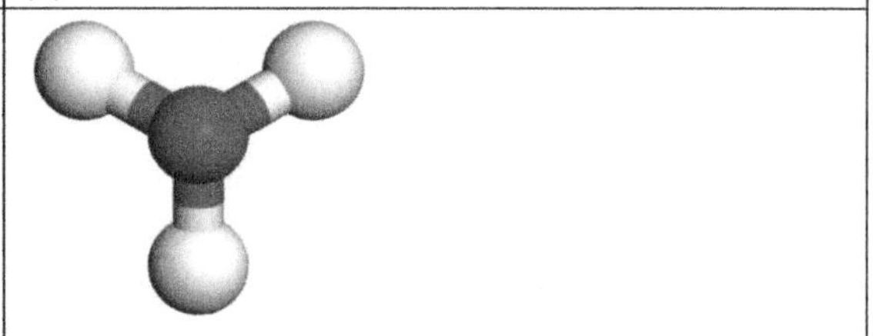
Boron trifluoride	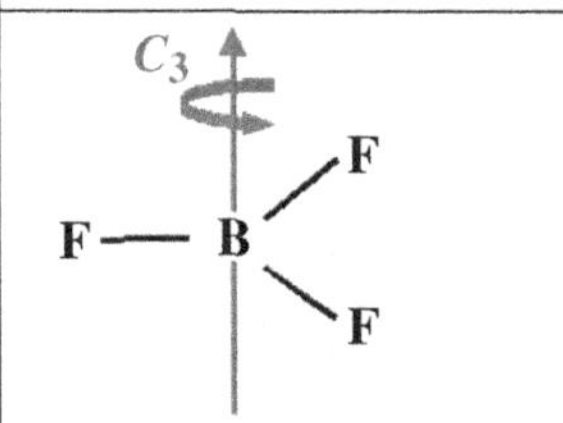	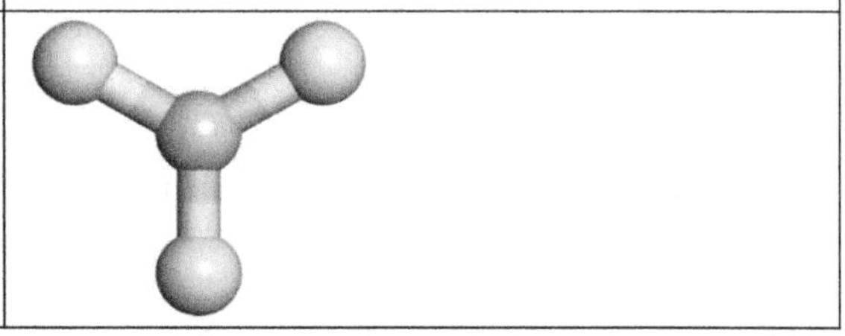

Figure 7.2 *Top*: C_{3v} symmetry of NH_3. *Bottom*: D_{3h} symmetry of BF_3. (a) Schematic showing C_3 axis; (b) optimized geometry from DFT (PBE/DNP) viewed along the C_3 axis.

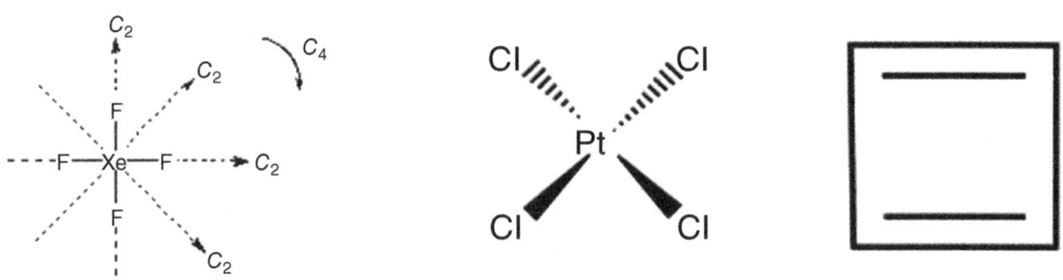

Figure 7.3 Different axes of symmetry in the XeF_4 molecule, $[PtCl_4]^{2-}$, and cyclobutadiene.

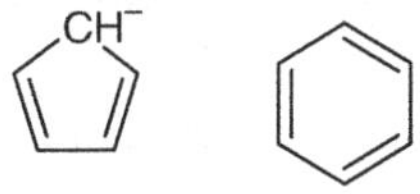

Figure 7.4 Examples of molecules possessing C_5 and C_6 axes of symmetry.

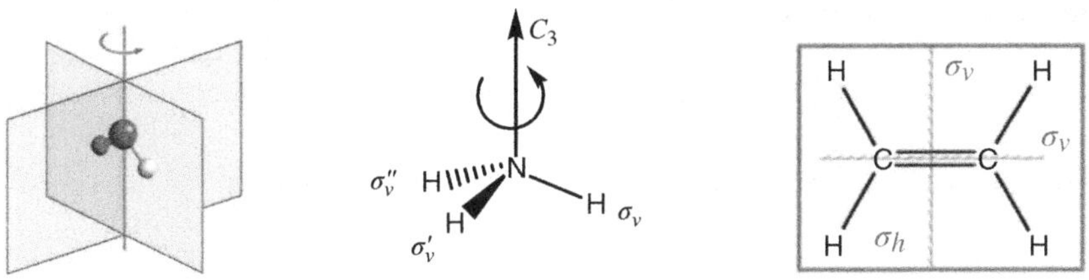

Figure 7.5 Reflection planes (*left to right*: water, ammonia, and ethylene).

F. Molecules with C_∞ axes

- Finally, **all linear molecules** possess a C_∞ axis, since rotation about the bond axis by *any* angle leaves the molecule indistinguishable from itself.

Examples include N_2, O_2, HCl, CO, linear triatomics such as CO_2, N_2O, and OCS, and longer linear molecules such as acetylene.

⚠ Red Flag 7.1

Many students confuse the **total** *number of symmetry operations* **in the group** with the **order** (n) of the rotation axis. A C_n axis generates n operations: rotation by $k \times (360°/n)$ for $k = 1, 2, \ldots, n$, where the case $k = n$ gives the identity E, but n itself refers to the *order of rotation*—not to the total number of symmetry elements.

7.3.3.2 Plane of Symmetry (σ)

A **plane of symmetry** is an imaginary plane that divides a molecule such that reflection in this plane leaves the appearance of the molecule unchanged.

Examples include:

- **Water (H_2O):** Two reflection planes—one containing the molecule and one perpendicular to it, both passing through the C_2 axis.
- **Ammonia (NH_3):** Three vertical planes, each containing one N—H bond and the C_3 axis.
- **Ethylene (C_2H_4):** Three mutually perpendicular reflection planes (Figure 7.5).

A **reflection** operation transforms a molecule through a plane of symmetry (a mirror plane). Each atom is moved to the opposite side of the plane, perpendicular to it, and at an equal distance.

We will use the example of **dichloromethane** again. The molecule has a vertical mirror plane, σ, that contains the C and two Cl atoms. Another plane contains the C atom and bisects the Cl—C—Cl angle. Reflecting across this plane will swap the two hydrogen atoms.

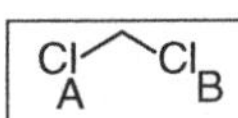	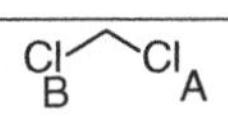
Before σ reflection	**After σ reflection**

Reflection through the plane swaps the positions of the two hydrogen atoms, but the resulting configuration is indistinguishable from the original.

Reflection planes are classified as **vertical** (σ_v) or **horizontal** (σ_h) depending on their orientation with respect to the **principal rotation axis**:

- A plane containing the principal axis is **vertical** (σ_v).
- A plane perpendicular to it is **horizontal** (σ_h).

Fact Box 7.1

Not all σ_h-planes are strictly horizontal, and not all σ_v-planes are strictly vertical.

For example:

- In H_2O and NH_3, the principal axes are C_2 and C_3, respectively; since all reflection planes include these axes, they are vertical.
- In BF_3, the C_3 axis is perpendicular to the molecular plane, and the plane of the molecule itself is the **horizontal plane σ_h**, while the three planes passing through the B—F bonds are **vertical planes σ_v**.

Red Flag 7.2: Common Misconception

Students—and sometimes even instructors—often assume that the *plane of the molecule* in H_2O is a **horizontal plane**, simply because it *appears* horizontal on the page. In fact, the criterion is **geometrical**, not visual. The plane of the H_2O molecule contains the C_2 axis, and therefore it is a **vertical plane** (σ_v). The second plane, which is perpendicular to it and also contains the C_2 axis, is likewise **vertical**. H_2O has **no horizontal mirror plane**.

Fact Box 7.2: The "σ_v Versus σ_h" Demystifier

If the principal axis is C_n (z-axis), then a vertical plane (σ_v) is one that contains the z-axis. It stands up, like a wall. A horizontal plane (σ_h) is one that is perpendicular to the z-axis. It is like a tabletop or a floor.

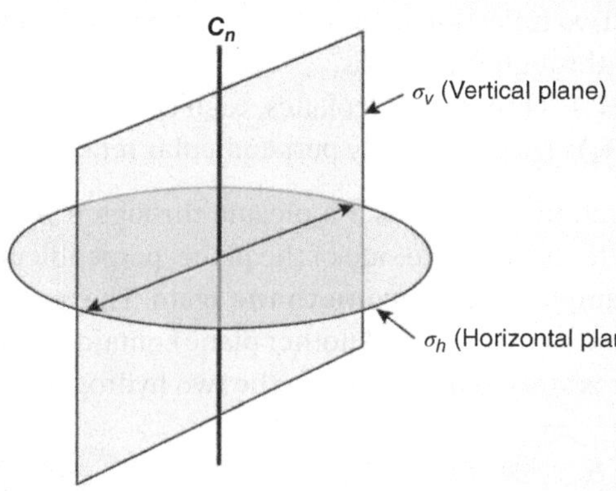

The Quick Check: Ask: "Does this mirror plane contain my main rotation axis?"
YES → It is σ_v (or σ_d).
NO → It is σ_h.
This instantly clarifies that the molecular plane of H_2O, which contains the C_2 axis, is a vertical plane.

When a molecule has more than one rotation axis, the **highest-order axis** is taken as the principal axis.

For naphthalene or ethylene, all three axes are C_2: the principal one is defined as that which leaves the **largest number of atoms unmoved** by the C_2 operation. In ethylene, this axis lies along the C=C bond.

Conventionally, the coordinate system is chosen such that:

- the ***z*-axis** lies along the principal axis,
- the ***yz*-plane** defines the molecular plane, and
- the ***x*-axis** is perpendicular to the molecular plane.

Hence, in ethylene:

- σ_{xz} and σ_{yz} are **vertical planes** (σ_v),
- σ_{xy} is the **horizontal plane** (σ_h).

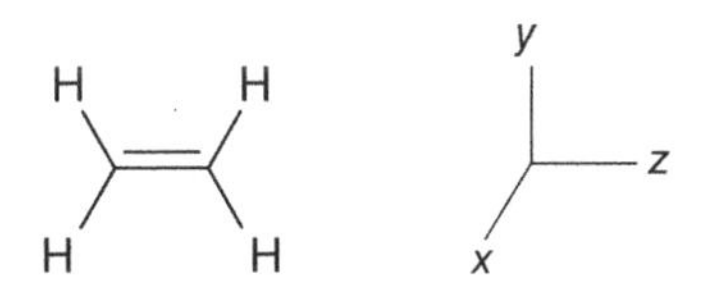

A third type, the **dihedral plane** (σ_d), contains the principal axis and bisects the angle between two secondary C_2 axes perpendicular to it.

A classic example is **allene**, which has three mutually perpendicular C_2 axes; the plane passing through each CH_2 group bisects the secondary axes and is thus a dihedral plane.

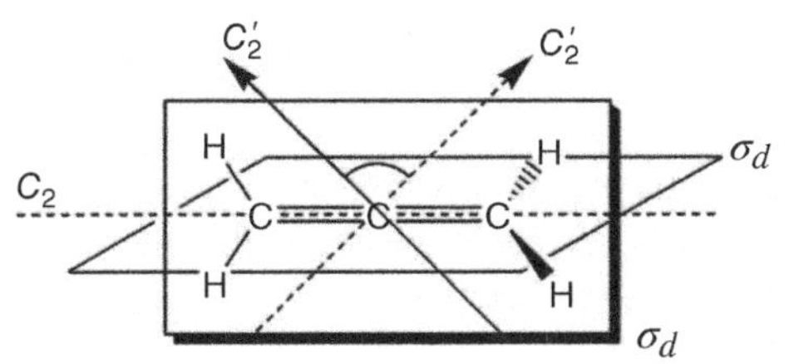

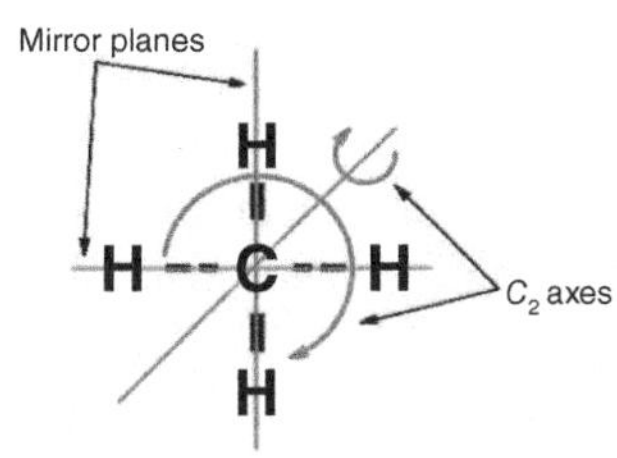

Some molecules possess only a plane of symmetry but no rotation axis. Such molecules are simply described as having a σ-plane, as in **aniline**, where the hydrogens of the —NH_2 group are out of the aromatic plane, yet a mirror plane bisects the group.

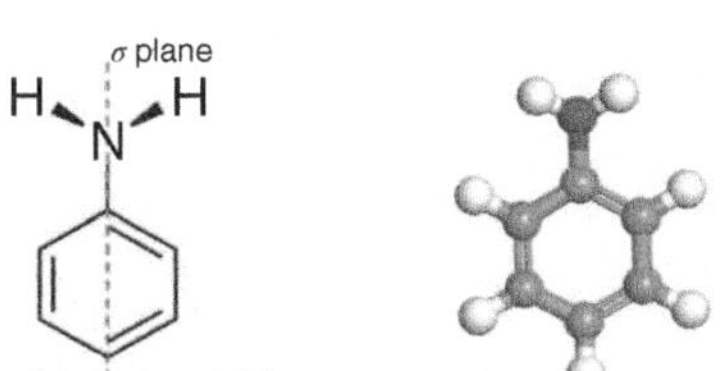

Mirror plane bisecting NH_2 group

Red Flag 7.3

A plane of symmetry must pass through the *entire molecule* and not merely divide one part of it. Students often mistake a local mirror plane (e.g., through one substituent) for a molecular plane of symmetry.

7.3.3.3 Center of Symmetry (i)

A **center of symmetry** (or **inversion center**) is a point within a molecule such that for every atom at a distance r in one direction, there exists an identical atom at the same distance r in the opposite direction.

Inversion is a transformation through a single point at the center of the molecule, called the inversion center. Each atom is projected through this center to an equal distance on the opposite side.

Equivalently, inversion through this point leaves the molecule unchanged.

A molecule can have only one center of symmetry.

An example is the square-planar complex $[\mathbf{PtCl_4}]^{2-}$.

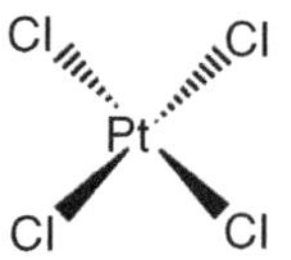

***trans*-1,2-Dichloroethylene** has an inversion center at the midpoint of the C=C bond.

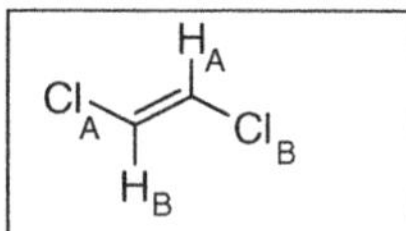	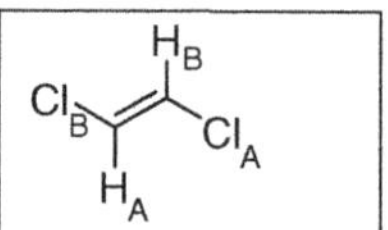
Before *i* inversion	**After *i* inversion**

Red Flag 7.4

Having an inversion center is not the same as having a mirror plane. A molecule can possess one without the other. A classic example is one specific staggered conformer of 1, 2-dichloro-1, 2-difluoroethane (ClFCH—CHFCl), which possesses an inversion center but no mirror plane or rotational axis other than C_1, placing it in the C_i point group.

1,2-Dichloro-1,2-Difluoroethane
(Staggered) C_i

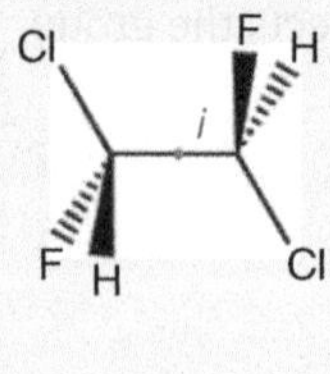

7.3.3.4 Identity Operation (E)

Every molecule possesses an **identity operation** (E), corresponding to doing nothing.

It leaves all points unchanged and serves as the neutral element in the group of symmetry operations.

Though trivial, it is essential for mathematical completeness—it ensures that the set of operations satisfies the closure and identity requirements of group theory.

! Red Flag 7.5

Do not overlook E in counting the total number of operations in a point group. In the C_1 group, it is the only operation, and it must always be included in group tables.

7.3.4 Combination of Symmetry Operations

In group theory, any two symmetry operations of a molecule can be performed successively, one after the other. The result of performing these operations in sequence is itself a valid symmetry operation of the same molecule. This property—that the *product* of any two operations is also an operation in the group—is called **closure** and is one of the defining characteristics of a group.

7.3.5 Rotation–Reflection (S_n)

A **rotation–reflection**, denoted S_n, is a *compound* or *improper* symmetry operation consisting of two successive steps:

1. **rotation** by an angle of $360°/n$ (or $2\pi/n$ radians) about an axis, followed by
2. **reflection** through a plane perpendicular to that axis.

Since symmetry operations do not necessarily commute, it is important to define the sequence unambiguously. The improper rotation S_n is defined as:

1. **rotation** C_n by $2\pi/n$ about an axis, followed by
2. **reflection** σ_h in a plane perpendicular to the rotation axis.

Symbolically,

$$S_n = \sigma_h C_n, \tag{7.1}$$

where the right-to-left notation means C_n is applied first and then σ_h. This specific sequence is part of the definition of the S_n element itself.

This is not a pure symmetry operation, because neither C_n nor σ_h alone leaves the molecule unchanged, but their *combination* may do so.

Examples:

- ***trans*-1,2-Dichloroethylene** possesses an S_2 operation. In this case, a 180° rotation about the C=C bond followed by reflection in the plane perpendicular to the bond is equivalent to inversion through the molecular center.

  Thus, for this molecule, $S_2 = i$.

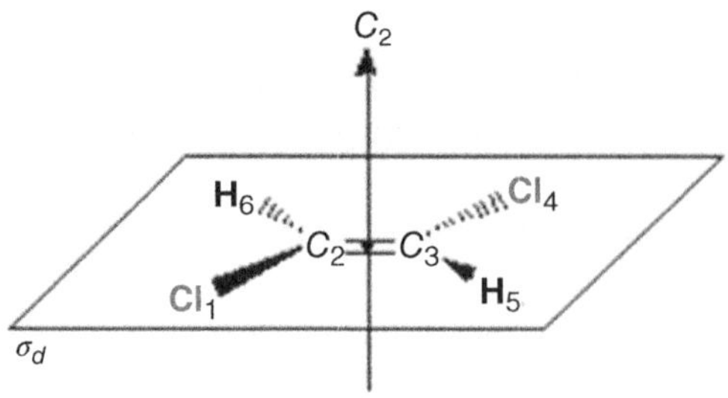

- **Methane (CH_4)** and other **tetrahedral molecules (T_d)** possess higher-order improper axes such as S_4 or S_6, which play an important role in defining their overall symmetry.

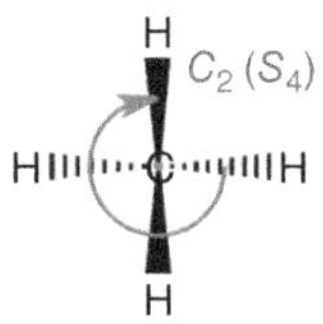

Methane (CH_4) possesses an S_4 axis. We first rotate counterclockwise by 90° (C_4) and then reflect through a perpendicular plane. The intermediate step after rotation is *not* an indistinguishable orientation, but the final result is.

Red Flag 7.6

It is a common mistake to treat S_n as if it were simply a rotation C_n followed by an arbitrary reflection. The defining feature is that the **reflection plane must be perpendicular to the rotation axis**, and both operations are applied successively in a fixed order.

7.3.6 Relationship Among Symmetry Operations

Some S_n operations correspond to simpler, already familiar operations:

$$S_1 = \sigma, \quad S_2 = i. \tag{7.2}$$

Thus, reflection and inversion can both be viewed as special cases of the general improper rotation.

If S_n is present and n is even, then $C_{n/2}$ is also present.

If S_n is present and n is odd, then C_n and σ_h also exist.

7.3.7 Conceptual Significance

The existence of such compound operations illustrates a key aspect of symmetry: combinations of operations are not merely abstract—they often reveal *hidden* symmetries of molecules that are not obvious from geometry alone. For example, the equivalence of S_2 and inversion in *trans*-1, 2-dichloroethylene demonstrates how distinct geometric actions (rotation + reflection) can result in the same final configuration.

Summary of symmetry elements and operations.

Element of symmetry	Symbol	Operation description	Example molecule(s)	Remarks/key points
Identity	E	Leaves the molecule unchanged. Equivalent to "doing nothing"	All molecules	Ensures mathematical completeness of the group Mathematical requirement for the group Corresponds to "doing nothing"
Proper rotation axis	C_n	Rotation by $360°/n$ about an axis through the molecule.	H_2O (C_2), NH_3 (C_3), BF_3 (C_3), XeF_4 (C_4), benzene (C_6)	Every molecule has at least a C_1 axis The axis with the highest n is the **principal axis**
Plane of symmetry	σ	Reflection through a plane dividing the molecule into two mirror-image halves	H_2O (two σ_v), NH_3 (three σ_v), BF_3 (one σ_h, three σ_v)	Planes are classified as σ_v, σ_h, or σ_d depending on orientation relative to the principal axis
Center of symmetry (inversion center)	i	Every atom at (x, y, z) is mapped to an identical atom at $(-x, -y, -z)$	$[PtCl_4]^{2-}$, *trans*-$[Co(en)_2Cl_2]^+$, *trans*-1,2-dichloroethylene	Only one possible per molecule Equivalent to S_2
Improper rotation axis (rotation–reflection)	S_n	Rotation by $360°/n$ followed by reflection in a plane perpendicular to the rotation axis.	*trans*-1,2-dichloroethylene ($S_2 = i$), CH_4 (S_4), benzene (S_6)	Combines proper rotation and reflection Not a pure operation
Principal axis	–	Highest-order rotation axis present in the molecule.	$NH_3 \rightarrow C_3$, $XeF_4 \rightarrow C_4$	Defines the molecular coordinate system (z-axis)

Red Flag 7.7: Notes

- The plane of H_2O is not horizontal—both reflection planes contain the C_2 axis and are therefore vertical (σ_v).
- $S_1 = \sigma$ and $S_2 = i$: these are not distinct symmetry elements but special cases of improper rotations.
- A molecule may possess a plane without a rotation axis (e.g., aniline), but not vice versa for nonlinear molecules.
- Identity (E) must always be included when listing symmetry operations, even though it produces no visible change.

Conceptual Note

- The E, C_n, σ, i, and S_n operations of a molecule are not just geometric features—they form a mathematical group obeying these properties.
- This group structure is what allows chemists to represent symmetry operations by matrices, leading to character tables and selection rules.

7.4 Mathematical Foundations: What Is a Group?

The set of all symmetry operations of a molecule forms a **group** in the mathematical sense. A *group* is a collection of elements (in this case, symmetry operations) that satisfy certain rules when combined. Understanding these properties helps explain why molecular symmetry can be treated mathematically.

7.4.1 Identity Element

A group must contain an **identity operation** (E) that leaves all parts of the molecule unchanged.
Performing E before or after any operation A has no effect:

$$AE = EA = A.$$

In molecular symmetry, E corresponds to "doing nothing." Every molecule, regardless of its symmetry, possesses E.

7.4.2 Closure

If A and B are two symmetry operations of a molecule, their product must also be a symmetry operation of the same molecule. By convention, the product $\boldsymbol{AB}$ signifies that operation $\boldsymbol{B}$ **is performed first, followed by operation** $\boldsymbol{A}$.

This property ensures that all possible combinations of operations still belong to the same set.

Example For H_2O, performing C_2 followed by σ_v gives another σ_v of the same molecule—therefore, the set of operations is **closed**.

Illustration: Closure of Operations in H_2O

Molecule: H_2O (C_{2v} symmetry).
Operations: E, $C_2(z)$, $\sigma_v(xz)$, $\sigma_v'(yz)$.

We'll consider C_2 followed by σ_v'.

Step 1: Define the geometry

Place the molecule in the yz-plane (the molecular plane).
The C_2 axis passes through O and bisects the H—O—H angle, along z.
$\sigma_v(xz)$ = a plane perpendicular to the molecular plane, containing the O atom and bisecting the H—H line.
$\sigma_v'(yz)$ = the molecular plane itself (contains both H atoms).

Step 2: Apply the operations

Start from the original configuration.
Apply $C_2(z)$ (rotation by 180°):
Each hydrogen moves to the position of the other.
Now apply $\sigma_v'(yz)$ (reflection in the molecular plane):
Reflection through $\sigma_v'(yz)$ leaves them in the same position.

Result:
The final configuration is indistinguishable from the one produced by $\sigma_v(xz)$ (reflection in the perpendicular plane, which also exchanges the two hydrogens).

Step 3: Conclusion

Thus, $\sigma_v'(yz)C_2(z) = \sigma_v(xz)$, which is another member of the same group (C_{2v}). Hence, the set is closed—combining any two operations yields another operation within the group.

⚠ Red Flag 7.8: CONVENTION ALERT: The Order of Operations in Group Theory

This is one of the most critical and often confusing points in group theory. The convention is:
In the product AB, the operation on the RIGHT is applied FIRST.
Think of it as function composition: AB means $A(B(\text{molecule}))$. You apply B, then apply A to the result.

Right-to-Left Reading: Always read the product AB as "B then A."
Matrix Multiplication: This convention ensures consistency with matrix representations.
Quantum Mechanics: This is the standard convention for operator algebra in quantum mechanics. An operator like $\widehat{A}\widehat{S}$ acting on a wavefunction, $\widehat{A}\widehat{S}\psi$, means $\widehat{S}$ acts first on ψ, followed by $\widehat{A}$.
Multiplication Tables: In a group multiplication table, the product is calculated as (Row Element) × (Column Element). This means you apply the column operation first, followed by the row operation.
Incorrect Assumption: AB means "A then B."
Correct Convention: AB means "B then A."

Always double-check your intuition against this rule!

7.4.3 Associative Law

The order in which operations are grouped does not affect the final result:

$$(AB)C = A(BC).$$

This is always true for symmetry operations because each can be represented by matrix multiplication, which obeys associativity.

7.4.4 Inverse Element

For every operation A in a group, there exists an **inverse operation** A^{-1} that reverses its effect:

$$AA^{-1} = A^{-1}A = E.$$

For example, the inverse of a rotation by +120° is a rotation by −120° (or by +240°). The inverse of σ or i is the same operation itself, since performing them twice restores the original configuration.

7.4.5 Commutative (or Abelian) Property

Two operations A and B may or may not give the same result when performed in reverse order.

If

$$AB = BA$$

for all elements in the group, the group is **Abelian** or **commutative**.

If not, it is **non-Abelian**.

Examples:

- C_{2v} (H_2O) is **Abelian**, since all operations commute.
- D_{3h} (BF_3) is **non-Abelian**, since a rotation followed by reflection gives a different result than when the order is reversed.

Summary of group properties.

Property	Definition	Molecular example/implication
Identity (*E*)	Doing nothing leaves the molecule unchanged	All molecules
Closure	Product of any two operations is also a valid operation	$\sigma'_v \times C_2 = \sigma_v$ (H_2O)
Associativity	$(AB)C = A(BC)$	Always true for matrix operations
inverse (A^{-1})	Every operation has an opposite, restoring the molecule	$C_3^{-1} = C_3^2$
Commutativity	$AB = BA$ for all operations	True for C_{2v} (Abelian), false for D_{3h} (non-Abelian)

7.5 Group Multiplication Table

The combination of symmetry operations follows definite rules that can be summarized in a **multiplication table**. This table lists all possible successive operations (*products*) among the symmetry elements of a molecule. For the C_{2v} point group, Table 7.1 shows that the set is **closed**.

Worked Example 1: C_{2v} Point Group (Water Molecule)

The symmetry operations of H_2O are

$$E, C_2, \sigma_v(xz), \sigma'_v(yz).$$

The **multiplication table** (Table 7.1) is constructed by performing each operation in sequence (row × column).

For example, applying C_2 followed by σ_v gives σ'_v.

Table 7.1 Group multiplication table for C_{2v}.

C_{2v}	E	C_2	σ_v	σ_v'
E	E	C_2	σ_v	σ_v'
C_2	C_2	E	σ_v'	σ_v
σ_v	σ_v	σ_v'	E	C_2
σ_v'	σ_v'	σ_v	C_2	E

Observations

1. The table is **closed**—every product is one of the four operations already present in the group.
2. The **identity element** (E) acts as a neutral element in every row and column.
3. Each element appears exactly once in every row and column—this is the hallmark of a group.
4. The order of operations does not matter for C_{2v}:

$$C_2\,\sigma_v = \sigma_v' = \sigma_v\,C_2.$$

Hence, C_{2v} is an **Abelian group** (all operations commute).

! Red Flag 7.9

When constructing a multiplication table, always be clear about the order of operations.
In group theory, the product AB usually means "perform B first, then A."
Many textbooks omit this clarification, leading to confusion when comparing results.
In the group multiplication table, it means row × column.

Worked Example 2: C_{3v} Point Group (Ammonia Molecule)

In ammonia, nitrogen is the fixed point and the molecule has a C_3 rotational axis. Both C_3 and C_3^2 operations take place around this same axis.

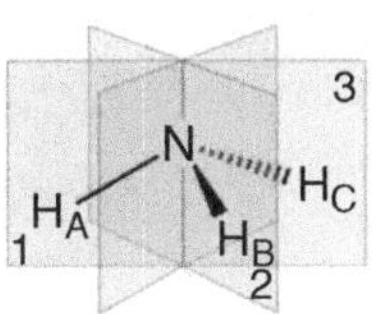

There are three reflection planes, each containing an NH bond and dividing the opposite HNH angle. Each plane is labeled by the hydrogen atom it contains; for example, σ_1 includes H_A.

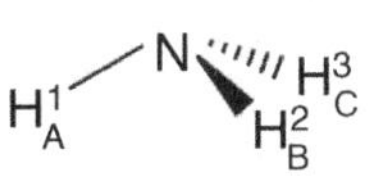

1. The σ_1 reflection swaps atoms H_B and H_C, keeping H_A and the numbers in place. Symmetry elements should stay in their original positions during atom transformations. Previously, we labeled reflection planes using fixed external axes, but that is not always practical (remember: numbers don't move, only letters do. Use of double labels when referencing external axes is inconvenient. Whether you fix numbers or letters is immaterial—just maintain consistency).

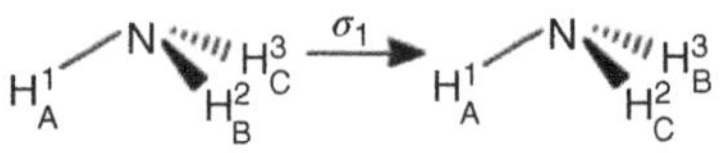

Since $\sigma_1^2 = \sigma_2^2 = \sigma_3^2 = E$, each reflection is self-inverse. For rotations, C_3^2 is the inverse of C_3 because $C_3^3 = C_3^2 C_3 = C_3 C_3^2 = E$. A C_3 rotation means a 120° counterclockwise turn, while C_3^2 represents a 240° counterclockwise (or 120°, clockwise) rotation C_3^{-1}.

In the following examples, we have shown the projection of the hydrogens and labeled them as A, B, and C.

Example A: Rotation, $C_3 \times C_3 = C_3^2$, for ammonia:

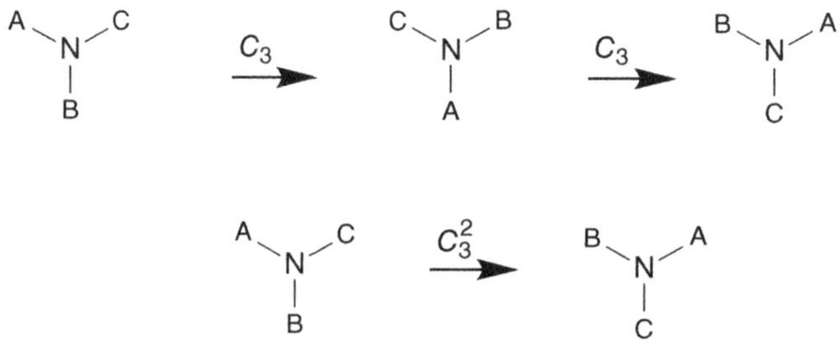

Example B: Reflection, $\sigma_2 \times \sigma_3 = C_3$, for ammonia:

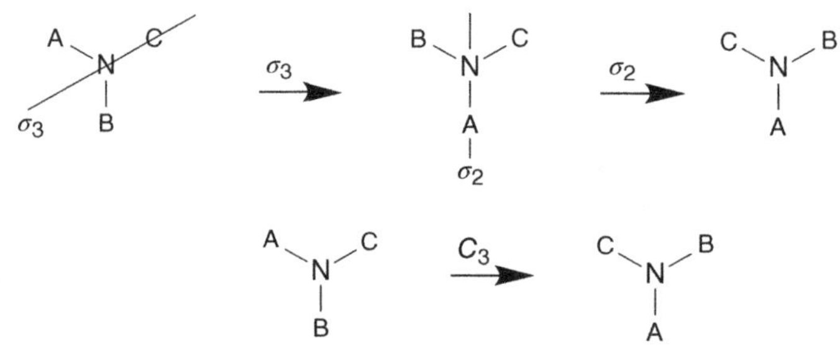

Example C: Both operations combined, $\sigma_1 \times C_3 = \sigma_2$, for ammonia:

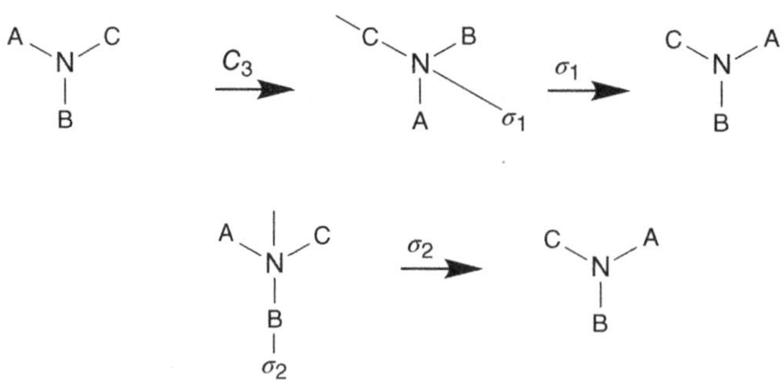

Table 7.2 Group multiplication table for C_{3v}.

C_{3v}	E	C_3	C_3^2	σ_v	σ_v'	σ_v''
E	E	C_3	C_3^2	σ_v	σ_v'	σ_v''
C_3	C_3	C_3^2	E	σ_v''	σ_v	σ_v'
C_3^2	C_3^2	E	C_3	σ_v'	σ_v''	σ_v
σ_v	σ_v	σ_v'	σ_v''	E	C_3	C_3^2
σ_v'	σ_v'	σ_v''	σ_v	C_3^2	E	C_3
σ_v''	σ_v''	σ_v	σ_v'	C_3	C_3^2	E

Note that:

i. The C_3 axis rotation is in a counterclockwise direction.
ii. The product is calculated as (Row Operation) × (Column Operation). Following the standard convention, the **column operation is applied first**, followed by the row operation.
iii. Each row or column is a permutation of another row or column.
iv. No symmetry element occurs more than once in any given row or column.
v. Labels for reflection planes are consistent with those assigned in the original molecule, regardless of updates to the numbering system following an operation (refer to examples B and C).

Based on these definitions and considerations, the group multiplication table can be constructed (Table 7.2). As per convention, the σ-planes are labeled σ_v, σ_v', and σ_v'' respectively.

7.5.1 Chemical Significance

Such tables confirm that symmetry operations form a **group** and also reveal patterns useful later in **character tables** and **representations**.

Fortunately, you don't need to create a group multiplication table each time to determine a molecule's point group. Identifying specific symmetry elements (C_n or σ) is enough. Once the point group is known (see below), its character table provides all the group operations. For character tables of common groups, see Appendix A.

7.6 Classification of Molecules into Point Groups

Every molecule can be assigned to a **point group**, which represents the complete set of symmetry operations that leave at least one point (usually the center of mass) unchanged. Point group classification provides a systematic way to describe molecular symmetry and serves as the foundation for applying group theory to spectroscopy, bonding, and crystal structure.

7.6.1 Step 1: Identify the Symmetry Elements

The first step is to determine all symmetry elements present in the molecule: E, C_n, σ, i, S_n, and their combinations. Remember that all molecules possess E (the identity operation), and that the **principal axis** is the one with the highest value of n.

7.6.2 Step 2: Determine the Principal Axis

Once the rotation axes are identified, the **axis of highest order** becomes the **principal axis** (commonly taken as z). Other symmetry elements—planes and secondary axes—are then described with respect to this axis.

7.6.3 Step 3: Identify Additional Elements

After locating the principal axis, look for:

- **vertical planes** (σ_v) containing the principal axis,
- **horizontal planes** (σ_h) perpendicular to it,
- **dihedral planes** (σ_d) bisecting secondary C_2 axes,
- **inversion centers** (i),
- **improper axes** (S_n).

7.6.4 Step 4: Assign the Point Group

Using these observations, molecules are classified according to the **combinations of symmetry elements** they possess:

1. **Single rotation axis only**
   If a molecule has an n-fold rotation axis (C_n) and no other symmetry elements, the molecule itself belongs to the C_n group.
   Example: H_2O_2 (C_2), $P(Me)_3$ (C_3).

   Hydrogen Peroxide C_2 Triphenylphosphine C_3

   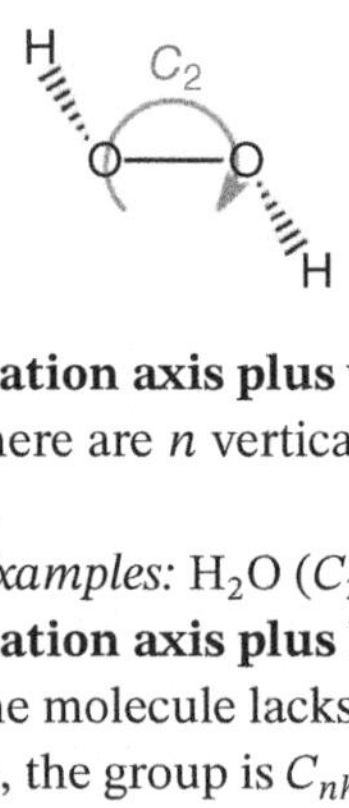

   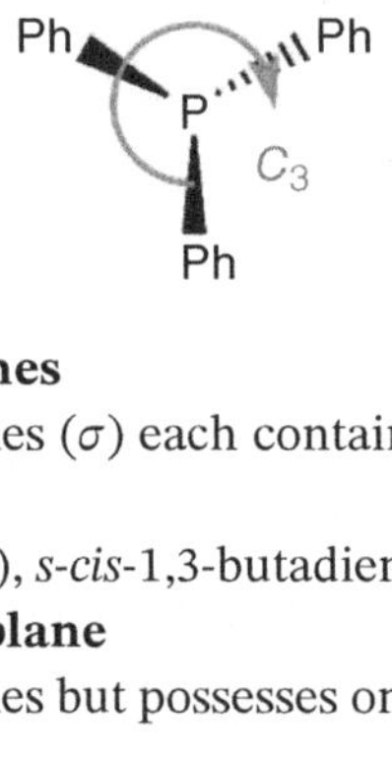

2. **Rotation axis plus vertical planes**
   If there are n vertical mirror planes (σ) each containing the C_n axis, the group is designated C_{nv}.
   Examples: H_2O (C_{2v}), NH_3 (C_{3v}), *s-cis*-1,3-butadiene (C_{2v}).
3. **Rotation axis plus horizontal plane**
   If the molecule lacks vertical planes but possesses one mirror plane **perpendicular** to the C_n axis, the group is C_{nh}.
   Example: s-*trans*-1,3-butadiene (C_{2h}).
   When n is even, inversion (i) is implied—for instance, C_{2h}, C_{4h}, C_{6h}.
4. **Presence of additional C_2 axes**
   If, in addition to the principal C_n axis, there are n twofold axes perpendicular to it, the molecule belongs to the D_n family.
   Depending on accompanying mirror planes:
   a. If a **horizontal plane** (σ_h) is present → D_{nh} (e.g., ethylene, D_{2h}, benzene, D_{6h}).
   b. If **dihedral planes** (σ_d) bisect the angles between secondary C_2 axes → D_{nd} (e.g., allene, D_{2d}).

5. Infinite rotation axes

If the highest-order axis is infinite (C_∞), the group corresponds to linear molecules:

a. Linear molecules with a center of symmetry belong to $D_{\infty h}$—for example, N_2, O_2, C_2H_2 (acetylene). All homonuclear diatomic molecules belong to this point group.

b. Linear molecules without inversion belong to $C_{\infty v}$—for example, HCl, CO, HCN.

Figure 7.6 presents a flowchart for the assignment of a point group to a molecule.

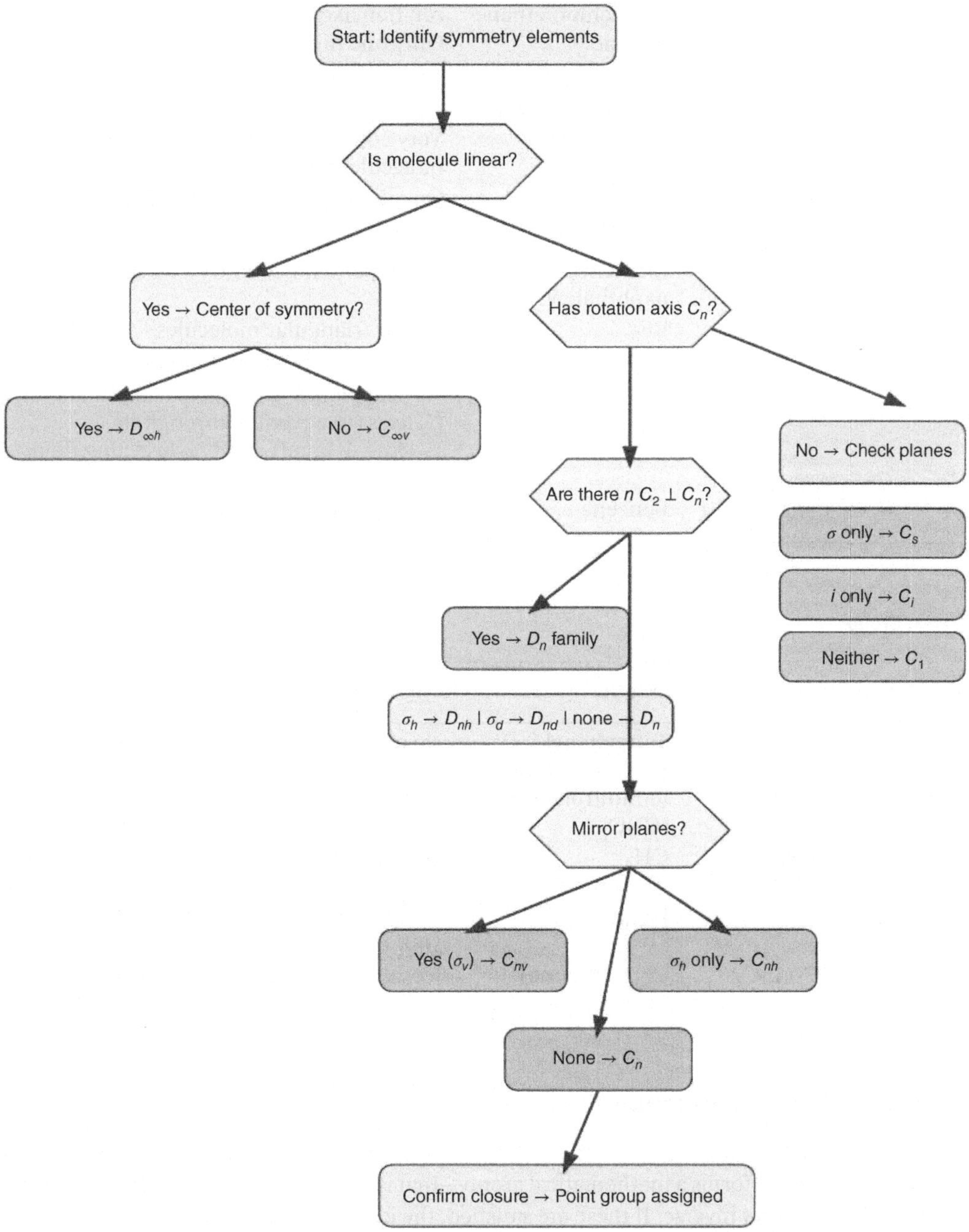

Figure 7.6 Flowchart for point group assignment.

The commonly encountered groups are summarized below.

a. High-symmetry Groups

Group	Characteristic elements	Example molecules	Comments
C_1	Only E	CHBrClF	Asymmetric molecules; no symmetry other than identity
C_i	E, i	*meso*-1,2-Dibromo-1,2-dichloroethane	Centrosymmetric; no mirror planes or rotation axes
C_s	E, σ	Aniline	Only one mirror plane; no rotation axis
C_2	E, C_2	H_2O_2 (gauche), hydrazine (N_2H_4)	Single twofold axis; no planes
C_{2v}	$E, C_2, 2\sigma_v$	H_2O, CH_2Cl_2	Very common; planar or pyramidal molecules with two vertical planes
C_{3v}	$E, C_3, 3\sigma_v$	NH_3, PCl_3	One threefold axis, several vertical planes
C_{4v}	$E, C_4, C_2, 4\sigma_v$	BrF_5, $XeOF_4$	Fourfold rotation, vertical planes
D_{2h}	$E, 3C_2, i, 3\sigma$	Ethylene, naphthalene	Three perpendicular C_2 axes and multiple planes
D_{3h}	$E, C_3, 3C_2, \sigma_h, 3\sigma_v$	BF_3	Planar triangular molecules
D_{4h}	$E, C_4, C_2, i, \sigma_h, 4\sigma_v$	XeF_4	High symmetry; both horizontal and vertical planes. C_4 principal axis and four C_2 axes (two passing through the F—Xe—F bonds and two bisecting them). Square-planar molecules
D_{6h}	$E, C_6, C_2, \sigma_h, 6\sigma_v, i$	Benzene	Hexagonal planar molecules

b. Special Groups

Group	Characteristic elements	Example	Comments
S_4	E, C_2, S_4, S_4^3	Tetraphenylmethane ($C(Ph)_4$), tetranitromethane ($(CNO_2)_4$)	No mirror plane but improper axis present
T_d	$8C_3, 3C_2, 6S_4, 6\sigma_d$	CH_4	Tetrahedral symmetry; highly degenerate energy levels
O_h	$3C_4, 4C_3, 6C_2, i, 9\sigma$	$[Mn(CO)_6]^+$	Octahedral; highest molecular symmetry after icosahedral
I_h	$6C_5, 10C_3, 15C_2, i$	C_{60} (fullerene)	Icosahedral; rare, very high symmetry

7.6.5 Step 5: Verify Group Closure

All symmetry operations identified must satisfy the **group postulates.** The set of all symmetry operations of a molecule forms a mathematical group—that is, they satisfy closure, have an identity element, and each has an inverse. If these are satisfied, the classification is consistent.

These properties have already been discussed earlier and need not be explicitly verified when assigning the point group.

Conceptual Notes

- Molecules with a **single rotation axis** only belong to the C_n family.
- If they also possess **vertical mirror planes**, they belong to C_{nv}.
- If they possess a **horizontal mirror plane**, they belong to C_{nh}.
- If the molecule has two or more C_2 axes perpendicular to the principal axis, it belongs to a D_n family.
- The presence of σ_h or i then gives D_{nh}; the presence of σ_d gives D_{nd}.
- There are no D_{nv}, only D_{nh} or D_{nd}.

Chemistry Connection

Point group classification may appear geometrical, but it is profoundly algebraic: the set of all operations that preserve molecular identity forms a mathematical group. These groups, once identified, allow chemists to predict degeneracies, infrared/Raman activity, and selection rules—turning geometry into quantitative science (see the next chapter).

Before proceeding further, let us walk through the classification of some common molecules that we shall frequently encounter. Referring to the flowchart:

Carbon dioxide	Water	Ammonia	Boron trifluoride
O=C=O Is the molecule linear? Yes Center of symmetry? Yes Conclusion: $\boldsymbol{D_{\infty h}}$	H-O-H Is the molecule linear? No Has rotation axis C_n? Yes, C_2 Are there $nC_2 \perp C_n$? No Mirror planes? Yes, σ_v Conclusion: $\boldsymbol{C_{2v}}$	H, N–H, H (Pyramidal) Is the molecule linear? No Has rotation axis C_n? Yes, C_3 Are there $nC_2 \perp C_n$? No Mirror planes? Yes, σ_v Conclusion $\boldsymbol{C_{3v}}$	F, F-B-F (Planar) Is the molecule linear? No Has rotation axis C_n? Yes C_3 Are there $3C_2 \perp C_3$? Yes, D_n family Mirror planes? Yes, σ_h Conclusion $\boldsymbol{D_{3h}}$

Example 7.1: Point Group Classification of SF_4 (See-saw Shape)

Task: Determine the point group of sulfur tetrafluoride (SF_4) in its common see-saw geometry.

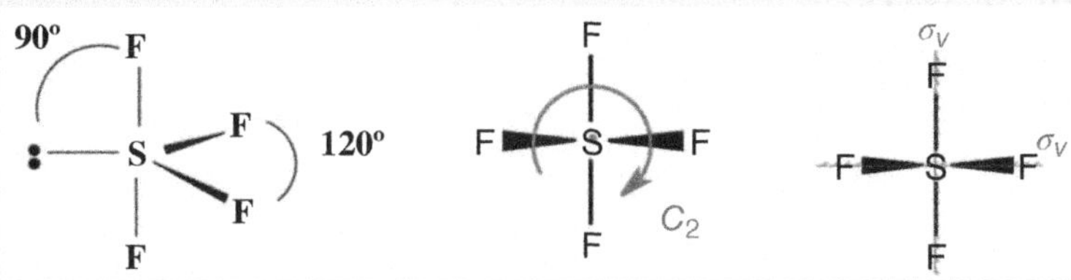

Step-by-step Solution:

Is the molecule linear? No

Has a rotation axis? Yes

Identify the Principal Axis:

Look at the see-saw structure. There is one axis that passes through the S atom and bisects the angle between the two equatorial fluorine atoms. A 180° rotation about this axis brings the molecule into an equivalent orientation.

This is a C_2 axis.

Are there two other C_2 axes perpendicular to this C_2? No

Look for Mirror Planes:

Vertical Mirror Planes (σ_v): There are two vertical mirror planes.

One contains the C_2 axis and the two axial fluorine atoms.

The other contains the C_2 axis and the two equatorial fluorine atoms.

Assign the Point Group:

We have a C_2 principal axis.

We have two vertical mirror planes.

The presence of vertical planes without a horizontal plane places the molecule in the C_{2v} point group.

Conclusion:

SF_4 (see-saw) belongs to the C_{2v} point group.

Note that the lone pair is ignored.

Since this is where students find it hard to assign point groups for all but the simplest molecules, here is an even simpler shortcut.

Shortcut 7.1: The "Quick and Dirty" Point Group Flowchart (for Chemists)

Instead of memorizing the full decision tree, this is a simplified, chemical-intuition-based approach:

Is it linear?

Yes → Does it have a center of inversion?

Yes → $D_{\infty h}$

No → $C_{\infty v}$

Is it a "special shape" you recognize?

Tetrahedral (e.g., CH_4) → T_d

Octahedral (e.g., SF_6) → O_h

Icosahedral (e.g., C_{60}) → I_h

Square planar (e.g., XeF_4) → D_{4h}

For everything else, find the "Best Axis":
Look for the highest-order rotation axis (C_n).

No C_n axis?
Is there a mirror plane (σ)? → C_s
Is there an inversion center (i)? → C_i
If not → C_1

You have a C_n axis. Now, look around it.

Are there nC_2 axes perpendicular to your C_n?
No→ It is a C-type group.
Yes → It is a D-type group.

Now, add the "mirror plane modifier":
For C-type:

Has a σ_h-plane? → C_{nh}
Has n σ_v-planes? → C_{nv}
Has no mirror planes? → C_n

For D-type:

Has a σ_h-plane? → D_{nh}
Has n σ_d-planes (bisecting the C_2 axes)? → D_{nd}
Has no mirror planes? → D_n

Why it works: This focuses on the three questions that matter: linear? Special shape? Best axis and its neighbors? It turns classification from a memorization task into a pattern-recognition game.

Example 7.2: The Quick Point Group Classification Drill

Task: Rapidly assign point groups to the following molecules: CH_2Cl_2, $CHCl_3$, $[Ni(CN)_4]^{2-}$ (square planar), and staggered ethane.

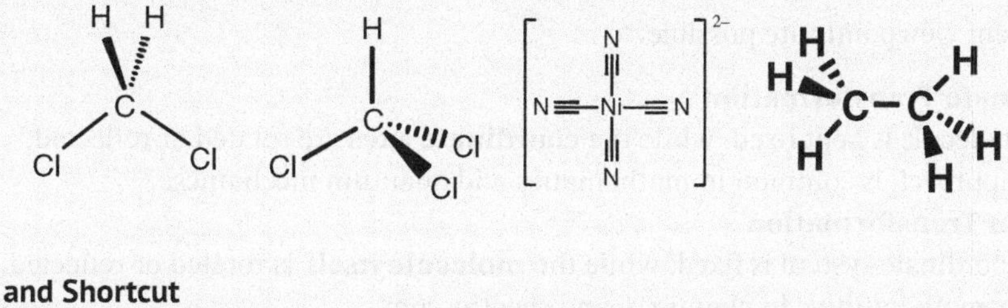

Solution and Shortcut

CH_2Cl_2 (dichloromethane):

Best Axis: It has a C_2 axis bisecting the H—C—H and Cl—C—Cl angles.

Mirror Planes? It has two vertical mirror planes (σ_v) that contain this C_2 axis.
Conclusion: C_2 axis + $2\sigma_v$ planes = C_{2v}.

$CHCl_3$ (chloroform):

Best Axis: It has a C_3 axis along the C—H bond.
Mirror Planes? It has three vertical mirror planes (σ_v), each containing the C_3 axis and one Cl atom.
Conclusion: C_3 axis + $3\sigma_v$ planes = C_{3v}.

$[Ni(CN)_4]^{2-}$ (square planar):

Special Shape? Yes, square planar.
Shortcut: Square planar complexes are almost always D_{4h}. (Has C_4, C_2's, σ_h, σ_v, i.)

Staggered Ethane (C_2H_6):

Best Axis: The C—C bond is a C_3 axis.
Perpendicular C_2's? Yes, there are three C_2 axes perpendicular to the main C_3 axis.
Mirror Planes? No horizontal mirror plane (σ_h). Three dihedral planes (σ_d) that bisect the H—C—H angles.
Conclusion: C_3 axis + 3 perpendicular C_2's + $3\sigma_d$ = D_{3d}.

Each distinct operation corresponds to a transformation that can be represented **algebraically**—and this naturally leads to **matrix representations**, the topic of the next section.

7.7 Matrix Representation of Symmetry Operations

Every symmetry operation can be expressed mathematically as a **linear transformation** acting on the coordinates (or on functions of coordinates) of the molecule, as discussed in previous chapters.

A matrix provides a convenient and precise way to represent such transformations.

7.7.1 Coordinate Versus Position Transformation

Two equivalent viewpoints are possible:

1. **Coordinate Transformation**
   The molecule is kept fixed, while the **coordinate axes** are rotated or reflected.
   This approach is common in mathematics and quantum mechanics.
2. **Position Transformation**
   The coordinate system is fixed, while the **molecule itself** is rotated or reflected.
   This is more intuitive in chemistry and spectroscopy.

Both viewpoints yield the same mathematical results—they differ only in convention. We'll follow the second here.

Worked Example 3: Rotation About the *z*-axis

A rotation by an angle θ about the z-axis is represented by the matrix (see Chapter 3):

$$R_z(\theta) = \begin{bmatrix} \cos\theta & -\sin\theta & 0 \\ \sin\theta & \cos\theta & 0 \\ 0 & 0 & 1 \end{bmatrix}. \tag{7.3}$$

This matrix transforms a coordinate vector (x, y, z) to (x', y', z') after rotation.

Worked Example 4: Reflection in the *xz*-plane

A reflection in the xz-plane changes the sign of y, while x and z remain unchanged:

$$\sigma_{xz} = \begin{bmatrix} 1 & 0 & 0 \\ 0 & -1 & 0 \\ 0 & 0 & 1 \end{bmatrix}. \tag{7.4}$$

Similarly, reflection in the yz-plane changes $x \to -x$.

Worked Example 5: Inversion

Inversion through the center of symmetry multiplies all coordinates by -1:

$$i = \begin{bmatrix} -1 & 0 & 0 \\ 0 & -1 & 0 \\ 0 & 0 & -1 \end{bmatrix}. \tag{7.5}$$

7.7.2 Physical Meaning

Each matrix expresses how the coordinates of a point, vector, or orbital transform under a symmetry operation. These **matrices form a representation of the group**, obeying the same multiplication rules as the group itself:

$$D(A)D(B) = D(AB),$$

where D denotes the matrix representation of operation A or B.

Conceptual Note

When we later apply these matrices to atomic orbitals, vibrational coordinates, or electronic wavefunctions, we obtain character tables that link mathematical symmetry to observable properties such as IR or Raman activity, orbital degeneracies, and selection rules.

As we have seen, every symmetry operation can be represented by a **matrix** that acts on a coordinate vector or on a function of coordinates. These matrices obey the same multiplication laws as the corresponding symmetry operations, and their use forms the basis for the study of **group representations**.

7.7.3 Nature of Transformation Matrices

Most of the transformation matrices encountered in molecular symmetry are **orthogonal** (for purely real operations) or, more generally, **unitary** (for complex transformations such as quantum-mechanical operators).

A unitary matrix U satisfies

$$U^{\dagger}U = I,$$

which ensures that lengths, angles, and inner products are preserved.

Thus, symmetry operations correspond to **rigid rotations or reflections** in coordinate space—they do not distort molecular geometry.

7.7.4 Successive Operations and Matrix Products

When several operations are performed in succession, the corresponding matrices are multiplied **in reverse order** of application. If operations A, B, C, D are applied successively to a molecule, the resulting transformation is

$$D(CBA)\,\mathbf{x} = \mathbf{X},$$

or, equivalently,

$$\mathbf{X} = DCBA\,\mathbf{x},$$

where $\mathbf{x}$ is the original coordinate vector.

Thus, the **last operation written acts first**. This convention is essential when comparing algebraic and geometric sequences of operations.

Example 7.3 uses matrix representations to demonstrate the equivalence of S_2 and i operations.

Example 7.3: Equivalence of S_2 and Inversion

Consider a rotation by 180° (C_2) about the z-axis followed by reflection in the xy-plane.

Rotation matrix ($\theta = 180°$) in Equation (7.3):

$$C_2(z) = \begin{bmatrix} -1 & 0 & 0 \\ 0 & -1 & 0 \\ 0 & 0 & 1 \end{bmatrix}.$$

Reflection matrix:

$$\sigma_{xy} = \begin{bmatrix} 1 & 0 & 0 \\ 0 & 1 & 0 \\ 0 & 0 & -1 \end{bmatrix}.$$

The product (reflection × rotation) gives

$$S_2 = \sigma_{xy}C_2(z) = \begin{bmatrix} 1 & 0 & 0 \\ 0 & 1 & 0 \\ 0 & 0 & -1 \end{bmatrix}\begin{bmatrix} -1 & 0 & 0 \\ 0 & -1 & 0 \\ 0 & 0 & 1 \end{bmatrix} = \begin{bmatrix} -1 & 0 & 0 \\ 0 & -1 & 0 \\ 0 & 0 & -1 \end{bmatrix} = i.$$

Hence, an S_2 operation (rotation + reflection) is equivalent to inversion, confirming the earlier geometric argument for *trans*-1, 2-dichloroethylene.

Key Takeaways

- Symmetry operations can be treated as mathematical transformations that preserve the metric properties of space.
- Their matrices are orthogonal or unitary, guaranteeing invariance of physical quantities such as distances and scalar products.
- Products of matrices correspond exactly to successive symmetry operations.
- These matrices provide the foundation for constructing reducible and irreducible representations, which in turn lead to character tables and the prediction of spectroscopic selection rules.

7.8 Matrix Representation of a Group

So far, we have seen that each symmetry operation of a molecule can be expressed as a **matrix** that performs a definite transformation on the coordinates or on functions of coordinates.

The complete set of these matrices, when combined under matrix multiplication, behaves exactly like the symmetry operations themselves.

This set of matrices is called a **representation of the group**.

7.8.1 Representation of a Group

A **representation** of a group is a set of matrices that correspond to the symmetry operations of that group and reproduce its multiplication table through matrix multiplication.

If $R(A)$ and $R(B)$ are the matrices representing operations A and B, then

$$R(A)R(B) = R(AB)$$

for all operations A, B in the group.

This ensures that the matrices "behave" like the symmetry operations they represent.

Each representation is thus a *matrix realization* of the abstract group.

7.8.2 Basis of a Representation

To build a representation, we need to specify **what is being transformed**—that is, the **basis**. A basis may consist of coordinates, vectors, orbitals, atomic displacements, or any set of quantities that transform in a definite way under the symmetry operations.

Examples of Common Bases:

- Cartesian coordinates $(x, y, z) \rightarrow$ used for rotational and reflection operations.
- Atomic orbitals $(s, p_x, p_y, p_z) \rightarrow$ used in molecular orbital theory.
- Vibrational displacements $\rightarrow$ used in infrared and Raman spectroscopy.

The number of basis functions determines the **dimension of the representation**. A representation using three coordinates (x, y, z) is a **three-dimensional representation**.

7.8.3 Matrices for Common Point Groups in Cartesian Coordinates

a. C_{2v} (H_2O)

Molecule in the yz-plane.

> **Red Flag 7.10**
>
> In C_{2v}, the notation σ_v and σ_v' is confusing. One of the vertical planes is the molecular plane, and the other is perpendicular to it. In some of the older textbooks, the molecular plane was chosen as xz, so σ_v became the molecular plane. The modern convention is to place the molecule in the yz-plane, so σ_v' is the molecular plane.

Operations: $E, C_2(z), \sigma_v(xz), \sigma_v'(yz)$.

$C_2(z)$: diag$(-1, -1, 1)$ (changes $x \to -x, y \to -y$).

$\sigma_v'(yz)$ is the molecular plane: diag$(-1, 1, 1)$ (changes $x \to -x$).

$\sigma_v(xz)$ is perpendicular to the molecular plane: diag$(1, -1, 1)$ (changes $y \to -y$).

$$\begin{aligned} E &= \begin{bmatrix} 1 & 0 & 0 \\ 0 & 1 & 0 \\ 0 & 0 & 1 \end{bmatrix}, C_2(z) = \begin{bmatrix} -1 & 0 & 0 \\ 0 & -1 & 0 \\ 0 & 0 & 1 \end{bmatrix}, \\ \sigma_v(xz) &= \begin{bmatrix} 1 & 0 & 0 \\ 0 & -1 & 0 \\ 0 & 0 & 1 \end{bmatrix}, \sigma_v'(yz) = \begin{bmatrix} -1 & 0 & 0 \\ 0 & 1 & 0 \\ 0 & 0 & 1 \end{bmatrix}. \end{aligned} \tag{7.6}$$

Each matrix represents how the coordinates (x, y, z) transform under the respective symmetry operation.

All are orthogonal, real, and their products reproduce the C_{2v} multiplication table.

b. C_{3v} (NH_3)

Operations: $E, C_3, C_3^2, 3\sigma_v$.

For a rotation by 120° about the z-axis (Equation 7.3),

$$C_3 = \begin{bmatrix} -\frac{1}{2} & -\frac{\sqrt{3}}{2} & 0 \\ \frac{\sqrt{3}}{2} & -\frac{1}{2} & 0 \\ 0 & 0 & 1 \end{bmatrix}, \ C_3^2 = \begin{bmatrix} -\frac{1}{2} & \frac{\sqrt{3}}{2} & 0 \\ -\frac{\sqrt{3}}{2} & -\frac{1}{2} & 0 \\ 0 & 0 & 1 \end{bmatrix}. \tag{7.7}$$

A reflection through the plane $xz(\sigma_v)$ changes $y \to -y$:

$$\sigma_v = \begin{bmatrix} 1 & 0 & 0 \\ 0 & -1 & 0 \\ 0 & 0 & 1 \end{bmatrix}. \tag{7.8}$$

The other two σ_v-planes are obtained by rotating this matrix by $\pm 120°$, yielding three equivalent reflection operations.

7.9 Order, Similarity Transformations, and Matrix Characters

Before analyzing reducible and irreducible representations, it is useful to establish a few key mathematical ideas that underpin the group–matrix connection.

7.9.1 Order of a Group and of an Element

- The **order of a group** (h) is the total number of symmetry operations it contains.
- For example:

Molecule	Point Group	Order (h)
H_2O	C_{2v}	4
NH_3	C_{3v}	6
BF_3	D_{3h}	12
Benzene	D_{6h}	24

Thus, the order of a group equals the number of distinct elements in its multiplication table.

- The **order of an element (operation)** is the number of times it must be applied successively to return the molecule to its original configuration.

For instance:

$C_2^2 = E \rightarrow$ order $= 2$.

$C_3^3 = E \rightarrow$ order $= 3$.

σ and i each has order 2.

The order of every element divides the order of the group.

7.9.2 Conjugate Elements and Classes

Two operations A and B are said to be **conjugate** if there exists another operation C in the same group such that

$$C^{-1}AC = B.$$

Conjugate operations are similar in their effect and are therefore said to belong to the same **class**. All operations belonging to the same class (e.g., all vertical mirror planes in ammonia) are equivalent from the standpoint of the group's symmetry properties and share the same character in character tables.

Example: In C_{3v}, the three vertical planes σ_v, σ_v', σ_v'' are conjugate to one another and form one class.

Likewise, C_3 and C_3^2 form another class.

Fact Box 7.3:

1 Conjugacy of the three σ_v-planes

Let us take σ_v as the reference operation (say, the plane passing through one vertex of a trigonal pyramid like NH_3).

Let us apply conjugation by C_3:

$$C_3^{-1}\,\sigma_v\,C_3 = ?.$$

From the group multiplication table (Table 7.2),

$$C_3^{-1}\,\sigma_v\,C_3 = C_3^{-1}\,(\sigma_v\,C_3) = C_3^{-1}\,\sigma'_v = \sigma''_v.$$

2 Conjugacy of C_3 and C_3^2

Now test C_3 and C_3^2.

Let $A = C_3$. We must find C such that:

$$C^{-1}\,C_3\,C = C_3^2.$$

Let's choose $C = \sigma_v$. Then:

$$\sigma_v^{-1}\,C_3\,\sigma_v = ?.$$

Using Table 7.2,

$$\sigma_v^{-1}(C_3\,\sigma_v) = \sigma_v^{-1}\,\sigma''_v = C_3^2.$$

3 Identity element

The identity element E always forms a class by itself:

Class 3: (E).

✅ Final Classification of Classes in C_{3v}.

Class	Operations	Number of elements
I	E	1
II	C_3, C_3^2	2
III	$\sigma_v, \sigma'_v, \sigma''_v$	3

Therefore:

$\sigma_v, \sigma'_v, \sigma''_v$ are mutually conjugate because rotation (C_3) cyclically permutes them.

C_3 and C_3^2 are conjugate because reflection (σ_v) interchanges them.

For a detailed geometric proof of these class relationships, see Appendix J.4.

7.9.3 Conjugation in C_{3v}: Matrix Proof

We define the principal axis C_3 along the z-axis and take the yz-plane as the reference σ_v-plane,

$$C_3 = \begin{bmatrix} -\frac{1}{2} & -\frac{\sqrt{3}}{2} & 0 \\ \frac{\sqrt{3}}{2} & -\frac{1}{2} & 0 \\ 0 & 0 & 1 \end{bmatrix}, \quad C_3^{-1} = \begin{bmatrix} -\frac{1}{2} & \frac{\sqrt{3}}{2} & 0 \\ -\frac{\sqrt{3}}{2} & -\frac{1}{2} & 0 \\ 0 & 0 & 1 \end{bmatrix},$$

$$\sigma_v = \begin{bmatrix} -1 & 0 & 0 \\ 0 & 1 & 0 \\ 0 & 0 & 1 \end{bmatrix}.$$

Then

$$C_3^{-1}\,\sigma_v\,C_3 = \begin{bmatrix} \frac{1}{2} & -\frac{\sqrt{3}}{2} & 0 \\ -\frac{\sqrt{3}}{2} & -\frac{1}{2} & 0 \\ 0 & 0 & 1 \end{bmatrix}.$$

This matrix represents reflection through a plane rotated by 120° relative to σ_v, that is, σ_v''. Hence,

$$C_3^{-1}\sigma_v\,C_3 = \sigma_v''.$$

Repeating the operation cyclically yields

$$C_3^{-1}\sigma_v' C_3 = \sigma_v'', C_3^{-1}\sigma_v'' C_3 = \sigma_v'.$$

Key Takeaways

1. The inverse C_3^{-1} is essential because conjugation represents a **change of basis**, not a physical rotation.
2. The expression $A^{-1}XA$ transforms an operation X into its equivalent form in the original coordinate frame.
3. The three σ_v operations of C_{3v} are mutually conjugate since rotation cyclically permutes them.

7.9.4 The Identity Operation and Classes

The **identity operation** (***E***) is always **by itself** in its own class, because no other operation satisfies the relation $C^{-1}EC = A$ unless $A = E$.

Hence, every group has at least one class—the ***E*** **class**.

7.9.5 Similarity Transformations

We have seen in previous chapters that two matrices A and B are said to be **similar** if there exists a nonsingular matrix P such that

$$B = P^{-1}AP.$$

Such matrices represent the same operation in **different coordinate systems or bases**.

Hence, similarity transformation corresponds to a **change of basis**—the physical operation remains the same, only the frame of reference changes.

Example: The C_2 operation in water may be represented by one matrix in Cartesian coordinates (x, y, z) and by another in symmetry-adapted coordinates $(a_1,\ b_1,\ b_2)$ (Chapter 8); the two matrices are *similar*.

Conceptual Note

Similarity transformations preserve the eigenvalues and trace of a matrix—and hence preserve the character, a quantity that uniquely identifies a representation.

7.9.6 Trace and Character of a Matrix

The **trace** of a square matrix is the sum of its diagonal elements. For a 3×3 matrix $D(R)$ representing symmetry operation R:

$$\mathrm{Tr}[D(R)] = D_{11} + D_{22} + D_{33}.$$

The **character** (χ) of an operation in a given representation is defined as the trace of its matrix:

$$\chi(R) = \mathrm{Tr}[D(R)].$$

Characters provide a compact way of comparing representations, since:

- They remain unchanged under similarity transformations.
- They are additive: for a **reducible representation**, the character equals the sum of the characters of its components.
- The complete set of characters for all operations forms a **character table**.

Example: Characters for C_{2v} using (x, y, z) basis.

Operation	Matrix	Character (Trace)
E	diag(1, 1, 1)	3
$C_2(z)$	diag(−1, −1, 1)	−1
$\sigma_v(xz)$	diag(1, −1, 1)	1
$\sigma_v'(yz)$	diag(−1, 1, 1)	1

Thus, the character set for this three-dimensional representation is (3, −1, 1, 1).

Summary of Key Terms

Term	Meaning	Example/remark
Representation	Set of matrices corresponding to group operations	Rotation matrices for C_{3v}
Basis	Quantities transformed by the operations (x, y, z; orbitals; vibrations)	Cartesian or orbital basis
Dimension of representation	Number of basis functions	3 for (x, y, z)
Order (h)	Total number of symmetry operations in the group	$h(C_{2v}) = 4$
Conjugate elements	Related by $C^{-1}AC = B$	$\sigma_v, \sigma_v', \sigma_v''$ in C_{3v}
Class	Set of mutually conjugate elements	Each class ↔ one irreducible representation
***E* Class**	Identity element's class	Always single-member

Conceptual Note

While symmetry operations act geometrically, representations act algebraically.

Each representation provides a mathematical lens through which we can study how molecular orbitals, vibrations, or electronic transitions respond to symmetry.

In subsequent sections, these ideas lead directly to the character table, where each irreducible representation corresponds to a distinct physical behavior.

7.10 Reducible and Irreducible Representations

Now that we can describe symmetry operations by matrices, we can analyze whether a given representation can be simplified or "reduced" into smaller independent parts.

7.10.1 Reducible Representation

A representation is said to be **reducible** if there exists a similarity transformation that converts all its matrices simultaneously into **block-diagonal form**:

$$P^{-1}D(R)P = \begin{bmatrix} D_1(R) & 0 \\ 0 & D_2(R) \end{bmatrix}.$$

Each smaller block ($D_1, D_2, \ldots$) behaves independently under all group operations.

Such a reducible representation can therefore be expressed as a **direct sum** of smaller representations.

Example: The three-dimensional (x, y, z) representation of C_{3v} can be decomposed into one 1D representation ($z \to A_1$) and one 2D representation ($x, y \to E$).

7.10.2 Irreducible Representation

If no such similarity transformation exists—that is, if the matrices cannot be simultaneously block-diagonalized—the representation is **irreducible**.

These are the *fundamental building blocks* of group theory, just as prime numbers are for integers.

Every reducible representation can be written as a **sum of irreducible representations**:

$$\Gamma_{\text{reducible}} = a_1\Gamma_1 + a_2\Gamma_2 + \cdots + a_n\Gamma_n,$$

where a_i is the number of times irreducible representation Γ_i occurs.

7.10.3 Relationship Between Classes and Representations

A fundamental theorem of group theory states that:

*The number of **irreducible representations** of a group equals the number of its **classes**.*

This means that in a group like C_{3v}, which has three classes (E, $2C_3$, and $3\sigma_v$), there are **three irreducible representations**—conventionally labeled A_1, A_2, and E.

7.10.4 Physical Meaning

- In molecular orbital theory, irreducible representations correspond to **orbital symmetries** (A_1, A_2, E, T_1, etc.). Remember, in crystal field theory, the d-orbitals of an octahedral complex are split into e_g and t_{2g}. These symbols come from group theory.
- In spectroscopy, they determine **allowed** or **forbidden** transitions.
- In vibrational analysis, each normal mode transforms as an irreducible representation of the molecular point group.

7.10.5 The Order and Character Connection

Since characters are invariant under similarity transformations, they serve as powerful tools for:

- Distinguishing nonequivalent irreducible representations.
- Decomposing reducible ones into irreducible components using the **reduction formula** (next section).

Summary Table

Concept	Definition	Significance
Order of group (h)	Number of symmetry operations	Used in normalization and reduction formula
Order of element	Number of times an operation must be repeated to give E	$C_2 \to 2, C_3 \to 3$
Similarity transformation	$B = P^{-1}AP$	Changes basis, preserves character
Character (χ)	Trace of transformation matrix	Unique, basis-independent identifier
Reducible representation	Can be simplified into blocks	Represents composite behavior
Irreducible representation	Cannot be simplified	Fundamental symmetry type

7.11 The Great Orthogonality Theorem (GOT)

The **Great Orthogonality Theorem** is the cornerstone of group theory. It provides the mathematical framework that governs all the relationships among the irreducible representations of a group.

7.11.1 Statement of the Theorem

For a group of order h with elements $R_1, R_2, \ldots, R_h$, let $D^{(\alpha)}(R)$ and $D^{(\beta)}(R)$ be the matrix representations of the operation R in the irreducible representations α and β, respectively, each of dimensions l_α and l_β.

The theorem states that the matrix elements of these irreducible representations obey the **orthogonality condition**

$$\sum_R D_{ij}^{(\alpha)}(R)\, D_{mn}^{(\beta)}(R)^* = \frac{h}{l_\alpha}\, \delta_{\alpha\beta}\, \delta_{im}\, \delta_{jn},$$

where δ is the Kronecker delta and h is the order of the group.

7.11.2 Physical Interpretation

- The theorem ensures that **different irreducible representations are orthogonal**—they represent *independent symmetry behaviors.*
- It also implies that **matrix elements within the same irreducible representation** are orthogonal if they belong to different rows or columns.
- These orthogonality relationships allow one to derive **characters, character tables**, and **projection operators** used in quantum chemistry.

7.11.3 Simplified Character Orthogonality

Since characters (χ) are traces of the matrices, the theorem simplifies to

$$\sum_R \chi^{(\alpha)}(R)\,\chi^{(\beta)}(R) = h\,\delta_{\alpha\beta}, \tag{7.9}$$

or, for classes:

$$\sum_i n_i\,\chi_i^{(\alpha)}\,\chi_i^{(\beta)} = h\,\delta_{\alpha\beta},$$

where n_i is the number of operations in class i.

This simpler relation is used to construct and verify **character tables.**

⚠ Red Flag 7.11: The "Class Size" Trap

When using the orthogonality formula (Equation 7.9) or reducing representations, **do not forget** the factor n_i (the number of operations in the class).

In character tables, this number appears as the coefficient in the top row (e.g., in C_{3v}, the headers are E, $2C_3$, $3\sigma_v$).

Wrong: Summing $\chi^{(\alpha)} \times \chi^{(\beta)}$ directly from the table.
Right: Summing $n_i \times \chi^{(\alpha)} \times \chi^{(\beta)}$.

If you simply sum the product of the characters without weighting them by the class size (n_i), the result will not equal the order of the group (h) or zero, and the theorem will appear to fail.

The use of this theorem will become clear when we study some examples in the next chapter.

7.12 Character Tables

The **character table** condenses all the essential information about the symmetry of a point group in a single, compact format (Appendix A). Each row corresponds to an **irreducible representation**, and each column corresponds to a **class of symmetry operations**. The entries are the **characters (traces)** for those operations.

- **Classes (columns):** Each unique type of symmetry operation.
- **Irreducible representations (rows):** Distinct symmetry types (A, B, E, T, etc.).
- **Characters:** Traces of representative matrices.
- **Last column:** It shows how typical physical quantities (coordinates, rotations, orbitals) transform.

Structure of a character table.

C_{2v}	E	C_2	$\sigma_v(xz)$	$\sigma'_v(yz)$	Linear & quadratic functions/rotations
A_1	1	1	1	1	z, x^2, y^2, z^2
A_2	1	1	−1	−1	R_z
B_1	1	−1	1	−1	x, R_y
B_2	1	−1	−1	1	y, R_x

Observe that z is categorized under A_1. Referring to the matrices for C_{2v} in Equation (7.6), it is evident that the z-coordinate consistently remains unchanged (with a character of +1), thereby confirming this classification.

Key Observations

1. The **number of irreducible representations = number of classes.**
2. The **sum of the squares of dimensions of irreducible representations** equals the **order of the group (h)**:

$$\sum_i l_i^2 = h.$$

   One-dimensional representations (A, B) are common in small groups; multidimensional ones (E, T) occur in highly symmetric molecules like CH_4 or BF_3.
3. The first row in any character table is the totally symmetric representation (A_1, Σ_1 [for linear molecules] or A_g or Σ_g [for linear types]), where all characters are 1.

7.12.1 Mulliken Symbols

Irreducible representations are designated using the **Mulliken notation**, which provides information about their symmetry properties.

Symbol	Meaning
A	One-dimensional representation, symmetric with respect to the principal axis (rotation by 360°/n)
B	One-dimensional, antisymmetric with respect to the principal axis
E	Two-dimensional representation
T	Three-dimensional representation
Subscripts 1, 2	Differentiate symmetric (1) and antisymmetric (2) with respect to $C_2 \perp$ principal axis or σ_v
g/u	"gerade" (even) or "ungerade" (odd)—symmetric or antisymmetric with respect to inversion

Examples:

- $C_{2v} \rightarrow A_1,\ A_2,\ B_1,\ B_2$
- $D_{3h} \rightarrow A'_1,\ A'_2,\ E',\ A''_1,\ A''_2,\ E''$
- $O_h \rightarrow A_{1g},\ A_{2u},\ E_g,\ T_{2g}$, and so on

Character tables of some important groups are provided in Appendix A.

Summary Thought

The Great Orthogonality Theorem provides the mathematical skeleton, character tables supply the symmetry map, and Mulliken symbols offer the chemical vocabulary through which group theory translates geometry into quantum mechanics.

7.13 Applications of Group Theory in Chemistry

Though the next chapter is wholly devoted to applications, before we wind up this chapter, some common applications are listed here. This will also answer the question of why we have devoted this chapter and the next to group theory. Group theory has wide-ranging applications across all branches of chemistry—from molecular structure to spectroscopy and bonding. Here is a sampler of what all group theory can do.

7.13.1 Vibrational Spectroscopy

Each normal mode of vibration transforms according to an irreducible representation. By consulting the character table, one can determine:

- the **number and symmetry types** of normal modes,
- which are **IR active** (transform like x, y, z), and
- which are **Raman active** (transform like quadratic functions x^2, y^2, z^2, xy, etc.).

7.13.2 Prediction of Spectroscopic Activity

A transition (infrared, Raman, or electronic) is **allowed** only if the direct product of representations of the initial and final states (and the transition moment operator) contains the totally symmetric representation. In the next chapter, we will explain why.

Example:

- In C_{2v}, z transforms as A_1, so only transitions between states whose product includes A_1 are IR active along the z-axis.

7.13.3 Molecular Orbital Theory

Group theory helps:

- Determine the **symmetry of atomic orbitals** in a molecule.
- Construct **symmetry-adapted linear combinations** of atomic orbitals.
- Label molecular orbitals by their **irreducible representations** (A_1, B_2, E, etc.).

Example: In NH_3 (C_{3v}), the nitrogen **2p**-orbitals split into A_1 ($\mathbf{p}_z$) and E $\left(\mathbf{p}_x, \mathbf{p}_y\right)$ combinations.

7.13.4 Optical Activity and Chirality

A molecule is **optically active** only if it belongs to a point group **lacking improper rotation axes** ($\boldsymbol{S_n}$). Thus, a molecule is optically active only if its point group contains **only proper rotations** (C_n, D_n) and lacks all improper symmetry elements (σ, i, S_n).

Thus, C_1, C_2, and D_2 groups can show optical activity, while C_{2v} and D_{3h} cannot.

7.13.5 Crystal Field and Ligand Field Theory

In complexes like $[Cr(H_2O)_6]^{3+}$ (O_h symmetry), group theory predicts:

- Splitting of d-orbitals into e_g and t_{2g} sets.
- Selection rules for d–d transitions.
- The symmetry labels of electronic states.

7.13.6 Chemical Bonding and Reaction Pathways

Symmetry arguments (using *group orbitals* and *correlation diagrams*) help explain:

- Why certain reactions are **symmetry-allowed or forbidden** (Woodward–Hoffmann rules).
- The **orientation and degeneracy** of frontier molecular orbitals in pericyclic reactions.

Before we end this chapter, we will briefly mention symmetry in periodic systems. The symmetry elements discussed above (rotation, reflection, inversion) exchange only the positions of atoms, keeping at least one point fixed (e.g., the center of mass for a molecule, or a point on a rotation axis). In crystals, there is also translational symmetry, because of which there are extra symmetry elements. We will discuss these briefly in the next section.

7.14 Symmetry in Periodic Structures: A Minimal Mathematical Note

Many important chemical systems—crystals, polymers, extended inorganic frameworks, e.g., MOFs, are **periodic** rather than finite. Unlike a molecule, which has a fixed number of atoms, a periodic pattern repeats indefinitely in space. To extend the group-theoretical ideas of this chapter to periodic systems, it is useful to first visualize how such repetition works.

7.14.1 A Geometric Picture of a Lattice

Consider a simple repeating arrangement of identical motifs (atoms, ions, or a small cluster) (Figure 7.7).

This geometric idea generalizes immediately to three dimensions (Figure 7.7b).

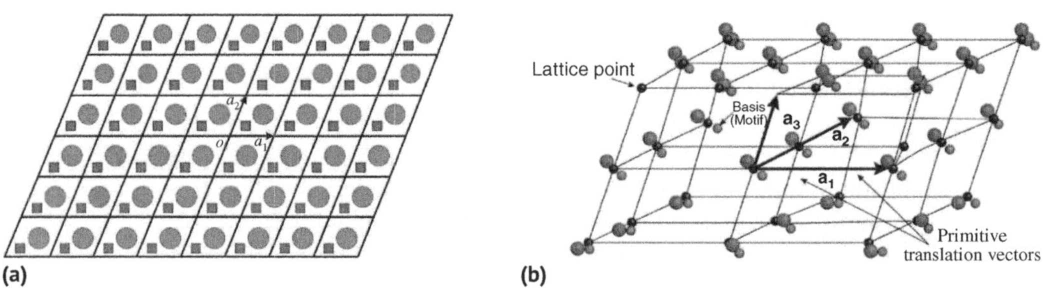

Figure 7.7 (a) A 2D periodic pattern of motifs. Two arrows, a_1 and a_2, connect one unit cell to the next identical unit cell. These two translations generate the entire infinite pattern. (b) Schematic representation of a three-dimensional lattice showing the primitive translation vectors $\mathbf{a}_1$, $\mathbf{a}_2$, $\mathbf{a}_3$. A basis attached to each lattice point generates the full crystal structure.

From Figure 7.7a:

- choosing a point as an origin (O), and
- identifying the two directions in which the pattern repeats, we can generate the entire infinite arrangement.

7.14.2 Primitive Translations and the Lattice

Let

$$\mathbf{a}_1, \mathbf{a}_2, \mathbf{a}_3$$

be three independent **primitive translation vectors** (Figure 7.7b). Every equivalent point in the periodic structure is obtained by

$$\mathbf{R} = n_1\mathbf{a}_1 + n_2\mathbf{a}_2 + n_3\mathbf{a}_3, n_1, n_2, n_3 \in \mathbb{Z}.$$

The set of all such translation vectors forms the **lattice**.

Math Box 7.1: The Translation Group

$$L = \{n_1\mathbf{a}_1 + n_2\mathbf{a}_2 + n_3\mathbf{a}_3 \mid n_i \in \mathbb{Z}\}$$

is an Abelian group under vector addition.

Any symmetry operation of a periodic structure must preserve this set.

7.14.3 Allowed Rotational Symmetry in Periodic Patterns

A fundamental geometric result: **only one-, two-, three-, four-, and sixfold rotations** can coexist with translational periodicity. Rotations of order 5, 7, or higher cannot tile space without gaps or overlaps in a periodic, long-range order. This restriction explains why crystallographic point groups form a **proper subset** of the molecular point groups discussed earlier.

Concept Box: Why Only These Orders?

A rotation must map the unit cell onto itself. Only one-, two-, three-, four-, and sixfold rotational divisions allow identical units to fill 2D/3D space without distortion.

Why Fivefold Rotational Symmetry Does Not Exist in a Periodic Crystal Lattice

This is a fundamental question in crystallography! The reason fivefold rotational symmetry does not exist in a periodic crystal lattice is that shapes with this symmetry cannot tile 2D or 3D space without leaving gaps.

Let us break this down with a visual example.

1. **What Is Fivefold Rotational Symmetry?**
   An object has n-fold rotational symmetry if it looks the same after being rotated by $360°/n$. For fivefold symmetry (C_5), the object must look identical after a $360°/5 = 72°$ rotation. A perfect example of a shape with fivefold symmetry is a regular pentagon.

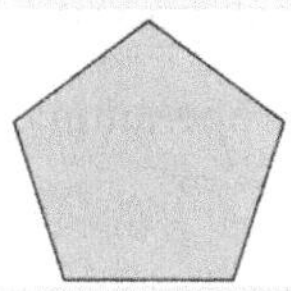

2. **The Tiling Problem**
   A crystal lattice is a periodic arrangement of points (or unit cells) that repeats infinitely in all directions, completely filling space without any gaps or overlaps.
   When we try to arrange identical pentagons to tile a 2D plane, we find that they cannot fit together perfectly. As you can see in the image below, placing them edge-to-edge inevitably creates gaps that cannot be filled by another pentagon. This is because the interior angle of a regular pentagon is 108°, and 360° is not a multiple of 108°. Therefore, you cannot fit an integer number of pentagons around a single point.

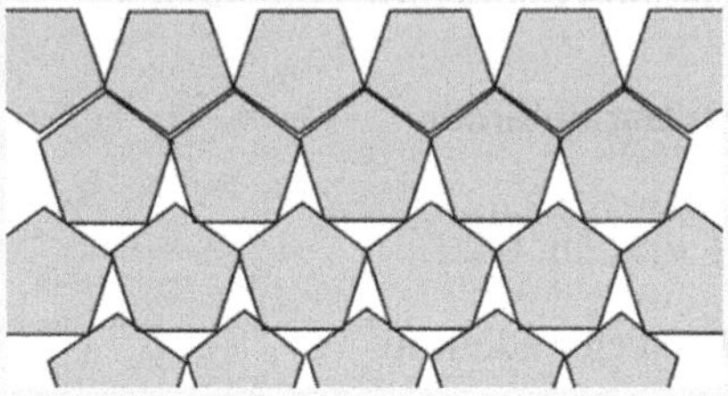

3. **A Contrast with an "Allowed" Symmetry**
   To contrast this with a symmetry that *is* allowed, let us look at sixfold symmetry. A regular hexagon has sixfold rotational symmetry. Its interior angle is 120°. Since $3 \times 120° = 360°$, three hexagons can perfectly meet at a single point, allowing them to tile a plane without any gaps.

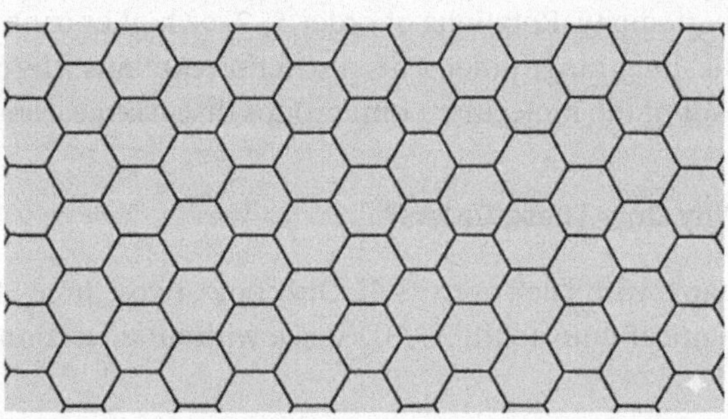

Because fivefold rotational symmetry does not allow for a seamless, periodic tiling of space, it is not a possible symmetry operation for a crystal lattice.

7.14.4 New Types of Symmetry Operations: Screws and Glides

In a molecule, translation cannot be a symmetry operation. In a periodic structure, however, combining a point group operation (proper rotation, mirror plane, inversion) with a **fractional translation** may still map the entire lattice onto itself.

This produces new symmetry operations:

- **screw axis:** rotation + fractional translation
- **glide plane:** reflection + fractional translation

Mathematically, a general symmetry operation can be written as

$$\mathbf{r}' = R\mathbf{r} + \mathbf{t},$$

where R is a rotation or reflection, and $\mathbf{t}$ may be a fraction of a lattice vector.

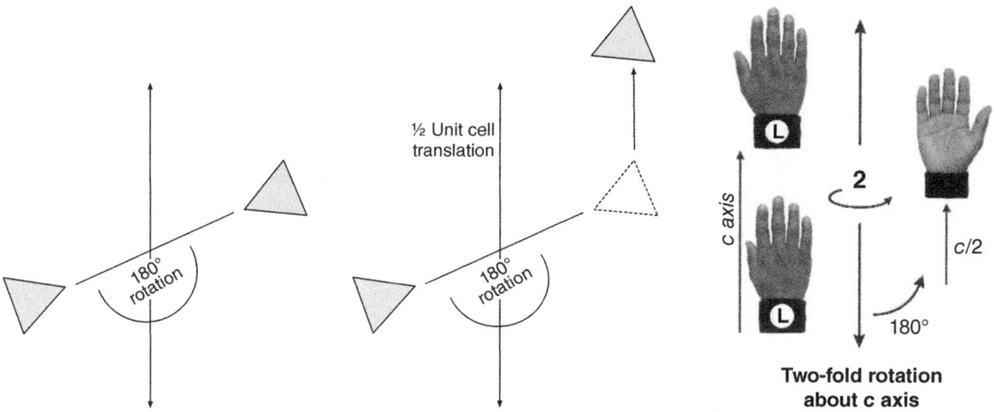

Figure 7.8 Geometry of screw axes n_m: rotation by $360°/n$ followed by the translation of m/n of the unit-cell length parallel to the axis. Here, the 2_1 screw axis is shown.

A symmetry operation combining a point group operation R with a translation $\mathbf{t}$ is written compactly as $\{R|\mathbf{t}\}$, where $\{R|\mathbf{t}\}\mathbf{r} = R\mathbf{r} + \mathbf{t}$.

7.14.4.1 Translational Symmetry Elements

a. **Screw Axes n_m**

Rotation by $360°/n$ followed by a translation of (m/n) of the unit-cell length in the direction of the axis (Figure 7.8).

Students understand this more effectively using the right hand, as illustrated for the 2_1 screw axis. A 2_1 screw axis is a symmetry operation in crystallography that combines a 180° rotation (C_2) with a translation of half the unit-cell length ($c/2$) along the axis. First, place your right hand at the bottom position with the back facing you. Rotate it by 180° (C_2 rotation) so that the palm faces you, then move it upward by half the unit-cell length ($c/2$). Next, rotate your hand again so the back faces you, and move it up by another $c/2$. This sequence results in your hand being translated by a full unit length (c) along the axis.

Examples: $2_1, 4_1, 4_3, 6_1, 6_3$.
Note that m must be less than n ($m < n$). Cases like 2_1, 4_2, or 6_3 represent rotations combined with a half-unit-cell translation.

b. **Glide Planes**

Reflection + translation within the plane.

A **glide plane** is a symmetry operation that combines a reflection across a plane with a translation parallel to that plane. A **c-glide plane** specifically involves a translation along the c-axis of the unit cell.

Here is a step-by-step illustration of a c-glide plane operation using a human hand (Figure 7.9):

1. **Initial State**
   First, consider a right hand placed in front of a mirror plane. This is our starting object.
2. **Reflection**
   The first step of the glide operation is a reflection across the mirror plane. As shown in Figure 7.9, the right hand is reflected to the other side of the plane, resulting in a left hand.

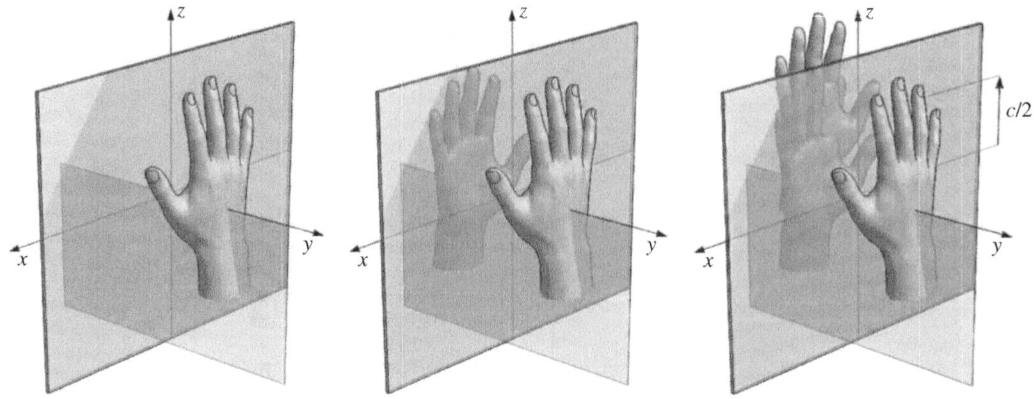

Figure 7.9 Step-by-step illustration of a c-glide plane operation using a human hand.

3. **Translation**
   The final step is a translation of the reflected object parallel to the plane. For a c-glide, this translation is along the c-axis (the z-axis in our diagram). Figure 7.9 shows the reflected left hand being translated upward by a distance of $c/2$.
   The final image illustrates the complete c-glide operation: the initial right hand is transformed into a translated left hand. This process demonstrates how glide planes are a combination of reflection and translation. Another repetition of these steps will result in the right hand being translated by c, that is, into the next unit cell.

Chemistry Connection

- Helical polymer chains and layered inorganic frameworks often reflect screw or glide symmetry.
- The detailed treatment of these operations belongs naturally to X-ray diffraction and solid-state spectroscopy.

7.14.5 Space Groups: Only the Mathematical Idea

A **space group** is the full symmetry group of a periodic pattern. It contains:

1. all lattice translations,
2. all allowed rotations and reflections, and
3. when necessary, screw and glide operations.

While there are 230 such groups, their classification is a crystallographic topic and lies outside the scope of this book.

As opposed to point groups, space groups also involve translation.

Key Insight

Molecular symmetry → point groups.

Periodic symmetry → space groups (point group operations + translations).

7.15 Summary

Group theory provides the mathematical language for symmetry in molecules.

Every molecule possesses symmetry operations—rotation, reflection, inversion, and identity—that form a **point group** obeying the rules of a **mathematical group**.

Matrix representations translate these geometric operations into algebraic form, allowing the use of traces (characters) to describe how orbitals, coordinates, and vibrations transform.

The **order** of a group equals the number of its symmetry operations. Operations related by **conjugation** form **classes**; each class corresponds to an **irreducible representation**, which cannot be simplified further.

The **Great Orthogonality Theorem** ensures that these irreducible representations are mutually independent and form the foundation of **character tables**, where Mulliken symbols provide their chemical labels (A_1, E, T_{2g}, etc.).

Group theory thus bridges geometry and quantum mechanics, explaining degeneracies, selection rules, orbital splitting, and optical activity.

In this chapter, we have built the complete machinery of group theory—from identifying symmetry elements to understanding irreducible representations and character tables. In the next chapter, we will apply this powerful toolkit to solve concrete chemical problems, including vibrational analysis of polyatomic molecules, construction of molecular orbitals, and prediction of spectroscopic selection rules.

Further Reading

N. L. Rosi et al., *Hydrogen storage in microporous metal-organic frameworks*, Science, 300, 1127–1129 (2003). https://www.science.org/doi/abs/10.1126/science.1083440.

F. A. Cotton, *Chemical Applications of Group Theory*, 3rd ed., Wiley-Interscience, 1990.

D. M. Bishop, *Group Theory and Chemistry*, Clarendon Press, Oxford, 1973.

R. L. Carter, *Molecular Symmetry and Group Theory*, Wiley, 1998.

P. Atkins and R. Friedman, *Molecular Quantum Mechanics*, 5th ed., Oxford University Press, 2010.

P. Atkins, J. de Paula, *Physical Chemistry*, 11th ed., Oxford University Press, 2023.

G. L. Miessler, P. J. Fischer, D. A. Tarr, *Inorganic Chemistry*, 6th ed., Pearson, 2024.

Nobel Foundation, Nobel Prize in Chemistry 2025—*for the design and discovery of metal-organic frameworks (MOFs)*. Official press release. https://www.nobelprize.org.

Exercises

A. Conceptual Questions

1. Define *symmetry operation* and *symmetry element* with one example each.

2. Distinguish between a *symmetry operation* and a *symmetry element* with examples from H_2O and NH_3.

3. What is meant by *equivalent atoms* in a molecule? How does symmetry make them equivalent?

4. Why do improper rotations (S_n) combine rotation and reflection? Give an example where S_4 symmetry arises.

5. Explain why all molecules belong to at least the C_1 point group.

6. Distinguish between σ_v-, σ_h-, and σ_d-planes with examples.

7. What is meant by the *order of a group* and the *order of an element*?

8. Why must the identity operation be included in every point group?

9. State the difference between *reducible* and *irreducible* representations.

10. What is the physical meaning of the *character* of a matrix?

11. Why do operations that are conjugate always have the same character?

12. What is the significance of Mulliken symbols A, B, E, and T?

13. Explain how group theory predicts whether a molecule can be optically active.

14. Explain geometrically why a fivefold rotation cannot be compatible with translational periodicity.

15. Why is translation a symmetry operation in a periodic structure but not in a molecule?

16. Give a clear distinction between a point group symmetry operation and a space-group operation.

17. Why do screw axes and glide planes not appear in molecular symmetry?

18. Consider the operation $\mathbf{r}' = R\mathbf{r} + \mathbf{t}$. What is the difference between $\mathbf{t}$ being a full lattice translation and a fractional translation?

B. Short-answer/Problem-solving

1. Write all symmetry elements of the following molecules and assign their point groups:
   a. BF_3.
   b. *trans*-1,2-dichloroethylene.
   c. CCl_4.
   d. HCN.
   e. eclipsed ethane.
   f. ethylene.

2. Construct the multiplication table for the operations of C_{2v} and verify closure.

3. Determine the characters of the three coordinate functions (x, y, z) for C_{2v} and identify the corresponding irreducible representations.

4. For the C_{3v} group, list its classes and explain why $\sigma_v, \sigma_v', \sigma_v''$ form one class.

5. The 3D Cartesian representation of C_{2v} has characters (3, −1, 1, 1). Decompose it into irreducible representations using the reduction formula.

6. A 2D lattice is generated by

$$\mathbf{a}_1 = (1, 0), \ \mathbf{a}_2 = \left(\frac{1}{2}, \frac{\sqrt{3}}{2}\right).$$

   a. Sketch the lattice.
   b. Does a threefold rotation about the origin preserve the lattice?
   c. Would a fourfold rotation preserve it? Justify.

7. Show that if a rotation R is a symmetry of a periodic structure, then

$$R\mathbf{a}_i = n_{i1}\mathbf{a}_1 + n_{i2}\mathbf{a}_2 + n_{i3}\mathbf{a}_3$$

   for integers n_{ij}.

8. In two dimensions, consider an operation made of a 180° rotation, followed by a translation of $\left(\frac{1}{2}, 0\right)$.
   a. Apply the operation to the origin.
   b. Show that performing it twice gives a pure translation.
   c. Explain why such an operation requires periodicity.

9. In two dimensions, consider an operation made of a reflection across the x-axis followed by a translation of $\left(\frac{1}{2}, 0\right)$. Show that performing it twice gives a pure translation.

C. Higher-order/Application-based

1. Show that the S_2 operation is equivalent to inversion by explicit matrix multiplication.

2. Use matrix representations to prove that the symmetry operations of H_2O form an Abelian group.

3. Explain how the presence of an inversion center influences the IR and Raman activity of a molecule.

4. Prove that for any periodic structure, the rotational part R of a symmetry operation must satisfy $RL = L$; that is, it must map each lattice vector to another lattice vector.

5. Show that an improper rotation S_n cannot be a space-group symmetry operation without an accompanying translation component.
(Hint: the reflection component must be compatible with the lattice.)

6. Prove that a glide plane operation $g = \{\sigma|\mathbf{t}\}$ performed twice must equal a primitive lattice translation, implying t must be half of a lattice vector.

8

Applications of Group Theory

C'est la dissymétrie qui crée le phénomène. **(It is asymmetry which creates the phenomenon.)**

—Pierre Curie (1894)

Learning Outcomes

After studying this chapter, students should be able to:

1. **Construct** reducible and irreducible representations and **use** them to obtain **vibrational modes**.
2. **Predict** infrared and Raman activity using the symmetry of μ vector and α tensor.
3. **Classify** molecular orbitals and hybridization patterns by symmetry.
4. **Explain** π-system behavior using SALCs and degeneracy.
5. **Apply** symmetry principles to determine whether a molecule is polar or chiral.
6. **Use** group theory qualitatively to interpret electronic spectra.

8.1 Introduction and Motivation

In the preceding chapter, we developed the formal machinery of symmetry operations, groups, and representations. We now move from abstraction to application. Group theory enables chemists to determine which vibrations of a molecule are infrared active or Raman active, to explain degeneracies in molecular orbitals (MOs), to classify normal modes, and to simplify quantum-mechanical calculations.

This chapter will cover vibrational analysis, orbital symmetry, and spectroscopic applications, emphasizing how group theory simplifies these problems. Many textbooks introduce these results as ready-made "rules." In this chapter, we will go further and explain *why* they work, tracing each selection rule back to its mathematical and physical origin. By the end, you should be able to not only use character tables but also understand the logic behind every entry.

8.2 Vibrational Analysis: General Principles

Every molecule with N atoms possesses $3N$ degrees of freedom, corresponding to the three Cartesian coordinates of each atom. Of these, three describe translational motion of the entire molecule, and, for nonlinear molecules, three describe rotation. The remaining $3N - 6$ degrees of freedom correspond to vibrational motions, called **normal modes** of vibration; for linear molecules, the number is $3N - 5$ because rotation about the bond axis is not a true rotational degree of freedom (negligible moment of inertia).

8.2.1 Choice of Coordinate System

It is essential to align the molecular coordinate system correctly: the highest-order rotation axis is chosen as z; for planar molecules, the molecular plane defines yz; and x is perpendicular to it. A careless choice of axes may interchange symmetry labels (e.g., B_1 and B_2 in C_{2v}). This is a common source of error—even among teachers.

8.2.2 Building the Total Representation

Each atom contributes a vector (x, y, z), and the collective $3N$-dimensional basis transforms as a reducible representation Γ_{total} or Γ_{3N}.

For each symmetry operation R, the *character* $\chi(R)$ equals the trace of the transformation matrix acting on all $3N$ coordinates. Only coordinates of atoms that remain fixed by R contribute to $\chi(R)$; when an atom moves to a different position, its submatrix becomes off-diagonal and does not affect the trace.

Typical characters for Cartesian coordinates of a single atom are:

Operation	Transformation	$\chi(R)$
E	$(x, y, z) \rightarrow (x, y, z)$	3
$C_2(z)$ (180°)	$(x, y, z) \rightarrow (-x, -y, z)$	−1
σ_{xz} (reflection)	$(x, y, z) \rightarrow (x, -y, z)$	1
i (inversion)	$(x, y, z) \rightarrow (-x, -y, -z)$	−3

In general, for a proper rotation C_n through an angle θ about the z-axis,

$$\chi = 1 + 2\cos\theta,$$

while for an improper rotation S_n (the same rotation followed by reflection through a plane perpendicular to the axis),

$$\chi = -1 + 2\cos\theta$$

for the three-dimensional Cartesian vector representation.

Note that E and C_n belong to the category of proper rotations with $\theta = 0°$ for E and $360°/n$ for C_n. S_n, $\sigma(\equiv S_1)$, and i (S_2) are improper rotations with $\theta = 360°/n$, $0°$, and π, respectively.

It is also useful to remember that the characters of E, σ, and i are always 3, 1, and −3, respectively, for the (x, y, z) basis of a single atom, regardless of whether the reflection plane is vertical, horizontal, or dihedral.

Proper and Improper Rotations: Quick Reference

Operation (R)	Formula for $\chi(R)$	Example
E	3	Identity
C_n	$1 + 2\cos(360°/n)$	Proper rotation
S_n	$-1 + 2\cos(360°/n)$	Improper rotation
σ	1	Reflection plane
i	-3	Inversion

These expressions apply specifically to the three-dimensional Cartesian vector representation (x, y, z).

8.3 Vibrational Analysis: The Water Molecule (C_{2v})

Let us now apply the ideas of symmetry and group theory to a real molecular system—water. H_2O is a bent molecule belonging to the C_{2v} point group, whose operations are $E, C_2(z), \sigma_v(xz)$, and $\sigma_v'(yz)$. Each atom contributes three Cartesian coordinates (x, y, z), giving a total of $3 \times 3 = 9$ degrees of freedom.

8.3.1 Choice of Coordinate System

The molecule is placed in the yz-plane, with the z-axis as the C_2 axis passing through the oxygen atom and bisecting the H—O—H bond angle. This is the standard orientation used in this book and in most character tables; other choices interchange B_1 and B_2 labels.

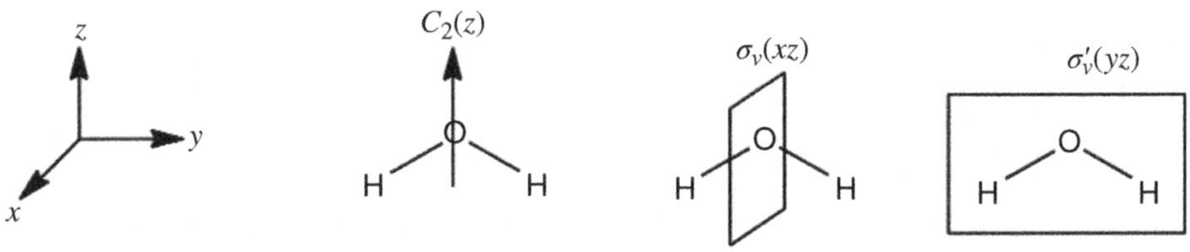

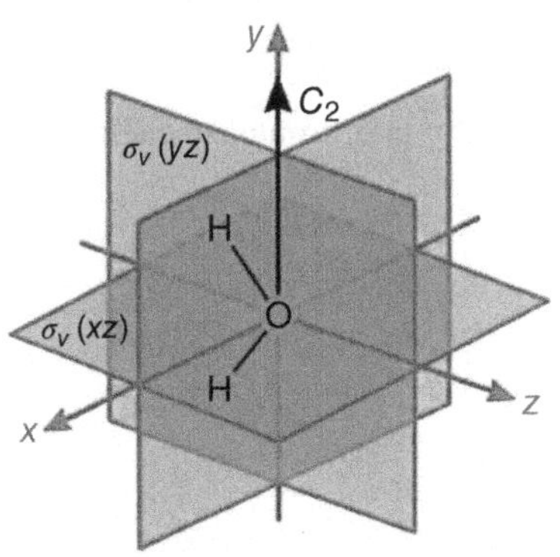

⚠ Red Flag 8.1

Choosing the wrong coordinate system can alter results. If the molecular plane is taken as xz, the representations B_1 and B_2 will be interchanged. Note that this is a labeling change, not a physical change.

8.3.2 Transformation of Coordinates

We now see how each symmetry operation acts on the set of nine coordinates:

$$(x_O, y_O, z_O, x_A, y_A, z_A, x_B, y_B, z_B),$$

where A and B denote the two hydrogen atoms.

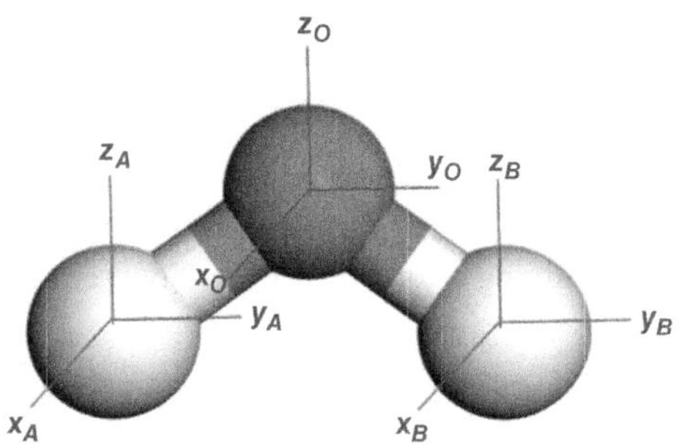

i. **Identity (E)**
   Each atom remains in its position, and each coordinate maps to itself,

$$E = \begin{bmatrix} I_3 & 0 & 0 \\ 0 & I_3 & 0 \\ 0 & 0 & I_3 \end{bmatrix}.$$

   Here, there are three blocks of identity matrices, each block corresponding to an atom, O, H_A, H_B. These blocks lie along the main diagonal and contribute to the trace.
   Trace $\chi(E) = 3 + 3 + 3 = 9$.

ii. **180° Rotation about the z-axis (C_2)**
   Hydrogen A goes to the position of B and vice versa. The x- and y-coordinates of each atom change sign, while z remains unchanged.
   The transformation is:

$$(x_O, y_O, z_O, x_A, y_A, z_A, x_B, y_B, z_B) \rightarrow (-x_O, -y_O, z_O, -x_B, -y_B, z_B, -x_A, -y_A, z_A).$$

   This gives the following block-structured matrix:

$$\Gamma(C_2) = \begin{bmatrix} C_2(O) & 0 & 0 \\ 0 & 0 & C_2(H_B) \\ 0 & C_2(H_A) & 0 \end{bmatrix}$$

Only the central atom (oxygen) remains in place, and for it, x and y change sign while z does not.

Hence, its 3×3 submatrix is

$$\begin{bmatrix} -1 & 0 & 0 \\ 0 & -1 & 0 \\ 0 & 0 & 1 \end{bmatrix}.$$

Hence, $\chi(C_2) = (-1) + (-1) + (1) = -1$.

All the elements corresponding to interchanged atoms (A ↔ B) are **off-diagonal**, so they do not contribute to the trace.

iii. Reflection through the *xz*-plane (σ_v)

This plane bisects the H—O—H bond and interchanges the two hydrogens. The y-coordinates change sign; x and z remain unchanged,

$$(x_O, y_O, z_O, x_A, y_A, z_A, x_B, y_B, z_B) \to (x_O, -y_O, z_O, x_B, -y_B, z_B, x_A, -y_A, z_A).$$

For each atom that is exchanged, the matrix elements become off-diagonal.

The oxygen atom remains fixed, and its y-coordinate changes sign:

$$\sigma_v(O) = \begin{bmatrix} 1 & 0 & 0 \\ 0 & -1 & 0 \\ 0 & 0 & 1 \end{bmatrix} \Rightarrow \chi(\sigma_v) = 1.$$

iv. Reflection through the *yz*-plane (σ_v')

This plane passes through all three atoms (the molecular plane). All atoms remain fixed, but the x-coordinate changes sign:

$$(x, y, z) \to (-x, y, z),$$

$$\sigma_v' = \begin{bmatrix} -1 & 0 & 0 \\ 0 & 1 & 0 \\ 0 & 0 & 1 \end{bmatrix} \Rightarrow \chi(\sigma_v') = 1,$$

$$\Gamma(\sigma_v') = \begin{bmatrix} \sigma_v'(O) & 0 & 0 \\ 0 & \sigma_v'(H_A) & 0 \\ 0 & 0 & \sigma_v'(H_B) \end{bmatrix}.$$

Trace $\chi(\sigma_v') = 1 + 1 + 1 = 3$.

8.3.3 Constructing the Reducible Representation

The characters for the four operations are:

$$\Gamma_{3N} = (9,\ -1,\ 1,\ 3).$$

The first entry corresponds to the total number of coordinates, and subsequent entries result from the invariance analysis above. Note that Γ_{3N} should be $3N$-dimensional, i.e., the character of E should be $3N$ (9 in this case).

8.3.4 Deriving the Vibrational Representation

Table 8.1 Character table of C_{2v}.

Operation	E	C_2	$\sigma_v(xz)$	$\sigma'_v(yz)$	Linear (x, y, z)	Rotation (R_x, R_y, R_z)	Quadratic functions
A_1	1	1	1	1	z	—	x^2, y^2, z^2
A_2	1	1	−1	−1	—	R_z	xy
B_1	1	−1	1	−1	x	R_y	xz
B_2	1	−1	−1	1	y	R_x	yz

From the character table of C_{2v} (Table 8.1) and using the reduction formula

$$a_i = \frac{1}{h}\sum_R \chi(R)\chi_i(R),$$

with $h = 4$, we find:

- **For A_1**: $a_{A_1} = \frac{1}{4}[(9)(1) + (-1)(1) + (1)(1) + (3)(1)] = \frac{1}{4}[9 - 1 + 1 + 3] = \frac{12}{4} = 3;$
- **For A_2**: $a_{A_2} = \frac{1}{4}[(9)(1) + (-1)(1) + (1)(-1) + (3)(-1)] = \frac{1}{4}[9 - 1 - 1 - 3] = \frac{4}{4} = 1;$
- **For B_1**: $a_{B_1} = \frac{1}{4}[(9)(1) + (-1)(-1) + (1)(1) + (3)(-1)] = \frac{1}{4}[9 + 1 + 1 - 3] = \frac{8}{4} = 2;$
- **For B_2**: $a_{B_2} = \frac{1}{4}[(9)(1) + (-1)(-1) + (1)(-1) + (3)(1)] = \frac{1}{4}[9 + 1 - 1 + 3] = \frac{12}{4} = 3;$

that is,

$$\Gamma_{3N} = 3A_1 \oplus A_2 \oplus 2B_1 \oplus 3B_2.$$

Notation Alert: ⊕ and ⊗ in This Book

In this book, the symbols ⊕ and ⊗ are used deliberately and consistently to distinguish **different mathematical operations on representations.**

Direct sum: ⊕

The symbol ⊕ denotes the **direct sum of representations**. For example,

$$\Gamma_{\text{vib}} = 2A_1 \oplus B_2$$

means that the vibrational representation decomposes into two independent parts transforming as A_1 and one as B_2.

The direct sum indicates a **listing of independent symmetry species**, not a multiplication or interaction between them.

Direct (tensor) product: ⊗

The symbol ⊗ denotes the direct (tensor) product of representations, for example,

$$A_1 \otimes B_2 = B_2.$$

Direct products describe how **two functions or operators combine**, and are central to:

- selection rules,
- orbital overlap,
- transition moment analysis.

Comparison with Common Chemistry Notation
Many chemistry texts write "+" for direct sums and "×" for direct products; the circled symbols ⊕ and ⊗ used here represent the *same operations* but make the distinction from ordinary arithmetic explicit.

Key Distinction (One Line to Remember)
⊕ lists **independent symmetry species**, whereas ⊗ combines symmetries to produce a new one.

Convention Adopted in This Book
Throughout this book, ⊕ denotes **direct sums** and ⊗ denotes **direct products** of representations.

8.3.5 Subtracting Translational and Rotational Contributions

The characters for translation are the same as x, y, z, and those for rotation are R_x, R_y, R_z. From the character table for C_{2v}:

$$\Gamma_{\text{trans}} = A_1 \oplus B_1 \oplus B_2, \Gamma_{\text{rot}} = A_2 \oplus B_1 \oplus B_2.$$

Subtracting these from Γ_{3N}:

$$\begin{aligned}\Gamma_{\text{vib}} = \Gamma_{3N} - \Gamma_{\text{trans}} - \Gamma_{\text{rot}} &= (3A_1 \oplus A_2 \oplus 2B_1 \oplus 3B_2) - (A_1 \oplus A_2 \oplus 2B_1 \oplus 2B_2)\\ &= 2A_1 \oplus B_2.\end{aligned}$$

Hence, H_2O has **three normal modes**: $2A_1$ and B_2.

At this stage, check that there are $3N - 6$ vibrational modes (for nonlinear molecules). The presence of three modes for water confirms that our calculations are correct.

Shortcut 8.1: The "Character From Unmoved Atoms" Shortcut for Γ_{vib}

When finding the reducible representation for vibrational analysis (Γ_{3N}), students struggle to construct full $3N \times 3N$ matrices. We have introduced this shortcut:
For any symmetry operation R:

$$\begin{aligned}\boldsymbol{\Gamma_{3N}(R)} &= \textbf{(Number of atoms unmoved by } \boldsymbol{R}\textbf{)} \times \textbf{(Contribution per unmoved atom)}\\ &= \boldsymbol{n_R \times \chi(R)},\end{aligned}$$

where the "contribution per unmoved atom ($\chi(R)$)" is as follows:
For a Proper Rotation (C_n): $1 + 2\cos\theta$:

For E ($\theta = 0°$): $1 + 2(1) = \mathbf{3}$;

For C_2($\theta = 180°$): $1 + 2(-1) = \mathbf{-1}$;

For C_3 ($\theta = 120°$): $1 + 2(-1/2) = \mathbf{0}$.

For an Improper Rotation (S_n)/Reflection (σ): $-1 + 2\cos\theta$:

For i ($S_2, \theta = 180°$): $-1 + 2(-1) = \mathbf{-3}$;

For σ ($S_1, \theta = 0°$): $-1 + 2(1) = 1$;

For S_3 ($\theta = 120°$): $-1 + 2(-1/2) = -2$.

Example for H_2O (C_{2v}):

E : Unmoved atoms = 3. Contribution = 3. $\chi(E) = 3 \times 3 = \mathbf{9}$.

C_2 : Unmoved atoms = 1 (O). Contribution = −1. $\chi(C_2) = 1 \times -1 = \mathbf{-1}$.

$\sigma_v(xz)$: Unmoved atoms = 1 (O). Contribution = 1. $\chi(\sigma_v) = 1 \times 1 = \mathbf{1}$.

$\sigma'_v(yz)$: Unmoved atoms = 3. Contribution = 1. $\chi(\sigma'_v) = 3 \times 1 = \mathbf{3}$.

This gives $\Gamma_{3N} = (9, -1, 1, 3)$ instantly, without any matrices.

Instead of separating translational and rotational components explicitly, we note that translational motion transforms like the coordinates (x, y, z) themselves, and rotations like the angular momenta (R_x, R_y, R_z). Also, the characters for angular momenta are the same as those for the coordinates in the case of proper rotations and the sign is opposite for the improper rotations.

Here is the complete shortcut solution for water (Table 8.2).

Using the Great Orthogonality Theorem, Γ_{vib} reduces to $2A_1 \oplus B_2$ as before. Another simplification is possible. For simple AB_n molecules with equivalent peripheral atoms, like H_2O, NH_3, BF_3, CO_2, where there is a central atom connected to n similar atoms, the number of bonds and hence the number of stretching frequencies is $n = N - 1$. The character of the stretches is obtained by

Table 8.2 Normal modes of water.

		$1+2\cos\theta$		$-1+2\cos\theta$
C_{2v}	E	$C_2(z)$	$\sigma_v\,(xz)$	$\sigma'_v\,(yz)$
$\chi(R)$	3	−1	1	1
n_R	3	1	1	3
Γ_{3N}	9	−1	1	3
$\Gamma_{trans} + \Gamma_{rot}^{a}$	6	−2	0	0
Γ_{vib}	3	1	1	3
$\Gamma_{stretch}$	2	0	0	2

a. You can check that $\Gamma_{trans} = A_1 \oplus B_1 \oplus B_2 = (3, -1, 1, 1)$, $\Gamma_{rot} = A_2 \oplus B_1 \oplus B_2 = (3, -1, -1, -1)$, and $\Gamma_{trans} + \Gamma_{rot} = (6, -2, 0, 0)$.

subtracting 1 from each of the characters of unshifted atoms [i.e., subtract A_1 from $\Gamma(n_R)$]. For H_2O this gives $A_1 \oplus B_2$.

8.3.6 Classification of Vibrations

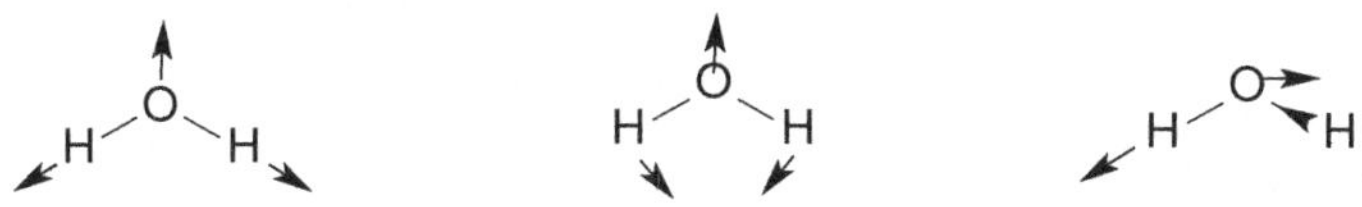

ν_1 ν_2 ν_3

Mode	Symmetry	Description
ν_1	A_1	Symmetric stretch: both O—H bonds lengthen/shorten together
ν_2	A_1	Bend (scissoring): bond angle changes within the molecular plane
ν_3	B_2	Asymmetric stretch: one bond lengthens while the other shortens

We can verify these assignments by examining how each mode behaves under the symmetry operations of C_{2v}:

- For the **symmetric stretch**, both bonds move identically → unchanged (A_1).
- For the **bending mode**, displacements remain symmetric within the plane → unchanged (A_1).
- For the **asymmetric stretch**, displacements change sign under C_2 and σ_v → antisymmetric (B_2).

Key Point

The symmetry labels (A_1, B_2) directly determine the physical motion of atoms.

By checking how the displacement vectors behave under the symmetry operations of the point group, we can identify which motion corresponds to which irreducible representation—without relying on visualization alone.

Key Learning Point

The full 9 × 9 matrices demonstrate that *symmetry operations do not act atom by atom but coordinate by coordinate.*

An operation that moves an atom to another equivalent position introduces off-diagonal couplings, and those are what group theory "filters out" through the trace.

This makes the concept of the character physically meaningful: it counts how much of the molecule remains invariant in space under each symmetry operation.

8.4 Infrared and Raman Activity

Once the normal modes of vibration have been identified, the next question is which of them can be **observed experimentally**.

The answer depends on the way each mode changes the **molecular dipole moment** and the **polarizability** during vibration.

8.4.1 Transition Moment Integral and Symmetry Requirement

A vibrational transition is possible only if the transition dipole moment

$$M = \int \psi' \mu \psi'' \, d\tau \tag{8.1}$$

is nonzero, where

- ψ'' and ψ' are the initial (ground) and final (excited) vibrational wavefunctions, and
- $\mu = (\mu_x, \mu_y, \mu_z)$ is the dipole-moment operator.

By group theory, this integral will be nonzero only if the **direct product**

$$\Gamma(\psi') \otimes \Gamma(\mu) \otimes \Gamma(\psi'') \tag{8.2}$$

contains the **totally symmetric irreducible representation** of the point group. If it does not, the integral vanishes because the integrand becomes an **odd function** over a symmetric region—just as the integral of an odd function from $-a$ to $+a$ is zero. Hence, the physical meaning: *the wavefunction overlap must remain symmetric under all symmetry operations.*

Direct Product

When two or more representations ($\Gamma_1, \Gamma_2, ...$) of a group act simultaneously on the same set of basis functions, their combined action is represented by the *direct product* of these representations. Mathematically, if $D^{(1)}(R)$ and $D^{(2)}(R)$ are the matrices representing a symmetry operation R in representations Γ_1 and Γ_2, then the direct product representation $D(R)$ is formed by all possible products of the corresponding basis functions:

$$D(R) = D^{(1)}(R) \otimes D^{(2)}(R).$$

The characters of the direct product representation are given by the product of the characters of the individual representations:

$$\chi_{\Gamma_1 \otimes \Gamma_2}(R) = \chi_{\Gamma_1}(R) \times \chi_{\Gamma_2}(R).$$

This concept is especially important for determining allowed transitions and interaction terms in spectroscopy and quantum mechanics.

Example: Direct Product in C_{2v}

Consider the $\boldsymbol{C_{2v}}$ point group, which has four irreducible representations: A_1, A_2, B_1, and B_2. Let us determine the result of the direct product $\boldsymbol{B_1 \otimes B_2}$.

From the C_{2v} character table:

Operation	E	C_2	$\sigma_v(xz)$	$\sigma'_v(yz)$
B_1	1	−1	1	−1
B_2	1	−1	−1	1

The character of the direct product $\Gamma = B_1 \otimes B_2$ is obtained by multiplying the corresponding characters for each operation:

Operation	$\chi(B_1)$	$\chi(B_2)$	$\chi(\Gamma) = \chi(B_1) \times \chi(B_2)$
E	1	1	1
C_2	−1	−1	1
$\sigma_v(xz)$	1	−1	−1
$\sigma'_v(yz)$	−1	1	−1

The resulting character set (1, 1, −1, −1) corresponds to A_2.
Hence:

$$B_1 \otimes B_2 = A_2.$$

Physical significance:
In this example, if a transition moment transforms as B_1 and a vibrational mode as B_2, their combined symmetry ($B_1 \otimes B_2 = A_2$) indicates that the resulting product transforms as A_2. Only if this product contains the **totally symmetric representation** (A_1 in C_{2v}) will the transition be *allowed*. This is the basis for using direct products in selection rules for infrared and Raman spectroscopy.

8.4.2 Change in Dipole Moment

For a **fundamental vibration** ($v'' = 0 \rightarrow v' = 1$),

- the ground-state wavefunction ψ_0 transforms as the totally symmetric representation (A_1 for C_{2v}),
- the excited state ψ_1 transforms into the same species as the normal-coordinate Q.

Therefore, Equation (8.2) reduces to

$$\Gamma(Q) \otimes \Gamma(\mu) \supset A_1. \tag{8.3}$$

A vibration is **infrared active** if its irreducible representation $\Gamma(Q)$ is the same as that of x, y, or z in the character table—that is, if the vibration produces a *change in dipole moment* along any Cartesian axis.

8.4.3 Change in Polarizability

In Raman spectroscopy the quantity of interest is the **polarizability tensor** α with components $xx, xy, xz, yx, yy, yz, zx, zy, zz$. A vibration is **Raman active** if it changes the polarizability—that is, if the direct product

$$\Gamma(Q) \otimes \Gamma(\alpha) \supset A_1 \tag{8.4}$$

is satisfied. Here, $\Gamma(\alpha)$ is the representation of the polarizability tensor α.

Each tensor component transforms according to the quadratic functions (x^2, xy, yz, etc.) listed in the character table.

8.4.4 Application to Water (C_{2v})

From the previous section,

$$\Gamma_{\text{vib}} = 2A_1 \oplus B_2,$$

and the character table (Table 8.1) shows

Species	Linear functions	Quadratic functions
A_1	z	x^2, y^2, z^2
A_2	R_z	xy
B_1	x, R_y	xz
B_2	y, R_x	yz

a. **Infrared Activity**
   Infrared (IR) activity requires overlap with x, y, or z.
   Thus:

Mode	Symmetry	Direction of dipole change	IR activity
ν_1 Symmetric stretch	A_1	Along z	Active
ν_2 Bending	A_1	Along z	Active
ν_3 Asymmetric stretch	B_2	Along y	Active

   Each mode produces a change in dipole moment along one of the molecular axes; hence, all three are IR active.

b. **Raman Activity**
   Raman activity requires overlap with quadratic functions.

Mode	Symmetry	Polarizability change	Raman activity
$\nu_1(A_1)$	Changes x^2, y^2, z^2	Strong	Raman active
$\nu_2(A_1)$	Same as above	Strong	Raman active
$\nu_3(B_2)$	Changes yz	Weak	Raman active

   Thus, all three vibrations are Raman active as well, in agreement with experiment.

8.4.5 Physical Interpretation

- **Infrared spectroscopy** detects those vibrations that cause an *oscillating dipole moment*, enabling interaction with the electric field of IR light.
- **Raman spectroscopy** probes vibrations that cause a *change in polarizability*—the molecule's ability to distort its electron cloud in response to light.

Mathematically, IR activity arises from first-order terms in the dipole expansion, whereas Raman activity comes from first-order changes in the polarizability tensor. Both obey the same symmetry rule: the product of the relevant functions must contain the **totally symmetric representation** (A_1 for C_{2v}).

Shortcut 8.2: The "IR and Raman Activity at a Glance" From the Character Table

Instead of memorizing rules, you can read the answer directly from the right-hand column of the character table.

IR Active? Check the "Linear Functions" column.

If your irreducible representation (e.g., B_1) is listed next to $\boldsymbol{x}$, $\boldsymbol{y}$, or $\boldsymbol{z}$, it is IR active.

Reason: The transition moment operator transforms as (x, y, z).

Raman Active? Check the "Quadratic Functions" column (often combined with Linear/ Rotations).

If your irreducible representation is listed next to $\boldsymbol{x^2}$, $\boldsymbol{y^2}$, $\boldsymbol{z^2}$, $\boldsymbol{xy}$, $\boldsymbol{xz}$, or $\boldsymbol{yz}$, it is Raman active.

Reason: The polarizability tensor transforms like these quadratic functions.

Example using C_{2v} table:

A_1 is listed with z and x^2, y^2, z^2 → **IR and Raman Active.**

B_2 is listed with y and yz → **IR and Raman Active.**

A_2 is listed with R_z and xy → In the C_{2v} table, A_2 has no linear functions listed, so it is inactive in IR.

This makes determining activity a 5-second visual task.

8.4.6 Summary Table for H_2O

Mode	Symmetry	IR activity	Raman activity	Polarization[a]
ν_1 Symmetric stretch	A_1	✓(z)	✓(Strong)	Parallel
ν_2 Bending	A_1	✓(z)	✓(Strong)	Parallel
ν_3 Asymmetric stretch	B_2	✓(y)	✓(Weak)	Perpendicular

a. Parallel bands are those for which the coordinate along the principal axis (z) excites the vibration, and perpendicular bands are those where a coordinate perpendicular to the principal coordinate excites the vibration. The polarization influences the shape of the IR band.

This explicit reasoning shows why we look at x, y, z for IR activity and at quadratic terms for Raman. It also reinforces the underlying idea: *only totally symmetric combinations survive integration; all others vanish because they are odd under at least one symmetry operation.*

Having now seen, through explicit matrices, how the symmetry operations of C_{2v} act on the Cartesian coordinates of H_2O, we no longer need to repeat $3N \times 3N$ matrices for every molecule. The principle is established:

Only those coordinates that remain unchanged under a symmetry operation contribute to the trace; atoms that shift to new positions give off-diagonal terms that do not contribute.

In the next example, we will therefore determine the characters directly by identifying the number of unshifted atoms and multiplying by the characteristic value for the operation ($\chi = 1 + 2\cos\theta$ for proper rotations, $\chi = -1 + 2\cos\theta$ for improper ones). This approach is equally rigorous but far more efficient.

8.4.7 Vibrational Analysis of NH_3 (C_{3v})

Ammonia is a trigonal-pyramidal molecule (point group C_{3v}) with four atoms → $3N = 12$ degrees of freedom. Three correspond to translation and three to rotation, so there are **six vibrational modes** ($3N - 6 = 6$).

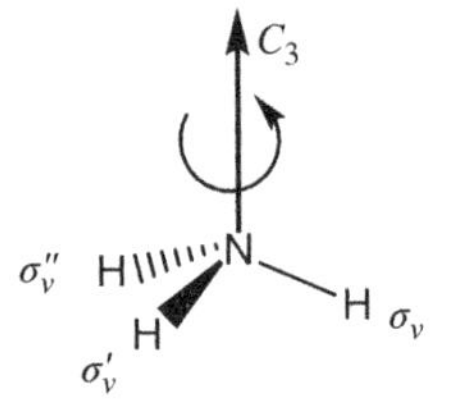

Step 1: Total Representation

In the same manner as for water (Table 8.2), we find the number of unshifted atoms (n_R) for every symmetry operation and form the 12-dimensional reducible representation of all Cartesian displacements (Γ_{3N}) (Table 8.3).

Table 8.3 Character table (C_{3v}) and vibrational analysis of ammonia.

C_{3v}	E	$2C_3$	$3\sigma_v$		
A_1	1	1	1	z	$x^2 + y^2, z^2$
A_2	1	1	−1	R_z	
E	2	−1	0	$(x, y), (R_x, R_y)$	$(x^2 - y^2, xy)(xz, yz)$
$\chi(R)$	3	0	1		
n_R	4	1	2		
Γ_{3N}	12	0	2		
$\Gamma_{trans} + \Gamma_{rot}$	6	0	0		
$\Gamma_{vib} = \Gamma_{3N} - (\Gamma_{trans} + \Gamma_{rot})$	6	0	2		
$\Gamma_{stretch} = \Gamma(n_R) - \Gamma(A_1)$	3	0	1		

As noted previously, the characters for rotation are the same as those for translation for the proper rotations and opposite in sign for the improper rotations, so the characters for $\Gamma_{trans} + \Gamma_{rot}$ are obtained by doubling the characters in $\chi(R)$ for proper rotations (E and C_3) and setting zero for the improper rotation (σ_v). This applies to Cartesian vector representations. To confirm,

- $\Gamma_{trans}(x, y, z) = A_1 \oplus E \rightarrow$ characters: $(3, 0, 1)$.
- $\Gamma_{rot}(R_x, R_y, R_z) = A_2 \oplus E \rightarrow$ characters: $(3, 0, -1)$.
- $\Gamma_{trans} + \Gamma_{rot} \rightarrow$ characters: $(6, 0, 0)$.

After subtraction of translations and rotations ($A_1 \oplus A_2 \oplus 2E$), the vibrational representation becomes:

$$a(A_1) = \frac{1}{6}(6 \times 1 + 2 \times 0 \times 1 + 3 \times 2 \times 1) = 2,$$

$$a(A_2) = \frac{1}{6}(6 \times 1 + 2 \times 0 \times 1 + 3 \times 2 \times (-1)) = 0,$$

$$a(E) = \frac{1}{6}(6 \times 2 + 2 \times 0 \times (-1) + 3 \times 2 \times 0) = 2,$$

$$\Gamma_{\text{vib}} = 2A_1 \oplus 2E.$$

Thus, there are **four fundamental frequencies**: two nondegenerate A_1 and two doubly degenerate E modes. Note that there are six vibrational modes, but only four fundamental bands will be observed because two of these are doubly degenerate (same frequency).

Step 2: Assigning Physical Meaning

The stretching vibrations (Γ_{stretch}) correspond to $A_1 \oplus E$.

Mode	Symmetry	Description	Degeneracy
ν_1	A_1	Symmetric stretch	1
ν_2	A_1	Symmetric bend ("umbrella")	1
ν_3	E	Degenerate asymmetric stretches	2
ν_4	E	Degenerate asymmetric bends	2

These modes are shown in Figure 8.1.

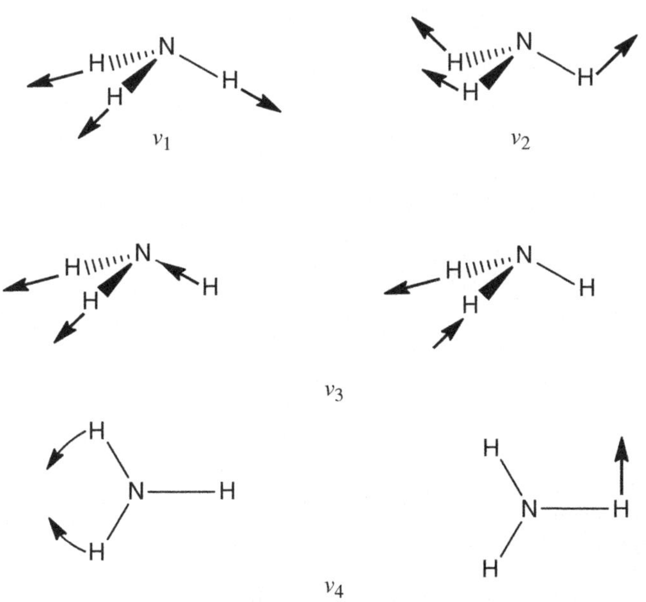

Figure 8.1 Normal modes of vibration of ammonia (C_{3v}).

Step 3: IR and Raman Activity

- **IR:** Active if transforms as $x, y, z \rightarrow A_1 \oplus E$.
- **Raman:** Active if transforms as quadratic functions $\rightarrow A_1 \oplus E$.

Hence, **all NH_3 vibrations are both IR and Raman active**, though intensities differ.

The doubly degenerate E modes give a characteristic pair of peaks in spectra; the umbrella (A_1) bending mode around 950 cm^{-1} is diagnostic.

Before winding up this section, we will briefly discuss the linear triatomic molecule, carbon dioxide.

8.4.8 The Carbon Dioxide Molecule

Carbon dioxide is a linear triatomic molecule whose symmetry group is $D_{\infty h}$. Infinite groups like $D_{\infty h}$ and $C_{\infty v}$ cannot be treated in the usual way because they are of infinite order, and hence the Great Orthogonality Theorem cannot be applied to them.

Its *normal-coordinate analysis* was already carried out in Chapter 6 (Appendixes J.2 and J.3) along the bond axis, yielding the stretching motions. We now consider those results from the standpoint of symmetry reasoning alone.

Each atom contributes three coordinates $\rightarrow 3 \times 3 = 9$ degrees of freedom. Three describe translation of the molecule as a whole, and two correspond to rotation about perpendicular axes (rotation about the bond axis has negligible moment of inertia). Hence, CO_2 has **four vibrational degrees of freedom**.

In analogy with water, carbon dioxide has the following modes. The bending mode is, however, doubly degenerate, in both directions perpendicular to the bond axis.

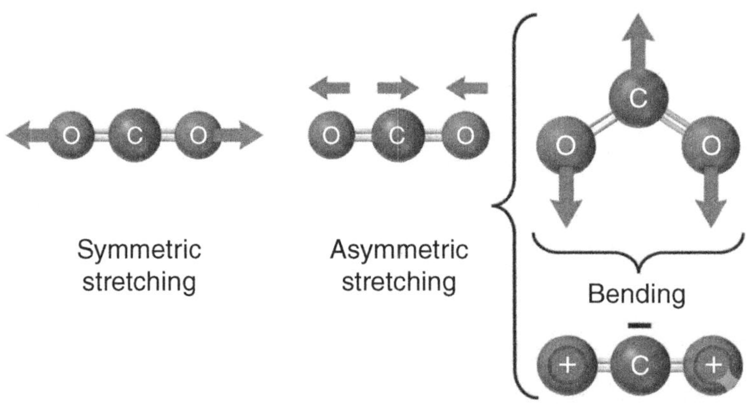

Mode	Symmetry	Description
ν_1	Σ_g^+	**Symmetric stretch**—both C═O bonds elongate/contract together; molecule remains centrosymmetric
ν_2	Π_u	**Degenerate bending mode**—molecule bends in two perpendicular planes
ν_3	Σ_u^+	**Asymmetric stretch**—one bond lengthens while the other shortens; this motion breaks the centrosymmetry and creates an oscillating dipole moment

Only the symmetric stretch (Σ_g^+) is **Raman active** since $x^2 + y^2$ and z^2 transform as Σ_g^+ (Appendix A.6).

The asymmetric stretch (Σ_u^+) is **IR active** because z transforms as Σ_u^+.

The bending mode (Π_u) is also **IR active** (x, y) and doubly degenerate, corresponding to bending in mutually perpendicular planes.

The u (*ungerade*) modes are IR active because they are unsymmetrical and correspond to transformations like x, y, z. In contrast, g (*gerade*) modes are symmetric, involve quadratic functions (g), and are Raman active. This leads to the **Rule of Mutual Exclusion**, according to which, in centrosymmetric molecules, a mode cannot be both IR and Raman active.

8.5 MOs and Symmetry

We now turn from vibrations to **MOs**, which obey the same mathematical laws of symmetry. Each MO must transform as one of the **irreducible representations** of the molecule's point group. This requirement arises from the commutation of the Hamiltonian with the symmetry operations of the molecule.

8.5.1 Symmetry and the Hamiltonian

For any molecule, the electronic Hamiltonian $\hat{H}$ is invariant under all symmetry operations $\hat{R}$ of its point group:

$$[\hat{R}, \hat{H}] = \hat{R}\hat{H} - \hat{H}\hat{R} = 0. \tag{8.5}$$

This means that performing a symmetry operation on the molecule before or after applying $\hat{H}$ yields the same result.

Mathematically, if ψ is an eigenfunction of $\hat{H}$ with eigenvalue E,

$$\hat{H}\psi = E\psi, \tag{8.6}$$

then operating by $\hat{R}$ on both sides gives

$$\hat{R}\hat{H}\psi = \hat{R}E\psi \Longrightarrow \hat{H}\hat{R}\psi = E\hat{R}\psi, \tag{8.7}$$

showing that $\hat{R}\psi$ is also an eigenfunction with the same eigenvalue E.

Hence, ψ and $\hat{R}\psi$ belong to the same **irreducible representation** of the point group.

If the eigenvalue is degenerate, all linearly independent functions belonging to that representation (e.g., the E set in C_{3v}) form a basis.

8.5.2 Consequences for MOs

This principle ensures that:

- Every MO, atomic hybrid, or combination of atomic orbitals (AOs) must transform as an **irreducible representation** of the molecule's point group.
- Orbitals of different symmetry **cannot mix**.
- Only orbitals of the **same symmetry species** can combine to form bonding or antibonding combinations.

Hence, group theory not only classifies vibrations and transitions—it also dictates how bonds form.

8.5.3 Symmetry-adapted Linear Combinations

In a polyatomic molecule, we construct MOs from **linear combinations of AOs** that have compatible symmetry. The simplest illustration is the formation of **σ-bonding orbitals** in water or **π-systems** in ethylene or benzene.

Suppose we have several ligand orbitals $\{\varphi_1, \varphi_2, \varphi_3, \ldots\}$ surrounding a central atom. We can form **symmetry-adapted linear combinations (SALCs))** using the **projection operator formula**:

$$P_i = \frac{l_i}{h} \sum_R \chi_i(R) R, \tag{8.8}$$

where:

- l_i is the dimension of the irreducible representation,
- h is the order of the group,
- the sum runs over all symmetry operations R of the group,
- $\chi_i(R)$ is the character of that operation in the ith representation.

Applying P_i to a trial orbital φ_{trial} extracts the part of that orbital that transforms as the irreducible representation Γ_i.

Example 8.1: Decomposing a Reducible Representation (The Core Skill)

The set of three p-orbitals $(\mathrm{p}_x, \mathrm{p}_y, \mathrm{p}_z)$ on the nitrogen atom in ammonia (NH_3, C_{3v} point group) forms a reducible representation, Γ_p. Decompose Γ_p into its irreducible components.

Solution:

Let's choose:

- z-axis along the $\boldsymbol{C_3}$ principal axis,
- x and y in the $\boldsymbol{H}$**-plane**.

Step-by-step Solution:

1. **Determine the character (χ) for each class of operations:** We use the "unmoved orbitals" trick but note that an orbital's sign flips if it is inverted.
   - $\boldsymbol{E}$: All three orbitals are unchanged. $\boldsymbol{\chi(E) = 3}$.
   - $\boldsymbol{C_3}$: Under C_3, the p_z-orbital is unchanged (+1), but the $(\mathrm{p}_x, \mathrm{p}_y)$ pair is rotated, contributing a character of $2\cos(120°) = -1$ to the trace. Thus, the total character is $1 + (-1) = 0$; $\chi(C_3) = \mathbf{0}$.
   - $\boldsymbol{\sigma_v}$: Consider the σ_v mirror plane that contains p_z and one of the in-plane orbitals (say p_x). This plane:
     - Leaves p_z unchanged (+1).
     - Leaves p_x unchanged (+1).
     - Reflects p_y to $-\mathrm{p}_y$ (−1).
   - **Total character**: $1 + 1 + (-1) = \mathbf{1}$.

So, the reducible representation is: $\boldsymbol{\Gamma_p} = (\mathbf{3, 0, 1})$ for the classes $(E, 2C_3, 3\sigma_v)$.

2. **Use the reduction formula:** The order of the group $h = 6$:
   - $a(A_1) = (1/6) \times [(1)(3)(1) + (2)(0)(1) + (3)(1)(1)] = (1/6)(3 + 0 + 3) = \mathbf{1}$,
   - $a(A_2) = (1/6) \times [(1)(3)(1) + (2)(0)(1) + (3)(1)(-1)] = (1/6)(3 + 0 - 3) = \mathbf{0}$,
   - $a(E) = (1/6) \times [(1)(3)(2) + (2)(0)(-1) + (3)(1)(0)] = (1/6)(6 + 0 + 0) = \mathbf{1}$.
3. **Conclusion:**

$$\boldsymbol{\Gamma_p = A_1 \oplus E}:$$

This tells us that the three p-orbitals split into a nondegenerate orbital (A_1) and a pair of degenerate orbitals (E). By checking the C_{3v} character table, we see that z (p_z) transforms as A_1, and the pair (p_x, p_y) transforms as E. Group theory predicts this splitting without any quantum mechanics!

Transformation of p-orbitals Under Symmetry Operations

When determining the transformation properties of a p-orbital, the sign change is decided by whether the **symmetry operation contains the orbital axis:**

- If the symmetry operation **contains the axis of the p-orbital** (e.g., $C_2(x)$ or $\sigma(xz)$ for a p_x-orbital), the orbital is returned with the **same orientation** and **does not change sign**.
- If the symmetry operation **does not contain the orbital axis**, the orbital orientation is reversed and the wavefunction **changes sign**.

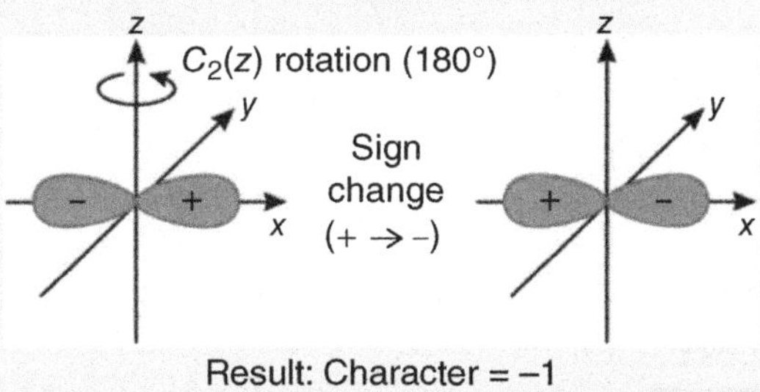

This rule applies to rotations and reflections alike and greatly simplifies the assignment of characters for p-orbitals.

Worked Example 1: σ SALCs in H_2O

Objective: To derive the shape of the MOs in water using the projection operator method.

System Setup:

- **Point Group:** C_{2v}.
- **Coordinate System:** By convention, we align the C_2 axis with the z-axis. We place the molecule in the ***yz*-plane**:
  - $\sigma_v(xz)$: Perpendicular to the molecular plane.
  - $\sigma_v'(yz)$: The molecular plane itself.
- **Basis Set:** The two hydrogen 1s-orbitals, labeled φ_A and φ_B.

Step 1: Generate the Reducible Representation (Γ_H)

We examine how the pair $\{\varphi_A, \varphi_B\}$ transforms under each symmetry operation.

	E	$C_2(z)$	$\sigma_v(xz)$	$\sigma'_v(yz)$
Operation Effect	Identity	Rotation 180°	Reflection ⊥	Reflection ||
Movement	No change	$H_A \leftrightarrow H_B$	$H_A \leftrightarrow H_B$	No change
Unmoved Atoms	2	0	0	2
Character ($\chi^{(\Gamma_H)}$)(R)	**2**	**0**	**0**	**2**

Step 2: Decompose Into Irreducible Representations

Looking at the C_{2v} character table, we can decompose Γ_H by inspection or using the reduction formula:

$$\Gamma_H = A_1 \oplus B_2.$$

(*Check:* $A_1(1, 1, 1, 1) + B_2(1, -1, -1, 1) = (2, 0, 0, 2)$. *This matches* Γ_H.)

Step 3: Construct SALCs Using Projection Operators

We apply the projection operator formula to one of the basis functions (φ_A):

$$\psi^{(\Gamma)} \propto \sum_R \chi^{(\Gamma)}(R) \cdot \hat{R}(\varphi_A).$$

First, determine the effect of operations on φ_A:

- $E\varphi_A = \varphi_A$
- $C_2\varphi_A = \varphi_B$
- $\sigma_v(xz)\varphi_A = \varphi_B$ (swaps atoms)
- $\sigma'_v(yz)\varphi_A = \varphi_A$ (in the plane)

C_{2v}	E	$C_2(z)$	$\sigma_v(xz)$	$\sigma'_v(yz)$
φ_A	φ_A	φ_B	φ_B	φ_A
φ_B	φ_B	φ_A	φ_A	φ_B
χ_H	**2**	**0**	**0**	**2**

For A_1 symmetry (Characters: 1, 1, 1, 1):

$$\psi(A_1) \propto (1)\varphi_A + (1)\varphi_B + (1)\varphi_B + (1)\varphi_A$$

$$\psi(A_1) \propto 2(\varphi_A + \varphi_B).$$

Normalization: Multiplying by $N = 1/\sqrt{2}$ (assuming negligible overlap for the normalization constant derivation):

$$\Psi_{\text{SALC}}(A_1) = \frac{1}{\sqrt{2}}(\varphi_A + \varphi_B).$$

(This corresponds to the hydrogens oscillating in phase.)

For B_2 symmetry (Characters: 1, −1, −1, 1):

$$\psi(B_2) \propto (1)\varphi_A + (-1)\varphi_B + (-1)\varphi_B + (1)\varphi_A$$

$$\psi(B_2) \propto 2(\varphi_A - \varphi_B).$$

Normalization:

$$\Psi_{\text{SALC}}(B_2) = \frac{1}{\sqrt{2}}(\varphi_A - \varphi_B).$$

(This corresponds to the hydrogens oscillating out of phase.)

Step 4: Orbital Interactions (Formation of MOs)

We now match these SALCs with oxygen valence orbitals of the same symmetry.

1. **A_1 Interaction:**
   - **SALC:** $\frac{1}{\sqrt{2}}(\varphi_A + \varphi_B)$.
   - **Oxygen Orbitals:** 2s and $2p_z$ both have A_1 symmetry.
   - **Result:** The SALC mixes with both, forming a deep bonding σ-orbital, a nonbonding (mostly lone pair) orbital, and an antibonding orbital.
2. **B_2 Interaction:**
   - **SALC:** $\frac{1}{\sqrt{2}}(\varphi_A - \varphi_B)$.
   - **Oxygen Orbital:** $2p_y$ has B_2 symmetry (matches the antisymmetry with respect to the xz-plane).
   - **Result:** Forms a bonding σ-orbital and an antibonding orbital.
3. **B_1 Noninteraction:**
   - **Oxygen Orbital:** $2p_x$ is perpendicular to the molecular plane.
   - **SALC:** None. There is no linear combination of hydrogen s-orbitals that has B_1 symmetry.
   - **Result:** The $O(2p_x)$ remains a strictly **nonbonding** pure p-orbital (a lone pair).

Based on this, Figure 8.2 depicts a schematic MO energy level diagram for H_2O.

8.5.4 Physical Interpretation and VSEPR

Group theory provides a rigorous description of the "Canonical Molecular Orbitals."

- **Bonding:** We have two bonding orbitals ($1a_1$ and $1b_2$) responsible for the O—H bonds.
- **Lone Pairs:** We have two nonbonding electrons.
  - One is pure p-character (b_1, from p_x).
  - One is mixed s/p-character (a_1, from s/p_z mixing).

Note: This differs from the elementary VSEPR "sp^3 rabbit ear" model where the two lone pairs are equivalent. However, experimentally (via photoelectron spectroscopy), the two lone pairs of water **do** have different energies, confirming the group theory prediction over the simple hybridization model.

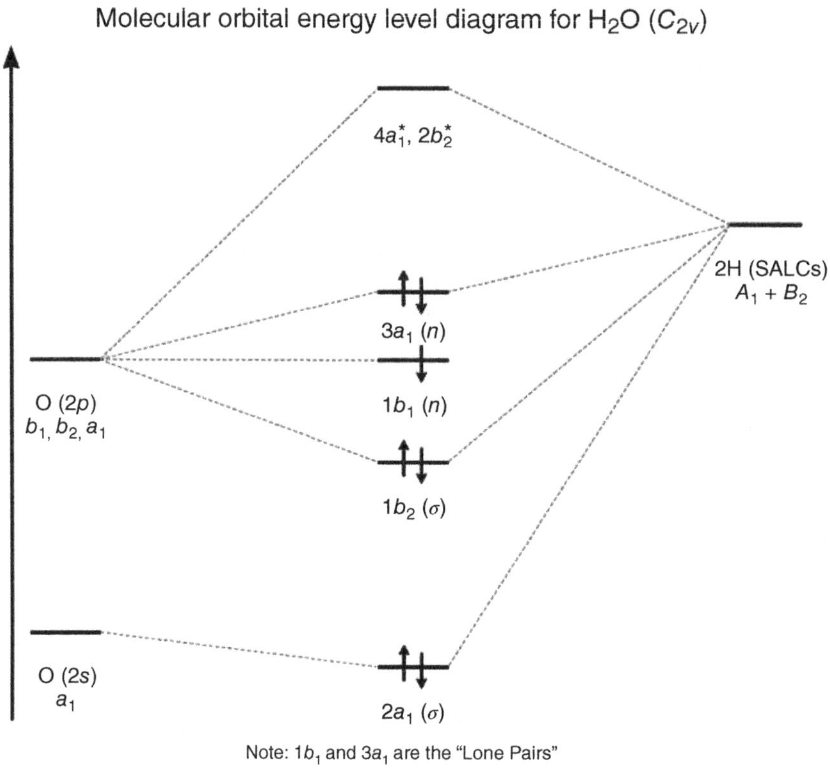

Figure 8.2 MO diagram for H_2O (C_{2v}).

8.5.5 SALCs and Hybridization

Hybridization itself is a symmetry phenomenon. The central atom's valence orbitals (s, p, d) form combinations that transform as irreducible representations of the molecular point group. The foundational principle is that only orbitals of the same symmetry can mix.

Geometry	Point group	Basis orbitals	Symmetry species	Hybridization
Linear	$D_{\infty h}$	s, p_z	$\Sigma_g^+ \oplus \Sigma_u^+$	sp
Trigonal planar	D_{3h}	s, p_x, p_y	$A_1' \oplus E'$	sp^2
Tetrahedral	T_d	s, p_x, p_y, p_z	$A_1 \oplus T_2$	sp^3
Trigonal bipyramidal	D_{3h}	s, p, d	$2A_1' \oplus E' \oplus A_2''$	dsp^3
Octahedral	O_h	s, p, d	$A_{1g} \oplus E_g \oplus T_{1u}$	d^2sp^3

Thus, hybrid orbitals can be viewed as **SALCs of the central atom's AOs** under the symmetry operations of the molecule.

Conceptual Note

Group theory not only classifies vibrations and electronic states but also **predicts the type and orientation of hybrid orbitals.** Hybridization, often taught empirically, is in fact a direct consequence of molecular symmetry and representation theory.

Worked Example 2: Symmetry of the Three Hydrogen 1s-orbitals in NH_3 and Construction of SALCs.

We now analyze how the three hydrogen 1*s* orbitals in ammonia combine to form **SALCs** that interact with nitrogen's valence orbitals.

Step 1: Define the Basis and Coordinate System

Ammonia (NH_3) belongs to the C_{3v} point group. The three H atoms occupy equivalent positions at the corners of a trigonal pyramid with the nitrogen atom at the apex (see Worked Example 2, Section 7.5),

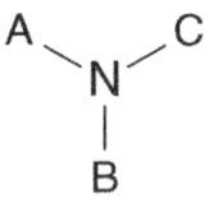

Let the three hydrogen orbitals be $\{\varphi_A, \varphi_B, \varphi_C\}$. We now examine how this *set* of orbitals transforms under each symmetry operation.

C_{3v}	E	$C_3^{(1)}$	$C_3^{(2)}$	$\sigma_v^{(1)}$	$\sigma_v^{(2)}$	$\sigma_v^{(3)}$
φ_A	φ_A	φ_C	φ_B	φ_A	φ_C	φ_B
φ_B	φ_B	φ_A	φ_C	φ_C	φ_B	φ_A
φ_C	φ_C	φ_B	φ_A	φ_B	φ_A	φ_C
$\chi(R)$	3	0	0	1	1	1

Step 2: Construct the Reducible Representation Γ_H

For each operation, count how many hydrogen 1s-orbitals remain unmoved (or mapped onto equivalent positions).

Operation	# of operations	Effect on H atoms	Character $\chi(\Gamma_H)$
E	1	All three remain in place.	3
$C_3(\pm 120°)$	2	All atoms are **cycled** (none fixed).	0
σ_v (3 planes)	3	One H lies in the plane (unchanged), the other two are interchanged.	1

Thus,

$$\Gamma_H = (3, 0, 1).$$

Step 3: Decompose Γ_H Into Irreducible Representations

Using the C_{3v} character table:

Representation	E	$2C_3$	$3\sigma_v$
A_1	1	1	1
A_2	1	1	−1
E	2	−1	0

The group order $h = 6$,

$$a_i = \frac{1}{6} \sum_R n_R \, \chi(\Gamma_H)_R \, \chi_i(R).$$

Compute for each irreducible representation:

Irrep	Calculation	Result
A_1	$(1/6)[(1)(3)(1)+(2)(0)(1)+(3)(1)(1)]=(1/6)(3+0+3)$	1
A_2	$(1/6)[(1)(3)(1)+(2)(0)(1)+(3)(1)(-1)]=(1/6)(3+0-3)$	0
E	$(1/6)[(1)(3)(2)+(2)(0)(-1)+(3)(1)(0)]=(1/6)(6+0+0)$	1

Thus,

$$\Gamma_H = A_1 \oplus E.$$

Step 4: Construct the SALCs

For A_1 symmetry (Characters: 1, 1, 1):

$$\psi(A_1) \propto (1)\varphi_A + (1)\varphi_C + (1)\varphi_B + (1)\varphi_A + (1)\varphi_C + (1)\varphi_B$$

$$\psi(A_1) \propto 2(\varphi_A + \varphi_B + \varphi_C).$$

Normalization: Multiplying by $N = 1/\sqrt{3}$:

$$\Psi_{\text{SALC}}(A_1) = \frac{1}{\sqrt{3}}(\varphi_A + \varphi_B + \varphi_C).$$

For the first orbital of E symmetry (Characters: 2, −1, 0):

$$\psi_1 \propto (2)\varphi_A + (-1)\varphi_C + (-1)\varphi_B.$$

The second and third rows give

$$\psi_2 \propto (2)\varphi_B + (-1)\varphi_A + (-1)\varphi_C,$$

$$\psi_3 \propto (2)\varphi_C + (-1)\varphi_B + (-1)\varphi_A.$$

The three are linearly dependent since their sum is zero. We are looking for two mutually orthogonal SALCs since E is doubly degenerate. We keep ψ_1 as such ($\psi(E_1) = \psi_1$) and take the sum and difference of the other two. The sum $\psi_2 + \psi_3$ gives just ψ_1 with the signs reversed, and hence no new linear combination is obtained. The difference gives

$$\psi(E_2) \propto 3(\varphi_B - \varphi_C).$$

Thus, the projection operator method gives the combinations:

$$\Psi(A_1) = \frac{1}{\sqrt{3}}(\varphi_A + \varphi_B + \varphi_C),$$
$$\Psi(E_1) = \frac{1}{\sqrt{6}}(2\varphi_A - \varphi_B - \varphi_C),$$
$$\Psi(E_2) = \frac{1}{\sqrt{2}}(\varphi_B - \varphi_C).$$

$\boldsymbol{\Psi(A_1)}$: All three 1s-orbitals in phase—totally symmetric combination.

- $\boldsymbol{\Psi(E_1)}$, $\boldsymbol{\Psi(E_2)}$: Two degenerate combinations with opposite phases on different hydrogens.

Step 5: Match with Nitrogen Orbitals

From Example 8.1, the nitrogen valence orbitals transform as

$$\Gamma_{(s+p)} = A_1(2s, p_z) \oplus E(p_x, p_y).$$

We now match the symmetries:

Hydrogen SALCs	Nitrogen orbitals	Type of interaction
A_1	2s, p_z	σ bonding (along N—H)
E	p_x, p_y	σ bonding (off-axis)

Thus, N's A_1 and E orbitals combine with H SALCs of the same symmetry to form **three equivalent N—H bonding orbitals**. The remaining $\mathbf{A_1}$ **(nonbonding)** combination on N corresponds to the **lone pair**.

Final Picture:

$$\text{Bonding scheme: } (A_1 \oplus E)_{\mathrm{H}} \leftrightarrow (A_1 \oplus E)_{\mathrm{N}}.$$

Three equivalent N—H bonds ($E \oplus A_1$).

- One nonbonding lone pair (A_1).

Step 6: Interpretation

This group-theoretical analysis reproduces the familiar **sp^3 hybridization model**, but with mathematical justification:

- The A_1 orbital (2s + p_z) and the E pair (p_x, p_y) on nitrogen mix to match the symmetry of the H SALCs.
- The combination of one A_1 and one E set yields four hybrid orbitals—three directed toward H atoms and one holding the lone pair.

Summary Insight:

Hybridization is not an arbitrary mixing of orbitals but a direct consequence of molecular symmetry. Group theory tells us *which orbitals are allowed to mix*, ensuring that each MO belongs to a definite irreducible representation.

Let us take another example, that of BH_3.

Example 8.2: sp^2 Hybridization in BH_3 (D_{3h})

Group theory explains the hybridization of atomic orbitals by combining them according to the symmetry of the molecule.

Step 1: Determine the Point Group

BH_3 is planar trigonal → D_{3h} symmetry.

The central boron uses one 2s- and three 2p-orbitals: (s, p_x, p_y, p_z).

We find how these orbitals transform under D_{3h}.

Step 2: Character Table for D_{3h} (extract)

D_{3h}	E	$2C_3$	$3C_2'$	σ_h	$2S_3$	$3\sigma_v$	Example functions
A_1'	1	1	1	1	1	1	$z^2, x^2 + y^2$
A_2'	1	1	−1	1	1	−1	R_z
E'	2	−1	0	2	−1	0	(x, y); (p_x, p_y)
A_1''	1	1	1	−1	−1	−1	—
A_2''	1	1	−1	−1	−1	1	z
E''	2	−1	0	−2	1	0	(R_x, R_y)

Step 3: Assign Symmetry of Boron Atomic Orbitals

Orbital	Transformation	Representation
s	Totally symmetric	A_1'
(p_x, p_y)	Like x, y	E'
p_z	Like z	A_2''

Step 4: Construct Hybrid Orbitals

Only orbitals of the same symmetry species can mix.

Hence, $A_1'(s)$ and $E'(p_x, p_y)$ combine to form **three equivalent sp² hybrid orbitals.**

These hybrids lie in the plane of the molecule (σ_h), 120° apart, and directed toward the three hydrogens.

The remaining A_2'' orbital (p_z) remains **unhybridized**, available for π-type interactions with adjacent atoms or ligands.

Step 5: Interpretation

Symmetry species	Orbitals involved	Result
$A_1' \oplus E'$	$s + (p_x, p_y)$	sp² hybrids in plane
A_2''	p_z	Unhybridized p-orbital (perpendicular to plane)

Thus, group theory **justifies the concept of sp² hybridization** as arising from symmetry-allowed linear combinations of A_1' and E orbitals of boron.

8.5.6 Hybridization as a Symmetry Problem

Thus, hybridization corresponds to forming **SALCs of central-atom orbitals**, which transform according to the molecule's irreducible representations.

Molecule	Point group	Symmetry of central-atom Aos	Hybrid type
CH_4	T_d	$A_1 \oplus T_2$	sp^3
NH_3	C_{3v}	$A_1 \oplus E$	sp^3 (one nonbonding)
BF_3	D_{3h}	$A_1' \oplus E'$	sp^2
$BeCl_2$	$D_{\infty h}$	$\Sigma_g^+ \oplus \Sigma_u^+$	sp

8.6 The π-systems

8.6.1 Why π-orbitals Deserve Special Treatment

The π-orbitals are formed by **p-orbitals perpendicular to the molecular plane** and therefore remain orthogonal to the σ-framework. Because they extend above and below the plane, their overlap is smaller but their delocalization is greater.

Group theory provides a powerful way to describe these orbitals without solving the Schrödinger equation explicitly. We simply treat the set of perpendicular p-orbitals as **basis functions** and find how they transform under the operations of the molecule's point group. This yields the symmetry species of the MOs directly, indicating which combinations are bonding, antibonding, or degenerate.

8.6.2 The π-systems and Symmetry

The same reasoning as that for σ orbitals applies to π-orbitals. For example, in **ethylene (D_{2h})**, we place the molecule in the yz-plane, with the z-axis along the C—C bond axis. Then the $2p_x$ orbitals of the two carbons are involved in π-bond formation.

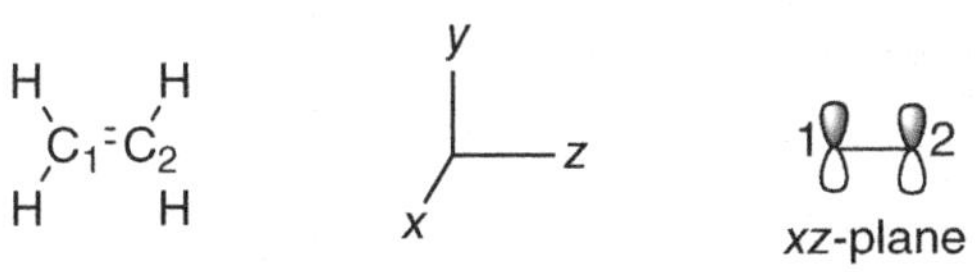

Red Flag 8.2

All π-orbitals do not arise from p_z orbitals. Depending on the symmetry, they may be p_x, p_y, or p_z.

Many organic chemistry texts label π-orbitals as p_z by adopting a different axis convention. The physics is unchanged; only the coordinate labels differ.

In this book, the identity of the π-orbital (p_x, p_y, or p_z) is determined strictly by the chosen coordinate system and the requirement that it be perpendicular to the molecular plane.

The two carbon $2p_x$ orbitals transform as a reducible representation:

D_{2h}	E	$C_2(z)$	$C_2(y)$	$C_2(x)$	i	$\sigma(xy)$	$\sigma(xz)$	$\sigma(yz)$	Linear, rotations
A_g	1	1	1	1	1	1	1	1	
B_{1g}	1	1	−1	−1	1	1	−1	−1	R_z
B_{2g}	1	−1	1	−1	1	−1	1	−1	R_y

(Continued)

(Continued)

D_{2h}	E	$C_2(z)$	$C_2(y)$	$C_2(x)$	i	$\sigma(xy)$	$\sigma(xz)$	$\sigma(yz)$	Linear, rotations
B_{3g}	1	−1	−1	1	1	−1	−1	1	R_x
A_u	1	1	1	1	−1	−1	−1	−1	
B_{1u}	1	1	−1	−1	−1	−1	1	1	z
B_{2u}	1	−1	1	−1	−1	1	−1	1	y
B_{3u}	1	−1	−1	1	−1	1	1	−1	x
φ_1	φ_1	$-\varphi_1$	$-\varphi_2$	φ_2	$-\varphi_2$	φ_2	φ_1	$-\varphi_1$	
φ_2	φ_2	$-\varphi_2$	$-\varphi_1$	φ_1	$-\varphi_1$	φ_1	φ_2	$-\varphi_2$	
$\Gamma(\pi)$	2	−2	0	0	0	0	2	−2	

$$\Gamma(\pi) = (2, -2, 0, 0, 0, 0, 2, -2) = b_{3u} \oplus b_{2g}.$$

This results in one bonding (π, b_{3u}) and one antibonding (π^*, b_{2g}) combination. AOs are typically denoted using lowercase letters, while uppercase notations are used for molecular states.

Concept Check: How Do π-orbitals "Change Sign" in SALCs?

In constructing symmetry-adapted linear combinations (SALCs) for π-electron systems, the phrase "change of sign" refers **not to electric charge** but to the **phase of the orbital wavefunction.**

A p-orbital consists of **two lobes of opposite phase**, conventionally labeled + and −. These signs indicate the algebraic sign of the wavefunction, not positive or negative charge.

When a symmetry operation is applied:

- if a lobe maps onto a lobe of the **same phase**, the orbital is unchanged;
- if a lobe maps onto a lobe of **opposite phase**, the orbital acquires a minus sign.

Remember: A p-orbital retains its sign if the symmetry operation contains its axis; otherwise, its sign is reversed.

Taking the first row for φ_1 (one representative operation per pair; the remaining four duplicate these), the SALC of b_{3u} symmetry is $(1)(\varphi_1) + (-1)(-\varphi_1) + (-1)(-\varphi_2) + (1)(\varphi_2) = 2(\varphi_1 + \varphi_2)$. This is a positive overlap, so it is a bonding orbital. Similarly, the b_{2g} SALC (from the same representative subset) is proportional to $\varphi_1 - \varphi_2$ and is an antibonding orbital. These belong to different symmetry species, and hence transitions between them ($\pi \rightarrow \pi^*$) will be symmetry-allowed if the transition dipole moment contains the totally symmetric representation.

Key Takeaway for SALCs

In forming SALCs:

- orbitals that transform **with the same sign** under a symmetry operation are added,
- orbitals that transform **with opposite signs** are subtracted.

This phase behavior determines whether a particular linear combination is allowed by symmetry.

8.6.2.1 Symmetry and Selection Rules

The electronic transition moment is governed by the same criterion as vibrational transitions:

$$M = \int \psi_f \, \mu \, \psi_i \, d\tau,$$

and the direct product rule applies:

$$\Gamma(\psi_i) \otimes \Gamma(\mu) \otimes \Gamma(\psi_f) \supset A_1.$$

For example:

- In ethylene (D_{2h}), $\pi(b_{3u}) \rightarrow \pi^*(b_{2g})$ transition is allowed because $B_{3u} \otimes B_{2g} = B_{1u}$. On looking up the D_{2h} character table, we find that z transforms as B_{1u}. Therefore, $B_{1u} \otimes B_{3u} \otimes B_{2g} = A_g$; that is, it contains the totally symmetric representation.
- Hence, the transition $\pi(b_{3u}) \rightarrow \pi^*(b_{2g})$ is **allowed** and corresponds to the observed strong $\pi \rightarrow \pi^*$ absorption band around 165 nm.

Thus, **group theory unifies IR, Raman, and electronic spectroscopy under one symmetry principle**.

8.6.3 The π-system of Benzene (D_{6h})

Benzene's six carbon p_z-orbitals form the archetypal delocalized π-system. The point group D_{6h} contains 12 symmetry operations, but only a few affect p_z signs and phases.

8.6.3.1 Constructing the Basis

Take the six p_z-orbitals as basis functions $\{\varphi_1, \ldots, \varphi_6\}$. Under successive 60° rotations (C_6), they transform cyclically. Applying the projection operator or simply using rotation symmetry gives six orthogonal SALCs (Table 8.4).

Table 8.4 Symmetry species of benzene SALCs.

Orbital label	Nodal planes	Symmetry (D_{6h})	Degeneracy
ψ_1	0	a_{2u}	1
ψ_2, ψ_3	1	e_{1g}	2
ψ_4, ψ_5	2	e_{2u}	2
ψ_6	3	b_{2g}	1

These alternate gerade/ungerade, reflecting the inversion symmetry of benzene.

The three lowest orbitals (a_{2u} and e_{1g}) are filled, giving a **closed shell with 6π-electrons**, consistent with aromatic stability.

8.6.3.2 Electronic Transitions and Selection Rules

For simple two-orbital π-systems such as ethylene, orbital-level symmetry arguments correctly reproduce the symmetry of the lowest excited state. For polyatomic π-systems with degeneracy, such as benzene, selection rules must be formulated in terms of electronic states rather than individual orbitals.

Let us first formulate this for ethylene:

For a closed-shell system like the ground state of ethylene or benzene, with all electrons paired, the ground state is ^1A_g, the 1 standing for singlet, since all electrons are paired. We'll ignore the spin multiplicity.

For ethylene,

- π (bonding) orbital: $\Gamma(\pi) = b_{3u}$;
- π^* (antibonding) orbital: $\Gamma(\pi^*) = b_{2g}$.

Ground-state electron configuration $(\pi)^2 = (b_{3u})^2$:

$$\Gamma_{\text{ground}} = B_{3u} \otimes B_{3u} = A_g.$$

The lowest excitation is effectively:

$$(\pi)^2 \to (\pi)^1(\pi^*)^1$$

The excited state configuration is $(\pi)^1(\pi^*)^1$.

The **electronic state symmetry** is therefore

$$\Gamma_{\text{excited}} = B_{3u} \otimes B_{2g}.$$

Using the D_{2h} multiplication table:

$$B_{3u} \otimes B_{2g} = B_{1u}.$$

Allowed transition condition:

$$A_g \otimes \Gamma(\mu) \otimes \Gamma_{\text{excited}} \supset A_g,$$

with

$$\Gamma_{\text{excited}} = B_{1u}.$$

This is satisfied for the **z-polarized transition**, exactly as observed.

Similarly, the ground state of benzene $(^1A_{1g})$ can undergo transitions to excited states if

$$\Gamma_i \otimes \Gamma(\mu) \otimes \Gamma_f \supset A_{1g}.$$

The dipole operator transforms as E_{1u} (x, y components) $\oplus$ A_{2u} (z component). Therefore, the lowest energy $\pi \to \pi^*$ transition corresponds to an excitation from the highest occupied molecular orbital (HOMO) (e_{1g}) to the lowest unoccupied molecular orbital (LUMO) (e_{2u}). The direct product of these representations is

$$E_{1g} \otimes E_{2u} = B_{1u} \oplus B_{2u} \oplus E_{1u}.$$

Since the dipole-moment operator transforms as E_{1u} (for x, y polarization), the $A_{1g} \to E_{1u}$ transition is symmetry-allowed. Transitions to B_{1u} and B_{2u} states are symmetry forbidden.

8.6.4 Symmetry Breaking and Correlation Tables

A powerful application of group theory is the analysis of **symmetry breaking**—predicting what happens to degenerate energy levels when the physical symmetry of a molecule is lowered (e.g., by substitution or geometric distortion).

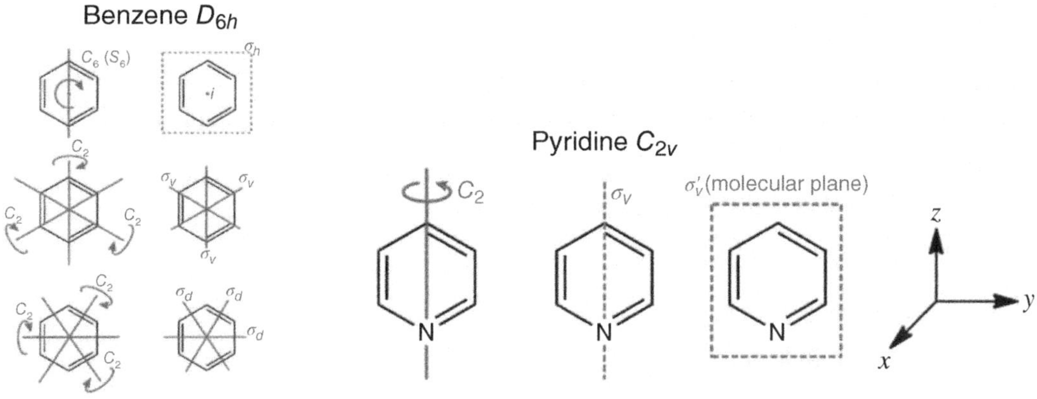

Figure 8.3 Symmetry elements of benzene and pyridine.

Mathematically, this corresponds to restricting the domain of the group operations from a parent group $\mathcal{G}$ to a subgroup $\mathcal{H}$. When we do this, an irreducible representation of $\mathcal{G}$ may become reducible in $\mathcal{H}$.

a. **Descent in Symmetry: Benzene (D_{6h}) to Pyridine (C_{2v})**
   Consider the HOMO of benzene, which we identified in the previous section as the doubly degenerate pair transforming as e_{1g}.
   If we replace one CH group with a nitrogen atom (forming pyridine) or attach a substituent (monosubstituted benzene), the D_{6h} symmetry is destroyed. The only remaining symmetry elements are the molecular plane, a C_2 axis passing through the substituent, and a vertical mirror plane perpendicular to the molecular plane (Figure 8.3). The point group descends from D_{6h} to C_{2v}.
   To determine the fate of the degenerate e_{1g} orbitals, we use a **Correlation Table**, which maps the irreducible representations of a group to those of its subgroups. Alternatively, we can derive the result by inspecting the characters of the preserved operations.

Step 1: Identify the Preserved Operations. In the descent D_{6h} to C_{2v}, the operations preserved are:

- E (Identity),
- C_2' (axis passing through atoms → becomes $C_2(z)$ in C_{2v}),
- σ_v (plane perpendicular to ring → becomes $\sigma_v = \sigma_{xz}$ in C_{2v}),
- σ_h (ring plane → becomes $\sigma_v' = \sigma_{yz}$ in C_{2v}).

Step 2: Determine Characters in the Subgroup. We look up the characters of the E_{1g} representation in the original D_{6h} table for these specific operations:

Operation in D_{6h}	E	C_2'	σ_v	σ_h
Character of E_{1g}	2	0	0	−2

Step 3: Reduce in the Subgroup (C_{2v}). We now treat these characters (2, 0, 0, −2) as a representation in the C_{2v} point group and reduce it using the standard reduction formula,

$$\Gamma_{\text{reduced}} = \frac{1}{h}\sum_R \chi_{\text{red}}(R)\chi_{\text{irr}}(R).$$

Since π-orbitals are antisymmetric to the molecular plane, only the A_2 and B_1 representations in the C_{2v} character table qualify because their character is -1 for σ_{yz} (molecular plane).

Using the C_{2v} character table:

- For A_2: $\frac{1}{4}[(1)(2) + (1)(0) + (-1)(0) + (-1)(-2)] = \frac{1}{4}(4) = 1$.
- For B_1: $\frac{1}{4}[(1)(2) + (-1)(0) + (1)(0) + (-1)(-2)] = 1$.

Performing the reduction yields:

$$E_{1g}(D_{6h}) \to A_2 \oplus B_1 \text{ (in } C_{2v}).$$

Physical Consequence: The irreducible representation E_{1g} has split into two one-dimensional representations, A_2 and B_1.

- **Mathematically:** The invariant subspace of dimension 2 has broken into two invariant subspaces of dimension 1.
- **Physically:** The degeneracy is lifted. The two orbitals, which had identical energy in benzene, now have different energies. This is the group-theoretical justification for **perturbation theory**: a change in the Hamiltonian (substitution) lowers the symmetry, splitting the eigenvalues.

Monosubstituted Benzene

Similarly, a single substituent lowers $D_{6h} \to C_{2v}$ (or C_1 if asymmetric). The degenerate e_{1g} and e_{2u} orbitals split into nondegenerate a and b types. This symmetry breaking explains the doublets observed in UV spectra of monosubstituted benzenes.

b. Disubstituted Benzenes

Type	Approx. group	Consequence
Para	D_{2h}	Minimal splitting, often symmetric bands
Meta	C_{2v}	Multiple nondegenerate transitions
Ortho	C_{2v}	Similar splitting, but stronger interaction between substituents

As symmetry decreases, previously forbidden transitions gain weak intensity because the dipole operator now contains components of lower symmetry.

8.6.5 The Benzene Cation and the Jahn–Teller Effect

A profound consequence of the degeneracy we just derived is observed when we ionize benzene.

Ground State (C_6H_6): The neutral molecule has 6π-electrons. Filling the orbitals from the bottom up (see Table 8.4), we obtain the configuration:

$$(a_{2u})^2(e_{1g})^4.$$

Since all shells are filled (closed shell), the total electronic state transforms as the totally symmetric representation $^1A_{1g}$. The molecule is stable and retains perfect hexagonal symmetry (D_{6h}).

Cation State ($C_6H_6^+$): If we remove one electron to form the radical cation, the configuration becomes

$$(a_{2u})^2(e_{1g})^3.$$

This corresponds to a "hole" in the degenerate e_{1g} orbitals. Consequently, the total electronic wavefunction of the cation possesses $^2E_{1g}$ symmetry.

The Jahn–Teller Theorem: This triggers a fundamental theorem in quantum mechanics: **"Any nonlinear molecule in a spatially degenerate electronic state is unstable."**

Because the electronic state is degenerate (E_{1g}), the molecule will spontaneously distort along a vibrational coordinate to lower its symmetry to D_{2h}. This distortion splits the degenerate e_{1g} level into two nondegenerate levels (typically b_{2g} and b_{3g} in D_{2h}), stabilizing the single electron (or hole).

Result: The benzene cation is **not** a regular hexagon. It distorts into a "squeezed" or "stretched" hexagon (lowering symmetry from $D_{6h} \rightarrow D_{2h}$) to lift the degeneracy and lower the total energy. This explains the complex structure observed in the photoelectron spectrum of benzene, which shows vibrational progressions corresponding to this distortion.

A fundamental principle of quantum mechanics is that degeneracy is a direct consequence of symmetry: when the symmetry of a system is reduced, degeneracies associated with the higher symmetry are lifted.

Key Concepts and Takeaways

- Each π-orbital or SALC must belong to an **irreducible representation**.
- Symmetry determines which combinations are bonding or antibonding and which transitions are allowed.
- **Degeneracy** arises from multidimensional irreducible representations (E-type).
- **Loss of symmetry** removes degeneracy and allows forbidden transitions to appear.
- Aromaticity of benzene is fundamentally a symmetry-imposed closed-shell condition.

8.7 Additional Conceptual Applications

a. Dipole Moment and Symmetry

A molecule possesses a **permanent dipole moment** only if it has components (x, y, z) that transform as *basis functions* for the totally symmetric irreducible representation of the molecule's point group. These components correspond to translational motion (T_x, T_y, T_z) in its point group. Thus, if the molecule belongs to a group containing an inversion center, its dipole must be zero—because a vector quantity changes sign under inversion.

How to Check With Symmetry (the Fast Way)—Questions to Ask Yourself:

1. **Does the molecule have a center of symmetry (inversion center)?** If **YES**, it **cannot** have a dipole moment (e.g., CO_2, benzene).
2. **Does the molecule have multiple high-order symmetry axes (like two or more C_3 or C_2 axes)?** If **YES**, it **cannot** have a dipole moment (e.g., CH_4, SF_6).

3. If the answer to both the above is **NO**, the molecule **can** have a dipole moment (e.g., H_2O, NH_3, HCl).

Dipole moments are found in molecules with **low or no symmetry**, where positive and negative charges are pulled to one side. High-symmetry forces any potential dipole to cancel itself out. Here are some examples:

Molecule	Point group	Dipole components present	Dipole?
H_2O	C_{2v}	$z(A_1) + x(B_1) + y(B_2) \rightarrow A_1$ along z	✓
NH_3	C_{3v}	$z(A_1) \rightarrow$ along C_3 axis	✓
CO_2	$D_{\infty h}$	Inversion center → all components cancel	✗
BF_3	D_{3h}	Planar with $\sigma_h \rightarrow z = 0$	✗
CH_2Cl_2	C_{2v}	No symmetry canceling vectors	✓
CCl_4	T_d	Dipole components cancel by symmetry	✗
trans-1,2-Dichloroethene	C_{2h}	Inversion center	✗
cis-1,2-Dichloroethene	C_{2v}	Vector sum $\neq 0$	✓

Key rule:
A molecule **cannot** have a permanent dipole moment if it belongs to any group containing an inversion center (i), any D group (e.g., D_n, D_{nh}, D_{nd}), or the high-symmetry cubic groups T_d, O_h, or I_h.

Conversely, a molecule **can** have a permanent dipole moment only if it belongs to the point groups C_n, C_{nv}, or C_s.

Note that the C_s point group contains a single mirror plane as its only symmetry element. Although this might seem to preclude a dipole (since the mirror plane is sometimes called σ_h), the dipole vector lies *within* the plane and is therefore unchanged by reflection through it. CH_2FCl is a good example: it belongs to C_s and is polar.

A simple but inexact shortcut sometimes seen in textbooks—"a molecule is polar if it lacks i, σ, and S_n"—fails precisely for this reason and should be used with caution.

Example 8.3: Predicting Molecular Properties From Symmetry

Use group theory to explain why:

a. H_2O has a permanent dipole moment.
b. Benzene has a zero dipole moment.
c. BF_3 has a zero dipole moment.

Solution and Shortcut:

a. H_2O (C_{2v}) Dipole Moment:

Shortcut: A molecule can have a permanent dipole moment if its symmetry allows a dipole vector component to be totally symmetric.

Check: In the C_{2v} character table, the translations (T_x, T_y, T_z) transform as B_1, B_2, and A_1, respectively. The dipole-moment vector itself must transform as the totally symmetric representation of the group. For a vector to be totally symmetric, it must be unchanged by all operations. The dipole-moment vector lies along the z-axis, which is the axis of the C_2 operation, and lies within the $\sigma_v'(yz)$ plane. Therefore, this vector is **invariant (unchanged)** under *all four* operations of the C_{2v} group (E, C_2, $\sigma_v(xz)$, $\sigma_v'(yz)$). Because it transforms as

(1, 1, 1, 1), it belongs to the totally symmetric A_1 representation, and a permanent dipole is allowed.

Conclusion: The dipole-moment vector is symmetric under all operations of the C_{2v} point group, so H_2O **has** a dipole.

b. **Benzene (D_{6h}) Dipole Moment:**

Shortcut: Any molecule with a center of inversion (i) or with multiple high-order axes cannot have a permanent dipole. The dipole-moment vector would not be symmetric under all these operations.

Check: Benzene is D_{6h}, which has a center of inversion and several C_2 axes perpendicular to the principal axis. Any proposed dipole vector would be reversed by a C_2 rotation or inversion, proving it must be zero.

Conclusion: Benzene **has no** permanent dipole moment.

c. **BF_3 (D_{3h}) Dipole Moment:**

Shortcut: Any molecule with multiple high-order symmetry axes cannot have a permanent dipole moment.

Check: The principal axis (C_3) is perpendicular to the plane of the molecule. In the D_{3h} character table, there are three C_2 axes perpendicular to the C_3 axis.

Conclusion: BF_3 **has no** permanent dipole moment.

b. **Optical Activity and Chirality**

Optical activity is a direct consequence of the **absence of improper symmetry elements** (i, σ, S_n). A molecule is **chiral** if it cannot be superimposed on its mirror image.

Condition	Consequence
Contains i, σ, or S_n	Achiral, optically inactive
Lacks i, σ, S_n	Chiral, optically active

Examples:

Molecule	Point group	Chiral?	Optical activity
$[Co(en)_3]^{3+}$	D_3	✓	Active
$[Ni(en)_3]^{2+}$	D_3	✓	Active
CHBrClF	C_1	✓	Active
trans-$[Co(NH_3)_4Cl_2]^+$	D_{4h}	✗	Inactive
cis-$[Co(NH_3)_4Cl_2]^+$	C_{2v}	✗	Inactive
$[Ru(bipy)_3]^{2+}$	D_3	✓	Active

Mnemonic:

Only molecules belonging to point groups containing only proper rotation axes (C_1, C_n, D_n) can be chiral. The presence of any improper rotation (S_n), including $S_1 = \sigma$ and $S_2 = i$, makes the molecule achiral.

Example 8.4: Symmetry and Optical Activity (Chirality)

Which of the following point groups implies that a molecule is chiral (optically active)? C_1, C_2, C_s, C_{2v}, C_{2h}, D_2, D_{3d}.

Solution and Shortcut:

- **The Shortcut Rule:** A molecule is chiral **if and only if** it does not possess any **improper rotation axis (S_n)**. This includes $S_1 = \sigma$ (a mirror plane) and $S_2 = i$ (an inversion center).
- **Apply the Rule:**
  - C_1: No symmetry elements. **Chiral.**
  - C_2: Only a C_2 axis. No S_n. **Chiral.**
  - C_s: Has a mirror plane (S_1). **Achiral.**
  - C_{2v}: Has mirror planes. **Achiral.**
  - C_{2h}: Has a mirror plane and an inversion center. **Achiral.**
  - D_2: Has three perpendicular C_2 axes. No S_n. **Chiral.** (This is the "chiral D_2" case, e.g., twisted allene or certain BINOL derivatives.)
  - D_{3d}: Has an S_2 axis (inversion center) and S_6 axes. **Achiral.**

Conclusion: The chiral point groups are those containing only proper rotation axes, such as C_n ($C_1, C_2, C_3, \ldots$) and D_n ($D_2, D_3, \ldots$).

8.8 Summary

- Symmetry determines bonding, degeneracy, and selection rules for all types of transitions.
- The π-systems of conjugated molecules are excellent demonstrations of symmetry-imposed nodal structure.
- Hybridization and orbital mixing are specific examples of **symmetry adaptation**.
- Even "forbidden" transitions can gain weak intensity when symmetry is dynamically lowered.
- Understanding the **why** behind these rules—commutation, invariance, and representations—transforms group theory from an abstract tool into a predictive framework for chemistry.

8.9 Practice Problems

1. Classify the symmetry of each normal mode of CO_2 and predict which are IR or Raman active.
2. Determine the point group of XeF_4 and derive its vibrational modes.
3. Determine the irreducible representations for the π-system of C_6H_6 and $C_6H_6^+$; discuss how ionization affects allowed transitions.

Further Reading

F. A. Cotton, *Chemical Applications of Group Theory*, 3rd ed., Wiley, 1990.

A. Vincent, *Molecular Symmetry and Group Theory: A Programmed Introduction to Chemical Applications*, Wiley, 2013.

J. S. Ogden, *Introduction to Molecular Symmetry*, Oxford University Press, 2001.
N. Fleming (ed.), *Group-Theoretical Approach to Molecular Orbitals*, LibreTexts Chemistry, online resource.
D. C. Harris and M. D. Bertolucci, *Symmetry and Spectroscopy: An Introduction to Vibrational and Electronic Spectroscopy*, Oxford University Press (1978), Dover Publications reprint, 1989.
R. Kakkar, *Atomic and Molecular Spectroscopy*, Cambridge University Press, 2015.

Exercises

A. Conceptual Questions

1. Why is it necessary that a vibrational integral contain the *totally symmetric representation* for a transition to be allowed? Explain using the parity (odd/even) property of integrals.

2. Why are degenerate vibrations always orthogonal?

3. Why do some molecules have vibrations that are IR active but not Raman active, and vice versa?

4. The C—H stretch in methane appears as a single band in the IR spectrum. Explain how group theory accounts for this.

5. Why can hybrid orbitals such as sp^2 and sp^3 be regarded as *symmetry-adapted linear combinations (SALCs)*?

6. Why are linear molecules special cases where group theory (based on finite point groups) cannot be directly applied?

7. **The Origin of Selection Rules**
   Explain why an electronic transition is only allowed if the direct product of the initial state, the transition moment operator, and the final state contains the totally symmetric representation (Γ_{TS}, e.g., A_1). How does this relate to the conservation of energy and momentum?

8. **Infrared versus Raman Activity**
   State the symmetry conditions required for a vibrational mode to be:
   a. Infrared active
   b. Raman active
      i. Why is the "symmetric stretch" of CO_2 Raman active but IR inactive, while the same mode in H_2O is active in both?
      ii. Using the C_{3v} character table, predict which normal modes of NH_3 are IR and Raman active.

9. **The Mutual Exclusion Principle**
   "For a molecule with a center of inversion (i), no vibrational mode can be both IR and Raman active."

Using the properties of the inversion operator on vector coordinates (x, y, z) and quadratic functions $(xy, x^2$, etc.), explain why this rule holds true.

10. **Chirality and Improper Rotations**
    A common misconception is that chiral molecules must lack a mirror plane. The rigorous condition is the absence of an improper rotation axis (S_n). Explain the difference, giving an example of a molecule that lacks a mirror plane but is still achiral due to an S_n axis.

11. **Projection Operators**
    Describe the physical function of a projection operator $\left(\widehat{P^{(\Gamma)}}\right)$. When applied to an AO basis set, what does the resulting linear combination represent in terms of Molecular Orbital Theory?

B. Problem-solving Exercises

1. Construct the reducible representation for the three p-orbitals $\left(p_x, p_y, p_z\right)$ in NH_3 (C_{3v}) and decompose it into irreducible representations. Identify which combinations correspond to hybrid orbitals.

2. Using the C_{2v} character table, determine which of the nine Cartesian displacements of H_2O remain invariant under each symmetry operation, and hence derive its three normal modes of vibration.

3. Identify which of the following molecules possess a dipole moment: (a) CO_2, (b) H_2O, (c) BF_3, (d) XeF_4, (e) CH_2Cl_2, (f) cis-1,2-dichloroethylene, (g) trans-1,2-dichloroethylene. Justify each answer based on symmetry.

4. Which of the following molecules can be optically active? (a) CHBrClF, (b) *cis*-$[Co(en)_2Cl_2]^+$, (c) *trans*-$[Co(en)_2Cl_2]^+$, (d) CH_2Cl_2.

5. Verify by direct multiplication that in C_{2v}:
   $B_1 \otimes B_1 = A_1, B_2 \otimes B_2 = A_1, B_1 \otimes B_2 = A_2$.
   Use this result to justify IR selection rules for H_2O.

6. **Vibrational Analysis of Boron Trifluoride (BF_3)**
   The BF_3 molecule belongs to the D_{3h} point group.
   a. Determine the number of degrees of freedom and the expected number of vibrational modes.
   b. Using the "Unmoved Atoms" shortcut, determine the reducible representation Γ_{vib} for the vibrational modes.
   c. Decompose Γ_{vib} into its irreducible components.
   d. Identify which modes are IR active and which are Raman active using the character table.

7. **Ammonia SALCs (C_{3v})**
   Consider the ammonia molecule (NH_3).
   a. Generate the reducible representation (Γ_H) for the three hydrogen 1s-orbitals.

b. Apply the projection operator to construct the symmetry-adapted linear combinations (SALCs) for the A_1 and E representations.
c. Sketch the resulting shapes of these group orbitals.

8. **Hybridization in Methane**
   Methane (CH_4) has T_d symmetry.
   a. Write down the characters (χ) for the reducible representation (Γ_{valence}) spanned by the four carbon valence orbitals $(2s, 2p_x, 2p_y, 2p_z)$.
   b. Show that Γ_{valence} decomposes into $A_1 \oplus T_2$.
   c. Explain how this decomposition confirms the sp^3 hybridization model.

C. Advanced/Extension Questions

1. Show that for any molecule, the **sum of the squares** of the dimensions of all irreducible representations equals the order of the group.

2. For the staggered and eclipsed conformations of ethane (D_{3d} and D_{3h}, respectively), predict which normal modes are IR active and which are Raman active.

3. Discuss how symmetry breaking in monosubstituted and disubstituted benzenes affects the degeneracy of π-molecular orbitals.

4. Derive the symmetry species of π-orbitals in benzene (D_{6h}) using the C_6 axis and the plane of the molecule as symmetry elements.

5. Prove that the commutation of the symmetry operator $\hat{R}$ with the Hamiltonian $\hat{H}$ implies that molecular orbitals belong to irreducible representations of the point group.

6. Using the direct product table for C_{3v}, show that electronic transitions $A_1 \rightarrow E$ and $E \rightarrow A_1$ are allowed, while $A_1 \rightarrow A_2$ is forbidden.

7. Derive the character for a rotation operation (C_n) and an improper rotation (S_n), starting from the transformation matrix for rotation about the z-axis.

9

Limits, Series, and Power-series Expansions

So what difference does it make, whether it's 20 minutes or 20 years, since neither amounts to the faintest echo of the tiniest whisper in the thunder of time.

—Dr. Paul Wayne

In this chapter, we build the foundations necessary for understanding power-series and Fourier expansions. We begin with limits and indeterminate forms, develop Taylor/Maclaurin expansions, study convergence tests, and finally introduce Fourier series with applications in spectroscopy, signal processing, and quantum mechanics.

Learning Outcomes

After completing this chapter, students will be able to:

1. **Evaluate** limits of functions, **recognize** indeterminate forms, and **apply** the mean value theorem and l'Hôpital's rule to resolve them.
2. **Construct** Taylor and Maclaurin series for analytic functions and **understand** their domains of validity and error behavior.
3. **Use** the generalized binomial expansion for real or complex exponents and **identify** convergence conditions.
4. **Apply** convergence tests—ratio, root, alternating, and comparison—to determine when infinite series converge or diverge.
5. **Formulate** Fourier series expansions (trigonometric, exponential, and half-range) and **compute** the corresponding coefficients using orthogonality.
6. **Interpret** applications of Fourier and power series in chemistry and physics, including spectroscopy, signal processing, and quantum mechanical wavefunctions.

9.1 Limit of a Function

Many times, we are interested in examining the behavior of a function, say, $f(x)$, at $x = 0$, $x = \infty$, or some intermediate value. Some of the deepest ideas in chemistry and physics—such as the Heisenberg Uncertainty Principle or the conditions for constructive interference in X-ray diffraction—arise from analyzing what happens to mathematical expressions as *a variable approaches a limiting value.*

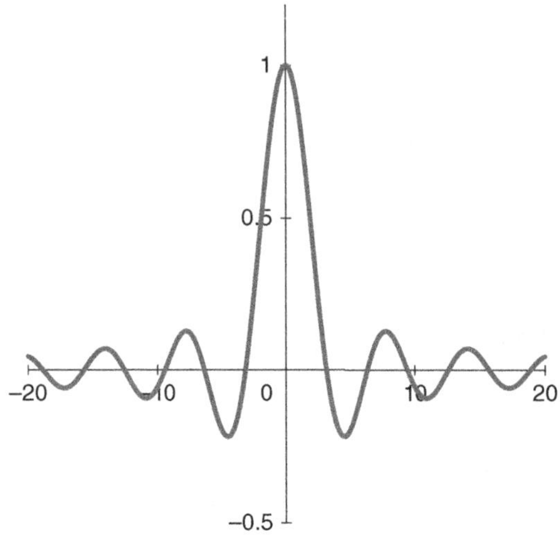

Figure 9.1 Plot of $\text{sinc}(x) = \sin(x)/(x)$ near $x = 0$.

For the uncertainty principle, the impossibility of measuring position and momentum to arbitrary precision emerges mathematically from the properties of wavefunctions under the limit process, as expressed in inequalities like $\Delta x \Delta p \geq \frac{h}{4\pi}$. Likewise, the distribution and intensity of X-ray diffraction patterns are predicted by analyzing what happens to the mathematical expressions for scattered waves as the variables approach resonant, or Bragg, conditions. Without the framework of limits, these foundational principles would lack formal mathematical support.

For example, the ratio $\sin(x)/x$ is encountered in several places, such as in the Fourier transform of the box function to demonstrate the Heisenberg Uncertainty Principle, in time-dependent perturbation theory to study the interaction of radiation with matter, and in the study of X-ray diffraction. Trouble comes when you try to evaluate the function at $x = 0$, because the ratio is $0/0$, which appears to be **indeterminate**. However, when you use plotting software and plot this function, you get the above result (Figure 9.1).

Clearly, in the limit as x tends to zero, the function $\sin(x)/x$ equals 1. Alternatively, we may write this as

$$\lim_{x\to 0}\left(\frac{\sin x}{x}\right) = 1. \tag{9.1}$$

The function within parentheses is so ubiquitous that it has been given a special name, sinc(x). Mathematical software programs such as **Mathematica®**, **Desmos Graphing Calculator**, or **GeoGebra** are optimized to calculate this ratio. However, if you write a simple computer program to evaluate this ratio at $x = 0$, it will return a "division by zero" or similar error.

The **indeterminate forms** include $0/0$, ∞/∞, $0\times\infty$, $\infty-\infty$, 1^∞, 0^0, and ∞^0. By contrast, the expression $0/\infty$ is unequivocally equal to zero. We will soon learn a general method for determining such limits.

Math Box 9.1: Indeterminate Forms in Calculus

An **indeterminate form** is an algebraic expression encountered in limits where direct substitution does not reveal a definite limit value. This is because the behavior of the expression cannot be determined solely from the limits of its individual parts; different limits may yield different results depending on the functions involved.

Common indeterminate forms: $\frac{0}{0}$, $\frac{\infty}{\infty}$, $0 \times \infty$, $\infty - \infty$, 1^{∞}, 0^0, ∞^0.

By contrast, an expression such as

$$\frac{0}{\infty} = 0$$

is *not* indeterminate.

Example: Evaluate $\lim_{x\to 0}\left(\frac{\ln(1+x)}{x}\right)$. Direct substitution gives $\frac{0}{0}$, so the limit is indeterminate and requires a different method (such as l'Hôpital's rule) to evaluate it.

Now, let us return to the question of determining *indeterminate* ratios such as $f(x)/g(x)$ at $x = a$, where both the numerator and denominator vanish. By the **mean value theorem**, for a function $f(x)$ continuous on (a, x) and differentiable in that interval, there exists a point x_1 between a and x such that

$$\frac{f(x) - f(a)}{x - a} = f'(x_1). \tag{9.2}$$

Substituting $x = b$ in Equation (9.2), we obtain

$$\frac{f(b) - f(a)}{b - a} = f'(x_1). \tag{9.3}$$

Let there be another function $g(x)$ for which a similar relation holds at x_2:

$$\frac{g(b) - g(a)}{b - a} = g'(x_2). \tag{9.4}$$

If both $f(a)$ and $g(a)$ are zero and we wish to evaluate $f(a)/g(a)$, we face the same situation as when finding $\lim_{x\to 0}(\sin x/x)$. Taking the ratio of Equations (9.3) and (9.4) gives

$$\frac{f(b) - f(a)}{g(b) - g(a)} = \frac{f'(x_1)}{g'(x_2)}. \tag{9.5}$$

Because both x_1 and x_2 lie between a and b, they approach a as $b \to a$, and we have

$$\lim_{x\to a} \frac{f(x)}{g(x)} = \lim_{x\to a} \frac{f'(x)}{g'(x)}. \tag{9.6}$$

This is known as **l'Hôpital's rule**. This rule applies provided $g'(x) \neq 0$ in a neighborhood of a (except possibly at a itself), and the limit on the right-hand side exists (or is $\pm\infty$), and $g'(x) \neq 0$ in a **punctured neighborhood** of a:

- The theorem requires differentiability in a punctured neighborhood.
- Both numerator and denominator must approach 0 or $\pm\infty$.

It was formulated by the French mathematician **Guillaume de l'Hôpital (1661–1704)**. In early texts, his name appeared as *l'Hospital* with a silent *s*; modern French orthography replaces the *s* with a circumflex accent (ô).

Historical Note 9.1: Guillaume de l'Hôpital and Johann Bernoulli

It is interesting that *l'Hôpital's rule* was not discovered by l'Hôpital himself.

In 1696, the French nobleman Guillaume de l'Hôpital published the first textbook on differential calculus, *Analyse des infiniment petits*. The rule for evaluating indeterminate forms, now bearing his name, actually originated from Johann Bernoulli, the younger brother of Jakob Bernoulli. l'Hôpital had paid Johann a stipend in exchange for exclusive access to his discoveries. Bernoulli's authorship became widely known only after his letters were published decades later. Despite this, the name persists because l'Hôpital ensured the technique reached a wider audience.

Now, we apply this rule to the ratio $\sin x/x$ as $x \to 0$. According to Equation (9.6),

$$\lim_{x\to 0} \frac{\sin x}{x} = \lim_{x\to 0} \frac{\cos x}{1} = 1. \tag{9.7}$$

Similarly,

$$\lim_{x\to 0}\left(\frac{\ln(1+x)}{x}\right) = \lim_{x\to 0} \frac{1}{1+x} = 1. \tag{9.8}$$

Example 9.1: Evaluating Limits

Evaluate $\lim_{x\to 0} \dfrac{\sin(kx)}{x}$.

Solution:

Using l'Hôpital's rule,

$$\lim_{x\to 0} \frac{\sin(kx)}{x} = \lim_{x\to 0} \frac{k\cos(kx)}{1} = k.$$

9.2 Taylor and Maclaurin Series

The general form of a **Taylor series** expansion of a function $f(x)$ about a point a is

$$f(x) = f(a) + f'(a)(x-a) + \frac{f''(a)}{2!}(x-a)^2 + \frac{f^{(3)}(a)}{3!}(x-a)^3 + \cdots. \tag{9.9}$$

The function must be **analytic** in the region of interest, its derivatives must exist, and the series must **converge** within that region.

Example: Expansion of ln *x* about *x* = 1

Let us illustrate this using an example. From Equation (9.9), the expansion of $\ln x$ around $x = 1$ is as follows:

$$\ln x = \ln(1) + \left(\frac{1}{x}\right)_{x=1}(x-1) + \left(\frac{-1}{x^2}\right)_{x=1}\frac{(x-1)^2}{2} + \left(\frac{2}{x^3}\right)_{x=1}\frac{(x-1)^3}{3!}$$
$$+ \left(\frac{-6}{x^4}\right)_{x=1}\frac{(x-1)^4}{4!} + \cdots$$

or

$$\ln x = (x-1) - \frac{(x-1)^2}{2} + \frac{(x-1)^3}{3} - \frac{(x-1)^4}{4} + \cdots. \tag{9.10}$$

Therefore,

$$\boxed{\ln x = \sum_{n=1}^{\infty} (-1)^{n+1} \frac{(x-1)^n}{n},}$$

which is valid for $0 < x \leq 2$.

(*Note:* The series also converges at $x = 2$, giving the alternating harmonic series.)

A special case arises when $a = 0$ in Equation (9.9); the series then becomes

$$f(x) = f(0) + f'(0)x + \frac{f''(0)}{2!}x^2 + \frac{f^{(3)}(0)}{3!}x^3 + \cdots. \tag{9.11}$$

This is known as the **Maclaurin series**, i.e., the Taylor series expansion about $x = 0$.

Historical Note 9.2: Brook Taylor and Colin Maclaurin

The idea of expanding a function in an infinite polynomial series was introduced by Brook Taylor in 1715. Taylor's contemporaries considered the method rather formal and even mysterious. Later, the Scottish mathematician Colin Maclaurin specialized the method by setting the expansion point to zero, producing what we now call the Maclaurin series. Their work predated the rigorous notion of convergence by nearly two centuries; Cauchy, Weierstrass, and others later provided the formal foundations that justify such expansions.

Some common Maclaurin series are:

$$\sin x = x - \frac{x^3}{3!} + \frac{x^5}{5!} - \cdots, \tag{9.12}$$

$$\cos x = 1 - \frac{x^2}{2!} + \frac{x^4}{4!} - \cdots, \tag{9.13}$$

$$e^x = 1 + \frac{x}{1!} + \frac{x^2}{2!} + \frac{x^3}{3!} + \cdots. \tag{9.14}$$

All of these can be obtained from Equation (9.11).

Consider Equation (9.12). It is clear that if $|x| \ll 1$, the successive terms become smaller, and the series rapidly converges to $\sin x$. A computer program cannot evaluate an infinite number of terms, but it can be designed to compute the sum up to a specified precision.

The summation may be terminated once the desired accuracy is achieved.

The difference between a truncated series and the exact function is called the remainder or **error term**; its magnitude decreases as higher-order terms are included.

A comparison of the Taylor and Maclaurin series is given in Table 9.1.

Table 9.1 Taylor versus Maclaurin series.

Feature	Taylor series	Maclaurin series
Expansion point	Any point a	Fixed at $a = 0$
General form	$f(x) = \sum_{n=0}^{\infty} \frac{f^{(n)}(a)}{n!}(x-a)^n$	$f(x) = \sum_{n=0}^{\infty} \frac{f^{(n)}(0)}{n!}x^n$
Usefulness	Suitable when behavior around a specific point a is needed	Useful for functions naturally defined at 0 (e.g., $\sin x$, $\cos x$, e^x)
Typical domain of validity	Depends on the distance from the expansion point to the nearest singularity	Same rule, but singularities measured from 0
Special case?	The Maclaurin series is a **special case** of the Taylor series	Yes—specialized at $a = 0$
Common applications	Approximations near arbitrary points; shifts in potential energy curves	Spectroscopy, perturbation expansions, computational chemistry algorithms

9.3 Binomial Expansion

Let us now develop a function $f(x) = (x+1)^{\alpha}$ as a **Maclaurin series**. It is easy to show for $|x| < 1$ that

$$(1+x)^{\alpha} = 1 + \alpha x + \frac{\alpha(\alpha-1)}{2!}x^2 + \frac{\alpha(\alpha-1)(\alpha-2)}{3!}x^3 + \cdots. \tag{9.15}$$

This is known as the **binomial series**, and the coefficients are the **binomial coefficients**, denoted by

$$\binom{\alpha}{k} = \frac{\alpha(\alpha-1)(\alpha-2)\cdots(\alpha-k+1)}{k!}. \tag{9.16}$$

Here, α can be any real or complex number and is not necessarily an integer.

For fractional or negative values of the exponent n, the binomial expansion becomes an infinite series, referred to as the **generalized binomial expansion**.

When α is a positive integer n, all coefficients beyond $k = n$ vanish. For example, for $k = n + 1$, Equation (9.16) becomes

$$\binom{n}{n+1} = 0 \tag{9.17}$$

because the numerator contains a zero factor. Hence, all subsequent coefficients also vanish, and the infinite series reduces to a polynomial of degree n:

$$(1+x)^n = \sum_{k=0}^{n} \binom{n}{k} x^k. \tag{9.18}$$

This is **Newton's Binomial Formula**.

Historical Note 9.3: Newton's Generalized Binomial Expansion

Although the binomial theorem is taught early in algebra, Isaac Newton discovered a much deeper fact: the algebraic formula

$$(1 + x)^n$$

makes sense even when n is fractional or negative.

In 1665–1666, during the famous "plague years," Newton developed the infinite binomial series while working in Woolsthorpe. This insight was revolutionary because it extended algebra beyond integers into analysis, paving the way for power-series methods that became indispensable in physics and chemistry.

Equation (9.15) finds numerous applications in physics and chemistry because, unlike Newton's finite formula, it applies equally well to fractional or negative values of α.

For example:

For $\alpha = 1/2$,

$$(1 + x)^{1/2} = 1 + \frac{1}{2}x - \frac{1}{8}x^2 + \frac{1}{16}x^3 - \cdots, \tag{9.19}$$

and for $\alpha = -1/2$,

$$(1 + x)^{-1/2} = 1 - \frac{1}{2}x + \frac{3}{8}x^2 - \frac{5}{16}x^3 + \cdots. \tag{9.20}$$

Such series are often used in computational approximations; the number of terms retained depends on the desired precision.

Example 9.2: Expansion of $(1 + y)^{-1/2}$

Expand $(1 - 2xt + t^2)^{-1/2}$ in powers of t and collect coefficients of t^0, t^1, and t^2.

Solution:

Let $y = -2xt + t^2$.

From Equation (9.20),

$$(1 + y)^{-1/2} = 1 - \frac{1}{2}y + \frac{3}{8}y^2 + \cdots.$$

Substituting $y = -2xt + t^2$ and simplifying, the coefficients are

$t^0 \to 1$, $t^1 \to x$, $t^2 \to (3x^2 - 1)/2$.

Before proceeding further, an important aspect of infinite series needs consideration—their **convergence**, i.e., whether the sum approaches a finite value as successive terms are added.

9.4 Tests of Series Convergence

A power series represents a function only within a specific interval, known as its **domain of validity**.

A power series may converge for some values of x and diverge for others. To determine this behavior, we use a set of classical tests.

9.4.1 Ratio (Cauchy–d'Alembert) Test

Cauchy's ratio test (also called **d'Alembert's ratio test**) is one of the most frequently used criteria for series convergence.

Let T_n and T_{n+1} denote the nth and $(n+1)$st terms of a series. Define

$$L = \lim_{n\to\infty}\left|\frac{T_{n+1}}{T_n}\right|.$$

If $L < 1$, the series **converges absolutely**. (**Absolute convergence** means the series would still converge if all its terms were replaced by their absolute values.) A series is said to be **conditionally convergent** if it converges but fails to converge absolutely:

- If $L > 1$ or L diverges, the series **diverges**.
- If $L = 1$ or the limit does not exist, the test is **inconclusive**.

Example 9.3: To Illustrate The Use Of The Ratio Test, Consider The Exponential Series

Apply the ratio test to the exponential series

$$e^x = 1 + \frac{x}{1!} + \frac{x^2}{2!} + \frac{x^3}{3!} + \cdots.$$

Here, $T_n = x^n/n!$.

Then

$$\left|\frac{T_{n+1}}{T_n}\right| = \frac{|x|}{n+1} \to 0$$

for all finite x; thus, the exponential series converges **for all values of x.**

Historical Note 9.4: Augustin-Louis Cauchy and Convergence

The modern notion of convergence originates largely from Cauchy (1821). Before this, series were manipulated with extraordinary boldness—Euler even summed divergent series to compute values of the zeta function! Cauchy introduced the rigorous distinction between convergence, absolute convergence, and divergence, and established criteria such as the ratio test. These developments transformed infinite series from a formal tool into a bedrock of mathematical physics.

Closely related to the ratio test is the **root test**, which examines the limit of the nth root of the absolute value of the terms of a series to determine convergence.

Table 9.2 lists some of the common series used in physics and chemistry.

Table 9.2 Frequently used Maclaurin series.[a]

Function	Series expansion	Validity
e^x	$1 + x + \frac{x^2}{2!} + \frac{x^3}{3!} + \cdots$	All x
$\sin x$	$x - \frac{x^3}{3!} + \frac{x^5}{5!} - \cdots$	All x
$\cos x$	$1 - \frac{x^2}{2!} + \frac{x^4}{4!} - \cdots$	All x
$\ln(1 + x)$	$x - \frac{x^2}{2} + \frac{x^3}{3} - \cdots$	$-1 < x \leq 1$
$(1 + x)^{\alpha}$	$1 + \alpha x + \frac{\alpha(\alpha - 1)}{2!}x^2 + \cdots$	
$\frac{1}{1-x}$	$1 + x + x^2 + x^3 + \cdots$	

a. All of these follow directly from the Taylor/Maclaurin formula and are used extensively in quantum mechanics, spectroscopy, and perturbation theory.

9.4.2 Alternating Series Test (Leibniz)

If a series has alternating signs, it is sufficient to check that the magnitude of successive terms decreases. A series

$$\sum_{n=0}^{\infty}(-1)^n a_n$$

converges if:

1. $a_{n+1} \leq a_n$,
2. $a_n \to 0$.

This applies to the series for $\sin x$ and $\ln(1 + x)$.

Consider the series for $\sin x$ from Equation (9.12):

$$T_n = \frac{(-1)^n x^{2n+1}}{(2n + 1)!}.$$

Alternate terms change sign; therefore, we need to consider only the absolute magnitudes of the terms. The **Alternating Series Test** also requires that the limit of the nth term as $n \to \infty$ is zero.

Since

$$\lim_{n\to\infty} |T_n| = \lim_{n\to\infty} \left|\frac{x^{2n+1}}{(2n + 1)!}\right| \to 0,$$

the series converges for all x.

Similarly, the $\cos x$ series (Equation 9.13) also converges everywhere because

$$\lim_{n\to\infty} |T_n| = \lim_{n\to\infty} \left|\frac{x^{2n}}{(2n)!}\right| \to 0.$$

Example 9.4: Convergence of Binomial Series

Show that the binomial series

$$(1+x)^\alpha = 1 + \alpha x + \frac{\alpha(\alpha-1)}{2!}x^2 + \cdots$$

converges for $|x| < 1$.

Solution:

From Equations (9.15) and (9.16), the nth term is

$$T_n = \frac{\alpha(\alpha-1)\cdots(\alpha-n+1)}{n!}x^n.$$

Hence,

$$\frac{T_{n+1}}{T_n} = \frac{(\alpha-n)}{n+1}x,$$

and taking the limit as $n \to \infty$, we get $|x| < 1$ for convergence. Thus, the series converges when $-1 < x < 1$.

At $x = \pm 1$, convergence depends on α: for $\alpha > -1$, the series converges at $x = -1$; at $x = 1$ the series always converges.

Sometimes, a series converges only within a specific range of x, as in Example 9.4; the **radius of convergence** must then be stated.

9.4.3 Advanced Example: Legendre-type Series

Consider a Legendre-type series with recurrence:

$$(n+2)(n+1)U_{n+2} = \lambda_n U_n, \text{ where } \lambda_n \sim n^2 \text{ as } n \to \infty.$$

Taking the limit of the ratio of successive nonzero terms:

$$\frac{U_{n+2}}{U_n} \to 1 (n \to \infty).$$

Thus

$$|x| < 1$$

is required for convergence.

At $x = \pm 1$, further analysis is needed: the series diverges unless λ_n corresponds to a finite Legendre polynomial.

Example 9.5: Convergence of the Legendre Series

The Legendre polynomials $P_l(x)$ arise as solutions to the Legendre differential equation. The power-series solution $y = \sum_{k=0}^{\infty} a_k x^k$ is governed by the recurrence relation

$$a_{k+2} = -\frac{(l-k)(l+k+1)}{(k+1)(k+2)}a_k,$$

where l is a nonnegative constant. Determine the interval of convergence for this series.

Solution:

To determine the radius of convergence R, we apply the **ratio test** to the consecutive nonzero terms of the series (which are two steps apart, k and $k+2$),

$$L = \lim_{k\to\infty}\left|\frac{a_{k+2}x^{k+2}}{a_k x^k}\right| = |x|^2 \cdot \lim_{k\to\infty}\left|\frac{(l-k)(l+k+1)}{(k+1)(k+2)}\right|.$$

For very large k, the constant l becomes negligible compared to k, so the numerator behaves like $-k^2$ and the denominator like k^2:

$$\lim_{k\to\infty}\left|\frac{-k^2}{k^2}\right| = 1.$$

Thus, $L = |x|^2 \cdot 1 = |x|^2$.

For the series to converge, the ratio test requires $L < 1$:

$$|x|^2 < 1 \Longrightarrow |x| < 1.$$

Conclusion:

The Legendre series converges for $-1 < x < 1$.

At the end points $x = \pm 1$, the series generally diverges unless l is an integer. If l is an integer (e.g., $l = 0, 1, 2$ ...), the numerator $l - k$ eventually becomes zero, causing the series to terminate and become a finite polynomial (a Legendre Polynomial), which converges everywhere.

Another useful criterion is the **comparison test**, in which the convergence or divergence of a series is inferred by comparing it with a second series whose behavior is already known.

The **comparison test** is particularly useful in physical applications, where series often arise from approximations to known functions. If $0 \le a_n \le b_n$ for all sufficiently large n, and the series $\sum b_n$ is known to converge, then the series $\sum a_n$ also converges. Conversely, if $\sum a_n$ diverges and $a_n \ge b_n \ge 0$, then $\sum b_n$ must also diverge. This test is frequently employed to establish the convergence of series encountered in quantum mechanical problems, such as those arising in the analysis of the harmonic oscillator.

Table 9.3 lists the convergence tests at a glance.

Table 9.3 Convergence of series.[a]

Type	What it means	Example	Test used
Absolute convergence	Series converges if all terms are replaced by their absolute values	$\sum \frac{(-1)^n}{n^2}$	Ratio/root test, p-series test
Conditional convergence	Series converges, but the absolute value series diverges	$\sum \frac{(-1)^n}{n}$	Alternating series test
Divergence	Series does not approach a finite sum	$\sum \frac{1}{n}$	Comparison test

a. A series that converges absolutely also converges conditionally. Ratio, root, comparison, and alternating series tests are common tools to determine convergence.

9.5 Fourier Series

Many physical quantities in chemistry, physics, and engineering vary periodically: lattice potentials in crystals, vibrational waveforms, electronic signals, and even the angular dependence of scattering amplitudes.

A **Fourier series** expresses a periodic function as a sum of sine and cosine functions of different frequencies. These form the "building blocks" for periodic motions, much as atomic orbitals form the building blocks of molecular wavefunctions.

It is named after the French mathematician **Joseph Fourier**. A function $f(x)$ can be expanded as a Fourier series provided it satisfies the **Dirichlet conditions**:

1. The function must be *periodic*.
2. It must be *single-valued* and continuous, except at a finite number of finite discontinuities.
3. It must have a finite number of maxima and minima within a single period.
4. The integral of $f(x)$ over one period must converge.

The Fourier series has numerous applications in science and engineering.

In chemistry, Fourier analysis is applied to **X-ray diffraction**, since the electron density in a crystal is a periodic function (atoms are repeated in each unit cell).

Historical Note 9.5: Fourier's Revolutionary Idea (1822)

Joseph Fourier proposed that *any* reasonable function—even one with jumps!—could be represented as a sum of sines and cosines.

This shocked the great mathematicians of his time.

Lagrange, Laplace, and Poisson initially rejected the idea.

Yet Fourier's insight, arising from the study of heat conduction, ultimately created an entirely new branch of mathematics: **harmonic analysis**. Today, Fourier's ideas underpin spectroscopy, quantum mechanics, signal processing, and crystallography.

A periodic function over time is shown in Figure 9.2.

For a periodic function with period $2\pi(-\pi \leq x \leq \pi)$, the **trigonometric series** is:

$$f(x) = \frac{a_0}{2} + \sum_{n=1}^{\infty}(a_n \cos nx + b_n \sin nx). \tag{9.21}$$

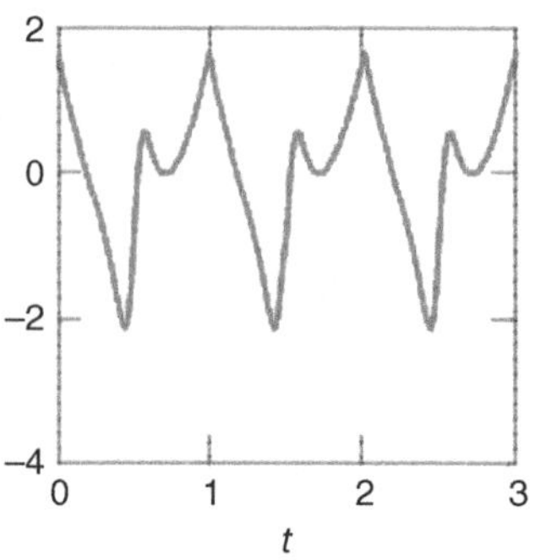

Figure 9.2 A periodic function having a unit period.

Since $\sin x$ is an *odd* function ($\sin 0 = 0$) and $\cos x$ is *even*,

- an **odd** function $f(x)$ with $f(0) = 0$ expands in sine terms only,
- an **even** function $f(x)$ expands in cosine terms only,
- a function that is neither even nor odd contains both.

Any function $f(x)$ can be decomposed into functions of **even** and **odd parity** as

$$f(x) = f_{\text{even}}(x) + f_{\text{odd}}(x), \tag{9.22}$$

where $f_{\text{even}}(x) = \frac{1}{2}[f(x) + f(-x)]$ and $f_{\text{odd}}(x) = \frac{1}{2}[f(x) - f(-x)]$, and f_{even} contributes to cosine terms and f_{odd} to sine terms.

To obtain the **Fourier coefficients**:

$$a_n = \frac{1}{\pi} \int_{-\pi}^{\pi} f(x) \cos(nx) dx, \tag{9.23}$$

$$b_n = \frac{1}{\pi} \int_{-\pi}^{\pi} f(x) \sin(nx) dx, \tag{9.24}$$

and

$$a_0 = \frac{1}{\pi} \int_{-\pi}^{\pi} f(x) dx. \tag{9.25}$$

The derivation uses **orthogonality relations** of trigonometric functions.

Geometrical Meaning of Orthogonality

Sine and cosine functions behave like perpendicular vectors in an infinite-dimensional function space. Their inner product—the integral over a full period—is zero unless they oscillate at the same frequency. Thus, they form an orthogonal basis, just like p_x and p_y orbitals in quantum chemistry.

Math Box 9.2: Orthogonality Relations

For integers n, m:

$$\int_{-\pi}^{\pi} \cos(nx) \cos(mx) dx = \begin{cases} 0, & n \neq m, \\ \pi, & n = m \neq 0, \end{cases} \tag{9.26}$$

$$\int_{-\pi}^{\pi} \sin(nx) \sin(mx) dx = \begin{cases} 0, & n \neq m, \\ \pi, & n = m \neq 0, \end{cases} \tag{9.27}$$

$$\int_{-\pi}^{\pi} \sin(nx)\cos(mx)dx = 0. \tag{9.28}$$

Math Box 9.3: Derivation of Equations (9.23)–(9.25)

Multiply Equation (9.21) by $\cos(mx)$ and integrate between $-\pi$ and π,

$$\int_{-\pi}^{\pi} f(x)\cos(mx)dx = \frac{a_0}{2}\int_{-\pi}^{\pi}\cos(mx)dx + \sum_{n=1}^{\infty} a_n \int_{-\pi}^{\pi}\cos(nx)\cos(mx)dx$$
$$+ \sum_{n=1}^{\infty} b_n \int_{-\pi}^{\pi}\sin(nx)\cos(mx)dx.$$

To solve this equation, we use the orthogonality relations (9.26)–(9.28) (n, m integers). These orthogonality relationships can be easily derived. For example, the first identity (9.26) can be obtained by using the trigonometric relation

$$\cos(A)\cos(B) = \frac{1}{2}[\cos(A-B) + \cos(A+B)].$$

Using this identity,

$$\int_{-\pi}^{\pi}\cos(nx)\cos(mx)dx = \frac{1}{2}\int_{-\pi}^{\pi}\cos((n-m)x)dx + \frac{1}{2}\int_{-\pi}^{\pi}\cos((n+m)x)dx.$$

For $n \neq m$, the first integral becomes

$$\frac{1}{2}\left[\frac{\sin((n-m))x)}{(n-m)}\right]_{-\pi}^{\pi} = \frac{\sin((n-m)\pi) - \sin((n-m)(-\pi))}{n-m} = 0,$$

as n and m are integers. The second integral is similarly zero.

For the special case, $n = m$, the first integral becomes $\frac{1}{2}\int_{-\pi}^{\pi} dx = \pi$. The second integral remains equal to zero.

Hence, Equation (9.23) becomes

$$\int_{-\pi}^{\pi} f(x)\cos(mx)dx = \sum_{n=1}^{\infty} a_n \int_{-\pi}^{\pi}\cos(nx)\cos(mx)dx = a_m\pi,$$

because only the $n = m$ term remains.

Therefore, we may write

$$a_n = \frac{1}{\pi}\int_{-\pi}^{\pi} f(x)\cos(nx)dx, \quad n \geq 0.$$

When $n = 0$, $\cos(nx) = 1$, and we may write

$$a_0 = \frac{1}{\pi}\int_{-\pi}^{\pi} f(x)dx.$$

Similarly,

$$b_n = \frac{1}{\pi}\int_{-\pi}^{\pi} f(x)\sin(nx)dx, \quad n > 0.$$

Because $\sin(nx) = 0$ when $n = 0$, b_0 does not exist.

Multiplication of Equation (9.21) by $\sin(mx)$ and integration between $-\pi$ and π similarly yields Equation (9.24).

An important consequence of the orthogonality of trigonometric functions is **Parseval's theorem**, which relates the sum of the squares of the Fourier coefficients to the mean square value of the function. In physical terms, it expresses the conservation of energy between a function and its Fourier components.

For a function expanded in a Fourier series, Parseval's theorem states that the total energy of the signal is equal to the sum of the energies of its Fourier modes.

$$\frac{1}{\pi}\int_{-\pi}^{\pi} [f(x)]^2 dx = \frac{a_0^2}{2} + \sum_{n=1}^{\infty}(a_n^2 + b_n^2).$$

We now use these ideas to expand certain functions as a Fourier series.

Example 9.6: Fourier Series for $f(x) = x$ on $(-\pi, \pi)$

Since $f(x) = x$ is an **odd** function, all $a_n = 0$ and only sine terms remain.

From Equation (9.24),

$$b_n = \frac{1}{\pi}\int_{-\pi}^{\pi} x\sin(nx)dx.$$

Let $u = nx$, then $du = ndx$. Integration by parts gives

$$\begin{aligned}
b_n &= \frac{1}{n^2\pi}\int_{-n\pi}^{n\pi} u\sin(u)du = \frac{1}{n^2\pi}[-u\cos(u) + \sin(u)]_{-n\pi}^{n\pi} \\
&= \frac{1}{n^2\pi}(-n\pi\cos(n\pi) + \sin(n\pi) - n\pi\cos(-n\pi) - \sin(-n\pi)) \\
&= \frac{1}{n^2\pi}(-2n\pi\cos(n\pi)) = -\frac{2}{n}(-1)^n = \frac{2(-1)^{n+1}}{n},
\end{aligned}$$

because $\sin(n\pi) = 0$ and $\cos(n\pi) = (-1)^n$ when n is an integer.

Hence,

$$f(x) = x = \sum_{n=1}^{\infty} \frac{2(-1)^{n+1}}{n} \sin(nx). \tag{9.29}$$

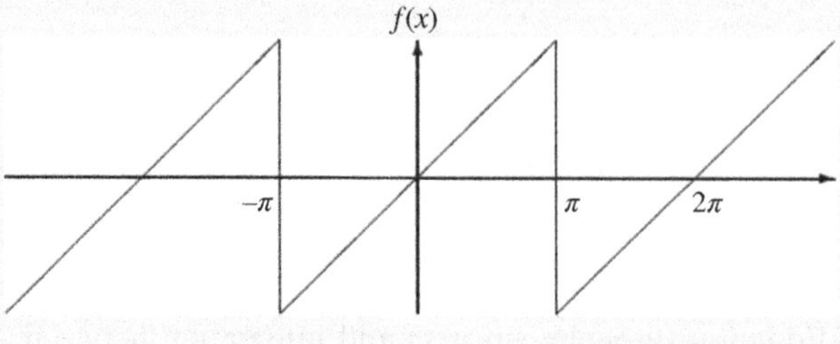

The periodic function $f(x) = x$ in $[-\pi, \pi]$.

This is the **sawtooth function**. Note that this series (9.29) consists of only sine terms because all a_n are equal to zero. This is because our function has odd parity.

Math Box 9.4: Parity and Symmetry in Functions

Parity refers to whether a function is even or odd, based on its symmetry:

- An **even function** (below left) satisfies $f(x) = f(-x)$ for all x in its domain. The graph of an even function has *reflectional symmetry* with respect to the y-axis (e.g., cos x or x^2).
- An **odd function** (below right) satisfies $f(-x) = -f(x)$ for all x in its domain. The graph of an odd function has *rotational symmetry* of order 2 about the origin (e.g., sin x or x^3).

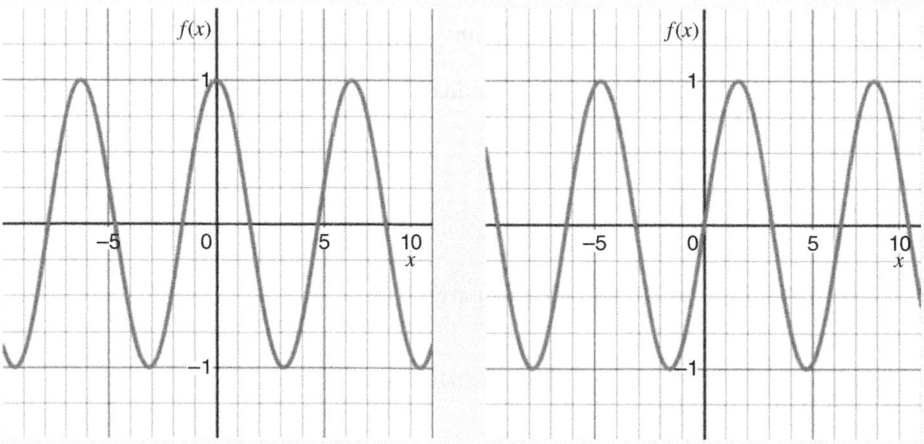

cos x (even) sin x (odd)

Implication for Fourier Series:

- Only cosine terms (which are even) appear in the expansion of even functions.
- Only sine terms (which are odd) appear in the expansion of odd functions.
- Functions that are neither even nor odd will include both sine and cosine terms.

This greatly simplifies computations.

As more terms are included, the Fourier approximation of Example 9.6 approaches the true waveform (Figure 9.3).

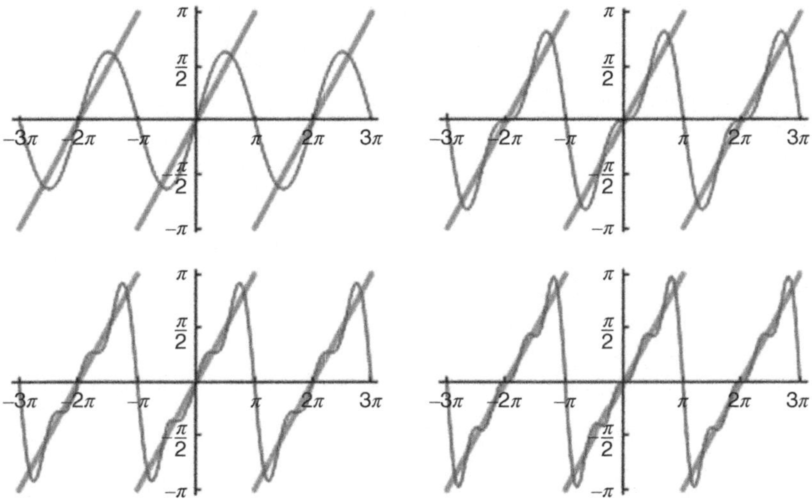

Figure 9.3 Successive approximations to a sawtooth function.

Many important relations can be derived from Fourier expansions.
At $x = \pi/2$, Equation (9.29) gives

$$\frac{\pi}{2} = 2\left(1 - \frac{1}{3} + \frac{1}{5} - \frac{1}{7} + \cdots\right)$$

or

$$1 - \frac{1}{3} + \frac{1}{5} - \frac{1}{7} + \cdots = \frac{\pi}{4}. \tag{9.30}$$

Example 9.7: Fourier Series for $f(x) = x^2$ on $(-\pi, \pi)$

Since $f(x) = x^2$ is **even**, all $b_n = 0$ and the series has cosine terms only,

$$a_n = \frac{1}{\pi}\int_{-\pi}^{\pi} x^2 \cos(nx)dx.$$

Again, we may substitute $u = nx$, $du = ndx$. Integrating twice by parts yields

$$a_n = \frac{1}{n^3\pi}\int_{-n\pi}^{n\pi} u^2\cos(u)du = \frac{1}{n^3\pi}\left[(u^2-2)\sin(u)+2u\cos(u)\right]_{-n\pi}^{n\pi}$$
$$= \frac{1}{n^3\pi}((n^2\pi^2-2)\sin(n\pi)+2n\pi\cos(n\pi)-(n^2\pi^2-2)\sin(-n\pi)$$
$$+2n\pi\cos(-n\pi))$$
$$= \frac{4(-1)^n}{n^2}.$$

Thus, $a_n = \frac{4(-1)^n}{n^2}$
and

$$a_0 = \frac{1}{\pi}\int_{-\pi}^{\pi} x^2 dx = \frac{2\pi^2}{3}.$$

Hence,

$$f(x) = x^2 = \frac{\pi^2}{3} + 4\sum_{n=1}^{\infty}\frac{(-1)^n}{n^2}\cos(nx). \tag{9.31}$$

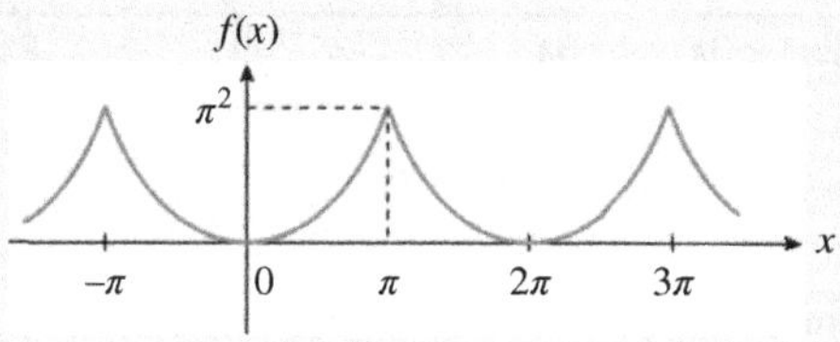

The periodic function $f(x) = x^2$ in $[-\pi, \pi]$.

Setting $x = 0$,

$$0 = \frac{\pi^2}{3} + 4\sum_{n=1}^{\infty}\frac{(-1)^n}{n^2}.$$

This gives the famous result:

$$\sum_{n=1}^{\infty}\frac{(-1)^{n+1}}{n^2} = \frac{\pi^2}{12}. \tag{9.32}$$

Example: Fourier Series of a Square Wave

Given:

$$f(x) = \begin{cases} 1, & 0 < x < \pi, \\ -1, & -\pi < x < 0. \end{cases}$$

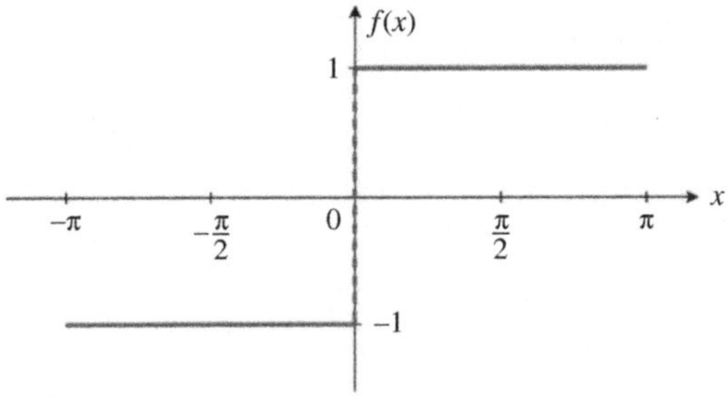

The function is **odd**, so only sine coefficients exist. From Equation (9.24),

$$b_n = \frac{2}{\pi}\int_0^{\pi} \sin(nx)dx = \frac{2}{n\pi}(1 - \cos n\pi) = \frac{2}{n\pi}(1 - (-1)^n).$$

Thus, nonzero terms occur only for **odd** n:

$$f(x) = \frac{4}{\pi}\left(\sin x + \frac{1}{3}\sin 3x + \frac{1}{5}\sin 5x + \cdots\right).$$

9.5.1 Exponential Fourier Series

While the trigonometric Fourier series expresses a periodic function as a sum of sines and cosines, the exponential Fourier series uses **complex** exponentials $\left(e^{inx}\right)$ for a more compact and general representation, as derived from **Euler's formula**.

Using Euler's identity,

$$e^{inx} = \cos(nx) + i\sin(nx), \tag{9.33}$$

we may write the Fourier series as

$$f(x) = \sum_{n=-\infty}^{\infty} c_n e^{inx}. \tag{9.34}$$

The coefficients are:

$$c_n = \frac{1}{2\pi}\int_{-\pi}^{\pi} f(x)e^{-inx}dx. \tag{9.35}$$

Math Box 9.5: Derivation of Exponential Fourier Series

Using Euler's identity, we may write

$$\cos(nx) = \frac{1}{2}(e^{inx} + e^{-inx}) \tag{1}$$

and

$$\sin(nx) = \frac{1}{2i}(e^{inx} - e^{-inx}). \tag{2}$$

On substitution of Equations (1) and (2) in Equation (9.21), we obtain

$$
\begin{aligned}
f(x) &= \frac{a_0}{2} + \sum_{n=1}^{\infty}\left(\frac{a_n}{2}(e^{inx} + e^{-inx}) + \frac{b_n}{2i}(e^{inx} - e^{-inx})\right)\\
&= \frac{a_0}{2} + \sum_{n=1}^{\infty}\left(\left(\frac{a_n}{2} + \frac{b_n}{2i}\right)e^{inx} + \left(\frac{a_n}{2} - \frac{b_n}{2i}\right)e^{-inx}\right)\\
&= \frac{a_0}{2} + \sum_{n=1}^{\infty}\left(\frac{1}{2}(a_n - ib_n)e^{inx} + \frac{1}{2}(a_n + ib_n)e^{-inx}\right)\\
&= \sum_{n=-\infty}^{\infty} c_n e^{inx},
\end{aligned}
$$

where we have allowed n to take both negative and positive values, including zero, and we have defined

$$
c_n = \begin{cases} \frac{1}{2}(a_n - ib_n) & (n > 0), \\ \frac{1}{2}(a_n + ib_n) & (n < 0), \\ \frac{1}{2}a_0 & (n = 0). \end{cases} \tag{3}
$$

We may therefore rewrite Equation (9.21) as

$$
f(x) = \sum_{n=-\infty}^{\infty} c_n e^{inx}, \quad -\pi \le x \le \pi. \tag{4}
$$

The mth coefficient is obtained by multiplying Equation (4) by e^{-imx} and integrating

$$
\int_{-\pi}^{\pi} f(x)e^{-imx}dx = \int_{-\pi}^{\pi} e^{-imx} \sum_{n=-\infty}^{\infty} c_n e^{inx}dx = \int_{-\pi}^{\pi} \sum_{n=-\infty}^{\infty} c_n e^{i(n-m)x}dx. \tag{5}
$$

The integral on the right-hand side is zero unless $n = m$. In case $n \neq m$,

$$
\begin{aligned}
\int_{-\pi}^{\pi} e^{i(n-m)x}dx &= \left[\frac{e^{i(n-m)x}}{i(n-m)}\right]_{-\pi}^{\pi} = \frac{1}{i(n-m)}(e^{i(n-m)\pi} - e^{-i(n-m)\pi})\\
&= \frac{2\sin((n-m)\pi)}{(n-m)} = 0,
\end{aligned}
$$

because the sine of the integral multiple of π is zero. Hence, all terms in the summation vanish except $n = m$, for which Equation (5) reduces to

$$
\int_{-\pi}^{\pi} f(x)e^{-imx}dx = \int_{-\pi}^{\pi} c_m dx = 2\pi c_m.
$$

Therefore,

$$
c_n = \frac{1}{2\pi}\int_{-\pi}^{\pi} f(x)e^{-inx}dx. \tag{6}
$$

Equations (4) and (6), therefore, define the exponential (or complex) Fourier series and its coefficients, respectively.

Example 9.8: Heaviside Step Function

Find the exponential Fourier series for the function

$$f(x) = \begin{cases} 0, & -\pi < x \leq 0, \\ 1, & 0 < x \leq \pi. \end{cases}$$

Solution:

The coefficients c_n are obtained as follows. Substituting the function in Equation (9.35),

$$c_n = \frac{1}{2\pi} \int_0^{\pi} e^{-inx} dx = \frac{1}{2\pi} \left[\frac{e^{-inx}}{-in} \right]_0^{\pi} = -\frac{1}{2\pi in}(e^{-in\pi} - 1) = \frac{1 - e^{-in\pi}}{2\pi in}.$$

Since $e^{-in\pi} = \cos(n\pi) - i\sin(n\pi) = (-1)^n$,

$$c_n = \frac{1 - e^{-in\pi}}{2\pi in} = \begin{cases} 0 & (n = \text{even}), \\ \dfrac{1}{in\pi} & (n = \text{odd}). \end{cases}$$

Also,

$$c_0 = \frac{1}{2\pi} \int_0^{\pi} dx = \frac{1}{2}.$$

Therefore, substitution in Equation (9.34) gives

$$f(x) = \frac{1}{2} + \frac{1}{i\pi} \sum_{n=\text{odd}}^{\infty} \frac{1}{n} e^{inx}.$$

9.5.2 Change of Interval

Until now, we have considered functions in the range $(-\pi, \pi)$ with a period of 2π. It is not necessary for this function to be within this range. The function may be temporal (dependent on time) with period T. If $f(t)$ has period T, we replace x by $\frac{2\pi t}{T}$.

Thus, the Fourier series becomes

$$f(t) = \frac{a_0}{2} + \sum_{n=1}^{\infty} \left[a_n \cos\left(\frac{2n\pi t}{T}\right) + b_n \sin\left(\frac{2n\pi t}{T}\right) \right], \tag{9.36}$$

with coefficients

$$a_n = \frac{2}{T} \int_{-T/2}^{T/2} f(t) \cos\left(\frac{2n\pi t}{T}\right) dt, \tag{9.37}$$

$$b_n = \frac{2}{T} \int_{-T/2}^{T/2} f(t) \sin\left(\frac{2n\pi t}{T}\right) dt, \tag{9.38}$$

and

$$a_0 = \frac{2}{T} \int_{-T/2}^{T/2} f(t) dt. \tag{9.39}$$

Example 9.9: Fourier Series of a Square Wave

Expand the square-wave function

$$f(t) = \begin{cases} -A, & -T/2 < t < 0, \\ +A, & T/2 > t \geq 0, \end{cases}$$

in a Fourier series.

Solution:

The function is as follows:

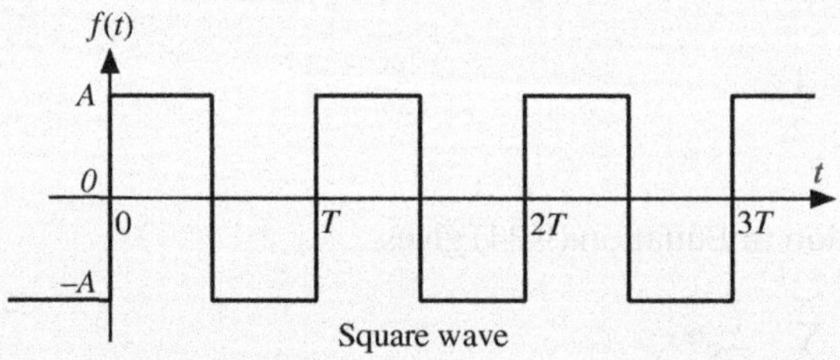

Square wave

The function has odd parity; therefore, we expect only the sine terms. Indeed, from Equation (9.39),

$$a_0 = \frac{2}{T}\left[\int_{-T/2}^{0} (-A)dt + \int_{0}^{T/2} (+A)dt\right] = \frac{2}{T}\left[-\frac{T}{2} + \frac{T}{2}\right] = 0.$$

Insertion of the function in Equation (9.37) similarly gives

$$a_n = \frac{2}{T}\left[\int_{-T/2}^{0} -A\cos(2n\pi t/T)dt + \int_{0}^{T/2} A\cos(2n\pi t/T)dt\right] = 0.$$

The sine coefficients are obtained using relation (9.38),

$$b_n = \frac{2}{T}\left[\int_{-T/2}^{0} -A\sin(2n\pi t/T)dt + \int_{0}^{T/2} A\sin(2n\pi t/T)dt\right]$$

$$= \frac{2}{T}\left[-\int_{T/2}^{0} A\sin(2n\pi t/T)dt + \int_{0}^{T/2} A\sin(2n\pi t/T)dt\right]$$

$$= \frac{2}{T}\left[2\int_{0}^{T/2} A\sin(2n\pi t/T)dt\right] = \begin{cases} \dfrac{4A}{n\pi} & \text{if } n = \text{odd}, \\ 0 & n = \text{even}. \end{cases}$$

Therefore, the Fourier series is

$$f(t) = \frac{4A}{\pi}\sum_{n=1(\text{odd})}^{\infty} \frac{1}{n}\sin\left(\frac{2n\pi t}{T}\right) \equiv \frac{4A}{\pi}\sum_{m=0}^{\infty}\frac{1}{2m+1}\sin\left(\frac{2(2m+1)\pi t}{T}\right)$$

$$= \frac{4A}{\pi}\sin\left(\frac{2\pi t}{T}\right) + \frac{4A}{3\pi}\sin\left(\frac{6\pi t}{T}\right) + \frac{4A}{5\pi}\sin\left(\frac{10\pi t}{T}\right) + \cdots$$

If the function is periodic on the length scale and has a period of $2L$, the Fourier series expression becomes

$$f(x) = \frac{a_0}{2} + \sum_{n=1}^{\infty}(a_n\cos(n\pi x/L) + b_n\sin(n\pi x/L)), \quad -L \le x \le L, \tag{9.40}$$

with

$$a_n = \frac{1}{L}\int_{-L}^{L} f(x)\cos(n\pi x/L)dx, \quad n \ge 0, \tag{9.41}$$

$$a_0 = \frac{1}{L}\int_{-L}^{L} f(x)dx, \tag{9.42}$$

$$b_n = \frac{1}{L}\int_{-L}^{L} f(x)\sin(n\pi x/L)dx, \quad n > 0. \tag{9.43}$$

Similarly, the exponential (or complex) Fourier series for a function of period $2L$ is given by Equations (9.44) and (9.45), respectively,

$$f(x) = \sum_{n=-\infty}^{\infty} c_n e^{in\pi x/L}, \quad -L \le x \le L, \tag{9.44}$$

$$c_n = \frac{1}{2L}\int_{-L}^{L} f(x)e^{-in\pi x/L}dx. \tag{9.45}$$

Example 9.10: Exponential Fourier Series for the Heaviside Step Function

$$f(x) = \begin{cases} 1, & 0 < x < L, \\ 0, & -L < x < 0. \end{cases}$$

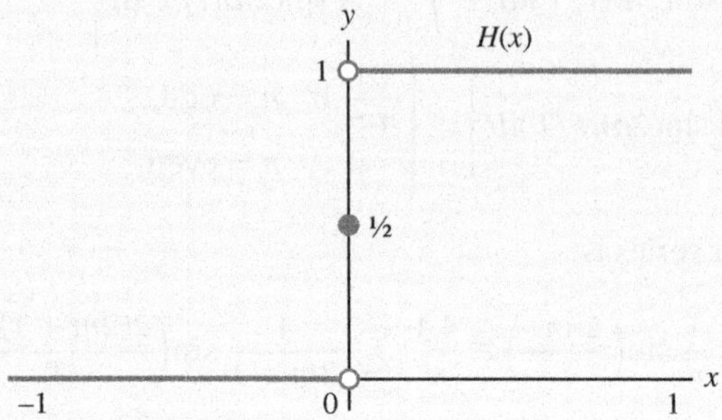

At $x = 0$, $f(0) = 1/2$ (by convention).
Then from Equation (9.45),

$$c_n = \frac{1}{2L} \int_0^L e^{-in\pi x/L}\, dx = \frac{1}{2L}\left[\frac{L}{-in\pi}(e^{-in\pi} - 1)\right] = \frac{1-(-1)^n}{2in\pi}.$$

The formula for c_n gives 0/0 for $n = 0$. The c_0 coefficient must be calculated separately:

$$c_0 = \frac{1}{2L} \int_0^L 1 dx = \frac{1}{2}.$$

9.5.3 Half-range Expansions

Sometimes a function is defined only on $(0, L)$. Depending on whether we choose an **odd** or **even** extension to $(-L, L)$, we obtain the following.

Half-range Sine Series (odd extension)
If $f(x)$ is an **odd** function of period $2L$, its **half-range sine series** is

$$f(x) = \sum_{n=1}^{\infty} b_n \sin\left(\frac{n\pi x}{L}\right), \tag{9.46}$$

where

$$b_n = \frac{2}{L} \int_0^L f(x) \sin\left(\frac{n\pi x}{L}\right) dx. \tag{9.47}$$

Half-range Cosine Series (even extension)
If $f(x)$ is **even**, its **half-range cosine series** is

$$f(x) = \frac{a_0}{2} + \sum_{n=1}^{\infty} a_n \cos\left(\frac{n\pi x}{L}\right), \tag{9.48}$$

with

$$a_n = \frac{2}{L} \int_0^L f(x) \cos\left(\frac{n\pi x}{L}\right) dx. \tag{9.49}$$

The sine half-range expansion effectively *extends* the function as an odd function over $(-L, L)$, while the cosine half-range extension is even.

Math Box 9.6: Orthogonality Relations in Fourier Analysis

The following integrals illustrate the orthogonality of sine and cosine functions used in the Fourier series. These properties ensure that each coefficient isolates only the respective frequency from a periodic function:

For integers n, m over interval $(-L, L)$,

$$\int_{-L}^{L} \cos\left(\frac{n\pi x}{L}\right) \cos\left(\frac{m\pi x}{L}\right) dx = \begin{cases} 0 & \text{if } n \neq m, \\ L & \text{if } n = m \neq 0, \\ 2L & \text{if } n = m = 0, \end{cases}$$

$$\int_{-L}^{L} \sin\left(\frac{n\pi x}{L}\right) \sin\left(\frac{m\pi x}{L}\right) dx = \begin{cases} 0 & \text{if } n \neq m, \\ L & \text{if } n = m, \end{cases}$$

$$\int_{-L}^{L} \cos\left(\frac{n\pi x}{L}\right) \sin\left(\frac{m\pi x}{L}\right) dx = 0.$$

These orthogonality relations are key to extracting individual coefficients when expanding functions as a sum of sines and cosines in the Fourier series.

Connection to Quantum Chemistry: Particle in a Box

The functions $\sin\left(\frac{n\pi x}{L}\right)$ are not only the building blocks for Fourier sine series over the interval $(0, L)$; they are also the normalized **stationary state solutions** (energy eigenfunctions) of the quantum mechanical "particle in a box" problem, i.e., solutions to the time-independent Schrödinger equation with infinite potential walls at $x = 0$ and $x = L$.

Expanding an arbitrary wavefunction in this basis is mathematically equivalent to writing a general quantum state as a linear combination of energy eigenstates.

This connection is fundamental to time-dependent quantum mechanics.

Math Box 9.7: Half-range Expansions for $f(x) = x$, $0 \le x \le \pi$

a. Cosine Half-range Series (Even Extension)

Since $f(x) = x$ for $0 \le x \le \pi$, extend evenly to $(-\pi, \pi)$. From Equation (9.49) with $L = \pi$,

$$a_n = \frac{2}{\pi}\int_0^{\pi} x\cos(nx)dx.$$

Integrating by parts,

$$a_n = \frac{2}{\pi}\left[\frac{x\sin(nx)}{n} + \frac{\cos(nx) - 1}{n^2}\right]_0^{\pi} = \frac{2}{\pi n^2}[(-1)^n - 1].$$

Also,

$$a_0 = \frac{2}{\pi}\int_0^{\pi} xdx = \pi.$$

Hence,

$$f(x) = \frac{\pi}{2} - \frac{4}{\pi}\left(\frac{\cos x}{1^2} + \frac{\cos 3x}{3^2} + \frac{\cos 5x}{5^2} + \cdots\right). \tag{1}$$

b. Sine Half-range Series (Odd Extension)

In this case, $a_n = 0$, and

$$b_n = \frac{2}{\pi}\int_0^{\pi} x\sin(nx)dx = \frac{2}{n}(-1)^{n+1}.$$

Hence,

$$f(x) = \sum_{n=1}^{\infty} \frac{2(-1)^{n+1}}{n}\sin(nx).$$

c. Although both series represent $f(x)$ on $0 < x < \pi$, the **cosine series converges more rapidly** because it is smoother at the boundaries.

Derived Relation:

Setting $x = \pi$ in Equation (1):

$$\frac{\pi}{2} = \frac{4}{\pi}\left(1 + \frac{1}{3^2} + \frac{1}{5^2} + \frac{1}{7^2} + \cdots\right),$$

or

$$1 + \frac{1}{3^2} + \frac{1}{5^2} + \frac{1}{7^2} + \cdots = \frac{\pi^2}{8}.$$

In connection with Math Box 9.7, it is instructive to see how the two half-range expansions extend the function.

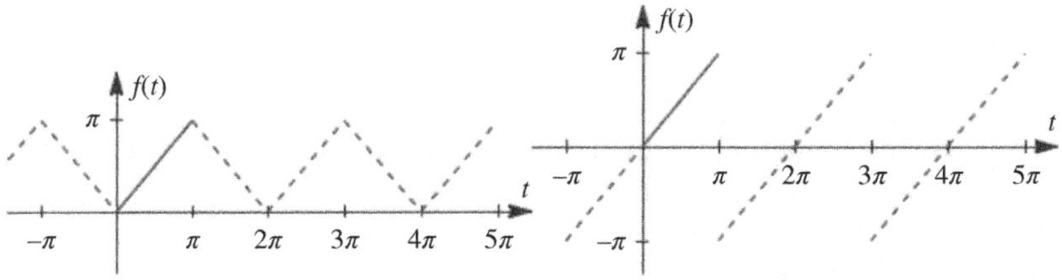

n		
0	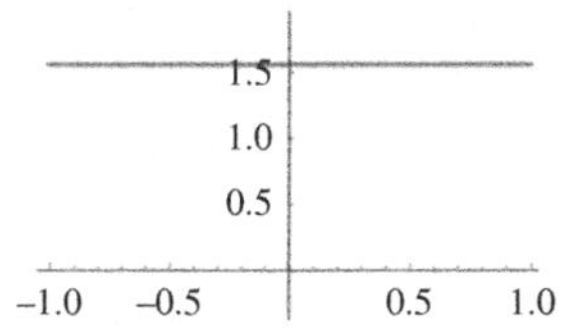	(x from –1 to 1)
1	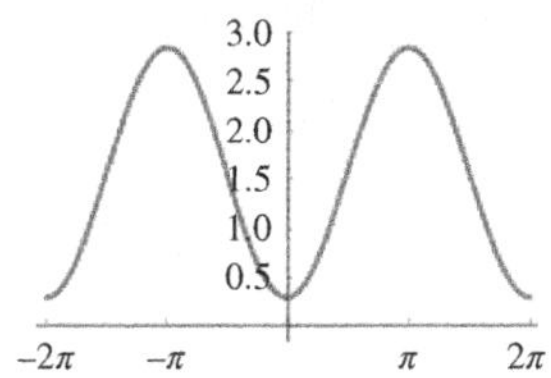	(x from –6.3 to 6.3)
2	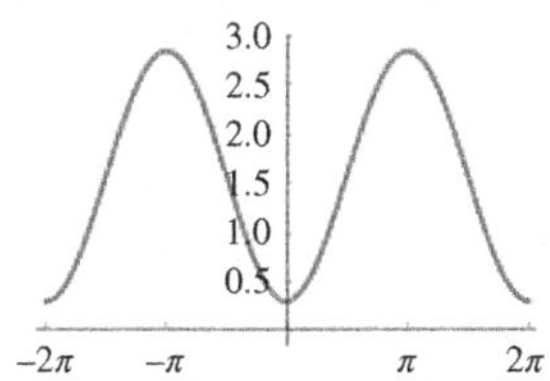	(x from –6.3 to 6.3)
3	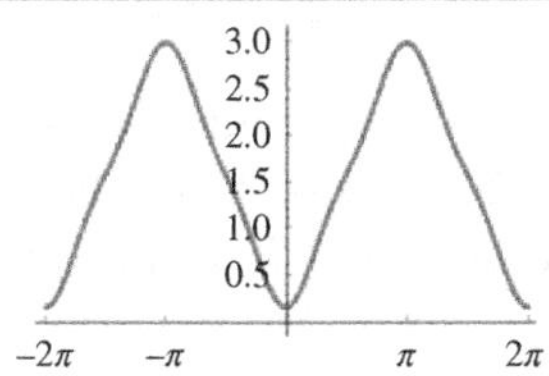	(x from –6.3 to 6.3)

n		
1	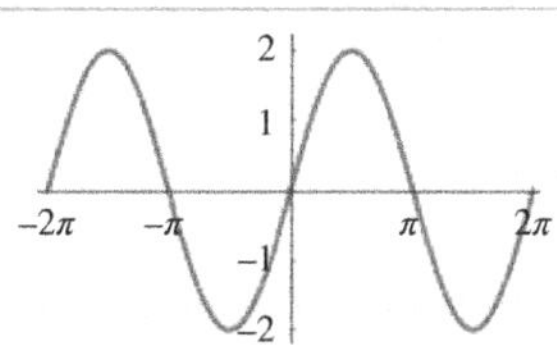	(x from –6.3 to 6.3)
2	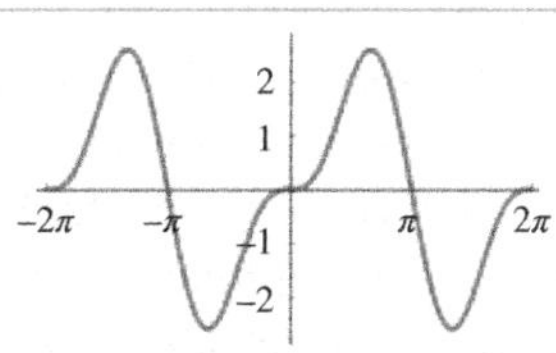	(x from –6.3 to 6.3)
3	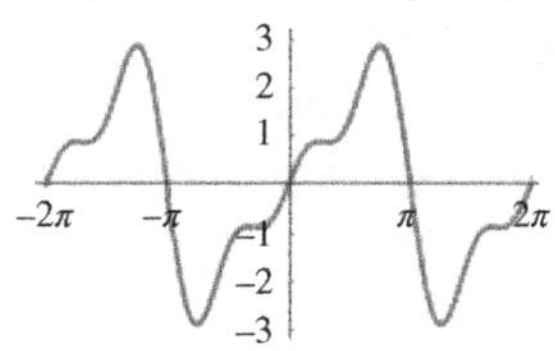	(x from –6.3 to 6.3)

On the left is shown an even extension (cosine half-range series), and on the right is shown the odd extension (sine half-range series).

In the figure above, the left panel shows how the cosine extension progresses as the number of terms in the expansion increases, whereas the panel on the right shows the same for the sine extension. The cosine function converges more rapidly.

9.5.4 Physical Applications

Worked Example 1: Signal Processing

A periodic electrical signal $f(t)$ can be analyzed using a Fourier series, breaking it into a sum of sine and cosine terms. For instance, a square wave used in digital electronics has the Fourier expansion:

$$f(t) = \frac{4}{\pi} \sum_{n=1,3,5,...}^{\infty} \frac{1}{n} \sin(n\omega t).$$

This decomposition reveals the fundamental frequency and all higher harmonics present in the signal, allowing engineers to design better filters and communication systems.

Worked Example 2: Quantum Mechanics

In quantum mechanics, the wavefunction $\psi(x)$ of a **particle in a box** is expanded in terms of sine and cosine functions—its Fourier series representation. The uncertainty in a particle's position and momentum can be explored using the spread and behavior of its Fourier coefficients, which directly links to the **Heisenberg Uncertainty Principle**:

$$\Delta x \Delta p \geq \frac{h}{4\pi}.$$

A localized wave packet requires combining many sinusoidal waves, and the Fourier series gives the mathematical machinery to study these combinations within a bounded domain, demonstrating why perfect localization is impossible in quantum systems.

9.6 Summary

In this chapter, we discussed the concept of **limits** and their importance in defining continuity and differentiability of functions.

The **mean value theorem** and **l'Hôpital's rule** were presented as essential tools for analyzing functions and evaluating indeterminate forms.

We then explored the **Taylor** and **Maclaurin series** as powerful methods for approximating analytic functions using polynomial expansions.

The **binomial series** was introduced as a special case, with its convergence criteria explained using **Cauchy's ratio test**.

Finally, we studied the **Fourier series**—expressing periodic functions as sums of sines and cosines—along with their **exponential**, **half-range sine**, and **cosine** forms, which have wide applications in mathematical physics and chemistry (e.g., in X-ray diffraction and quantum mechanics).

Further Reading

M. R. Spiegel, *Schaum's Outline of Theory and Problems of Fourier Analysis with Applications to Boundary-Value Problems*, McGraw-Hill, 1974—covers Fourier series, orthogonality, and convergence with solved problems.

R. Wrede and M. R. Spiegel, *Schaum's Outlines of Advanced Calculus*, 3rd ed., McGraw-Hill, 2010—classic source for Taylor, Maclaurin, and binomial series.

E. Kreyszig, *Advanced Engineering Mathematics*, 10th ed., Wiley, 2011—comprehensive reference for Fourier series, limits, and power series in applied sciences.

M. L. Boas, *Mathematical Methods in the Physical Sciences*, 3rd ed., Wiley, 2006—integrates series expansions and Fourier analysis with practical examples from physics and chemistry.

H. F. Davis, *Fourier Series and Orthogonal Functions*, Dover Publications, Inc., 1963—specialized treatment for students in mathematics, physics, and engineering.

J. W. Brown and R. V. Churchill, *Complex Variables and Applications*, 9th ed., McGraw-Hill, 2021—focuses on convergence and analytic continuation.
E. B. Saff and A. D. Snider, *Fundamentals of Complex Analysis with Applications to Engineering and Science*, Pearson, 3rd ed., 2003—detailed, applied discussions on series convergence.
P. Dennery and A. Krzywicki, *Mathematics for Physicists*, Dover Publications, 1996—detailed explanations of expansions used in physical modeling.
H. Margenau and G. M. Murphy, *The Mathematics of Physics and Chemistry*, 2nd ed., Van Nostrand, 1956—connects physical insight with rigorous series and integral formulations.

Exercises

A. Conceptual Questions

1. Explain what is meant by the *limit* of a function. How does it relate to continuity?
2. Distinguish between *stationary points* and *critical points* of a function.
3. Define the **mean value theorem** and explain its physical interpretation.
4. What are *indeterminate forms*? List six common types.
5. Discuss the conditions under which a function can be expanded as a Taylor series.
6. What is the difference between a **Taylor series** and a **Maclaurin series**?
7. Why is the binomial expansion for $(1 + x)^\alpha$ valid only for $|x| < 1$ when α is not an integer?
8. Explain *absolute convergence* and *conditional convergence* of a series.
9. State the **Dirichlet conditions** for a function to be expressible as a Fourier series.
10. Explain how Fourier series are used in chemistry or physics (choose one example).

B. Problems

1. Compute the following limits:
   a. $\lim_{x\to 0} \dfrac{\sin x}{x}$
   b. $\lim_{x\to 0} \dfrac{e^x - 1}{x}$
   c. $\lim_{x\to 0} \dfrac{\ln(1 + x)}{x}$
2. Derive the Taylor expansion of $\ln(1 + x)$ up to the x^4-term.

3. Expand $(1 - 3x)^{-2}$ as a binomial series up to x^3.

4. Determine the first four nonzero terms in the Fourier sine series for $f(x) = x$ on $0 < x < \pi$.

5. Find the Fourier cosine series for $f(x) = x^2$ on $0 < x < \pi$.

6. Apply Cauchy's ratio test to determine the convergence of $\sum_{n=1}^{\infty} \frac{x^n}{n!}$.

7. Determine the value of $\sum_{n=1}^{\infty} \frac{1}{n^2}$ using the Fourier series of $f(x) = x^2$ on $(-\pi, \pi)$.
(*Hint:* Use the Fourier cosine series for $f(x) = x^2$ on $(-\pi, \pi)$ [see Equation 9.31]. Then evaluate the series at a convenient value of x [such as $x = \pi$] to obtain a relation involving $\sum_{n=1}^{\infty} \frac{1}{n^2}$.)

8. Show that the binomial series for $(1 + x)^{\alpha}$ converges for $-1 < x < 1$ and diverges for $|x| > 1$.

9. Plot (using any plotting software) the first five partial sums of

$$f(x) = \frac{4}{\pi}\left(\sin x + \frac{1}{3}\sin 3x + \frac{1}{5}\sin 5x + \ldots\right)$$

and comment on how the approximation improves with more terms.

C. Advanced Problems

1. Prove l'Hôpital's rule rigorously, starting from Cauchy's mean value theorem.

2. Derive Euler's formula

$$e^{ix} = \cos x + i \sin x$$

using power-series expansions of e^x, $\sin x$, and $\cos x$.

3. Show that

$$\sum_{n=1}^{\infty} \frac{1}{(2n-1)^2} = \frac{\pi^2}{8}.$$

(*Hint:* Use the Fourier series of a square-wave function.)

4. Derive the relation between the Fourier coefficients a_n, b_n, and the exponential coefficients c_n.

5. Using the binomial series, obtain an approximate expression for

$$(1 - x)^{-1/3}$$

correct up to x^3.

6. Show that the Fourier series of $f(x) = x^2$ on $(-\pi, \pi)$ satisfies Parseval's theorem.

7. Compute the exponential Fourier coefficients c_n for $f(x) = |x|$ on $(-\pi, \pi)$.

8. Evaluate the improper integral

$$\int_0^\infty \frac{\sin x}{x} dx$$

using the Fourier transform of a rectangular pulse.

9. Using the Maclaurin series for $\sin x$, show that the error term when approximating $\sin x \approx x - x^3/6$ is less than 0.001 for $|x| < 0.5$.

10. Derive the general formula for the *radius of convergence* using the ratio test for $\sum a_n x^n$.

D. Additional Exercises

1. Prove that the binomial series for $(1 + x)^\alpha$,

$$(1 + x)^\alpha = 1 + \alpha x + \frac{\alpha(\alpha - 1)}{2!} x^2 + \cdots,$$

converges if $|x| < 1$.

2. Expand $\ln(1 + x)$ in a Taylor series and show that

$$\ln(1 + x) = x - \frac{x^2}{2} + \frac{x^3}{3} - \frac{x^4}{4} + \cdots.$$

The resulting series is useful for developing a second solution to **Legendre's differential equation** and for normalizing Legendre polynomials.

3. Find the half-range sine and cosine expansions of the **triangular function**

$$f(x) = \begin{cases} x, & 0 < x < \pi, \\ 2\pi - x, & \pi < x < 2\pi. \end{cases}$$

Plot both series and also determine and plot the full-range Fourier series.

4. Expand $f(x) = \cos(kx)$, where k is not an integer, into a Fourier series in the interval $(-\pi, \pi)$.

5. The Fourier series can be employed to evaluate important infinite sums. For instance, show that the Fourier series for the square-wave function $f(x)$ defined on $(-\pi, \pi)$ as

$$f(x) = \begin{cases} -\pi/2 \text{ if } -\pi < x < 0, \\ \pi/2 \text{ if } 0 < x < \pi \end{cases}$$

is

$$f(x) = 2\left(\sin x + \frac{1}{3} \sin 3x + \frac{1}{5} \sin 5x + \cdots\right).$$

Setting $x = \pi/2$, obtain the result of Euler:

$$1 - \frac{1}{3} + \frac{1}{5} - \frac{1}{7} + \cdots = \frac{\pi}{4}.$$

Extend the idea and prove that

$$1 + \frac{1}{2^2} + \frac{1}{3^2} + \frac{1}{4^2} + \cdots = \frac{\pi^2}{6}.$$

6. Expand the function $f(x) = h$ (constant) into a **complex Fourier series** in $(0, 2\pi)$.

7. Show that the set of functions

$$1, \cos\left(\frac{n\pi x}{L}\right), \sin\left(\frac{n\pi x}{L}\right)$$

forms an *orthogonal set* in the interval $(-L, L)$.
Determine the normalization constant so that the set is *orthonormal* in $(-L, L)$.

8. Expand the function

$$f(x) = \begin{cases} 0, & -\pi < x < 0, \\ 1, & 0 < x < \pi, \end{cases}$$

into a Fourier series and discuss its convergence near the discontinuities.

9. Find the exponential Fourier series coefficients c_n for the square-wave function defined in Example 9.9.

10

Fourier Transforms

Fourier's theorem is not only one of the most beautiful results of modern analysis, but it may be said to furnish an indispensable instrument in the treatment of nearly every recondite question in modern physics.

— Lord Kelvin (1867)

Learning Outcomes

After studying this chapter, you will be able to:

1. **Explain** the concept of the Fourier transformation and how it connects time (or coordinate) and frequency (or momentum) domains.
2. **Derive and apply** the Fourier transform pair for standard mathematical functions used in chemistry and physics.
3. **Interpret** the relationship between a function and its transform—particularly the width–reciprocal width behavior and its link to the Heisenberg uncertainty principle.
4. **Compute** Fourier, sine, and cosine transforms of simple functions and **identify** their spectral features.
5. **Recognize and discuss** major applications of Fourier transforms in spectroscopy, diffraction, image analysis, and signal processing.
6. **Relate** the concepts of lifetime, damping, and linewidth to experimental spectra and **understand** their Fourier connections.

10.1 Introduction

The Fourier theorem is one of the most elegant and powerful results of modern analysis. It provides a means to express an arbitrary periodic or aperiodic function as a superposition of sinusoidal components of different frequencies. This idea forms the mathematical foundation of many branches of physics and chemistry, particularly in spectroscopy, signal analysis, and quantum mechanics.

The **Fourier transform (FT)** and the related **Fourier series** are among the most important tools for analyzing periodic and nonperiodic phenomena. For a periodic function $f(x)$ of period $2L$, the Fourier series representation (Chapter 9) is

$$f(x) = \sum_{n=-\infty}^{\infty} c_n e^{in\pi x/L}. \tag{10.1}$$

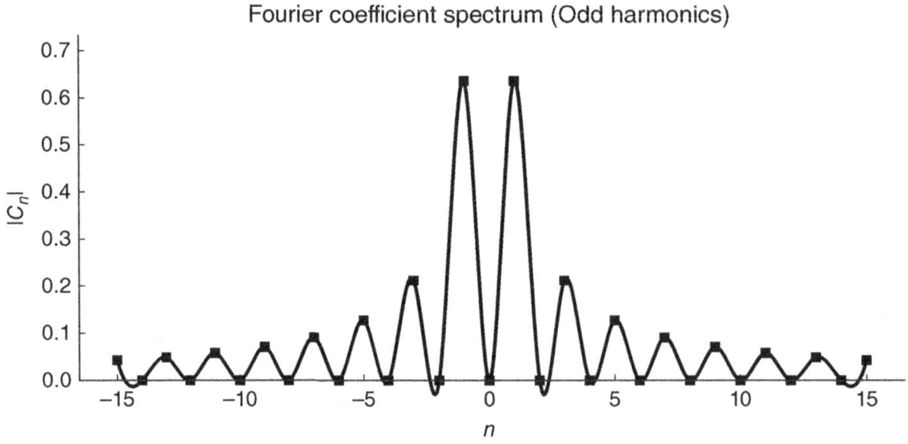

Figure 10.1 Magnitude of Fourier coefficients $|c_n|$ for the Heaviside step function (Example 9.8).

Here, the complex coefficients c_n are given by

$$c_n = \frac{1}{2L} \int_{-L}^{L} f(x) e^{-in\pi x/L}\, dx. \tag{10.2}$$

The coefficients c_n constitute the **Fourier spectrum** of $f(x)$—a representation of the function in terms of its frequency components.

When we plot the coefficients c_n for a given function, we obtain its **frequency spectrum** (Figure 10.1). For the Heaviside step function (Example 9.8), for example, the coefficients are imaginary; hence, their magnitudes are represented as $|c_n|$, obtained from the square root of $c_n c_n^*$.

10.1.1 From Discrete to Continuous Spectrum

The Fourier series is applicable to periodic functions. To analyze nonperiodic functions, we consider the limit where the period $2L$ tends to infinity. In this limit, the fundamental frequency π/L becomes infinitesimally small, and the discrete sum over frequencies evolves into a continuous integral.

Let us define the wavenumber k_n associated with each mode n as

$$k_n = \frac{n\pi}{L}.$$

The difference between successive wavenumbers is

$$\Delta k = k_{n+1} - k_n = \frac{\pi}{L}. \tag{10.3}$$

We now substitute the expression for c_n from (10.2) into the series (10.1):

$$f(x) = \sum_{n=-\infty}^{\infty} \left[\frac{1}{2L} \int_{-L}^{L} f(x') e^{-ik_n x'}\, dx' \right] e^{ik_n x}.$$

A key step is to incorporate the factor $\frac{\Delta k}{\pi}$ into the expression. Since $\Delta k = \frac{\pi}{L}$, it follows that $\frac{1}{2L} = \frac{\Delta k}{2\pi}$. Making this substitution yields

$$f(x) = \sum_{n=-\infty}^{\infty} \left[\frac{\Delta k}{2\pi} \int_{-L}^{L} f(x') e^{-ik_n x'} dx' \right] e^{ik_n x}.$$

We can now define a new function, $F_L(k_n)$, which depends on the continuous variable k_n,

$$F_L(k_n) = \frac{1}{\sqrt{2\pi}} \int_{-L}^{L} f(x') e^{-ik_n x'} dx'. \tag{10.4}$$

Notice that the quantity inside the square brackets in the previous expression for $f(x)$ is $\left(\frac{\Delta k}{\sqrt{2\pi}} \right) F_L(k_n)$. Therefore, we can write

$$f(x) = \frac{1}{\sqrt{2\pi}} \sum_{n=-\infty}^{\infty} F_L(k_n) e^{ik_n x} \Delta k. \tag{10.5}$$

We now take the limit $L \to \infty$. In this limit:

- The function $f(x)$ is no longer restricted to a finite interval, so the limits of integration in (10.4) become $\pm\infty$.
- The discrete wavenumbers k_n merge into a continuous variable k.
- The increment Δk becomes the differential dk.
- The sum over n in (10.5) becomes an integral over k.
- The function $F_L(k_n)$ approaches a limiting function $F(k)$.

Thus, in the limit $L \to \infty$, Equation (10.4) becomes the **FT** and Equation (10.5) becomes the **inverse FT**,

$$F(k) = \frac{1}{\sqrt{2\pi}} \int_{-\infty}^{\infty} f(x) e^{-ikx} \, dx, \tag{10.6}$$

$$f(x) = \frac{1}{\sqrt{2\pi}} \int_{-\infty}^{\infty} F(k) e^{ikx} \, dk. \tag{10.7}$$

Equations (10.6) and (10.7) form the **FT pair** in its symmetric (unitary), normalized form. This **symmetrical representation**, which treats the **forward and inverse transforms** on an equal footing, will be adopted throughout this chapter, unless otherwise stated.

Example 10.1: Fourier Transform of the Box Function

Find the Fourier transform of the rectangular (box) function:

$$f(x) = \begin{cases} 1, & |x| < a, \\ 0, & |x| > a. \end{cases}$$

Solution:

Using Equation (10.6),

$$F(k) = \frac{1}{\sqrt{2\pi}} \int_{-a}^{a} e^{-ikx} dx = \frac{1}{\sqrt{2\pi}} \cdot 2\frac{\sin(ka)}{k} = \sqrt{\frac{2}{\pi}} \frac{\sin(ka)}{k}.$$

The box function has a width of $2a$. The width of the central peak of its transform is inversely proportional to a, since $F(k)$ crosses the k-axis whenever $ka = \pm n\pi$. Thus, broad functions in real space correspond to narrow transforms in reciprocal space, and vice versa.

The function (left) and its Fourier transform (right) are plotted below for a value of $a = 1$.

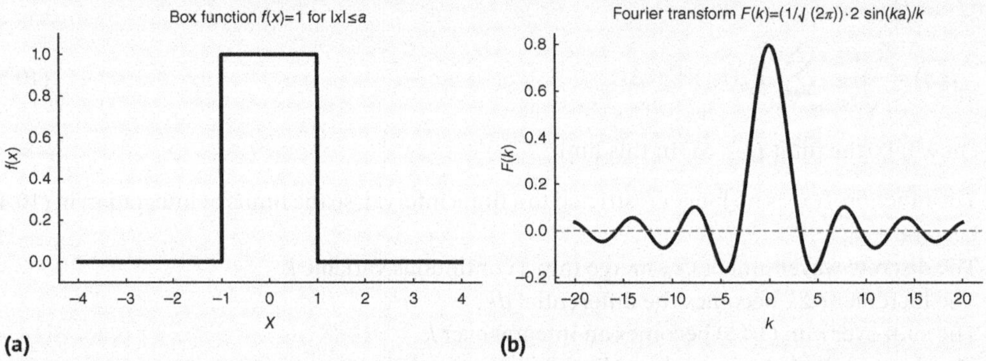

This inverse relationship between a function and its FT has high significance in quantum chemistry. We shall next find the FT of a **Gaussian function**, which is a function of the kind e^{-ay^2}. Such functions are bell-shaped, i.e., flat and rounded at the top. **Normalization** ensures that the total probability equals one.

Example 10.2: Fourier Transform of the Gaussian Function

Find the Fourier transform of the normalized Gaussian function

$$f(x) = Ne^{-\alpha x^2} = Ne^{-x^2/(2\sigma^2)},$$

where σ is the standard deviation.

Solution:

For this Gaussian,

$$\int_{-\infty}^{\infty} f(x)dx = 1$$

gives

$$\boxed{N = (2\pi\sigma^2)^{-1/2}}$$

(see Math Box 10.1).
Then, $N = (2\pi\sigma^2)^{-1/2}$ and $\alpha = 1/(2\sigma^2)$.
From Equation (10.6),

$$F(k) = \frac{1}{\sqrt{2\pi}} \int_{-\infty}^{\infty} N e^{-x^2/(2\sigma^2)} e^{-ikx}\, dx.$$

To solve this integral, we convert the exponent into a sum of squares thus

$$\alpha x^2 + ikx = \left(\sqrt{\alpha}x + \frac{ik}{2\sqrt{\alpha}}\right)^2 + \frac{k^2}{4\alpha}.$$

Therefore,

$$F(k) = \frac{N}{\sqrt{2\pi}} \int_{-\infty}^{\infty} e^{-\left(\sqrt{\alpha}x + ik/2\sqrt{\alpha}\right)^2} e^{-k^2/4\alpha} dx.$$

Let

$$u = \sqrt{\alpha}x + \frac{ik}{2\sqrt{\alpha}}.$$

Then $du = \sqrt{\alpha}dx$ and

$$F(k) = \frac{N}{\sqrt{2\pi\alpha}} e^{-k^2/4\alpha} \int_{-\infty}^{\infty} e^{-u^2} du = \frac{N}{\sqrt{2\alpha}} e^{-k^2/4\alpha}$$

using the standard integral

$$\int_{0}^{\infty} e^{-ay^2} dy = \frac{1}{2}\sqrt{\frac{\pi}{a}}.$$

Substitution for N and α gives

$$F(k) = \frac{1}{\sqrt{2\pi}} e^{-k^2\sigma^2/2}.$$

Hence, the **FT of a Gaussian** is another Gaussian (Figure 10.2).
We observe that:

1. The FT convention we used is the *unitary* FT convention. Under this convention, the $\boldsymbol{L^2}$**-norm** is preserved:

$$\int_{-\infty}^{\infty} |f(x)|^2 dx = \int_{-\infty}^{\infty} |F(k)|^2 dk.$$

However, the **area** of the function is *not* preserved. Thus $f(x)$ may be normalized, but $F(k)$ generally is not.

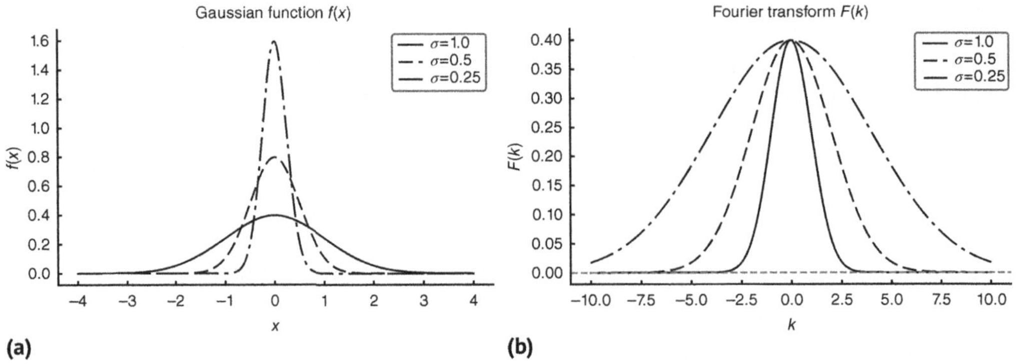

Figure 10.2 Gaussian function (a) and its Fourier transform (b). Curves for $\sigma = 1$ (continuous), 0.5 (dashed), and 0.25 (dash and dots).

2. $f(x)$ is narrow for small σ, while $F(k)$ is broad, and vice versa.
3. $\Delta x\, \Delta k = 1$. This **reciprocal relationship between widths** is the mathematical origin of uncertainty relations in conjugate variables.

Math Box 10.1: Gaussian Function and Its Fourier Transform

1. Normalization of the Gaussian $f(x)$

We verify

$$\int_{-\infty}^{\infty} f(x)\, dx = 1.$$

Substitute the explicit form:

$$\int_{-\infty}^{\infty} f(x)\, dx = \frac{1}{\sqrt{2\pi}\sigma} \int_{-\infty}^{\infty} \exp\left(-\frac{x^2}{2\sigma^2}\right) dx.$$

Let $x = \sigma u$ so that $dx = \sigma\, du$. Then

$$\int_{-\infty}^{\infty} f(x)\, dx = \frac{1}{\sqrt{2\pi}} \int_{-\infty}^{\infty} e^{-u^2/2}\, du.$$

Using the standard Gaussian integral

$$\int_{-\infty}^{\infty} e^{-u^2/2}\, du = \sqrt{2\pi},$$

we obtain

$$N = \left(2\pi\sigma^2\right)^{-1/2},$$

$$\int_{-\infty}^{\infty} f(x)\,dx = 1.$$

2. Normalization of the FT $F(k)$

The transform of a Gaussian is another Gaussian (Example 10.2):

$$F(k) = \frac{1}{\sqrt{2\pi}} \exp\left(-\frac{1}{2}\sigma^2 k^2\right).$$

Consider the integral: $\int_{-\infty}^{\infty} F(k)\,dk$.
Substitute

$$\int_{-\infty}^{\infty} F(k)\,dk = \frac{1}{\sqrt{2\pi}} \int_{-\infty}^{\infty} \exp\left(-\frac{1}{2}\sigma^2 k^2\right) dk.$$

Let $v = \sigma k$, $dk = dv/\sigma$. Then

$$\int_{-\infty}^{\infty} F(k)\,dk = \frac{1}{\sqrt{2\pi}\sigma} \int_{-\infty}^{\infty} e^{-v^2/2}\,dv = \frac{1}{\sqrt{2\pi}\sigma} \cdot \sqrt{2\pi} = \frac{1}{\sigma}.$$

Thus, the spatial and Fourier-domain Gaussians are not individually normalized. In particular, for the Gaussian pair used in this chapter,

$$\int_{-\infty}^{\infty} f(x)\,dx = 1, \quad \int_{-\infty}^{\infty} F(k)\,dk = \frac{1}{\sigma}.$$

So the correct statement is:

$$\boxed{\text{Unitary Fourier transforms preserve } L^2 \text{ normalization, not area normalization.}}$$

Fact Box 10.1: Gaussian–Fourier Pair

$$\boxed{\begin{aligned} f(x) &= \frac{1}{\sqrt{2\pi}\sigma} \exp\left(-\frac{x^2}{2\sigma^2}\right), \\ F(k) &= \frac{1}{\sqrt{2\pi}} \exp\left(-\frac{1}{2}\sigma^2 k^2\right). \end{aligned}}$$

- Widths obey

$$\sigma_x \sigma_k = 1.$$

- The Gaussian is the **unique** function that preserves its functional form under the Fourier transform.

Unitarity Check (Parseval)

- With our Fourier convention:

$$\int |f(x)|^2 dx = \int |F(k)|^2 dk.$$

- For our Gaussian:

$$\int |f(x)|^2 dx = \frac{1}{2\sqrt{\pi}\,\sigma} \text{ and } \int |F(k)|^2 dk = \frac{1}{2\sqrt{\pi}\,\sigma}.$$

✓ The norms match → **unitary transform**, as required.

⚠ Red Flag 10.1: Common Mistake—"Shouldn't the Fourier Transform Also Integrate to 1?"

No. $\int_{-\infty}^{\infty} f(x)\,dx = 1$ does not imply $\int_{-\infty}^{\infty} F(k)\,dk = 1.$

Under the **unitary Fourier transform**

$$F(k) = \frac{1}{\sqrt{2\pi}} \int f(x) e^{-ikx} dx,$$

the transform of a normalized Gaussian is

$$F(k) = \frac{1}{\sqrt{2\pi}} e^{-\sigma^2 k^2/2}.$$

Although $f(x)$ is normalized,

$$\int f(x) dx = 1,$$

its Fourier transform satisfies

$$\int F(k) dk = \frac{1}{\sigma}.$$

Area is not preserved.
What *is* preserved is the L^2**-norm** (energy):

$$\int |f(x)|^2 dx = \int |F(k)|^2 dk.$$

This distinction is important and often overlooked by students.

Historical Note 10.1: Gauss, Fourier, and the Ubiquity of the Gaussian

The Gaussian function was introduced by **Carl Friedrich Gauss (1777–1855)** in his study of observational errors in astronomy. Small, independent fluctuations were found to combine into a characteristic bell-shaped distribution. This insight, now formalized in the central limit theorem, made the Gaussian the fundamental model of noise and uncertainty.

Joseph Fourier (1768–1830), while analyzing heat conduction, discovered that even complicated functions can be decomposed into sums or integrals of sines and cosines. His work laid the foundation for the Fourier transform, which allows switching between the time/space domain and the frequency domain.

The pairing of Gauss and Fourier is especially elegant: the Fourier transform of a Gaussian is another Gaussian.

This invariance explains the appearance of Gaussian wave packets in quantum mechanics, Gaussian beams in optics, and the role of Gaussians as minimum-uncertainty states.

The Gaussian is therefore central not only to probability and statistics but also to spectroscopy, signal processing, molecular dynamics, and any field where frequency-domain analysis is used.

Why the Gaussian Is Special Under the Fourier Transform

- The Gaussian is the *only* function whose Fourier transform is another Gaussian of the same functional form.
- This invariance made the Gaussian (and its transform) crucial in:
  - quantum mechanics (ground-state wavefunctions of the harmonic oscillator);
  - probability theory (stable distributions, CLT);
  - signal processing (minimum-uncertainty wave packets);
  - optics (Gaussian beams).

Heisenberg and the Uncertainty Principle (1927)

- The relation

  $$\sigma_x \sigma_k = 1$$

  for the unitary transform we are using arises directly from the Fourier pair of Gaussians.
- Schrödinger later showed that **only Gaussians** achieve the minimum-uncertainty product.

In summary:
Gauss discovered the function; Fourier discovered the transform; and together they form one of the most elegant dualities in mathematics, chemistry, and physics.

10.2 The Dirac Delta Function

The Dirac delta function, introduced by Paul A. M. Dirac, is not an ordinary function but a *generalized function* (or *distribution*) used to represent an idealized point source or an infinitely narrow pulse.

In physics and chemistry, it plays an essential role in expressing orthogonality and completeness relations of continuous wavefunctions and in formulating transition probabilities and Green's functions.

The Dirac delta function, denoted by $\delta(x)$, is **defined** by the two fundamental properties:

$$\delta(x) = 0 \text{ for } x \neq 0$$

and

$$\int_{-\infty}^{\infty} \delta(x)\, dx = 1. \tag{10.8}$$

It possesses the ***sifting* (or *sampling*) property**:

$$\int_{-\infty}^{\infty} f(x)\, \delta(x - x_0)\, dx = f(x_0), \tag{10.9}$$

which means that $\delta(x - x_0)$ extracts the value of $f(x)$ at the point $x = x_0$. This property is widely used in quantum mechanics when normalizing continuous eigenfunctions.

10.2.1 Representations of the Delta Function

The delta function may be represented as the limiting form of a normalized function whose width tends to zero while its area remains unity. One common representation uses a Gaussian function

$$\delta(x) = \lim_{\sigma \to 0} \frac{1}{\sqrt{2\pi}\, \sigma} e^{-x^2/(2\sigma^2)}. \tag{10.10}$$

As $\sigma \to 0$, the Gaussian becomes infinitely sharp and tall, but its integral remains equal to 1. Another frequently used form is its **Fourier integral representation**

$$\delta(x) = \frac{1}{2\pi} \int_{-\infty}^{\infty} e^{ikx}\, dk. \tag{10.11}$$

This expression shows that $\delta(x)$ can be viewed as a superposition of sinusoidal waves of all possible wavenumbers k.

If we substitute this representation into the inverse FT (Equation 10.7), we obtain

$$f(x) = \int_{-\infty}^{\infty} F(k) e^{ikx}\, dk = \int_{-\infty}^{\infty} f(x')\, \delta(x - x')\, dx', \tag{10.12}$$

demonstrating that the delta function serves as the **kernel** of the FT—the function that allows exact reconstruction of $f(x)$ from its transform $F(k)$.

10.2.2 Orthogonality of Exponentials

The sifting property leads directly to the orthogonality of the complex exponential basis:

$$\int_{-\infty}^{\infty} e^{i(k-k')x}\, dx = 2\pi\, \delta(k - k'). \tag{10.13}$$

Equation (10.13) expresses the **completeness** of the exponential functions e^{ikx} in representing any square-integrable function.

In quantum mechanics, this corresponds to the orthogonality of momentum eigenfunctions, where

$$\langle k \mid k' \rangle = \delta(k - k'). \tag{10.14}$$

Thus, the Dirac delta function forms a fundamental link between **Fourier analysis** and **quantum theory**, appearing naturally in the normalization of continuous states, in Green's function formalism, and in transition amplitude calculations.

10.3 Time and Frequency Domains

The FT provides a direct correspondence between two complementary ways of describing a system: the **time (or space) domain** and the **frequency (or wavenumber) domain**. In chemistry and quantum mechanics, these domains often represent pairs of **conjugate variables** such as **time–energy** or **position–momentum**.

Let $f(t)$ be a time-dependent function. Its FT and inverse transform are defined as

$$F(\omega) = \frac{1}{\sqrt{2\pi}} \int_{-\infty}^{\infty} f(t)\, e^{-i\omega t}\, dt, \tag{10.15}$$

$$f(t) = \frac{1}{\sqrt{2\pi}} \int_{-\infty}^{\infty} F(\omega)\, e^{i\omega t}\, d\omega. \tag{10.16}$$

Here, ω is the **angular frequency**, which is related to the linear frequency ν by $\omega = 2\pi\nu$.

The function $f(t)$ describes how a quantity varies with time, while $F(\omega)$ represents how the same signal is distributed among different frequency components.

Because these domains are Fourier conjugates, localization in one implies delocalization in the other. If $f(t)$ is sharply confined in time, its transform $F(\omega)$ is broad, and, conversely, a broad $f(t)$ in time produces a narrow $F(\omega)$ in frequency. This reciprocal behavior is the mathematical basis of the **uncertainty principle**, expressed as

$$\Delta t\, \Delta\omega \geq \frac{1}{2}, \tag{10.17}$$

or, in quantum-mechanical (**position–momentum**) form,

$$\Delta x\, \Delta p \geq \frac{\hbar}{2}. \tag{10.18}$$

10.3.1 Interpretation and Physical Significance

The FT separates a complex signal into its constituent frequencies, showing how energy or intensity is distributed in the frequency domain. The magnitude $|F(\omega)|$ gives the **amplitude spectrum**, while $|F(\omega)|^2$ represents the **power spectrum**.

In spectroscopy, this distinction is critical:

- $f(t)$ may represent a **time-domain signal**, such as the free-induction decay (FID) in NMR or a transient absorption trace in ultrafast spectroscopy.

- $F(\omega)$ gives the **frequency-domain spectrum**, corresponding to the observed intensity as a function of frequency or energy.

Hence, the FT provides the mathematical bridge between **time-resolved** and **frequency-resolved** descriptions of a system. It enables chemists and physicists to analyze oscillatory phenomena, relaxation dynamics, and wave–particle duality within a unified mathematical framework.

10.3.2 Coordinate and Momentum Space

The FT provides a natural bridge between the **coordinate** and **momentum** representations of a wavefunction in quantum mechanics. Just as the FT relates the spatial and frequency domains of a signal, it also connects the position-space wavefunction $\psi(x)$ with its momentum-space counterpart $\phi(p)$.

10.3.3 Fourier Relation Between Coordinate and Momentum Representations

The **wavefunction $\psi(x)$ in coordinate space** and the corresponding **wavefunction $\phi(p)$ in momentum space** form an **FT pair**:

$$\phi(p) = \frac{1}{\sqrt{2\pi\hbar}} \int_{-\infty}^{\infty} \psi(x)\, e^{-ipx/\hbar}\, dx, \tag{10.19}$$

$$\psi(x) = \frac{1}{\sqrt{2\pi\hbar}} \int_{-\infty}^{\infty} \phi(p)\, e^{ipx/\hbar}\, dp. \tag{10.20}$$

The factor $1/\sqrt{2\pi\hbar}$ ensures that the transformation preserves **normalization**:

$$\int_{-\infty}^{\infty} |\psi(x)|^2\, dx = \int_{-\infty}^{\infty} |\phi(p)|^2\, dp = 1. \tag{10.21}$$

The position-space function $\psi(x)$ describes the probability amplitude of finding a particle at position x, while the momentum-space function $\phi(p)$ gives the probability amplitude for momentum p.

Both contain the same physical information but are expressed in conjugate variables.

10.3.4 Deriving the Quantum Mechanics Convention

The FT provides a natural bridge between the coordinate and momentum representations of a wavefunction. We can derive this relationship directly from our chapter's symmetric FT pair (Equations 10.6 and 10.7) and the fundamental de Broglie relation.

The **de Broglie relation** connects a particle's momentum p to its wavenumber $k(k = 2\pi/\lambda = 2\pi p/h = p/\hbar)$. This implies $k = p/\hbar$ and $dk = dp/\hbar$. Our goal is to relate the position-space wavefunction, $\psi(x)$, to the momentum-space wavefunction, $\phi(p)$.

1. **Forward Transform ($\psi(x) \to \phi(p)$)**
   First, let us apply our standard FT to $\psi(x)$ to get the wavenumber-space wavefunction, which we can call $\phi_k(k)$:

$$\phi_k(k) = \mathcal{F}\{\psi(x)\} = \frac{1}{\sqrt{2\pi}} \int_{-\infty}^{\infty} \psi(x) e^{-ikx} dx.$$

   The functions $\phi_k(k)$ and $\phi(p)$ describe the same physical state, so their probability densities must be equivalent. The probability of finding the particle with momentum between p and $p + dp$ must equal the probability of finding it with wavenumber between k and $k + dk$:

$$|\phi(p)|^2 dp = |\phi_k(k)|^2 dk.$$

   Substituting $dp = \hbar dk$, we get

$$|\phi(p)|^2 (\hbar\, dk) = |\phi_k(k)|^2 dk \Longrightarrow |\phi(p)|^2 = \frac{1}{\hbar} |\phi_k(k)|^2.$$

   This shows we must define our momentum-space wavefunction $\phi(p)$ as

$$\phi(p) = \frac{1}{\sqrt{\hbar}} \phi_k(k).$$

   Now, we substitute $k = p/\hbar$ and the expression for $\phi_k(k)$:

$$\phi(p) = \frac{1}{\sqrt{\hbar}} \left[\frac{1}{\sqrt{2\pi}} \int_{-\infty}^{\infty} \psi(x) e^{-i(p/\hbar)x} dx \right].$$

   Combining the constants, we arrive at the standard momentum-space transform:

$$\phi(p) = \frac{1}{\sqrt{2\pi\hbar}} \int_{-\infty}^{\infty} \psi(x) e^{-ipx/\hbar} dx.$$

   This is precisely Equation (10.19).
2. **Inverse Transform ($\phi(p) \to \psi(x)$)**
   To find the inverse, we start with our standard inverse transform:

$$\psi(x) = \mathcal{F}^{-1}\{\phi_k(k)\} = \frac{1}{\sqrt{2\pi}} \int_{-\infty}^{\infty} \phi_k(k) e^{ikx} dk.$$

   Now, we use our relationships $\phi_k(k) = \sqrt{\hbar}\phi(p)$ and $dk = dp/\hbar$:

$$\psi(x) = \frac{1}{\sqrt{2\pi}} \int_{-\infty}^{\infty} \left[\sqrt{\hbar}\phi(p)\right] e^{i(p/\hbar)x} \left[\frac{dp}{\hbar}\right].$$

   Grouping the constants:

$$\psi(x) = \frac{\sqrt{\hbar}}{\sqrt{2\pi} \cdot \hbar} \int_{-\infty}^{\infty} \phi(p) e^{ipx/\hbar} dp = \frac{1}{\sqrt{2\pi\hbar}} \int_{-\infty}^{\infty} \phi(p) e^{ipx/\hbar} dp.$$

   This is precisely Equation (10.20).

By using the de Broglie relation, we have shown that the position–momentum FT pair used in quantum mechanics is not a different convention but a direct physical consequence of the symmetric k-space transform defined at the start of this chapter. This derivation also confirms the symmetric normalization factor $\frac{1}{\sqrt{2\pi\hbar}}$ that ensures probability is conserved, as stated in Equation (10.21).

10.3.4.1 Example: Gaussian Wave Packet

A particularly important case is the Gaussian wave packet:

$$\psi(x) = \left(\frac{1}{\pi\sigma^2}\right)^{1/4} e^{-x^2/(2\sigma^2)}. \tag{10.22}$$

The FT of this function gives the momentum-space wavefunction:

$$\phi(p) = \frac{1}{\sqrt{2\pi\hbar}} \int_{-\infty}^{\infty} \psi(x) e^{-ipx/\hbar}\, dx = \left(\frac{\sigma^2}{\pi\hbar^2}\right)^{1/4} e^{-\sigma^2 p^2/(2\hbar^2)}. \tag{10.23}$$

Hence, the momentum-space function is also Gaussian, with a width inversely proportional to that in coordinate space:

$$\Delta x\, \Delta p = \hbar/2. \tag{10.24}$$

For a detailed derivation, please refer to Math Box 10.2. This expresses the **Heisenberg uncertainty principle**, which states that a particle cannot have both a precisely defined position and momentum simultaneously. A narrow wave packet in coordinate space (small σ) implies a broad distribution in momentum space, and vice versa.

This specific Gaussian packet is a "**minimum-uncertainty wave packet**" and exactly satisfies the limit of the Heisenberg uncertainty principle ($\Delta x\, \Delta p \geq \hbar/2$).

Math Box 10.2: Derivation of Equation (10.24): The Rigorous Calculation

In quantum mechanics, uncertainty is defined as the **standard deviation** of the operator:

- $(\Delta x)^2 = \langle x^2 \rangle - \langle x \rangle^2$;
- $(\Delta p)^2 = \langle p^2 \rangle - \langle p \rangle^2$.

1. **Finding Δx (Position Uncertainty)**
   1. **Wavefunction:** $\psi(x) = Ce^{-x^2/(2\sigma^2)}$.
   2. **Probability Density:** $|\psi(x)|^2 = C^2 e^{-x^2/\sigma^2}$. With the normalization constant $C = (\pi\sigma^2)^{-1/4}$, this is $|\psi(x)|^2 = (\pi\sigma^2)^{-1/2} e^{-x^2/\sigma^2}$.
   3. Find $\langle x \rangle$: $\langle x \rangle = \int_{-\infty}^{\infty} x|\psi(x)|^2 dx$.

      The integrand $x \cdot e^{-x^2/\sigma^2}$ is an odd function, so its integral from $-\infty$ to ∞ is 0.
   4. Find $\langle x^2 \rangle$: $\langle x^2 \rangle = \int_{-\infty}^{\infty} x^2 |\psi(x)|^2 dx = \frac{1}{\sqrt{\pi\sigma^2}} \int_{-\infty}^{\infty} x^2 e^{-x^2/\sigma^2} dx$.

      This is a standard Gaussian integral with the known result: $\int_{-\infty}^{\infty} x^2 e^{-ax^2} dx = \frac{1}{2a}\sqrt{\frac{\pi}{a}}$.
      - Here, $a = 1/\sigma^2$.

- $\int_{-\infty}^{\infty} x^2 e^{-x^2/\sigma^2}\,dx = \frac{1}{2(1/\sigma^2)}\sqrt{\frac{\pi}{1/\sigma^2}} = \frac{\sigma^2}{2}\sqrt{\pi\sigma^2}.$
- $\langle x^2\rangle = \frac{1}{\sqrt{\pi\sigma^2}}\left(\frac{\sigma^2}{2}\sqrt{\pi\sigma^2}\right) = \frac{\sigma^2}{2}.$

5. Calculate Δx:
   $(\Delta x)^2 = \langle x^2\rangle - \langle x\rangle^2 = \frac{\sigma^2}{2} - 0^2 = \frac{\sigma^2}{2}.$
   Therefore, $\Delta x = \frac{\sigma}{\sqrt{2}}.$

2. **Finding Δp (Momentum Uncertainty)**
   1. **Momentum Wavefunction:** $\phi(p) = De^{-\sigma^2 p^2/(2\hbar^2)}$.
   2. **Probability Density:** $|\phi(p)|^2 = D^2 e^{-\sigma^2 p^2/\hbar^2}$.
   3. Find $\langle p\rangle$:
      By symmetry, $\langle p\rangle = \int_{-\infty}^{\infty} p|\phi(p)|^2 dp = 0.$
   4. Find $\langle p^2\rangle$: $\langle p^2\rangle = \int_{-\infty}^{\infty} p^2|\phi(p)|^2 dp = D^2\int_{-\infty}^{\infty} p^2 e^{-\sigma^2 p^2/\hbar^2} dp.$
      - We use the same standard integral, this time with $a = \sigma^2/\hbar^2$.
      - $\int_{-\infty}^{\infty} p^2 e^{-(\sigma^2/\hbar^2)p^2} dp = \frac{1}{2(\sigma^2/\hbar^2)}\sqrt{\frac{\pi}{\sigma^2/\hbar^2}} = \frac{\hbar^2}{2\sigma^2}\sqrt{\frac{\pi\hbar^2}{\sigma^2}}.$
      - The normalization constant $D^2 = \sqrt{\sigma^2/(\pi\hbar^2)}$.
      - $\langle p^2\rangle = \sqrt{\frac{\sigma^2}{\pi\hbar^2}}\left(\frac{\hbar^2}{2\sigma^2}\sqrt{\frac{\pi\hbar^2}{\sigma^2}}\right) = \frac{\hbar^2}{2\sigma^2}.$
   5. Calculate Δp:
      $(\Delta p)^2 = \langle p^2\rangle - \langle p\rangle^2 = \frac{\hbar^2}{2\sigma^2} - 0^2 = \frac{\hbar^2}{2\sigma^2}.$
      Therefore, $\Delta p = \frac{\hbar}{\sqrt{2}\sigma}.$

3. **The Uncertainty Product**
   Now we multiply the two results:

$$\Delta x \cdot \Delta p = \left(\frac{\sigma}{\sqrt{2}}\right)\cdot\left(\frac{\hbar}{\sqrt{2}\sigma}\right) = \frac{\hbar}{2}.$$

10.3.4.2 Physical Interpretation

1. The FT in quantum mechanics converts between the *representation of the state* in position and momentum variables.
2. The operator $-i\hbar\frac{d}{dx}$ in coordinate space corresponds to multiplication by p in momentum space, showing that differentiation and multiplication are Fourier conjugates.
3. Plane waves $e^{ipx/\hbar}$ form the orthogonal basis functions of momentum eigenstates, satisfying

$$\langle p \mid p'\rangle = \delta(p - p').$$

The normalization condition ensures that the total probability remains unity in either representation.

This equivalence of coordinate and momentum descriptions is a cornerstone of **wave mechanics**, emphasizing that the two domains are not distinct realities but complementary aspects of the same quantum state.

10.4 Properties of the FT

The FT possesses several useful mathematical properties that make it an indispensable analytical tool in physics and chemistry. These properties simplify the solution of differential equations, the treatment of boundary conditions, and the analysis of experimental data such as spectroscopic signals.

Let $f(x)$ and $F(k)$ form an FT pair:

$$F(k) = \frac{1}{\sqrt{2\pi}} \int_{-\infty}^{\infty} f(x)\, e^{-ikx}\, dx, \; f(x) = \frac{1}{\sqrt{2\pi}} \int_{-\infty}^{\infty} F(k)\, e^{ikx}\, dk.$$

1. **Linearity**
   The FT is a linear operation. If $f_1(x) \leftrightarrow F_1(k)$ and $f_2(x) \leftrightarrow F_2(k)$, then for any constants a and b,

   $$af_1(x) + bf_2(x) \longleftrightarrow aF_1(k) + bF_2(k). \tag{10.25}$$

   This property follows directly from the linearity of the integral and is extensively used in representing linear combinations of wavefunctions or in the superposition of spectral components.
2. **Shifting (Translation) Property**
   A shift in the variable x results in a phase factor in the frequency domain.
   If $f(x - x_0)$ is the shifted function, then

   $$f(x - x_0) \longleftrightarrow e^{-ikx_0} F(k). \tag{10.26}$$

   Similarly, a shift in the frequency domain produces a phase change in real space:

   $$e^{ik_0x} f(x) \longleftrightarrow F(k - k_0). \tag{10.27}$$

3. **Scaling Property**
   If the function is stretched or compressed in one domain, its transform is inversely scaled in the other:

   $$f(ax) \longleftrightarrow \frac{1}{|a|} F\left(\frac{k}{a}\right), \tag{10.28}$$

   where a is a real, nonzero constant. This property underlies the reciprocal relationship between width in one domain and spread in the other, a key aspect of the uncertainty principle.
4. **Differentiation Property**
   Differentiation in one domain corresponds to multiplication by $(ik)^n$ in the other:

   $$\frac{d^n f(x)}{dx^n} \longleftrightarrow (ik)^n F(k). \tag{10.29}$$

   Conversely,

   $$x^n f(x) \longleftrightarrow i^n \frac{d^n F(k)}{dk^n}. \tag{10.30}$$

   These properties are particularly useful in solving linear differential equations and in expressing quantum-mechanical operators (such as momentum) in reciprocal space.

5. **Convolution Theorem**
The convolution of two functions in real space corresponds to the multiplication of their FTs. Under the **unitary FT convention** used in this chapter, the transform of a convolution becomes the product of the individual transforms, scaled by $\sqrt{2\pi}$:

$$(f * g)(x) = \int_{-\infty}^{\infty} f(\xi)g(x-\xi)d\xi \longleftrightarrow \sqrt{2\pi}F(k)G(k). \tag{10.31}$$

Conversely, the product of two functions in one domain corresponds to a convolution in the other:

$$f(x)g(x) \longleftrightarrow \frac{1}{\sqrt{2\pi}}(F * G)(k). \tag{10.32}$$

The constants in (10.31) and (10.32) are specific to the unitary choice and differ from engineering texts that place all 2π in one direction.
The convolution theorem has numerous applications in spectroscopy, such as in **line broadening**, instrumental response corrections, and **signal filtering**.
6. **Parseval's Theorem**
The total energy (or power) of a signal is conserved under the Fourier transformation:

$$\int_{-\infty}^{\infty} |f(x)|^2 dx = \int_{-\infty}^{\infty} |F(k)|^2 dk. \tag{10.33}$$

This relation is fundamental in physical chemistry and spectroscopy, where it guarantees that the total energy content of a signal remains invariant whether expressed in time or frequency domains.
7. **Complex Conjugation and Symmetry**
If $f(x)$ is real, then

$$F(-k) = F^*(k), \tag{10.34}$$

where the asterisk denotes complex conjugation. Hence, $|F(k)|^2$ is an even function of k.
If $f(x)$ is even, then $F(k)$ is also even and real; if $f(x)$ is odd, $F(k)$ is odd and purely imaginary. These symmetry relationships are helpful when analyzing signals with definite parity, such as vibrational or rotational modes.
Together, these properties make the FT a versatile mathematical framework for describing wave phenomena, spectral transitions, and linear systems in chemistry and physics.

10.5 Fourier Sine and Cosine Transforms

When a function $f(x)$ is defined only for $x \geq 0$ and satisfies appropriate boundary conditions, it is often more convenient to express it using **Fourier sine** and **cosine transforms**.

These real transforms are particularly useful in solving partial differential equations involving semi-infinite domains such as diffusion, heat conduction, or quantum wells.

10.5.1 Even and Odd Extensions

For a given function $f(x)$ defined for $x \geq 0$, the even and odd extensions are

$$f_{\text{even}}(x) = \begin{cases} f(x), & x \geq 0, \\ f(-x), & x < 0, \end{cases} \quad f_{\text{odd}}(x) = \begin{cases} f(x), & x \geq 0, \\ -f(-x), & x < 0. \end{cases} \tag{10.35}$$

The FT of $f_{\text{even}}(x)$ gives the **Fourier cosine transform (FCT)**, while that of $f_{\text{odd}}(x)$ gives the **Fourier sine transform (FST)**.

10.5.2 Fourier Cosine Transform

For an even function,

$$F_c(k) = \sqrt{\frac{2}{\pi}} \int_0^\infty f(x)\cos(kx)dx, \tag{10.36}$$

and the inverse relation is

$$f(x) = \sqrt{\frac{2}{\pi}} \int_0^\infty F_c(k)\cos(kx)dk. \tag{10.37}$$

The cosine transform is useful for symmetric boundary conditions, e.g., vibrations of a membrane fixed at both ends or a particle confined in a symmetric potential well.

10.5.3 Fourier Sine Transform

For an odd function,

$$F_s(k) = \sqrt{\frac{2}{\pi}} \int_0^\infty f(x)\sin(kx)dx, \tag{10.38}$$

and its inverse is

$$f(x) = \sqrt{\frac{2}{\pi}} \int_0^\infty F_s(k)\sin(kx)dk. \tag{10.39}$$

The sine transform applies to systems where the function vanishes at $x = 0$, such as diffusion from a boundary or a quantum particle in a rigid wall potential.

Example 10.3: Sine and Cosine Transforms of an Exponential

Find the Fourier sine and cosine transforms of $f(x) = e^{-ax}$ where $a > 0$.

Solution:

$$F_s(k) = \sqrt{\frac{2}{\pi}} \int_0^\infty e^{-ax} \sin(kx)dx = \sqrt{\frac{2}{\pi}} \frac{k}{a^2 + k^2} \tag{10.40}$$

and

$$F_c(k) = \sqrt{\frac{2}{\pi}} \int_0^\infty e^{-ax} \cos(kx)dx = \sqrt{\frac{2}{\pi}} \frac{a}{a^2 + k^2}. \tag{10.41}$$

These transforms illustrate that an exponentially decaying function in real space gives rise to a Lorentzian-type frequency distribution, a concept encountered in **spectral line broadening** and **NMR line-shape analysis**.

Remarks

1. Both transforms are **real-valued**, avoiding explicit complex exponentials.
2. They satisfy the same basic properties (linearity, scaling, differentiation, and convolution) as the general FT.
3. The choice between sine and cosine forms depends on the boundary conditions—sine transforms for *Dirichlet* ($f = 0$ at the boundary) and cosine transforms for *Neumann* ($df/dx = 0$) types.

10.6 Applications of Fourier Transforms

FTs find application in almost every branch of modern science and engineering—far beyond what Lord Kelvin could have imagined. The advent of high-speed computers and efficient numerical algorithms such as the **fast Fourier transform (FFT)**, and its later variant, the **sparse Fourier transform (SFT)** (10–100 times faster), has revolutionized data analysis. Before the development of FFT, the numerical evaluation of FTs was extremely cumbersome. In many modern applications, FFT has itself been supplemented or replaced by the **discrete cosine transform (DCT)**, a computationally efficient analog of the **FCT** (Equation 10.36).

All such applications are based on one simple but profound idea: **any waveform can be represented as a sum of simple sinusoidal functions of different frequencies**. An FT decomposes a complicated signal into its constituent frequency components.

In fields such as spectroscopy, crystallography, and NMR, FFT-based implementations are indispensable for modern data acquisition and signal reconstruction. While the unitary transformation convention (Equations 10.6 and 10.7) is maintained as the default framework throughout this chapter, normalization factors frequently scale differently depending on the specific physical context and historical traditions of the discipline. For a comparative overview of these alternative conventions and their corresponding Fourier transform pairs, refer to Table 10.1.

Table 10.1 Common Fourier transform conventions.[a]

Convention name	Forward transform $f(x) \to F(k)$	Inverse transform $F(k) \to f(x)$	Key properties/usage
Symmetric (unitary) (Chapter 10 standard)	$F(k) = \frac{1}{\sqrt{2\pi}} \int_{-\infty}^{\infty} f(x)e^{-ikx}\,dx$	$f(x) = \frac{1}{\sqrt{2\pi}} \int_{-\infty}^{\infty} F(k)e^{ikx}\,dk$	Preserves L^2-norm
Quantum mechanics (position–momentum)	$\phi(p) = \frac{1}{\sqrt{2\pi\hbar}} \int_{-\infty}^{\infty} \psi(x)e^{-ipx/\hbar}\,dx$	$\psi(x) = \frac{1}{\sqrt{2\pi\hbar}} \int_{-\infty}^{\infty} \phi(p)e^{ipx/\hbar}\,dp$	• Derived from symmetric convention via $k = p/\hbar$ • Preserves probability
X-ray crystallography (3D periodic)	$F_{hkl} = \int_{\text{cell}} \rho(\boldsymbol{r})e^{2\pi i\mathbf{h}\cdot\mathbf{r}}\,d\mathbf{r}$	$\rho(\boldsymbol{r}) = \frac{1}{V} \sum_{hkl} F_{hkl}\,e^{-2\pi i\mathbf{h}\cdot\mathbf{r}}$	• Asymmetric normalization (1 and $1/V$) • Forward sign in exponent is $+i$ *(Note the sign convention in crystallography uses $+i$ [forward] versus $-i$ [standard unitary])* • Used for a discrete reciprocal lattice
Spectroscopy (FTIR/NMR) (cosine/one-sided)	$I(\delta) = \int_0^{\infty} S(\nu)\cos(2\pi\nu\delta)d\nu$	$S(\nu) = 2\int_0^{\infty} I(\delta)\cos(2\pi\nu\delta)d\delta$	• Asymmetric, real-valued cosine transform • Used for interferograms (Equations 10.46 and 10.47) • Factor of 2 appears in inverse transform
General cosine/sine transforms (for $x \geq 0$)	$F_c(k) = \sqrt{\frac{2}{\pi}} \int_0^{\infty} f(x)\cos(kx)dx$ $F_s(k) = \sqrt{\frac{2}{\pi}} \int_0^{\infty} f(x)\sin(kx)dx$	$f(x) = \sqrt{\frac{2}{\pi}} \int_0^{\infty} F_c(k)\cos(kx)dk$ $f(x) = \sqrt{\frac{2}{\pi}} \int_0^{\infty} F_s(k)\sin(kx)dk$	• Real-valued transforms for even/odd extensions • Use symmetric prefactor $\sqrt{2/\pi}$ (Equations 10.36–10.39) • Useful for semi-infinite domain problems

a The choice of convention does not change the physical content but affects the normalization of the transformed function. Always check the prefactors and exponent signs when comparing formulas from different sources.

Math Box 10.3: Fourier Transform Conventions

Fourier transforms are defined in several conventions, and all are mathematically equivalent.

They differ only in how the factors of 2π are distributed between the forward and inverse transforms.

Three conventions appear most commonly in physics, engineering, and applied mathematics:

1. **Unitary (Symmetric) Convention**

$$F(\omega) = \frac{1}{\sqrt{2\pi}} \int_{-\infty}^{\infty} f(t)e^{-i\omega t}\,dt, \quad f(t) = \frac{1}{\sqrt{2\pi}} \int_{-\infty}^{\infty} F(\omega)e^{i\omega t}\,d\omega.$$

   This convention treats time and frequency **symmetrically** and preserves normalization:

$$\int |f(t)|^2 dt = \int |F(\omega)|^2 d\omega.$$

   It is widely used in quantum mechanics and in any context where Parseval's Theorem in symmetric form is preferred.

2. **Physics Convention (Asymmetric)**

$$F(\omega) = \int_{-\infty}^{\infty} f(t)e^{-i\omega t}\,dt, \quad f(t) = \frac{1}{2\pi} \int_{-\infty}^{\infty} F(\omega)e^{i\omega t}\,d\omega.$$

   Here, the forward transform carries no prefactor, while the inverse transform contains the entire factor of 2π.

   This is common in many analytical calculations and in older physics literature.

3. **Engineering/Signal-processing Convention**

$$F(f) = \int_{-\infty}^{\infty} f(t)e^{-i2\pi ft}\,dt, \quad f(t) = \int_{-\infty}^{\infty} F(f)e^{i2\pi ft}\,df.$$

   This version uses **ordinary frequency** f instead of angular frequency ω.

   It avoids writing 2π explicitly in the inverse transform and is convenient for discrete-time and digital signal analysis.

Key Idea:

All conventions describe the **same transform**, and the choice of prefactors is purely a matter of style.

Changing conventions rescales the Fourier transform but never its physical content.

In this textbook, we adopt the **unitary convention** for its symmetry and its compatibility with quantum-mechanical notation.

Worked Example 1: Fourier Transform of a Cosine Function

Consider

$$f(t) = \cos(\alpha t).$$

From the definition of the FT (Equation 10.6),

$$F(\omega) = \frac{1}{\sqrt{2\pi}} \int_{-\infty}^{\infty} \cos(\alpha t) e^{-i\omega t}\, dt. \tag{10.42}$$

Expressing cosine in exponential form,

$$\cos(\alpha t) = \frac{1}{2}(e^{i\alpha t} + e^{-i\alpha t}),$$

and using Equation (10.11), we obtain

$$F(\omega) = \sqrt{\left(\frac{\pi}{2}\right)}[\delta(\omega - \alpha) + \delta(\omega + \alpha)], \tag{10.43}$$

where $\delta(x)$ is the Dirac delta function (see Equation 10.9). The delta function is zero everywhere except at $\omega = \pm\alpha$; hence, the transform of $\cos(\alpha t)$ consists of two spikes located at these frequencies (Figure 10.3 [top right]).

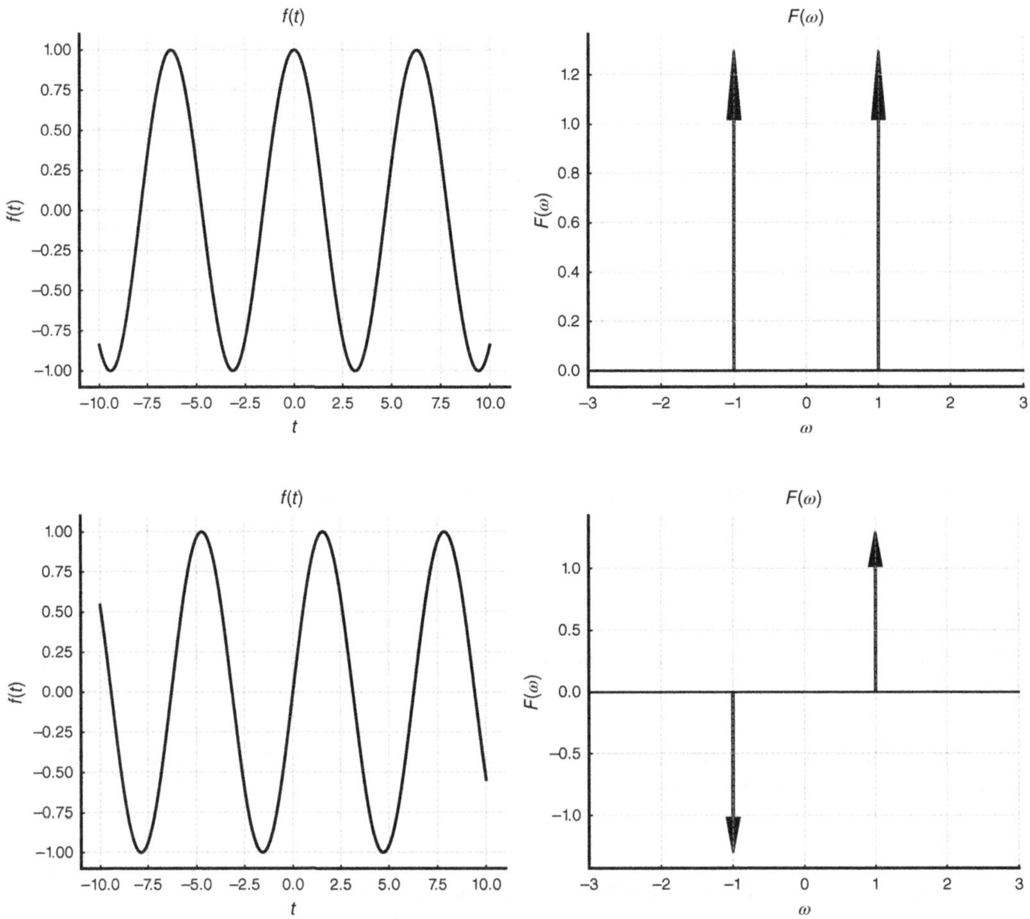

Figure 10.3 *Top:* Cosine: the time-domain cosine wave (left) and its Fourier transform (right) for $\alpha = 1$. *Bottom:* Sine: the time-domain sine wave (left) and its Fourier transform (right) for $\alpha = 1$.

Both graphs contain equivalent information—the left graph represents variation with time, while the right graph expresses the same periodicity as discrete frequency components at $\omega = \pm 1$ (see Figure 10.3).

The frequency of oscillation is $\alpha/2\pi$ s^{-1} (or α rad s^{-1}).

Worked Example 2: Fourier Transform of a Sine Function

Similarly,

$$f(t) = \sin(\alpha t) \Longrightarrow F(\omega) = i\sqrt{\left(\frac{\pi}{2}\right)}[\delta(\omega + \alpha) - \delta(\omega - \alpha)].$$

The FT again consists of two spikes at $\omega = \pm\alpha$, but they have imaginary components with opposite signs, indicating a phase difference (Figure 10.3 [bottom right]). This reflects the **odd-symmetry** of the sine transform.

Worked Example 3: Linear Combination of Frequencies

For a function containing two cosine components,

$$f(t) = \cos(\alpha t) + \cos(\beta t),$$

the transform yields

$$F(\omega) = \sqrt{\left(\frac{\pi}{2}\right)}[\delta(\omega - \alpha) + \delta(\omega + \alpha) + \delta(\omega - \beta) + \delta(\omega + \beta)],$$

showing that the signal contains frequencies at both α and β (Figure 10.4). Here, we have used $\alpha = 1, \beta = 2$). In general, any complex signal is composed of multiple such sinusoidal components.

Real-world waves are far more intricate than these idealized sine or cosine waves. For example, the waveform of a trumpet sound is highly complex.

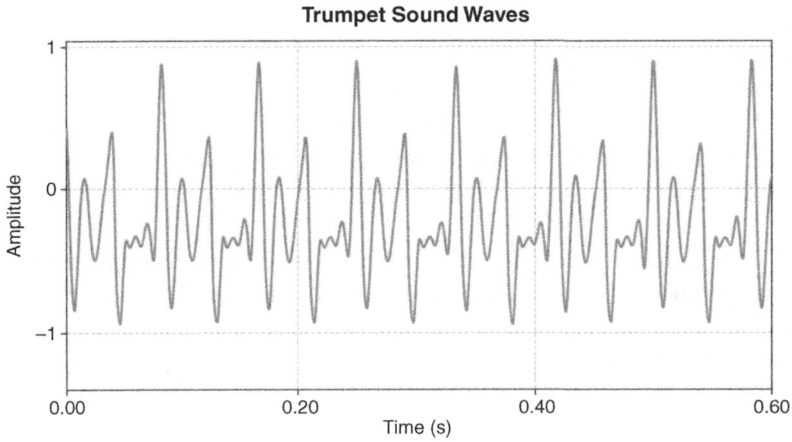

Yet, the **remarkable fact** is that every continuous waveform can be represented as a weighted sum of sine and cosine functions:

$$f(x) = a\cos(\alpha x) + b\sin(\beta x) + c\cos(\gamma x) + \cdots.$$

Performing an FT of such a wave yields its **frequency spectrum**, which displays peaks corresponding to each frequency component. Adding all these sinusoidal components together reconstructs the original waveform.

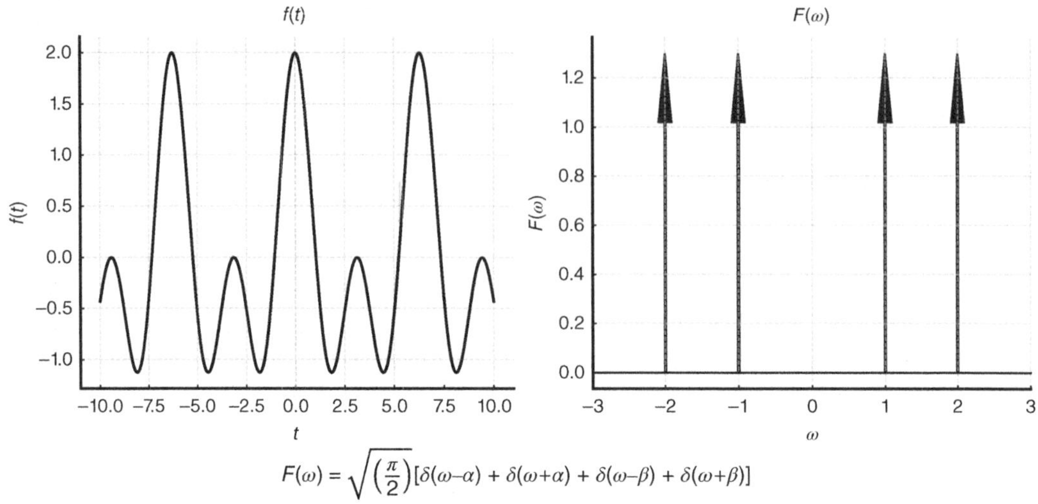

Figure 10.4 The time-domain $f(t) = \cos(\alpha t) + \cos(\beta t)$ wave (left) and its Fourier transform (right) for $\alpha = 1, \beta = 2$.

Thus, Fourier analysis serves as a universal method for translating a signal from the **time (or spatial) domain** into the **frequency (or wavenumber) domain**.

Table 10.2 lists the commonly used FTs under the *unitary* convention for easy reference:

$$F(\omega) = \frac{1}{\sqrt{2\pi}} \int_{-\infty}^{\infty} f(t)e^{-i\omega t}\, dt, \quad f(t) = \frac{1}{\sqrt{2\pi}} \int_{-\infty}^{\infty} F(\omega)e^{i\omega t}\, d\omega.$$

All entries below are **fully consistent** with this convention.

10.6.1 Practical and Scientific Importance

FTs underpin an enormous range of applications involving wave-like behavior:

- **Sound and Vibration:** Analysis of audio signals, musical tones, and acoustic noise.
- **Light and Optics:** Decomposition of electromagnetic fields, diffraction, and image processing.
- **Seismology:** Extraction of dominant frequencies in earthquake waves for structural design.
- **Radio and Telecommunications:** Modulation, filtering, and bandwidth optimization.
- **Data Compression:** MP3 audio and JPEG image compression employ Fourier-based transforms to reduce data size without perceptible loss.
- **Noise Reduction:** Filtering out high-frequency noise from experimental or astronomical data.
- **Machine Learning and Feature Extraction:** Predicting mechanical failure from vibration patterns, identifying specific human activities from smartphone data, or diagnosing heart conditions from the spectral signature of a heartbeat.

Let us explore a few of these applications in detail.

Table 10.2 Useful unitary Fourier transform pairs.

Time-domain function $f(t)$	Fourier transform $F(\omega)$	Notes
1	$\sqrt{2\pi}\,\delta(\omega)$	Constant ↔ delta
$\delta(t)$	$\frac{1}{\sqrt{2\pi}}$	Impulse ↔ constant
$\delta(t-t_0)$	$\frac{1}{\sqrt{2\pi}}e^{-i\omega t_0}$	Time shift → phase factor
$e^{i\alpha t}$	$\sqrt{(2\pi)}\delta(\omega-\alpha)$	Pure tone
$e^{-i\alpha t}$	$\sqrt{(2\pi)}\delta(\omega+\alpha)$	
$\cos(\alpha t)$	$\sqrt{\left(\frac{\pi}{2}\right)}[\delta(\omega-\alpha)+\delta(\omega+\alpha)]$	Even function
$\sin(\alpha t)$	$\left(\frac{1}{i}\right)\sqrt{\left(\frac{\pi}{2}\right)}[\delta(\omega-\alpha)-\delta(\omega+\alpha)]$	Odd function
$\text{rect}(t/2a)=1$ for $t\leq$ a, 0 for $t>a$	$\sqrt{\frac{2}{\pi}}\frac{\sin(a\omega)}{\omega}$	Transforms to a sinc function $\sqrt{\frac{2}{\pi}}a\ \sin c(a\omega)$; sharper time localization yields broader frequency spread
$\text{sinc}(t)$	$\sqrt{\frac{\pi}{2}}\ \text{rect}(\omega/2)$	Dual of above
$e^{-t^2/(2\sigma^2)}$	$\sigma\, e^{-\sigma^2\omega^2/2}$	Gaussian ↔ Gaussian
Normalized Gaussian: $f(t)=\frac{1}{\sqrt{2\pi\sigma}}e^{-t^2/(2\sigma^2)}$	$\frac{1}{\sqrt{2\pi}}e^{-\sigma^2\omega^2/2}$	The spatial Gaussian is normalized; the Fourier-domain Gaussian integrates to $1/\sigma$; the L^2-norms match in both domains
$e^{-\gamma t}(t\geq 0)$	$\frac{1}{\sqrt{2\pi}}\frac{1}{\gamma+i\omega}$	Exponential decay → Lorentzian spectrum
$u(t)$ (Heaviside step)	$\frac{1}{\sqrt{2\pi}}\left(\pi\delta(\omega)+\text{PV}\frac{1}{i\omega}\right)$	Requires distributions
$t\,f(t)$	$i\frac{dF}{d\omega}$	Time multiplication ↔ frequency differentiation
$\frac{df}{dt}$	$i\omega F(\omega)$	Differentiation property

Notes:

1. **All delta coefficients here are correct for the *unitary* definition,** and several rows carry a factor of $\sqrt{(2\pi)}$ or $\sqrt{(\pi/2)}$ — these come directly from the identity $\delta(x)=(1/2\pi)\int e^{ikx}dk$ (Equation 10.11) and are not typos.
2. **Normalization of the Gaussian pair is self-consistent:** $\int|f(t)|^2dt=\int|F(\omega)|^2d\omega=1$.
3. **The box–sinc and sinc–box transforms are often differently written in engineering books** (those use the nonunitary conventions). This table is corrected for our chosen convention.
4. The exponential decay function $e^{-\gamma t}(t\geq 0)$ is also written as $e^{-\gamma t}u(t)$, where $u(t)$ is the **unit step** function ensuring the function exists only for $t\geq 0$. This is standard in signal processing and physics.

10.6.2 Example: Fourier Transform in Sound and Image Processing

An analog record stores sound as physical grooves corresponding to waveform amplitudes. An MP3 file, by contrast, stores data digitally in the **frequency domain**—as lists of frequencies and amplitudes obtained via an FT. Because most information resides in low-frequency components, high-frequency contributions (beyond human hearing) can be truncated, greatly reducing file size. The same principle is used to remove unwanted background noise.

Images behave similarly, except that they vary in **two spatial dimensions**. Brightness and color changes correspond to **spatial frequencies**.

In the FT of an image, most information is concentrated near the center (low-frequency region). Removing data far from the center—a process known as **low-pass filtering**—reduces file size without appreciable loss of quality. For example, keeping only 50% of low-frequency components might reduce the total information content by merely 5%.

Figure 10.5 illustrates how most spatial information in a natural image is concentrated near the origin of its FT. The tiger cub's fine fur and whiskers correspond to high-frequency components lying away from the center. Removing these components produces a softened but still recognizable image, demonstrating that much of the essential structure resides in the low-frequency region.

Just as in audio processing, Fourier filtering can **clean noisy photographs**, e.g., removing speckles from under-exposed images or astronomical frames, thereby enhancing the focus on the region of interest.

In the preceding example (Figure 10.5), a natural image was used to show how spatial information can be decomposed into low- and high-frequency components. The same principle applies to scientific imaging, where FTs reveal periodicities and structural order in materials.

For instance, scanning electron microscopy (SEM) or transmission electron microscopy micrographs of crystalline or nanostructured materials contain information spanning a wide range of spatial frequencies. Applying an FT to such images converts the real-space data into reciprocal

Figure 10.5 *Top left:* Original grayscale image (spatial domain). *Top right:* Two-dimensional Fourier transform showing bright central low-frequency components and faint high-frequency edges. *Bottom left:* Reconstructed image after applying a low-pass filter that retains only the central region of the frequency domain; the result is smoother and slightly blurred. *Bottom right:* Frequency-domain mask illustrating the circular low-frequency region retained during filtering.

space, where regular lattice patterns appear as well-defined spots and surface roughness or disorder as diffuse scattering.

The concept is illustrated in Figure 10.6, where a simulated SEM micrograph and its FT are compared. The bright central region corresponds to low-frequency morphology, while the outer areas represent fine structural variations and defects. Such transformations are the foundation of X-ray diffraction (XRD), electron diffraction, and spectroscopic imaging techniques, where reciprocal space analysis provides deep insight into the atomic and electronic structure of materials.

In short, **Fourier analysis is among the most powerful mathematical tools in contemporary science**, enabling the translation between the time/spatial domain and the frequency/momentum domain—concepts that form the basis of quantum mechanics, spectroscopy, and solid-state physics.

The connection between real-space images and their Fourier representations extends naturally to diffraction phenomena. In essence, a diffraction pattern obtained from a crystal is the FT of the electron density distribution within that crystal. Peaks in reciprocal space correspond to the periodic arrangement of atoms in real space, and the intensity of each peak encodes information about the relative amplitudes of the scattered waves. Thus, just as Fourier analysis of micrographs reveals morphological features and defect structures, FTs in diffraction experiments provide direct insight into atomic-scale order and symmetry.

This principle forms the foundation of **XRD**, discussed in the following section, where the spatial periodicity of matter is mapped through its Fourier-related **reciprocal lattice**.

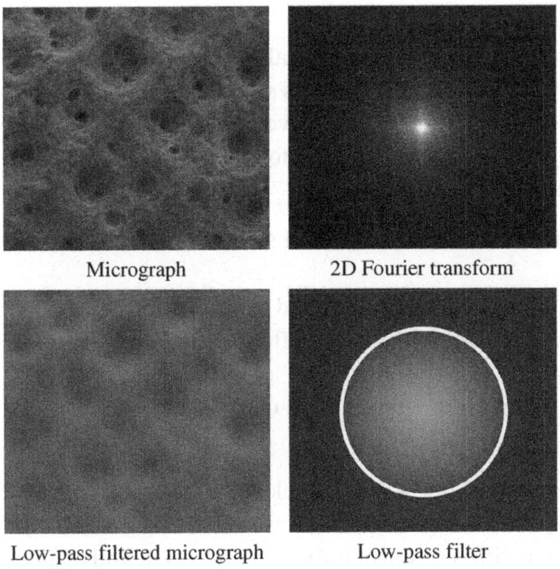

Figure 10.6 Fourier Transform of a scientific image (simulated SEM micrograph). *Top left:* Simulated scanning electron microscope (SEM)-like image of a textured material surface. *Top right:* Two-dimensional Fourier transform showing a bright central region corresponding to low spatial frequencies and diffuse high-frequency components representing fine structural features. *Bottom left:* Low-pass-filtered reconstruction of the image, in which high-frequency details are suppressed, and only large-scale morphology remains visible. *Bottom right:* Frequency-domain mask used to retain only the central (low-frequency) region while discarding higher frequencies.

10.6.3 X-ray Diffraction

Among the most significant applications of the FT in chemistry and materials science is its use in **XRD**, which reveals the atomic arrangement within crystals.

When a beam of X-rays scatters from the periodic **electron density**, $\rho(\mathbf{r})$, of a crystal, the scattered waves interfere to produce a diffraction pattern of discrete bright spots (reflections). This discrete pattern is a direct consequence of the crystal's 3D periodicity.

The intensity pattern observed experimentally is directly related to the **FT** of the crystal's **electron density distribution**, $\rho(\mathbf{r})$. Mathematically,

$$F(hkl) = \int \rho(\mathbf{r}) e^{2\pi i \mathbf{h}\cdot\mathbf{r}}\, dr, \tag{10.44}$$

where $F(hkl)$ is the **structure factor** corresponding to the reflection with Miller indices (h, k, l), and $\mathbf{h}$ is the reciprocal-lattice vector defined by $\mathbf{h} = h\mathbf{a}^* + k\mathbf{b}^* + l\mathbf{c}^*$.

Conversely, the electron density can be recovered by taking the **inverse FT**:

$$\rho(\mathbf{r}) = \frac{1}{V} \sum_{hkl} F(hkl) e^{-2\pi i \mathbf{h}\cdot\mathbf{r}}, \tag{10.45}$$

where V is the unit-cell volume.

Thus, the XRD pattern provides $F(hkl)$, the FT of $\rho(\mathbf{r})$. By performing the inverse transform, the **electron density map** of the crystal can be reconstructed. Each atom in the crystal appears as a peak in this electron density distribution.

10.6.3.1 Relating to the Crystallographic Convention

This periodicity means we must adapt our continuous Fourier integral (Equation 10.6) to a **3D Fourier series. In X-ray crystallography, a different convention is standard** for this reason.

The equations used in crystallography are the standard forward and inverse Fourier series for a 3D periodic function (Equations 10.44 and 10.45). These equations relate to our chapter's symmetric integral convention (Equations 10.6 and 10.7) as follows:

1. **Series Versus Integral:** We are using a discrete sum (a series) because the function is periodic, not a continuous integral.
2. **Normalization:** This is an **asymmetric convention**. The forward transform (for F_{hkl}) has a factor of 1, while the inverse (for $\rho(\mathbf{r})$) has a factor of $1/V$. This is different from our symmetric $1/\sqrt{2\pi}$ convention.
3. **Exponent:** The 2π is explicitly included in the exponent. The sign choice ($+i$ for the forward transform) is also a standard convention in crystallography.

Thus, the statement that the diffraction pattern is the FT of the electron density is correct; Equations (10.44) and (10.45) are simply the specific form of the FT pair used for 3D periodic systems. These formulas are still FT pairs but with a different normalization convention and discrete indices (Table 10.1).

Chemical and Historical Significance

This Fourier-transform-based method revolutionized structural chemistry. The same mathematical approach enabled **Watson and Crick's elucidation of the DNA double helix** in the 1950s, based on the X-ray diffraction data obtained by Rosalind Franklin. Today, X-ray crystallography—together with neutron and electron diffraction—provides atomic-level insights into complex structures such as **proteins, enzymes, catalysts, and even viruses**.

Thus, X-ray diffraction stands as a prime example of how the **Fourier transform bridges mathematical abstraction and chemical reality**, translating patterns of scattered intensity into real-space molecular architecture.

10.6.4 FT Spectroscopy

Another major application of the FT in chemistry is in **FT spectroscopy**, which has transformed experimental spectroscopy by enabling rapid and highly accurate acquisition of spectra.

FT methods are applied in a wide range of spectroscopic techniques—including **optical spectroscopy**, **infrared spectroscopy (FTIR)**, **nuclear magnetic resonance spectroscopy (FT-NMR)**, **magnetic resonance spectroscopic imaging**, **mass spectrometry**, and **electron spin resonance (ESR)** spectroscopy. Among these, FTIR remains the most widely used and best known.

10.6.4.1 Conventional Versus FT Spectroscopy

In a conventional (dispersive) spectrometer, radiation is separated by wavelength using a **monochromator**. Each wavelength band is measured sequentially, so that the complete spectrum of intensity versus wavelength is obtained one point at a time. This process is inherently slow and inefficient, as most of the incoming radiation is blocked during each measurement.

In **FT spectroscopy**, by contrast, *all frequencies are measured simultaneously*. The instrument uses a **Michelson interferometer**, in which one mirror is movable. As the mirror moves, it introduces an **optical path difference** (retardation, δ) between the two beams. The resulting interference pattern—intensity versus δ—is recorded as a function $I(\delta)$ called an **interferogram**.

Mathematically, the interferogram is related to the spectral intensity $S(\nu)$ by an **FCT**:

$$I(\delta) = \int_0^\infty S(\nu)\cos(2\pi\nu\delta)d\nu, \tag{10.46}$$

and the desired spectrum is obtained by performing the **inverse transform**:

$$S(\nu) = 2\int_0^\infty I(\delta)\cos(2\pi\nu\delta)d\delta. \tag{10.47}$$

In FT spectroscopy, the interferometer records intensity as a function of optical path difference, producing an interferogram. This time- or space-domain signal is mathematically equivalent to the sum of all frequency components of the radiation field. Applying an FT converts this to the frequency (or wavenumber) domain, yielding the spectrum. Figure 10.7 shows this relationship: the central burst in the interferogram corresponds to constructive interference at zero path difference, while the oscillations at larger retardations give rise to spectral features after transformation.

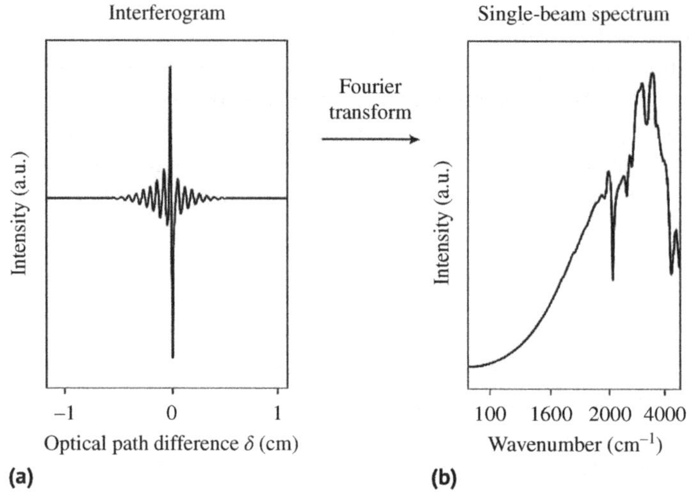

Figure 10.7 Relationship between the interferogram and its Fourier-transformed spectrum. (a) The interferogram, intensity versus retardation (δ), obtained directly from a Michelson interferometer. (b) The single-beam spectrum obtained by applying a Fourier transform to the interferogram. The broad spectral envelope corresponds to the source intensity distribution, while the finer modulations reflect molecular absorption features.

Note on convention: These two equations (Equations 10.46 and 10.47) are the standard asymmetric Fourier cosine transform pair used in spectroscopy (Table 10.1). They differ from the symmetric, $\sqrt{2/\pi}$ normalized convention defined in Section 10.5 (Equations 10.36 and 10.37) but are mathematically equivalent. This asymmetric form, with factors of 1 and 2, is the common standard in this specific field.

Historical Perspective

The conceptual basis of FT spectroscopy dates back to A. A. Michelson in the late 19th century, whose interferometer was later adapted for spectral measurements. The modern revival came in the mid-20th century with the advent of digital computers capable of performing the necessary Fourier transforms efficiently.

As summarized by B. F. Hochheimer (1967):

> *About ten years ago, when large digital computers became widely available, an old spectroscopic method was revived. An interferometer invented by A. A. Michelson is used in a method called Fourier transform spectroscopy. The interferometer processes radiant energy into a Fourier transform of the spectrum, which is then inverted by a digital computer.*
>
> *Because of the high light-gathering power of the interferometer and the efficient use of this energy, many problems are now being done which previously were considered possible only in theory.*

This passage elegantly captures how computation and optics combine to make Fourier transform spectroscopy possible.

Thus, an FT spectrometer is effectively a **computational spectrometer**—the interferometer collects a combined signal in the time (or retardation) domain, and a computer performs an FT to recover the frequency-domain spectrum.

10.6.4.2 FT-NMR and Related Techniques

In **FT-NMR spectroscopy**, a short pulse of radiofrequency (RF) radiation excites a broad range of nuclear spin transitions simultaneously. The nuclei then re-emit radiation as their magnetic moments relax back to equilibrium, producing a decaying oscillatory signal known as an **FID**. The FID is a function of time, and its FT yields the corresponding **frequency-domain NMR spectrum**:

$$S(\omega) = \int_0^{\infty} \mathrm{FID}(t)e^{-i\omega t}\,dt. \tag{10.48}$$

A Note on Convention: Because the FID signal has a definite start time $t = 0$, we use a **one-sided Fourier Transform** (integrating from 0 to ∞) to find its spectrum. This is a common and physically necessary modification of the standard two-sided transform for signals that begin at a specific time.

A similar approach is used in **ESR spectroscopy**, except that microwaves rather than RFs are employed to excite electron spins. In each case, the FT converts a time-dependent signal into a frequency-dependent spectrum.

10.6.4.3 Advantages of FT Methods

1. **Multiplex (Fellgett's) advantage**—All frequencies are measured simultaneously, improving signal-to-noise ratio.
2. **Throughput (Jacquinot) advantage**—The absence of a monochromator allows greater light throughput.
3. **Wavelength precision**—The laser reference and digital processing yield very high-frequency accuracy.
4. **Speed**—The use of FFT algorithms makes data acquisition and transformation nearly instantaneous.

In summary, FT spectroscopy exemplifies the power of the FT to connect **time (or path-difference) information** with **frequency information**, providing a unifying framework across all spectroscopic domains. Its widespread success in IR and NMR spectroscopy stands as a testament to how a mathematical concept can reshape entire branches of experimental chemistry.

10.6.5 Time Signals and Decay Information

One of the most important uses of the FT in spectroscopy is the extraction of **frequency information** from **time-dependent signals**. Experimental data are often recorded as a function of time—for example, the emission or absorption intensity following excitation—and the underlying frequencies must be determined from this temporal behavior.

An FT provides a natural way to convert such time-domain data into the frequency domain.

10.6.5.1 Single-frequency Example

Consider the exponentially damped oscillatory function

$$f(t) = e^{-\alpha t}\cos(\omega_0 t), \quad t > 0, \tag{10.49}$$

where α is the damping constant and ω_0 the natural angular frequency.

Figure 10.8a plots the **time-domain** function $f(t)$ against t for $\alpha = 0.10$ and $\omega_0 = 1$; Figure 10.8b shows its FT, the **frequency-domain** function, $F(\omega)$.

The transform clearly exhibits a single peak at $\omega = \omega_0$, corresponding to the single-frequency component present in the time record.

10.6.5.2 Composite-frequency Example

Now consider a function composed of a realistic, damped, multicomponent signal:

$$f(t) = A_1 e^{-a_1 t}\cos(\omega_1 t) + A_2 e^{-a_2 t}\cos(\omega_2 t) + A_3 e^{-a_3 t}\cos(\omega_3 t), \tag{10.50}$$

with

- $A_1 = 1.0,\ a_1 = 0.10, \omega_1 = 1.2;$
- $A_2 = 0.6,\ a_2 = 0.20,\ \omega_2 = 2.8;$
- $A_3 = 0.3,\ a_3 = 0.35,\ \omega_3 = 4.5$

plotted in Figure 10.9a. From inspection, it is difficult to discern how many frequencies are present or what their values are. However, the FT of this time-domain signal,

$$F_n(\omega) = \frac{A_i}{2\sqrt{2\pi}}\left[\frac{1}{a_i + i(\omega - \omega_i)} + \frac{1}{a_i + i(\omega + \omega_i)}\right] \tag{10.51}$$

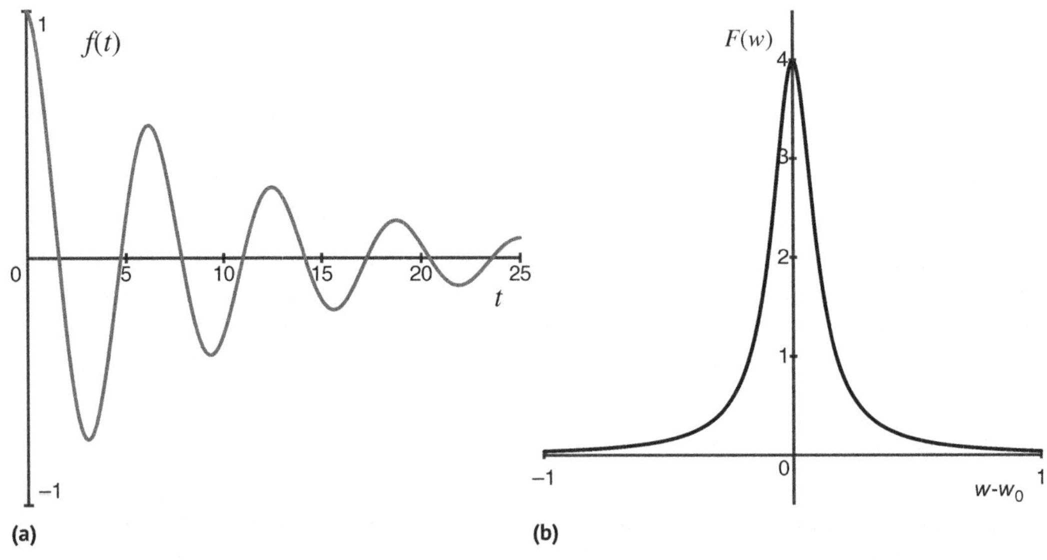

Figure 10.8 (a) The time-domain function $e^{-\alpha t}\cos\omega_0 t$ plotted against time for $\alpha = 0.10$ and $\omega_0 = 1$. (b) The corresponding frequency-domain function $\frac{1}{2\sqrt{2\pi}}\left(\frac{\alpha}{\alpha^2 + (\omega - \omega_0)^2}\right)$ plotted against $(\omega - \omega_0)$.

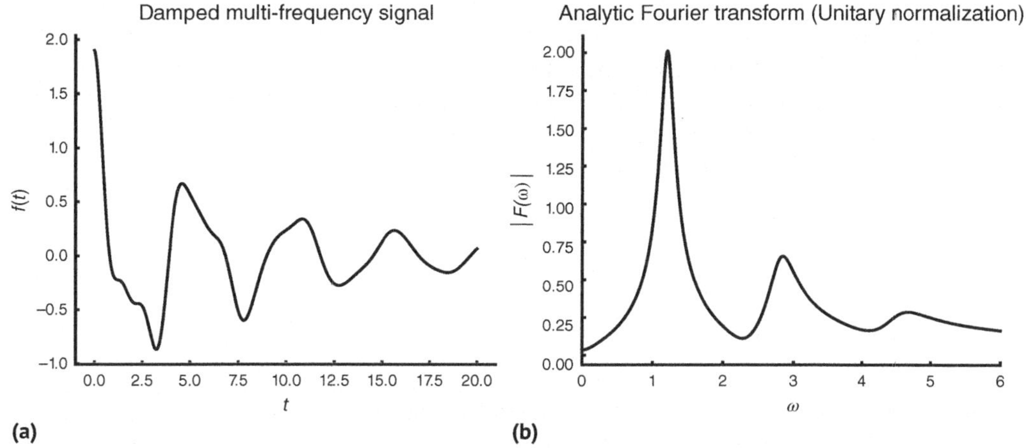

Figure 10.9 (a) Time domain: the function $f(t)$ given by Equation (10.50) plotted versus t. (b) Frequency domain: the Fourier transform of $f(t)$ (Equation 10.51), plotted against ω.

reveals three distinct peaks, as shown in Figure 10.9b, each corresponding to one of the component frequencies.

Thus, the FT cleanly decomposes a complicated time signal into its individual frequency contributions.

10.6.5.3 Numerical Implementation

In practice, time-domain data such as $f(t)$ are obtained as discrete numerical values. To evaluate their FTs efficiently, the **FFT** algorithm is used. FFT computes the discrete transform in $N \log_2 N$ operations instead of N^2, enabling real-time spectral analysis. This computational breakthrough is what made FT spectroscopy—particularly **FTIR** and **FT-NMR**—widely practical. The same principle underlies modern signal processing, vibration analysis, and even microwave spectroscopy.

Connection to Chemistry: Lifetime, Linewidth, and the Fourier Transform

A signal that decays in time cannot have a sharply defined frequency. If the field amplitude of an oscillation decreases as

$$f(t) = e^{-\gamma t} e^{i\omega_0 t}, t > 0,$$

where γ is the damping constant (inverse of lifetime τ) and ω_0 is the central transition frequency, its Fourier transform is a **Lorentzian** line shape,

$$I(\omega) \propto \frac{1}{(\omega - \omega_0)^2 + \gamma^2}.$$

The full width at half maximum is

$$\Delta\omega_{1/2} = 2\gamma,$$

showing that a **rapid temporal decay** (large γ) produces a **broad frequency distribution**, while a long-lived signal leads to a sharp spectral line.

This relationship is a direct reflection of the **time–frequency uncertainty principle**: a function localized in time cannot be localized with equal precision in frequency. In spectroscopy, this manifests as the intrinsic ("natural") linewidth of a transition: a short-lived excited state necessarily exhibits a larger energy uncertainty.

The energy uncertainty associated with a finite lifetime is

$$\Delta E = \hbar\,\Delta\omega_{1/2} \approx \frac{\hbar}{\tau},$$

so that

$$\boxed{\Delta E\,\tau \approx \hbar,} \tag{10.52}$$

a direct manifestation of the time–energy uncertainty principle $\Delta E\,\Delta t \geq \frac{\hbar}{2}$.

The apparent differences in numerical factors (½, 1, or 2) arise solely from the mathematical form assumed for the decay (exponential vs. Gaussian) and from whether the lifetime refers to the decay of **amplitude** or **intensity**.

In all cases, the physical meaning is the same: a short-lived excited state has a correspondingly large energy uncertainty, producing a broader spectral line.

Equation (10.52) is the **time–energy form of the Heisenberg uncertainty principle**. The result is consistent with this principle, as $\hbar > \hbar/2$. The exact numerical prefactor depends on the specific definitions of ΔE and Δt, but the inverse relationship between lifetime and energy uncertainty is the essential physical consequence.

Hence, a shorter-lived excited state (small τ) gives rise to a broader spectral line, while a long-lived state produces a sharper peak.

10.6.6 Machine Learning, Feature Extraction, and the Frequency Domain

In modern data science and artificial intelligence, the FT is a fundamental preprocessing tool for **time-series analysis**. Raw **sensor data**—from **accelerometers** in wearable devices, **vibration monitors** in industrial machinery, or **electrocardiogram (ECG) sensors** in medical diagnostics—is typically collected in the **time domain**. While time-domain signals can be visually inspected, their **frequency characteristics** often provide more discriminative and compact representations for **pattern recognition** and **classification**.

10.6.6.1 From Time Series to Discriminative Features

Consider a smartphone accelerometer recording motion during different activities (walking, running, climbing stairs). The raw acceleration versus time plot may appear as noisy, overlapping waveforms. Applying an **FT** converts this temporal signal into its **frequency spectrum**, revealing dominant oscillation modes:

- **Walking:** Strong low-frequency components (~1–2 Hz) from stride cycles.
- **Running:** Higher-frequency peaks (~2–3 Hz) with broader harmonics.
- **Climbing stairs:** Distinct impulsive patterns with specific harmonic content.

The **magnitudes of dominant frequency peaks** then serve as **numerical features** that can be fed into a **machine learning classifier** (e.g., a support vector machine or neural network). This

frequency-domain feature extraction often leads to higher accuracy and better generalization than using raw time-domain data directly.

10.6.6.2 Concrete Examples in Application

1. **Predictive Maintenance:** Vibration sensors on industrial motors produce time-series data. A developing fault (e.g., bearing wear) introduces specific high-frequency oscillations. The **Fourier spectrum** is monitored for emerging peaks; machine learning models can predict failure weeks in advance.
2. **Medical Diagnostics:** An ECG signal contains not just the heart rate (~1 Hz) but also finer structures (P, QRS, T waves) with characteristic frequency signatures. **Abnormal rhythms** (arrhythmias) alter this spectral profile. Fourier-based features enable automated detection of conditions like atrial fibrillation.
3. **Audio and Speech Recognition:** Apps like **Shazam** convert audio snippets into **spectrograms** (time–frequency maps via short-time FTs) and match spectral "fingerprints." Similarly, voice-recognition systems use **mel-frequency cepstral coefficients**, derived from FTs, to identify phonemes and speakers.

10.6.6.3 Why Frequency Features Work Well for Machine Learning

- **Invariance:** The overall spectral shape is often invariant to small time shifts or amplitude scaling.
- **Dimensionality Reduction:** A few spectral peaks can summarize seconds of high-sample-rate data.
- **Noise Robustness:** Low-pass filtering in the frequency domain can remove high-frequency noise without distorting the signal structure.
- **Physical Interpretability:** Frequency components often correspond to tangible physical processes (motor RPM, heart rate, resonant vibrations).

10.6.6.4 Broader Technological Impact

Fourier methods are embedded in countless technologies beyond ML:

- **Medical Imaging:** MRI reconstructs images from ***k*-space data**, which is essentially the spatial FT of the magnetization.
- **Communications:** Orthogonal frequency-division multiplexing uses FTs to encode data on multiple carrier frequencies in 4G/5G and Wi-Fi.
- **Audio Compression:** MP3 uses a modified DCT to discard inaudible frequency components.
- **Radio Astronomy:** Interferometric telescopes like the VLA measure Fourier components of sky brightness; images are reconstructed via inverse transforms.

Thus, the FT serves as a **bridge** between raw, temporal sensor data and the **compact, informative feature vectors** that power modern machine learning systems. Its ability to reveal the **hidden periodic structure** in signals makes it indispensable not only in traditional spectroscopy and diffraction but also in the era of data-driven discovery and intelligent systems.

10.7 Summary

FTs unify diverse physical and chemical phenomena under a single mathematical framework. They connect experimentally measurable quantities (spectra, diffraction patterns, time-domain signals) with the underlying theoretical descriptions of motion, structure, and interaction. Mastery of Fourier methods is therefore fundamental for understanding and interpreting modern quantum chemistry, spectroscopy, and materials science.

10.7.1 Key Relations

1. **FT Pair**

$$F(k) = \frac{1}{\sqrt{2\pi}} \int_{-\infty}^{\infty} f(x)e^{-ikx}\,dx, \quad f(x) = \frac{1}{\sqrt{2\pi}} \int_{-\infty}^{\infty} F(k)e^{ikx}\,dk.$$

The transform expresses a function $f(x)$ as a continuous superposition of exponential basis functions e^{ikx}.

2. **Fourier Sine and Cosine Transforms**

$$F_s(k) = \sqrt{\frac{2}{\pi}} \int_{0}^{\infty} f(x)\sin(kx)\,dx, \quad F_c(k) = \sqrt{\frac{2}{\pi}} \int_{0}^{\infty} f(x)\cos(kx)\,dx.$$

These are useful for functions defined on $x \geq 0$ with specific boundary conditions.

3. **Inverse Relations**

$$f(x) = \sqrt{\frac{2}{\pi}} \int_{0}^{\infty} F_s(k)\sin(kx)dk, \quad f(x) = \sqrt{\frac{2}{\pi}} \int_{0}^{\infty} F_c(k)\cos(kx)dk.$$

Dirac Delta Representation

$$\delta(x) = \frac{1}{2\pi} \int_{-\infty}^{\infty} e^{ikx}\,dk,$$

$$\int_{-\infty}^{\infty} f(x)\delta(x - x_0)\,dx = f(x_0).$$

The delta function serves as the kernel of the FT and expresses the completeness of the exponential basis.

4. **Key Properties**

Property	Mathematical form	Physical meaning
Linearity	$af_1 + bf_2 \leftrightarrow aF_1 + bF_2$	Superposition of waves/signals
Shift	$f(x - x_0) \leftrightarrow e^{-ikx_0}F(k)$	Phase shift in the spectrum
Scaling	$f(ax) \longleftrightarrow \frac{1}{\lvert a\rvert}F\left(\frac{k}{a}\right)$	Stretching and compression

Differentiation	$\frac{d^n f}{dx^n} \leftrightarrow (ik)^n F(k)$	Connection with operators
Convolution	$f * g \leftrightarrow F(k)G(k)$	Line broadening, correlation
Parseval's Theorem	$\int_{-\infty}^{\infty} \lvert f(x)\rvert^2\, dx = \int_{-\infty}^{\infty} \lvert F(k)\rvert^2\, dk$	Conservation of energy (or power) of a signal

5. **Time–Frequency Duality**

$$\Delta t\, \Delta\omega \geq \frac{1}{2}, \quad \Delta x\, \Delta p \geq \frac{\hbar}{2}.$$

This expresses the reciprocal localization of conjugate variables and forms the mathematical basis of the **Heisenberg uncertainty principle**.

6. **Applications**
   - Transformation between position and momentum space in quantum mechanics.
   - Conversion of time-domain experimental data (e.g., FID signals) to frequency-domain spectra.
   - Analysis of line broadening, correlation functions, and diffraction patterns.
   - Simplification of differential equations (diffusion, heat conduction, etc.).

Further Reading

R. N. Bracewell, *The Fourier Transform and Its Applications*, 3rd ed., McGraw-Hill, 1999.
E. Kreyszig, *Advanced Engineering Mathematics*, 10th ed., Wiley, 2011.
M. L. Boas, *Mathematical Methods in the Physical Sciences*, 3rd ed., Wiley, 2006.
A. Papoulis, *The Fourier Integral and Its Applications*, McGraw-Hill, 1962.
I. N. Sneddon, *The Use of Integral Transforms*, McGraw-Hill, 1972.
R. Kakkar, *Atomic and Molecular Spectroscopy*, Cambridge University Press, 2015.

Exercises

A. Conceptual Questions

1. Explain in your own words the physical meaning of an FT.

2. Differentiate between the time (or spatial) domain and the frequency (or wavenumber) domain.

3. What is the significance of the Dirac delta function in Fourier analysis?

4. How does the FT relate to the uncertainty principle?

5. Under what conditions can the Fourier series of a function be replaced by an FT?

6. Describe how the width of a function in real space is related to the width of its transform in reciprocal space.

7. Explain the difference between the Fourier, Fourier sine, and Fourier cosine transforms.

8. State and interpret Parseval's theorem in physical terms.

9. What are the advantages of using the sine and cosine transforms for semi-infinite boundary-value problems?

10. What does the phase factor e^{-ikx_0} represent in the shift property of the FT?

11. Why does narrowing a Gaussian in the x-domain cause its FT to broaden in the k-domain?

12. Explain why the FT preserves L^2-norm but not area.

13. What makes the Gaussian unique under the FT?

B. Problems

1. Find the Fourier transform of the following functions:
   a. $f(x) = e^{-|x|}$;
   b. $f(x) = e^{-a|x|}$, where $a > 0$;
   c. $f(x) = e^{-x^2}$.

2. Determine the Fourier cosine and sine transforms of $f(x) = e^{-ax}$, and verify that they satisfy the inversion relations.

3. Prove that the Fourier transform of the derivative $\frac{df(x)}{dx}$ is $ikF(k)$.

4. Show that the Fourier transform of a convolution $f(x) * g(x)$ equals the product $F(k)G(k)$.

5. Using the differentiation property, find the Fourier transform of $f(x) = xe^{-a|x|}$.

6. Obtain the Fourier transform of the box function:

$$f(x) = \begin{cases} 1, & |x| < a, \\ 0, & |x| > a. \end{cases}$$

7. Show that the Fourier transform of a Gaussian function is another Gaussian, and verify the relation

$$\Delta x \, \Delta k = 1.$$

8. Find the inverse Fourier transform of $F(k) = e^{-k^2}$.

9. Using the FST, solve the boundary-value problem:

$$\frac{\partial^2 u}{\partial x^2} = \frac{1}{D}\frac{\partial u}{\partial t}, \quad u(0,t) = 0, \quad u(x,0) = f(x).$$

10. Derive the Fourier transform of the first derivative of a Gaussian and discuss its symmetry.

11. Show that the Gaussian

$$g(x) = \exp(-\alpha x^2)$$

transforms to

$$G(k) = \sqrt{\frac{\pi}{\alpha}}\exp\left(-\frac{k^2}{4\alpha}\right)$$

under the unitary Fourier transform.
(*Hint:* No normalization factor is included in $g(x)$.)

12. Let

$$f(x) = \frac{1}{\sqrt{2\pi\sigma}}e^{-x^2/(2\sigma^2)}.$$

Compute the integrals

$$\int |f(x)|^2 dx, \quad \int |F(k)|^2 dk,$$

and verify that they are equal.
Verify Parseval's Theorem numerically using the values from Problem 11.

13. For the Gaussian in Problem 12, determine the full width at half maximum of $f(x)$ and of $F(k)$. Show explicitly that the widths satisfy

$$\sigma_x \sigma_k = 1.$$

C. Advanced/Applied Problems

1. Show that for the normalized Gaussian wavefunction

$$\psi(x) = \left(\frac{1}{\pi\sigma^2}\right)^{1/4} e^{-x^2/(2\sigma^2)},$$

the momentum-space wavefunction is

$$\phi(p) = \left(\frac{\sigma^2}{\pi\hbar^2}\right)^{1/4} e^{-\sigma^2 p^2/(2\hbar^2)}.$$

Discuss the relation $\Delta x\, \Delta p = \hbar/2$.

2. The absorption coefficient in spectroscopy is proportional to the Fourier transform of the dipole autocorrelation function:

$$\alpha(\omega) \propto \mathrm{Re}\left[\int_0^\infty \langle \mu(0)\mu(t)\rangle e^{-i\omega t}\,dt\right].$$

Explain the physical meaning of this expression and the role of the imaginary part.

3. Using Fourier transform methods, solve the diffusion equation:

$$\frac{\partial C}{\partial t} = D\frac{\partial^2 C}{\partial x^2}, \quad C(x,0) = \delta(x).$$

4. The structure factor in XRD is the Fourier transform of the electron density:

$$F(hkl) = \int \rho(\mathbf{r})e^{2\pi i\mathbf{h}\cdot\mathbf{r}}\,d\mathbf{r}.$$

Explain how this relation leads to the reconstruction of the crystal electron density and why the *phase problem* arises.

5. Show that the Fourier transform of the second derivative $\frac{d^2 f}{dx^2}$ is $-(k^2)F(k)$ and apply this result to a Schrödinger-type differential equation.

6. Prove that

$$\int_{-\infty}^{\infty} e^{i(k-k')x}\,dx = 2\pi\delta(k-k'),$$

and interpret this result in terms of the orthogonality of momentum eigenfunctions.

7. The transient signal from a laser pulse can be modeled as

$$f(t) = e^{-t/\tau}\sin(\omega_0 t), t > 0.$$

Obtain its Fourier transform and discuss how the line width depends on the lifetime τ.

8. Derive the Fourier sine transform of $f(x) = xe^{-ax}$ and interpret the result physically.

9. For a one-dimensional box of width L, use Fourier analysis to show how the allowed wavenumbers satisfy $k_n = n\pi/L$, leading to the quantization of energy levels.

10. Using the convolution theorem, explain how instrumental broadening affects the observed spectrum and how deconvolution can be used to recover the true spectral profile.

11

Laplace Transforms

Nature laughs at the difficulties of integration.

— Pierre-Simon Laplace

Learning Outcomes

After completing this chapter, the student should be able to:

1. **Define and compute** the Laplace transform of elementary functions.
2. **Apply** linearity and shifting theorems.
3. **Obtain** inverse transforms using partial fractions.
4. **Solve** initial-value differential equations using Laplace methods.
5. **Interpret** physical problems (decay, oscillation, kinetics) through Laplace analysis.

11.1 Introduction

You must have encountered operational calculus and been told that it works because of its close relationship to the Laplace transform. We now define the Laplace transform. Like the Fourier transform, the Laplace transform is an integral transform that takes a function of a real variable t (often in the time domain) to a function of a complex variable s (in the frequency domain). However, while the Fourier transform is a complex function of a *real* variable (frequency), the Laplace transform of a function is a complex function of a *complex* variable (typically written $s = \sigma + i\omega$).

A key difference is that the kernel e^{-st} contains exponential damping when $\mathrm{Re}\, s > 0$, which greatly improves convergence (Figure 11.1).

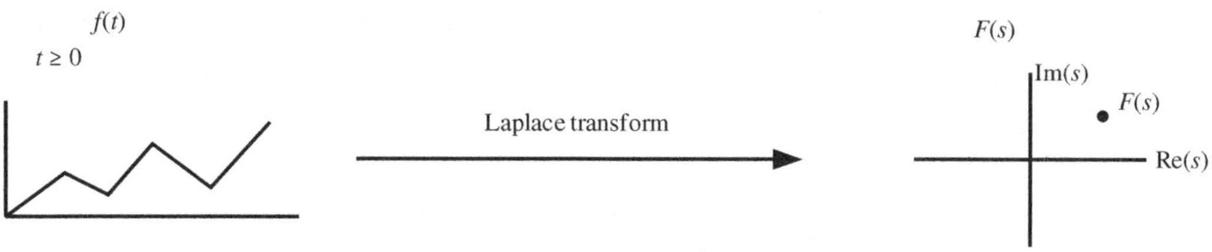

Figure 11.1 Mapping from the time domain to the s-domain.

Concept Box: Why Laplace Transforms Are Useful

- Convert differential equations into simple algebraic equations.
- Handle **initial conditions** automatically through the derivative formulas.
- Allow solution of **nonhomogeneous** equations without first finding complementary and particular solutions separately.
- Particularly suitable for kinetics, relaxation phenomena, oscillators, electrical circuits, and spectroscopy.
- The Laplace transform converges for a broader class of functions than the Fourier transform due to the damping factor e^{-st}.

Historical Note 11.1: Laplace and the Origins of Transform Methods

Pierre-Simon Laplace (1749–1827) introduced integrals of the form

$$\int_0^\infty e^{-st} f(t)dt$$

in his work on celestial mechanics and probability. His aim was to convert complicated differential equations into algebraic relations, anticipating the modern transform method. A century later, Oliver Heaviside employed similar ideas in electrical circuit theory, developing the operational calculus used to analyze linear systems. Their combined contributions form the basis of the Laplace transform as it is used today in physics, chemistry, and engineering, where it provides an elegant method for solving initial-value problems.

11.2 Laplace Transform

Definition: Let $f(t)$ be defined for $t \geq 0$. The **Laplace transform** of $f(t)$, denoted by $F(s)$ or $\mathscr{L}\{f(t)\}$, is

$$F(s) = \mathscr{L}\{f(t)\} = \int_0^\infty f(t)e^{-st}\, dt, \tag{11.1}$$

where s is complex, provided this improper integral converges. Since the kernel must be dimensionless, s has dimensions of reciprocal time; the variable s thus represents **complex frequency** (Figure 11.2).

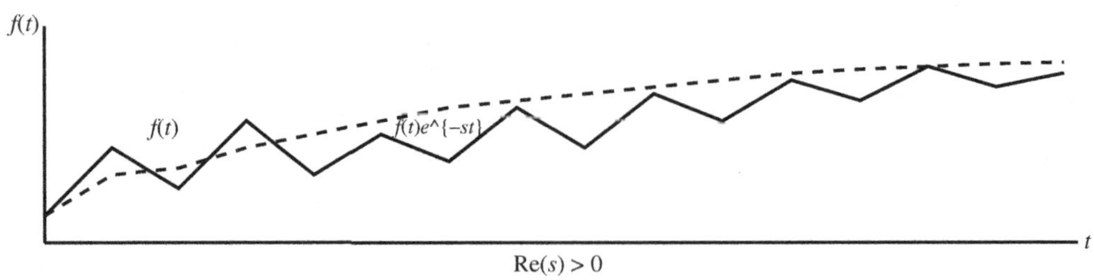

Figure 11.2 Effect of exponential damping.

Example 11.1: Laplace Transforms

Find the Laplace transform of (i) $f(t) = 1$, (ii) $f(t) = t$, (iii) $f(t) = e^{at}$

Solution:

i. Let $\boldsymbol{f(t) = 1}$. Show that $F(s) = 1/s$, $\text{Re}\, s > 0$,

$$F(s) = \int_0^\infty 1 \cdot e^{-st}\, dt = \frac{1}{s}, \quad \text{Re}\, s > 0.$$

The integral diverges for $s \leq 0$. If the frequency s is negative, the "damping" becomes "growth," and the integral of an exponentially growing function over infinite time is infinite.

ii. Laplace Transform of $\boldsymbol{f(t) = t}$,

$$F(s) = \int_0^\infty t\, e^{-st}\, dt = \frac{1}{s^2}, \quad s > 0.$$

iii. Laplace Transform of $\boldsymbol{f(t) = e^{at}}$,

$$F(s) = \int_0^\infty e^{at}\, e^{-st}\, dt = \frac{1}{s-a}, \quad s > a.$$

In a similar fashion, the Laplace transforms of other familiar functions can be computed. Here are some common Laplace transforms (Table 11.1).

In the Laplace transform framework, the locations of **poles** in the complex **s-plane** determine the **region of convergence (ROC)**, which specifies the values of s for which the transform exists.
The Laplace transform exists for functions that grow no faster than an exponential, a condition known as **exponential order**.

Table 11.1 Laplace transforms of some common functions.

S. no.	$f(t)$	$F(s)$	Condition
1	1	$1/s$	$\text{Re}\, s > 0$
2	t	$1/s^2$	$\text{Re}\, s > 0$
3	t^n	$n!/s^{n+1}$	$\text{Re}\, s > 0$
4	e^{at}	$1/(s-a)$	$\text{Re}\, s > a$
5	$\sin kt$	$k/(s^2 + k^2)$	$\text{Re}\, s > 0$
6	$\cos kt$	$s/(s^2 + k^2)$	$\text{Re}\, s > 0$

Note on Convergence: Region of Convergence

The Laplace transform

$$F(s) = \int_0^\infty f(t)e^{-st}\,dt$$

is defined as an **improper integral**, and it converges only for certain values of the complex variable $s = \sigma + i\omega$. The set of all s for which the integral converges absolutely is called the **region of convergence (ROC)**.

Key points about ROC:

1. **Exponential order condition:**
   If $|f(t)| \leq Me^{\alpha t}$ for some $M > 0$ and real α, then the Laplace transform converges for $\mathrm{Re}(s) > \alpha$.
   The ROC is a **right half-plane** in the complex s-plane: $\mathrm{Re}(s) > \sigma_0$, where σ_0 is the **abscissa of convergence**.
2. **Uniqueness of the inverse transform:**
   Two different time-domain functions can have the same Laplace transform $F(s)$, but with **different ROCs**. Therefore, the ROC must be specified to ensure a unique inverse Laplace transform.
3. **Poles and ROC:**
   For rational Laplace transforms, the ROC contains **no poles**. The ROC is bounded to the right by the rightmost pole but extends infinitely to the left unless the signal is left-sided. In this chapter, we mostly work with right-sided (causal) functions where ROC is of the form $\mathrm{Re}(s) > \sigma_0$.
4. **Practical implication:**
   In solving differential equations, we usually assume the ROC is the right half-plane large enough to include all poles of $F(s)$, so that the inverse transform corresponds to **causal** (right-sided) signals.

Illustrative Example

Consider $f(t) = e^{at}$. Its Laplace transform (Example 11.1[iii]) is

$$F(s) = \frac{1}{s-a}, \quad \text{ROC:} \quad \mathrm{Re}(s) > a.$$

The pole is at $s = a$, and the ROC is the half-plane to the right of this pole (Figure 11.3). If we instead consider a **left-sided** signal $g(t) = -e^{at}u(-t)$, its transform is also $\frac{1}{s-a}$, but with ROC: $\mathrm{Re}(s) < a$.

Thus, the same $F(s)$ corresponds to different time functions depending on the ROC.

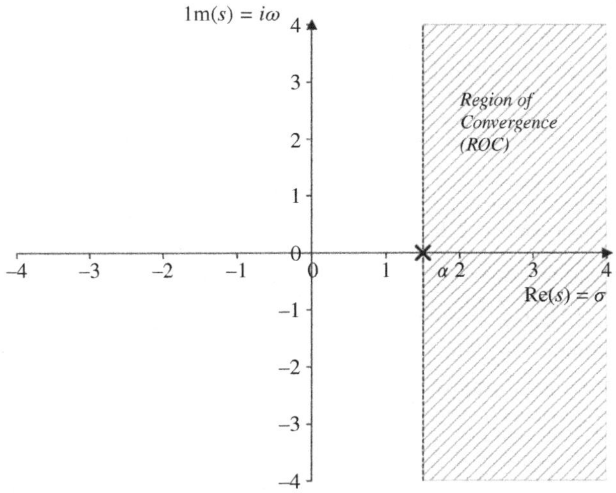

Figure 11.3 Pole–zero plot and region of convergence (ROC) for $f(t) = e^{at}$. The pole at $s = a$ (marked by ×) defines the boundary of the ROC. The shaded right-half plane, defined by $\text{Re}(s) > a$, represents the domain where the Laplace integral converges.

Chemist's Insight: The Physical Meaning of a Pole

In the standard chemical decay equation $N(t) = N_0\, e^{-kt}$, the concentration decays exponentially with a rate constant k. The Laplace transform of this function is

$$\mathcal{L}\{N_0 e^{-kt}\} = \frac{N_0}{s+k}.$$

Notice that the denominator becomes zero when $s = -k$. This point of singularity is called a **pole**.

While mathematicians view poles as convergence boundaries, chemists can view them as physical fingerprints of the system:

- A **real pole** at $s = -k$ represents a **reaction rate** (relaxation or decay).
- An **imaginary pole** at $s = \pm i\omega$ represents a **vibrational frequency** (oscillation).

Thus, the s-domain essentially acts as a "map" of all the decay rates and oscillation frequencies present in your chemical system. In this example the pole at $s = -k$ corresponds to the decay rate k; in general, the real part of a pole gives a decay (or growth) rate and the imaginary part gives an oscillation frequency.

Electrochemist's Insight: Poles as Relaxation Times

In electrochemical impedance spectroscopy (EIS), the impedance $Z(s)$ of an electrode interface is defined directly in the Laplace domain:

$$Z(s) = \frac{\mathcal{L}\{E(t)\}}{\mathcal{L}\{I(t)\}},$$

where $E(t)$ is applied potential and $I(t)$ is measured current.

For a simple RC cell (solution resistance R_s + double-layer capacitance C_{dl}):

$$Z(s) = R_s + \frac{1}{C_{dl}s}.$$

The pole at $s = 0$ (from the capacitor) gives the charging time constant $\tau = R_s C_{dl}$.

For a Randles circuit, add charge-transfer resistance R_{ct}.

Two important operational rules are the **shifting theorems**: multiplication of a function by an exponential factor shifts its Laplace transform along the s-axis, while time shifting of a function introduces an exponential factor in the transform domain.

Fact Box 11.1: Shifting Theorems

First Shifting Theorem

If

$$\mathcal{L}\{f(t)\} = F(s),$$

then

$$\mathcal{L}\{e^{at} f(t)\} = F(s - a).$$

This corresponds to a **horizontal shift** in the s-domain.

Second Shifting Theorem (Heaviside Step)

If $u(t - a)$ is the unit step function, then

$$\mathcal{L}\{u(t - a)f(t - a)\} = e^{-as}F(s).$$

This theorem is essential for piecewise and delayed signals.

Examples 11.2 and 11.5 illustrate the use of these theorems.

Example 11.2: Find Laplace Transforms

Find the Laplace transform of each function:

i. te^{5t};

ii. $e^{-t}\sin(2t)$;

iii. $e^{i\alpha t}$.

Solution:

i. Using the known integral:

$$\mathcal{L}\{te^{5t}\} = \frac{1}{(s-5)^2}, \qquad s > 5.$$

ii.

$$\mathscr{L}\{e^{-t}\sin(2t)\} = \int_0^\infty e^{-(s+1)t}\sin(2t)\,dt = \frac{2}{(s+1)^2+4}.$$

This illustrates the **first shifting theorem.**

iii.

$$\mathscr{L}\{e^{i\alpha t}\} = \int_0^\infty e^{-(s-i\alpha)t}dt = \frac{1}{s-i\alpha}, \quad \operatorname{Re} s > 0.$$

Using Euler's identity:
Real part → transform of $\cos \alpha t$.
Imaginary part → transform of $\sin \alpha t$.

Example 11.3: Recognizing Transforms by Inspection

Each function $F(s)$ below is defined by a definite integral. Without integrating, find an explicit expression for each $F(s)$:
(i) $\int_0^\infty e^{-(s+7)t}dt$; (ii) $\int_0^\infty t^2e^{-(s-3)t}dt$; (iii) $\int_0^\infty 4e^{-st}\sin(6t)dt$.

Solution:

i. This can be recognized as $\mathscr{L}\{e^{-7t}\}$. Hence, this is equal to $1/(s+7)$.
ii. This is $\mathscr{L}\{t^2e^{3t}\}$. By the first shifting theorem, this is equal to $2!/(s-3)^3 = 2/(s-3)^3$.
iii. This is equal to $4\mathscr{L}\{\sin(6t)\} = 4\dfrac{6}{s^2+36} = \dfrac{24}{s^2+36}$.

Example 11.4: F(0) as the Area Under the Curve

Let $f(t)$ be a continuous function and $F(s)$ be its Laplace transform. Suppose $f(t) \geq 0$ for $t \geq 0$.
a. Show that $F(0)$ is equal to the area of the region under the curve $y = f(t)$, $0 \leq t < \infty$.
b. Verify this fact by comparing the value of $F(0)$ with the area under the curve obtained the usual way, i.e., by integration, for each of the functions (i) e^{-t}; (ii) te^{-2t}; and (iii) t^2e^{-4t}.

Solution:

a. The definition $F(s) = \int_0^\infty e^{-st}f(t)dt \Rightarrow F(0) = \int_0^\infty f(t)e^{-0t}dt = \int_0^\infty f(t)dt = A$.
b. **i.** $\mathscr{L}\{e^{-t}\} = F(s) = \dfrac{1}{(s+1)} \Rightarrow F(0) = 1 = A$,

$$A = \int_0^\infty e^{-t}dt = \left[\frac{e^{-t}}{-1}\right]_0^\infty = \frac{0-1}{-1} = 1.$$

ii. $\mathscr{L}\{te^{-2t}\} = F(s) = \dfrac{1}{(s+2)^2} \Rightarrow F(0) = \dfrac{1}{4} = A,$

$$A = \int_0^\infty te^{-2t}\,dt = \frac{1}{2^2} = \frac{1}{4}.$$

iii. $\mathscr{L}\{t^2e^{-4t}\} = F(s) = 2/(s+4)^3 \Rightarrow F(0) = 1/32 = A.$

By direct integration

$$A = \int_0^\infty t^2e^{-4t}\,dt = \frac{2}{(4)^3} = \frac{1}{32}.$$

The **Heaviside (unit step) function** provides a convenient way to represent piecewise or delayed functions in the Laplace transform method (see Example 11.5).

Example 11.5: Second Shifting Theorem

Find the Laplace transform of

$$g(t) = \begin{cases} 0 & \text{for } 0 \le t < 2, \\ (t-2)^2 & \text{for } t \ge 2. \end{cases}$$

Solution:

We can write $g(t)$ using the Heaviside step function $u(t-a)$:

$$g(t) = (t-2)^2\,u(t-2).$$

Let $f(t) = t^2$. Then $f(t-2) = (t-2)^2$.

From Table 11.1,

$$\mathscr{L}\{t^2\} = F(s) = \frac{2!}{s^3} = \frac{2}{s^3}, \quad \operatorname{Re} s > 0.$$

By the **Second Shifting Theorem:**

$$\mathscr{L}\{u(t-a)f(t-a)\} = e^{-as}F(s).$$

Here $a = 2$ and $f(t-2) = (t-2)^2$, so

$$\mathscr{L}\{(t-2)^2u(t-2)\} = e^{-2s}\cdot\frac{2}{s^3}.$$

Thus,

$$\boxed{\mathscr{L}\{g(t)\} = \frac{2e^{-2s}}{s^3}, \quad \operatorname{Re} s > 0.}$$

11.3 Inverse Laplace Transforms

As stated earlier, if $F(s)$ is the Laplace transform of $f(t)$, then the two functions form a **Laplace transform pair**:

$$f(t) \overset{|\mathscr{L}|}{\longleftrightarrow} F(s).$$

We now see how to use Table 11.1 to find inverse transforms.

Example 11.6: Inverse Transform by Direct Recognition

Find $f(t)$ if

$$F(s) = \frac{3}{s+1} - \frac{2}{s+5}.$$

Solution:

Perform partial fractions (already in decomposed form):

$$F(s) = 3\frac{1}{s+1} - 2\frac{1}{s+5}.$$

Thus,

$$f(t) = 3e^{-t} - 2e^{-5t}.$$

Example 11.7: Partial Fractions with the Cover-Up Method

Find $f(t)$ if

$$F(s) = \frac{5s+7}{(s-1)(s+2)(s-6)}.$$

Solution:

We write

$$\frac{5s+7}{(s-1)(s+2)(s-6)} = \frac{a}{s-1} + \frac{b}{s+2} + \frac{c}{s-6}.$$

Using the cover-up method:
At $s = 1$,

$$5(1) + 7 = 12 = a(1+2)(1-6) = a(3)(-5) \Rightarrow a = -\frac{12}{15} = -\frac{4}{5}.$$

At $s = -2$,

$$5(-2) + 7 = -3 = b(-2-1)(-2-6) = b(-3)(-8) \Rightarrow b = -\frac{1}{8}.$$

At $s = 6$,

$$5(6) + 7 = 37 = c(6-1)(6+2) = c(5)(8) \Rightarrow c = \frac{37}{40}.$$

Thus,

$$F(s) = -\frac{4}{5}\frac{1}{s-1} - \frac{1}{8}\frac{1}{s+2} + \frac{37}{40}\frac{1}{s-6}.$$

Therefore,

$$f(t) = -\frac{4}{5}e^{t} - \frac{1}{8}e^{-2t} + \frac{37}{40}e^{6t}.$$

Example 11.8: Inverse Transform of a Sine–Cosine Combination

Find

$$\mathcal{L}^{-1}\left\{\frac{s+2}{s^2+1}\right\}.$$

Solution:

Decompose

$$\frac{s+2}{s^2+1} = \frac{s}{s^2+1} + \frac{2}{s^2+1}.$$

Using Table 11.1,

$$\mathcal{L}^{-1}\left\{\frac{s}{s^2+1}\right\} = \cos t,\ \mathcal{L}^{-1}\left\{\frac{2}{s^2+1}\right\} = 2\sin t.$$

Thus,

$$f(t) = \cos t + 2\sin t.$$

Example 11.9: Direct Lookup from the Transform Table

Find

$$\mathcal{L}^{-1}\left\{\frac{1}{s^2+1}\right\}.$$

Solution:

We recognize from Table 11.1 that

$$\mathcal{L}\{\sin t\} = \frac{1}{s^2+1}.$$

Hence,

$$f(t) = \sin t.$$

Example 11.10: Inverse Laplace Transforms

Find the inverse transforms of

i.

$$F(s) = \frac{2s+7}{s^2+4s+5};$$

ii.

$$F(s) = \frac{3}{(s+1)^3}.$$

Solution:

i. Complete the square:

$$s^2 + 4s + 5 = (s+2)^2 + 1.$$

Rewrite numerator:

$$2s + 7 = 2(s+2) + 3.$$

Thus,

$$F(s) = 2\frac{s+2}{(s+2)^2+1} + \frac{3}{(s+2)^2+1}.$$

Hence,

$$f(t) = 2e^{-2t}\cos t + 3e^{-2t}\sin t.$$

ii. Use the known inverse,

$$\mathcal{L}^{-1}\left\{\frac{n!}{(s+a)^{n+1}}\right\} = t^n e^{-at}.$$

Here,

$$\frac{3}{(s+1)^3} = \frac{2!}{(s+1)^3}\cdot\frac{3}{2} \Rightarrow f(t) = \frac{3}{2}t^2e^{-t}.$$

11.4 Some Properties of the Laplace Transform

Below are some important properties of the Laplace transform. These follow directly from linearity and differentiation under the integral sign.

1. **Linearity**

$$\mathcal{L}\{0\} = 0, \tag{11.2}$$

$$\mathcal{L}\{f(t) \pm g(t)\} = \mathcal{L}\{f(t)\} \pm \mathcal{L}\{g(t)\}, \tag{11.3}$$

$$\mathcal{L}\{c\,f(t)\} = c\,\mathcal{L}\{f(t)\},\ c \text{ a constant}. \tag{11.4}$$

2. **Differentiation in the s-domain**
   Let $F(s) = \mathscr{L}\{f(t)\}$. Then,

$$F'(s) = \mathscr{L}\{-t\, f(t)\}, \quad \mathscr{L}\{t f(t)\} = -F'(s). \tag{11.5}$$

   This can be generalized to higher powers:

$$F^{(n)}(s) = \mathscr{L}\{(-t)^n f(t)\}, \mathscr{L}\{t^n f(t)\} = (-1)^n F^{(n)}(s). \tag{11.6}$$

   These are extremely useful for constructing transforms of functions like te^{at}, $t^n e^{at}$, polynomials, and more. For convenience, they are tabulated in Table 11.2.

Another important property of the Laplace transform is the **convolution theorem**, which states that convolution in the time domain corresponds to multiplication in the transform domain.

! Red Flag 11.1: Warning

The Laplace transform is **not multiplicative**:

$$\mathscr{L}\{f(t)g(t)\} \neq \mathscr{L}\{f(t)\}\mathscr{L}\{g(t)\}.$$

This is a common beginner's misconception.

Table 11.2 Key properties of the Laplace transform. Let $\mathscr{L}\{f(t)\} = F(s)$ and $\mathscr{L}\{g(t)\} = G(s)$, with a, b, c constants and n a positive integer.

Property	Time domain	s-domain	Conditions
1. Linearity	$af(t) + bg(t)$	$aF(s) + bG(s)$	
2. First shifting	$e^{at} f(t)$	$F(s-a)$	$\mathrm{Re}(s) > \mathrm{Re}(a) + c$
3. Second shifting	$u(t-a)f(t-a)$	$e^{-as}F(s)$	$a \geq 0$
4. Derivative in t	$f'(t)$	$sF(s) - f(0)$	f continuous, f' piecewise
	$f^{(n)}(t)$	$s^n F(s) - s^{n-1} f(0) - \cdots - f^{(n-1)}(0)$	
5. Derivative in s	$t^n f(t)$	$(-1)^n F^{(n)}(s)$	
6. Integral in t	$\int_0^t f(\tau)\, d\tau$	$\frac{F(s)}{s}$	
7. Convolution	$(f \times g)(t) = \int_0^t f(\tau)g(t-\tau)\, d\tau$	$F(s)G(s)$	
8. Initial value	$\lim_{t\to 0^+} f(t)$	$\lim_{s\to\infty} sF(s)$	If limits exist
9. Final value	$\lim_{t\to\infty} f(t)$	$\lim_{s\to 0} sF(s)$	If poles of $sF(s)$ in LHP

Example 11.11: Differentiation in the s-Domain: A First Example

Using Property (11.5), show that $\mathcal{L}\{t^2\} = 2/s^3$.

Solution:

Here, $f(t) = t$, and from Table 11.1,

$$\mathcal{L}\{t\} = \frac{1}{s^2}.$$

Using (11.5),

$$\mathcal{L}\{t^2\} = -\frac{d}{ds}\left(\frac{1}{s^2}\right) = -\left(-\frac{2}{s^3}\right) = \frac{2}{s^3}.$$

This matches the general formula $n!/s^{n+1}$ for $n = 2$.

Example 11.12: Building Toward Higher Derivatives in the s-Domain

a. Use property (11.5) to show that

$$\mathcal{L}\{te^{at}\} = \frac{1}{(s-a)^2}.$$

b. Determine $\mathcal{L}\{t^2e^{at}\}$.

Solution:

a. From Table 11.1,

$$\mathcal{L}\{e^{at}\} = \frac{1}{s-a}.$$

Using $\mathcal{L}\{t^n f(t)\} = (-1)^n F^{(n)}(s)$,

$$\mathcal{L}\{te^{at}\} = -\frac{d}{ds}\left(\frac{1}{s-a}\right) = \frac{1}{(s-a)^2}.$$

b. From (11.6),

$$\mathcal{L}\{t^2e^{at}\} = (+1)\frac{d^2}{ds^2}\left(\frac{1}{s-a}\right) = \frac{2}{(s-a)^3}.$$

Example 11.13: Higher Derivatives in the s-domain

Given $\mathcal{L}\{e^{at}\} = \frac{1}{s-a}$, use the property

$$\mathcal{L}\{t^n f(t)\} = (-1)^n \frac{d^n}{ds^n} F(s)$$

to find $\mathcal{L}\{t^3 e^{at}\}$.

Solution:

Let $f(t) = e^{at}$, so that

$$F(s) = \frac{1}{s-a}.$$

We want $\mathcal{L}\{t^3 e^{at}\}$. Using Equation (11.6) with $n = 3$,

$$\mathcal{L}\{t^3 e^{at}\} = (-1)^3 \frac{d^3}{ds^3}\left(\frac{1}{s-a}\right).$$

Step 1–First derivative:

$$\frac{d}{ds}\left(\frac{1}{s-a}\right) = -\frac{1}{(s-a)^2}.$$

Step 2–Second derivative:

$$\frac{d^2}{ds^2}\left(\frac{1}{s-a}\right) = \frac{d}{ds}\left(-\frac{1}{(s-a)^2}\right) = \frac{2}{(s-a)^3}.$$

Step 3–Third derivative:

$$\frac{d^3}{ds^3}\left(\frac{1}{s-a}\right) = \frac{d}{ds}\left(\frac{2}{(s-a)^3}\right) = -\frac{6}{(s-a)^4}.$$

Step 4–Apply $(-1)^3$:

$$\mathcal{L}\{t^3 e^{at}\} = (-1)^3\left[-\frac{6}{(s-a)^4}\right] = \frac{6}{(s-a)^4}.$$

Thus,

$$\boxed{\mathcal{L}\{t^3 e^{at}\} = \frac{6}{(s-a)^4}, \quad \operatorname{Re} s > a.}$$

Verification:

This matches the known formula $\mathcal{L}\{t^n e^{at}\} = \frac{n!}{(s-a)^{n+1}}$ for $n = 3$, since $3! = 6$.

11.5 Solution of Initial-value Problems

One of the most powerful applications of the Laplace transform is the solution of **linear differential equations**. The transform converts differentiation in the time domain into multiplication by s, and it automatically incorporates initial conditions into algebraic expressions.

11.5.1 Theorem: Laplace Transform of Derivatives

Suppose f is of **exponential order** and is continuous on $(0, \infty)$, and $f'(t)$ is piecewise continuous on every finite interval. Then,

$$\mathscr{L}\{f'(t)\} = s\mathscr{L}\{f(t)\} - f(0).$$

Applying this repeatedly:

$$\mathscr{L}\{f''(t)\} = s^2\mathscr{L}\{f(t)\} - sf(0) - f'(0),$$

$$\mathscr{L}\{f^{(3)}(t)\} = s^3\mathscr{L}\{f(t)\} - s^2f(0) - sf'(0) - f''(0),$$

and, in general,

$$\mathscr{L}\{f^{(n)}(t)\} = s^n\mathscr{L}\{f(t)\} - s^{n-1}f(0) - s^{n-2}f'(0) - \cdots - sf^{(n-2)}(0) - f^{(n-1)}(0).$$

This property is what makes the Laplace transform an extremely efficient tool: **a differential equation becomes an algebraic equation**.

11.5.2 Solving Initial-value Problems Using Laplace Transforms

To solve a linear ODE using the Laplace transform, there are three basic steps:

1. **Take the Laplace transform of both sides** of the differential equation.
2. **Solve for** $Y(s) = \mathscr{L}\{y(t)\}$ algebraically.
3. **Take the inverse Laplace transform** to obtain $y(t)$.

This method works uniformly for both homogeneous and nonhomogeneous equations and avoids the need to find complementary and particular solutions separately.

An algorithm for the process is given in the following box, and Figure 11.4 gives a condensed version of the same.

◆ ▢→ Algorithm 11.1: Laplace Method for Linear IVPs

Given:
A linear ODE with constant coefficients and initial conditions:

$$a_n y^{(n)} + \cdots + a_1 y' + a_0 y = g(t), y(0) = y_0, \ y'(0) = y_1, \dots .$$

Step 1: Transform

Take the Laplace transform of both sides:

$$\mathscr{L}\{\text{ODE}\} \Longrightarrow \text{Algebraic equation in } Y(s).$$

Use derivative formulas:

$$\mathcal{L}\{y^{(k)}\} = s^k Y(s) - s^{k-1} y(0) - \cdots - y^{(k-1)}(0).$$

Step 2: Solve for $Y(s)$

Rearrange to isolate $Y(s)$:

$$Y(s) = \frac{\text{Terms from g}(t) + \text{Initial condition terms}}{a_n s^n + \cdots + a_0}.$$

The denominator $a_n s^n + \cdots + a_0$ is the characteristic polynomial and its roots are the system's poles.

Step 3: Simplify

- If $Y(s)$ is rational, use **partial fractions.**
- If $Y(s)$ contains e^{-as}, apply the **second shifting theorem.**
- If $Y(s)$ is a product, consider the **convolution theorem.**

Step 4: Invert

Use Table 11.1 and properties to find

$$y(t) = \mathcal{L}^{-1}\{Y(s)\}.$$

Step 5: Verify

Check that $y(t)$ satisfies the original ODE and initial conditions.

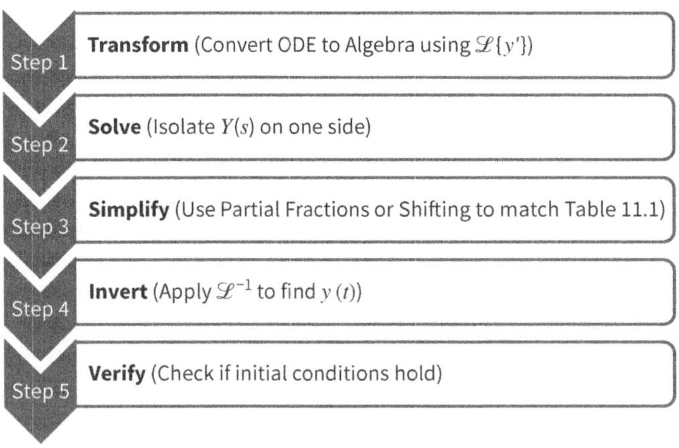

Figure 11.4 Laplace method for linear IVPs.

11.5.2.1 Example Walkthrough

Let us solve the following initial-value problem (IVP) using the process:

IVP: $y'' - y' - 2y = 0$, with $y(0) = 1$ and $y'(0) = 0$.

1. **Take Laplace Transform:**

$$\mathscr{L}\{y''\} - \mathscr{L}\{y'\} - 2\mathscr{L}\{y\} = \mathscr{L}\{0\},$$

$$\left(s^2Y - sy(0) - y'(0)\right) - (sY - y(0)) - 2Y = 0.$$

2. **Substitute Initial Conditions:**
   Substitute $y(0) = 1$ and $y'(0) = 0$:

$$\left(s^2Y - s(1) - 0\right) - (sY - 1) - 2Y = 0,$$

$$s^2Y - s - sY + 1 - 2Y = 0.$$

3. **Solve for $Y(s)$:**
   Group the terms with Y and move others to the right side:

$$Y\left(s^2 - s - 2\right) = s - 1,$$

$$Y(s) = \frac{s-1}{s^2 - s - 2}.$$

4. **Is $Y(s)$ rational?**
   Yes. The denominator can be factored: $s^2 - s - 2 = (s-2)(s+1)$.
   So, $Y(s) = \dfrac{s-1}{(s-2)(s+1)}$.
   Now, apply partial fraction decomposition:

$$\frac{s-1}{(s-2)(s+1)} = \frac{A}{s-2} + \frac{B}{s+1}.$$

   Solving for A and B (a separate algebraic step) gives $A = 1/3$ and $B = 2/3$.
   So, $Y(s) = \dfrac{1/3}{s-2} + \dfrac{2/3}{s+1}$.

5. **Inverse Laplace Transform:**
   Take the inverse transform of each term using the standard rule $\mathscr{L}^{-1}\left\{\frac{1}{s-a}\right\} = e^{at}$,

$$y(t) = \mathscr{L}^{-1}\left\{\frac{1/3}{s-2}\right\} + \mathscr{L}^{-1}\left\{\frac{2/3}{s+1}\right\},$$

$$y(t) = \frac{1}{3}e^{2t} + \frac{2}{3}e^{-t}.$$

   This is the final solution to the IVP.

As an example from electrochemistry, consider a simple RC cell.

Simple RC Cell: Potential Step

- Consider a simple electrochemical cell modeled as a resistor R in series with a capacitor C. A potential step $E(t) = E_0 u(t)$ is applied at $t = 0$; find the current $i(t)$.
- Governing equation (series RC):
  $E(t) = Ri(t) + \frac{1}{C}\int_0^t i(\tau)\,d\tau$, with $i(0) = \frac{E_0}{R}$ (or derive from $q = CV$, etc.).
- Take Laplace transforms, use $\mathcal{L}\left\{\int_0^t i(\tau)d\tau\right\} = I(s)/s$, solve for $I(s)$, and then invert to get the standard decaying exponential current.

Example 11.14: Solving an IVP - Laplace vs. the Classical Method

Solve

$$y'' - 6y' + 5y = 0, y(0) = 1, y'(0) = -3.$$

Solution:

Step 1—Transform both sides:

$$\mathcal{L}\{y''\} - 6\mathcal{L}\{y'\} + 5\mathcal{L}\{y\} = 0.$$

Using the formulas for derivatives:

$$(s^2Y - sy(0) - y'(0)) - 6(sY - y(0)) + 5Y = 0.$$

Substitute initial conditions:

$$(s^2Y - s + 3) - 6(sY - 1) + 5Y = 0.$$

Step 2—Solve algebraically for $Y(s)$:

$$s^2Y - s + 3 - 6sY + 6 + 5Y = 0,$$

$$(s^2 - 6s + 5)Y = s - 9.$$

Factor denominator:

$$(s - 1)(s - 5)Y = s - 9.$$

Thus,

$$Y(s) = \frac{s - 9}{(s - 1)(s - 5)}.$$

Step 3—Inverse transform via partial fractions:

Write

$$\frac{s - 9}{(s - 1)(s - 5)} = \frac{a}{s - 1} + \frac{b}{s - 5}.$$

Cover-up method:

- At $s = 5$: numerator is $5 - 9 = -4$, denominator factor: $5 - 1 = 4$;

$$b = \frac{-4}{4} = -1.$$

At $s = 1$: numerator is $1 - 9 = -8$, denominator factor: $1 - 5 = -4$;

$$a = \frac{-8}{-4} = 2.$$

Thus,

$$Y(s) = \frac{2}{s-1} - \frac{1}{s-5}.$$

Therefore,

$$y(t) = 2e^t - e^{5t}.$$

The same IVP can be solved by the standard method. The characteristic equation is

$$r^2 - 6r + 5 = 0 \Longrightarrow (r-1)(r-5) = 0 \Longrightarrow r = 1, 5.$$

General solution:

$$y(t) = Ae^t + Be^{5t}.$$

Initial conditions:

$$y(0) = A + B = 1,$$

$$y'(0) = A + 5B = -3.$$

Solving gives $A = 2, B = -1$, so

$$y(t) = 2e^t - e^{5t}.$$

Laplace advantage: Initial conditions are incorporated automatically during the algebraic step—no separate system of equations needed.

Example 11.15: A First-Order Linear IVP with Exponential Forcing

Solve

$$y' - y = e^{-x}, \quad y(0) = 0.$$

Solution:

Step 1–Transform:

We rewrite the equation in terms of the variable t': $y' - y = e^{-t}, y(0) = 0$ for consistency with the Laplace variable s,

$$\mathcal{L}\{y'\} - \mathcal{L}\{y\} = \mathcal{L}\{e^{-t}\},$$

$$(sY - y(0)) - Y = \frac{1}{s+1}.$$

Using $y(0) = 0$,

$$(s-1)Y = \frac{1}{s+1}.$$

Step 2—Solve for $Y(s)$:

$$Y(s) = \frac{1}{(s-1)(s+1)}.$$

Step 3—Partial fractions:

$$\frac{1}{(s-1)(s+1)} = \frac{a}{s-1} + \frac{b}{s+1}.$$

- At $s = -1$, $1 = -2b \Rightarrow b = -\frac{1}{2}$.
- At $s = 1$, $1 = 2a \Rightarrow a = \frac{1}{2}$.

Thus,

$$Y(s) = \frac{1}{2}\frac{1}{s-1} - \frac{1}{2}\frac{1}{s+1}.$$

Hence,

$$y(x) = \frac{1}{2}(e^{x} - e^{-x}).$$

Example 11.16: IVP with Polynomial-Exponential Forcing

Solve:

$$y' + 2y = 4te^{-2t}, \; y(0) = -3.$$

Solution:

Step 1—Transform:

$$(sY - y(0)) + 2Y = \mathcal{L}\{4te^{-2t}\}. \qquad (1)$$

Since

$$\mathcal{L}\{te^{-2t}\} = \frac{1}{(s+2)^2},$$

we have

$$\mathcal{L}\{4te^{-2t}\} = \frac{4}{(s+2)^2}.$$

Thus, substituting in Equation (1),

$$sY - (-3) + 2Y = \frac{4}{(s+2)^2},$$

or

$$sY + 3 + 2Y = \frac{4}{(s+2)^2}.$$

Step 2–Solve for $Y(s)$:

$$(s+2)Y = \frac{4}{(s+2)^2} - 3,$$

$$Y(s) = \frac{4}{(s+2)^3} - \frac{3}{s+2}.$$

Step 3–Inverse transform:

$$\mathcal{L}^{-1}\left\{\frac{4}{(s+2)^3}\right\} = 4 \cdot \frac{t^2}{2!}e^{-2t} = 2t^2e^{-2t},$$

$$\mathcal{L}^{-1}\left\{\frac{3}{s+2}\right\} = 3e^{-2t}.$$

Thus,

$$y(t) = 2t^2e^{-2t} - 3e^{-2t}.$$

Example 11.17: IVP with Exponential Forcing (Distinct Real Roots)

Solve:

$$y'' - 3y' + 2y = e^{3t}, y(0) = 1, y'(0) = 0.$$

Solution:

Step 1–Transform:

$$(s^2Y - s - 0) - 3(sY - 1) + 2Y = \frac{1}{s-3}.$$

Step 2–Solve for $Y(s)$:

$$(s^2 - 3s + 2)Y - s + 3 = \frac{1}{s-3},$$

$$(s-1)(s-2)Y = \frac{1}{s-3} + s - 3.$$

Do partial fractions:

$$Y(s) = \frac{1}{(s-3)(s-1)(s-2)} + \frac{s-3}{(s-1)(s-2)}.$$

Carrying out decomposition yields:

$$Y(s) = \frac{5}{2(s-1)} - \frac{2}{s-2} + \frac{1}{2(s-3)}.$$

Thus,

$$y(t) = \frac{5}{2}e^{t} - 2e^{2t} + \frac{1}{2}e^{3t}.$$

Example 11.18: IVP with Sinusoidal Forcing and Complex Roots

Solve

$$y'' - 2y' + 2y = \cos t, \quad y(0) = 1, y'(0) = 0.$$

Solution:

Step 1–Transform:

$$(s^2 Y - s - 0) - 2(sY - 1) + 2Y = \frac{s}{s^2+1}.$$

Step 2–Solve for $Y(s)$:

$$(s^2 - 2s + 2)Y - s + 2 = \frac{s}{s^2+1}.$$

$$Y(s) = \frac{s}{(s^2+1)(s^2-2s+2)} + \frac{s-2}{s^2-2s+2}. \tag{1}$$

Complete the square:

$$s^2 - 2s + 2 = (s-1)^2 + 1.$$

First Term of Equation (1)

$$\frac{s}{(s^2+1)\left((s-1)^2+1\right)}. \tag{2}$$

We use **partial fractions**.
Assume

$$\frac{s}{(s^2+1)\left((s-1)^2+1\right)} = \frac{As+B}{s^2+1} + \frac{C(s-1)+D}{(s-1)^2+1}.$$

Solving gives $A = 1/5, B = -2/5, C = -1/5, D = 3/5$.

Therefore, the first term is:

$$\frac{s}{(s^2+1)\left((s-1)^2+1\right)} = \frac{1}{5}\frac{s}{s^2+1} - \frac{2}{5}\frac{1}{s^2+1} - \frac{1}{5}\frac{s-1}{(s-1)^2+1} + \frac{3}{5}\frac{1}{(s-1)^2+1}. \quad (3)$$

Take Inverse Laplace Term-by-term

Use known transforms:

$$\mathcal{L}^{-1}\left\{\frac{s-1}{(s-1)^2+1}\right\} = e^t \cos t,$$

$$\mathcal{L}^{-1}\left\{\frac{1}{(s-1)^2+1}\right\} = e^t \sin t,$$

$$\mathcal{L}^{-1}\left\{\frac{s}{s^2+1}\right\} = \cos t,$$

$$\mathcal{L}^{-1}\left\{\frac{1}{s^2+1}\right\} = \sin t.$$

Apply to (3). This gives

$$\frac{1}{5}\cos t - \frac{2}{5}\sin t - \frac{1}{5}e^t \cos t + \frac{3}{5}e^t \sin t. \quad (4)$$

Second Term of Equation (1)

Rewrite $s-2$ as $s-2=(s-1)-1$.

So,

$$\frac{s-2}{(s-1)^2+1} = \frac{s-1}{(s-1)^2+1} - \frac{1}{(s-1)^2+1}.$$

These are known transforms. Thus, the second large term in (1) → in the time domain:

$$e^t \cos t - e^t \sin t.$$

Add the first term (4):

Final Answer:

$$\boxed{y(t) = \frac{1}{5}\cos t - \frac{2}{5}\sin t + \frac{4}{5}e^t \cos t - \frac{2}{5}e^t \sin t.}$$

11.6 Applications in Electrochemistry and Chemical Kinetics

While Laplace transforms are standard in engineering for circuits, they are a "secret weapon" in physical chemistry. Laplace transforms are particularly powerful for solving **kinetic schemes**—systems of coupled differential equations where species transform into one another (e.g., radioactive decay, consecutive reactions, or enzyme kinetics).

Instead of integrating factor methods or matrix diagonalization, Laplace transforms allow us to track the population of species algebraically.

Unified View: The Two-pole Pattern in Relaxation

Both the Bateman sequential decay (Example 11.19) and reversible kinetics (Example 11.20) yield rational Laplace transforms with two simple poles:

$$\overline{n_B}(s) = \frac{\lambda_A N_0}{(s+\lambda_A)(s+\lambda_B)}, \quad \overline{A}(s) \propto \frac{1}{s(s+k_f+k_b)}.$$

Inverse Laplace → bi-exponential time domain:

$$N_B(t) = A_1 e^{-\lambda_A t} + A_2 e^{-\lambda_B t}, \quad [A](t) = A_\infty + Be^{-(k_f+k_b)t}.$$

This two-pole → bi-exponential pattern unifies all linear relaxation problems—whether radioactive decay chains, reversible reactions, RC circuits, or spectroscopic line shapes.

Example 11.19: Sequential Radioactive Decay (The Bateman Equations)

Consider a radioactive decay chain where a parent nucleus A decays into an intermediate daughter B, which then decays into a stable isotope C. This is identical to a consecutive chemical reaction $A \rightarrow B \rightarrow C$,

$$A \xrightarrow{\lambda_A} B \xrightarrow{\lambda_B} C.$$

The rate equations are:

1. $\dfrac{dN_A}{dt} = -\lambda_A N_A$;
2. $\dfrac{dN_B}{dt} = \lambda_A N_A - \lambda_B N_B$.

Find the number of nuclei of B at time t, denoted $N_B(t)$, assuming we start with N_0 nuclei of A and none of B ($N_A(0) = N_0, N_B(0) = 0$).

Solution:

Step 1: Transform the System

We apply the Laplace transform to each equation. Let $\mathscr{L}\{N_A(t)\} = \overline{n_a}(s)$ and $\mathscr{L}\{N_B(t)\} = \overline{n_b}(s)$.

For equation (1):

$$s\overline{n_a} - N_A(0) = -\lambda_A \overline{n_a},$$

$$s\overline{n_a} - N_0 = -\lambda_A \overline{n_a} \Longrightarrow \overline{n_a}(s+\lambda_A) = N_0,$$

$$\overline{n_a} = \frac{N_0}{s+\lambda_A}.$$

For equation (2):

$$s\overline{n_b} - N_B(0) = \lambda_A \overline{n_a} - \lambda_B \overline{n_b}.$$

Since $N_B(0) = 0$, this simplifies to

$$(s + \lambda_B)\overline{n_b} = \lambda_A \overline{n_a}.$$

Step 2: Algebraic Substitution

Substitute the known expression for $\overline{n_a}$ into the equation for $\overline{n_b}$:

$$\overline{n_b} = \frac{\lambda_A}{s + \lambda_B} \cdot \overline{n_a} = \frac{\lambda_A N_0}{(s + \lambda_B)(s + \lambda_A)}.$$

Step 3: Partial Fraction Decomposition

This is a standard form (see Example 11.7). We separate the denominators:

$$\frac{1}{(s + \lambda_A)(s + \lambda_B)} = \frac{A}{s + \lambda_A} + \frac{B}{s + \lambda_B}.$$

Using the cover-up method:

- At $s = -\lambda_A$: $A = \dfrac{1}{-\lambda_A + \lambda_B} = \dfrac{1}{\lambda_B - \lambda_A}$.
- At $s = -\lambda_B$: $B = \dfrac{1}{-\lambda_B + \lambda_A} = \dfrac{-1}{\lambda_B - \lambda_A}$.

Thus:

$$\overline{n_b}(s) = \frac{\lambda_A N_0}{\lambda_B - \lambda_A}\left(\frac{1}{s + \lambda_A} - \frac{1}{s + \lambda_B}\right).$$

Step 4: Inverse Transform

Using the property $\mathscr{L}^{-1}\left\{\dfrac{1}{s + k}\right\} = e^{-kt}$,

$$N_B(t) = \frac{\lambda_A N_0}{\lambda_B - \lambda_A}(e^{-\lambda_A t} - e^{-\lambda_B t}).$$

Interpretation:

The population of B starts at 0, rises to a maximum as A decays, and then falls to 0 as B decays into C. The Laplace method handled the "coupling" automatically without needing complex integration techniques.

Example 11.20: Reversible Kinetics and Relaxation

Consider a reversible first-order reaction, common in conformational changes or isomerization:

$$A \underset{k_b}{\overset{k_f}{\rightleftharpoons}} B$$

The rate law for the concentration of A, denoted A, is

$$\frac{d[A]}{dt} = -k_f[A] + k_b[B].$$

Assume conservation of mass: $[A] + [B] = [A]_0$ (the total initial concentration). Solve for A as a function of time.

Solution:

Substitute $[B] = [A]_0 - [A]$ into the differential equation to get a linear ODE:

$$\frac{d[A]}{dt} = -k_f[A] + k_b([A]_0 - [A]),$$

$$\frac{d[A]}{dt} + (k_f + k_b)[A] = k_b[A]_0.$$

Step 1: Transform

$$s\overline{A} - [A]_0 + (k_f + k_b)\overline{A} = \frac{k_b[A]_0}{s}.$$

(*Note*: The transform of the constant $k_b[A]_0$ is $k_b[A]_0/s$.)

Step 2: Solve for A

Group terms with A:

$$\overline{A}(s + k_f + k_b) = [A]_0 + \frac{k_b[A]_0}{s} = [A]_0\left(\frac{s + k_b}{s}\right),$$

$$\overline{A} = [A]_0\frac{s + k_b}{s(s + k_{\text{sum}})},$$

where $k_{\text{sum}} = k_f + k_b$.

Step 3: Partial Fractions

$$\frac{s + k_b}{s(s + k_{\text{sum}})} = \frac{C_1}{s} + \frac{C_2}{s + k_{\text{sum}}}.$$

- At $s = 0$ (cover-up s): $C_1 = \frac{k_b}{k_{\text{sum}}}$.
- At $s = -k_{\text{sum}}$ (cover-up $s + k_{\text{sum}}$): $C_2 = \frac{-k_{\text{sum}} + k_b}{-k_{\text{sum}}} = \frac{-k_f}{-k_{\text{sum}}} = \frac{k_f}{k_{\text{sum}}}$.

Step 4: Inverse Transform

$$[A](t) = [A]_0\left(\frac{k_b}{k_{\text{sum}}}L^{-1}\left\{\frac{1}{s}\right\} + \frac{k_f}{k_{\text{sum}}}L^{-1}\left\{\frac{1}{s + k_{\text{sum}}}\right\}\right),$$

$$[A](t) = \frac{[A]_0}{k_f + k_b}\left(k_b + k_f e^{-(k_f+k_b)t}\right).$$

Result:

As $t \to \infty$, the exponential term vanishes, leaving the **equilibrium concentration** $[A]_{eq} = [A]_0 \dfrac{k_b}{k_f + k_b}$. The system "relaxes" to this state with a rate constant $k_{obs} = k_f + k_b$.

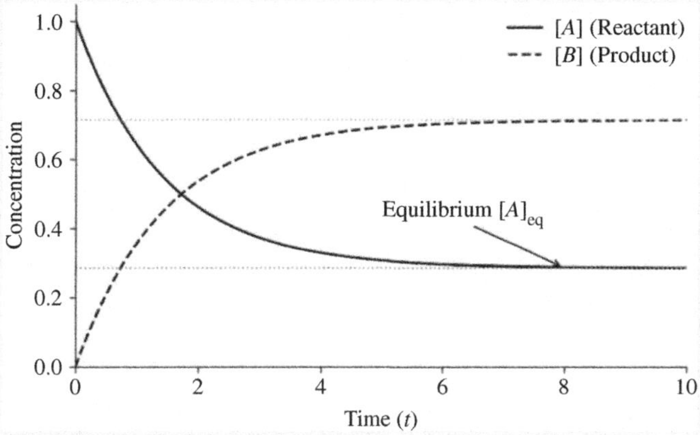

So far, we used Laplace transforms in chemical kinetics; now, we illustrate a closely related relaxation phenomenon in electrochemistry using an *RC* cell model of the electrode interface.

Example 11.21: Potential Step in an Electrochemical *RC* Cell

A simple electrochemical cell is modeled as a resistor R in series with a double-layer capacitor C. At $t = 0$, the potential is stepped from 0 to a constant value E_0:

$$E(t) = E_0 u(t).$$

Find the current $i(t)$ using the Laplace transform.

Solution:

The loop equation for the series *RC* cell is

$$E(t) = Ri(t) + V_C(t),$$

where $V_C(t)$ is the capacitor voltage. The capacitor relation is

$$i(t) = C\frac{dV_C}{dt}, V_C(0) = 0.$$

Take Laplace transforms (denote $\mathcal{L}\{i(t)\} = I(s)$, $\mathcal{L}\{V_C(t)\} = V_C(s)$):

$$\mathcal{L}\{E(t)\} = \mathcal{L}\{Ri(t) + V_C(t)\},$$

$$\frac{E_0}{s} = RI(s) + V_C(s),$$

and

$$I(s) = C[sV_C(s) - V_C(0)] = CsV_C(s).$$

Eliminate $V_C(s)$: from $I(s) = CsV_C(s)$ we get $V_C(s) = \frac{I(s)}{Cs}$. Substitute into the loop equation:

$$\frac{E_0}{s} = RI(s) + \frac{I(s)}{Cs} = I(s)\left(R + \frac{1}{Cs}\right).$$

Solve for $I(s)$:

$$I(s) = \frac{E_0/s}{R + 1/(Cs)} = \frac{E_0/s}{\frac{RCs+1}{Cs}} = \frac{E_0}{RCs + 1}.$$

Rewrite:

$$I(s) = \frac{E_0}{R}\frac{1}{s + 1/(RC)}.$$

Invert using $\mathcal{L}^{-1}\{1/(s+a)\} = e^{-at}$:

$$i(t) = \frac{E_0}{R}e^{-t/(RC)}, t \geq 0.$$

This shows that the current after a potential step in an RC electrochemical cell decays exponentially with time constant $\tau = RC$.

11.7 Summary

In this chapter, we introduced the Laplace transform as a tool for converting functions defined for $t \geq 0$ into functions of a complex frequency variable s. Beginning with the definition and evaluation of transforms of elementary functions, we then examined inverse transforms through partial fractions and the shifting theorems. An important feature of the Laplace transform is its handling of derivatives, which turns differential equations into algebraic equations while naturally incorporating initial conditions. This makes the Laplace method especially powerful for solving linear IVPs, including those arising in chemical kinetics, harmonic oscillators, damping phenomena, and forced vibrations. The chapter concludes with several applications illustrating the versatility and elegance of the Laplace transform technique.

Further Reading

D. G. Zill, W. Wright, and M. R. Culle, *Differential Equations with Boundary-Value Problems*, 8th ed., Cengage Learning, 2012.

R. L. Boylestad and L. Nashelsky, *Electronic Devices and Circuit Theory*, 11th ed., Pearson, 2013.

P. Atkins and J. de Paula, *Atkins' Physical Chemistry*, Oxford University Press, 2010 (applications to kinetics and relaxation processes).

J. W. Brown and R. V. Churchill, *Complex Variables and Applications*, 9th ed., McGraw-Hill, 2013 (background on transforms and complex analysis).

A. V. Oppenheim, A. S. Willsky, and S. H. Nawab, *Signals and Systems*, 2nd ed., Prentice Hall, 2015 (excellent coverage of Laplace transform properties and applications).

Exercises

A. Conceptual Questions

1. Explain why the Laplace transform converts differentiation into multiplication by s.

2. What is meant by a function being "of exponential order"? Why is this condition important for convergence?

3. Why does the Laplace transform automatically incorporate initial conditions?

4. Distinguish between the first and second shifting theorems.

5. Why does $\mathcal{L}\{f(t)g(t)\} \neq \mathcal{L}\{f(t)\}\mathcal{L}\{g(t)\}$?

B. Problems

1. **Laplace transforms of elementary functions**
   Find the Laplace transform of each function below:
   i. t^2;
   ii. te^{6t};
   iii. $\cos(3t)$;
   iv. $e^{-t}\sin(2t)$;
   v. $e^{i\alpha t}$, where α is a constant.

2. **Functions defined by integrals**
   Each function $F(s)$ below is defined by a definite integral. Without integrating, find an explicit expression for each $F(s)$:
   i. $\int_0^\infty e^{-st}e^{-7t}\,dt$
   ii. $\int_0^\infty t^2e^{-(s-3)t}\,dt$
   iii. $\int_0^\infty e^{-st}\cos(2t)\,dt$

3. **Area under a curve**
   Let $f(t)$ be continuous, $f(t) \geq 0$ for $t \geq 0$, and let $F(s)$ be its Laplace transform.
   a. Show that

$$F(0) = \int_0^\infty f(t)\,dt.$$

   b. Verify this for $f(t) = e^{-t}$, te^{-2t}, and t^2e^{-4t}.

4. Rational fraction decomposition

By performing the rational fraction decomposition, establish the results given below:

a. $\mathcal{L}^{-1}\left\{\frac{s+1}{s(s^2+1)}\right\} = 1 + \sin t - \cos t$

b. $\mathcal{L}^{-1}\left\{\frac{s+1}{s^2(s^2+s+1)}\right\} = t - \frac{2}{\sqrt{3}}e^{-t/2}\sin\frac{\sqrt{3}}{2}t$

(*Hint:* Complete square in denominator.)

5. Solve the following differential equations using Laplace transforms

a. $y' + y = e^{2t}, \quad y(0) = 0$

b. $y'' - 4y' + 4y = 0, \quad y(0) = 0, y'(0) = 1$

c. $y'' - y' - 2y = t^2, \quad y(0) = 1, y'(0) = 3$

d. $y'' + 4y = 0, \quad y(0) = 1, y\left(\frac{\pi}{4}\right) = -1$

e. $y' + 10y = t^2, \quad y(0) = 0$

f. $y'' + 2y' = te^{-t}, \quad y(0) = 6, y'(0) = -1$

g. $y''' + 3y'' + 3y' + y = 0, \quad y(0) = 7; y'(0) = -7, y''(0) = 11$

h. $y''' - y'' + 4y' - 4y = 26e^{3t}, \quad y(0) = -2; y'(0) = 3, y''(0) = 1$

6. Reversible reaction

Consider the reversible system $A \rightleftharpoons B$.

Write expressions for $d[A]/dt$ and $d[B]/dt$. Solve the coupled equations by Laplace transforms, given $[A]_0 = A_0$ and $[B]_0 = B_0$.

7. Forced harmonic oscillator

In a simplified chemical bond model, the vibration of an atom obeys

$$mx'' + kx = F_0 \cos(\omega t).$$

Solve for $x(t)$ in the case $m = 1$, and show that the amplitude is maximum when $\omega = \omega_0$. Also find $x(t)$ when $\omega = \omega_0$ (resonance).

8. Two masses and a spring

Two equal masses connected by a spring satisfy:

$$x_1'' = -k(x_1 - x_2), x_2'' = -k(x_2 - x_1).$$

Solve using Laplace transforms, assuming your original initial conditions.

9. Damped harmonic oscillator

Solve

$$x'' + 2\alpha x' + \omega_0^2 x = 0$$

for the cases:

i. underdamped $\omega_0^2 > \alpha^2$;

ii. critically damped $\omega_0^2 = \alpha^2$;

iii. overdamped $\omega_0^2 < \alpha^2$.

10. Forced damped oscillator

Solve Problem in Example 11.10 for the forcing condition

$$x'' + 2\alpha x' + \omega_0^2 x = F_0 \cos(\omega t),$$

with $\omega_0^2 > \alpha^2$.

C. Advanced Problems

1. Heaviside function (shifted forcing):

Solve

$$y'' + 4y = u(t-3), y(0) = 0, \; y'(0) = 0.$$

2. Impulse response:

$$y'' + y = \delta(t-\pi), y(0) = 0, \; y'(0) = 0.$$

3. Reversible–consecutive kinetics:

$$A \underset{k_2}{\overset{k_1}{\rightleftharpoons}} B \overset{k_3}{\longrightarrow} C.$$

Formulate and solve using Laplace transforms.

4. Overdamped oscillator decomposition:

Show that for

$$x'' + 2\alpha x' + \omega_0^2 x = 0,$$

with $\omega_0^2 < \alpha^2$, the solution is the sum of two exponentials and compute them explicitly using partial fractions.

5. Convolution theorem:

Verify that

$$\mathcal{L}\{f * g\} = \mathcal{L}\{f\}\mathcal{L}\{g\}$$

for $f(t) = e^{-t}$ and $g(t) = t$.

Compute the convolution directly and compare.

12

Linear Differential Equations

In order to solve this differential equation you look at it until a solution occurs to you.

—George Pólya

Learning Outcomes

After completing this chapter, students will be able to:

1. **Distinguish between ordinary points**, singular points, and regular singular points of linear differential equations.
2. **Use the Wronskian to test linear independence** of solutions and **understand** the structure of the solution space.
3. **Construct power-series solutions** of second-order linear differential equations and **derive** the associated recurrence relations.
4. **Apply the Frobenius method** at regular singular points and **interpret** the role of the indicial equation.
5. **Explain why**, for Legendre's equation, non-terminating series generally diverge at the endpoints $x = \pm 1$, and how polynomial truncation leads to physically acceptable solutions.
6. **Derive key properties of Legendre polynomials**, including their recurrence relations and orthogonality on [−1, 1].

12.1 Introduction

Linear second-order differential equations arise throughout physical chemistry and quantum mechanics because they naturally describe oscillations, wave motion, diffusion, heat flow, and the eigenvalue problems of quantum systems. Their solutions reveal much of the underlying physics, and different methods—Wronskian analysis, power-series expansions, Frobenius solutions near singular points, and the emergence of special functions—form a unified toolkit for handling these equations. In this chapter, we develop these methods systematically, beginning with linear independence and the Wronskian, followed by the power-series approach and its application to Legendre's equation. This provides the mathematical foundation needed for later chapters on special functions and quantum mechanical boundary-value problems.

Why This Matters

Second-order linear differential equations appear in every major area of physical chemistry and quantum mechanics: the harmonic oscillator, rotational and vibrational motion, diffusion, heat flow, spectroscopy, and the Schrödinger equation all reduce to this form. Their solutions introduce families of functions—Legendre, Hermite, Laguerre—that reappear throughout spectroscopy, molecular structure, and quantum theory. Understanding how these equations behave, how their solutions are constructed, and why certain series truncate is essential preparation for later chapters on special functions and for nearly all advanced courses in theoretical chemistry.

This chapter develops the **mathematical methods** for solving linear second-order differential equations: the Wronskian, power-series expansions, classification of ordinary/singular points, and the Frobenius approach. We illustrate these techniques using **Legendre's equation** as a primary example—not because we need all its properties now but because it clearly shows how **physical boundary conditions force series truncation**, leading to discrete, quantized solutions. The detailed properties of Legendre polynomials and other special functions (Hermite, Laguerre) will be treated systematically in the following chapter. Later in the chapter, we discuss the series-solution method for second-order linear differential equations, a method that will reappear in our study of quantum mechanics when we encounter special functions such as the Legendre, Laguerre, and Hermite functions.

The notations used in this chapter are provided in Appendix K.

12.2 Linear Independence and Wronskian

A linear homogeneous second-order differential equation of the type

$$y''(x) + p(x)y'(x) + q(x)y(x) = 0 \tag{12.1}$$

has two solutions, $y_1(x)$ and $y_2(x)$. It should be noted that $p(x)$ and $q(x)$ must be continuous over some interval for the theory to hold. A linear combination of these two solutions is also a solution; hence, the general solution of the differential equation is given by

$$y(x) = C_1 y_1(x) + C_2 y_2(x) \tag{12.2}$$

where $C_1$and C_2 are constants (not both zero). We now look more closely at the conditions of linear independence of the two solutions.

The two solutions $y_1(x)$ and $y_2(x)$ of Equation (12.1) are said to be **linearly independent** if

$$C_1\, y_1(x) + C_2\, y_2(x) = 0 \text{ for all } x$$

implies the **trivial case** $C_1 = C_2 = 0$.

If this is not so, then the two functions are linearly dependent. In the dependent case, one solution is proportional to the other:

$$y_2(x) = k\, y_1(x),$$

where k is a nonzero constant. Differentiation gives

$$y_2'(x) = k\, y_1'(x),$$

so that

$$\frac{y_2(x)}{y_1(x)} = \frac{y_2'(x)}{y_1'(x)} = k$$

or

$$y_1(x)\,y_2'(x) - y_2(x)\,y_1'(x) = 0.$$

Therefore, the condition for linear independence is

$$y_1(x)y_2'(x) - y_2(x)y_1'(x) \neq 0. \tag{12.3}$$

This expression is neatly written in terms of the **Wronskian**, defined as

$$W(y_1, y_2)(x) = y_1(x)y_2'(x) - y_2(x)y_1'(x). \tag{12.4}$$

Clearly, if $W = 0$, the two solutions are linearly dependent. Note that $W(x) \neq 0$ **at a single point** within the interval is sufficient for linear independence on that interval, given continuous coefficients.

The Wronskian may also be expressed in determinantal form:

$$W(y_1, y_2)(x) = \begin{vmatrix} y_1(x) & y_2(x) \\ y_1'(x) & y_2'(x) \end{vmatrix}. \tag{12.5}$$

If this determinant is nonzero, the two solutions are linearly independent, and any solution can be written uniquely in the form (12.2) for given constants C_1 and C_2.

In terms of $p(x)$, the Wronskian is given by

$$W(x) = C\exp\left(-\int p(x)dx\right), \tag{12.6}$$

where C is a constant.

Thus:

- If $p(x) = 0$, the Wronskian is constant.
- If the Wronskian is nonzero at **any** point, it is nonzero throughout the interval of continuity of $p(x)$ and $q(x)$, and the solutions are linearly independent everywhere in that interval.

Math Box 12.1: Derivation of the Wronskian Formula

Since both y_1 and y_2 satisfy Equation (12.1),

$$y_1'' + p(x)y_1' + q(x)y_1 = 0, \tag{i}$$

$$y_2'' + p(x)y_2' + q(x)y_2 = 0. \tag{ii}$$

Multiply Equation (i) by y_2 and (ii) by y_1:

$$y_2\,y_1'' + p(x)y_2\,y_1' + q(x)y_2\,y_1 = 0,$$

$$y_1\,y_2'' + p(x)y_1\,y_2' + q(x)y_1\,y_2 = 0.$$

Now subtract the second equation from the first:

$$y_2 y_1'' - y_1 y_2'' + p(x)(y_2 y_1' - y_1 y_2') = 0.$$

Recognizing the expression inside parentheses as the Wronskian-W (12.4), and the remaining term as its derivative (*Note that* $y_2y_1'' - y_1y_2'' = \frac{d}{dx}(y_2y_1' - y_1y_2') = W'(x)$), we obtain

$$W'(x) + p(x)W(x) = 0.$$

This is a first-order separable differential equation. Solving,

$$W(x) = C \exp\left(-\int p(x)dx\right),$$

where C is a constant.

12.2.1 General Solution of Homogeneous Equations

We can now express the steps for obtaining the general solution of a second-order linear homogeneous equation. In general, the solution is written as

$$y(x) = C_1\, y_1(x) + C_2\, y_2(x),$$

where $y_1(x)$ and $y_2(x)$ are linearly independent, as confirmed by a nonzero Wronskian (Equation 12.5).

To determine the **particular solution**, the constants C_1 and C_2 must be fixed using boundary or initial conditions. Often, the conditions are given as

$$y(x_0) = a,\, y'(x_0) = b,$$

at some point x_0. Substituting these conditions into Equation (12.2) and its derivative yields two simultaneous equations, which determine C_1 and C_2 uniquely.

$$a = C_1y_1(x_0) + C_2y_2(x_0)$$

$$b = C_1y_1'(x_0) + C_2y_2'(x_0)$$

In many problems, it is easy to identify one solution but not a second independent one. Obtaining the second, linearly independent solution then becomes a problem. A general method is outlined below.

12.2.2 Obtaining a Second Solution When One Is Known

Suppose one solution $y_1(x)$ is known. Consider the ratio y_2/y_1:

$$\frac{d}{dx}\left(\frac{y_2}{y_1}\right) = \frac{y_1(x)y_2'(x) - y_2(x)y_1'(x)}{y_1^2(x)} = \frac{W(x)}{y_1^2(x)}. \tag{12.7}$$

Integrating and substituting $W(x)$ from Equation (12.6):

$$\frac{y_2}{y_1} = \int \frac{C\, e^{-\int p(x)dx}}{y_1^2(x)}\, dx + K.$$

Therefore

$$y_2(x) = y_1(x)\left[K + C\int \frac{e^{-\int p(x)dx}}{y_1^2(x)}dx\right]. \tag{12.8}$$

This provides a general formula for constructing a **second linearly independent solution** when one solution is already known.

Though this method works well for some equations, for equations with variable coefficients such as the Legendre equation, it is often not possible to guess even a single simple solution. In such cases, we use the **power-series method**, which is introduced next.

12.3 Power-Series Method

Many linear second-order differential equations cannot be expressed in terms of elementary functions such as algebraic, trigonometric, exponential, or logarithmic functions. In such cases, a solution may be obtained by expressing the function as an **infinite power series**, usually a Taylor or Maclaurin series, and substituting it into the differential equation.

We begin with a first-order ODE.

Example 12.1: Power-series Method for a Simple First-order ODE

Objective: To illustrate the power-series method for a simple first-order ODE and show how recurrence relations arise.
Method Used: Substitute a Maclaurin series for y and equate coefficients to obtain a recurrence relation.
Problem: Solve the first-order homogeneous differential equation

$$y'(x) - y(x) = 0$$

by the power-series method.

Solution:

Let

$$y(x) = a_0 + a_1x + a_2x^2 + a_3x^3 + \ldots$$

Differentiating term by term, we obtain:

$$y'(x) = a_1 + 2a_2x + 3a_3x^2 + \ldots$$

The given differential equation can be written in the form $y'(x) = y(x)$
Comparing the constants in the two equations, we may write $a_1 = a_0$
Similarly, by comparison of the coefficients of x, we find that $2a_2 = a_1$, or $a_2 = a_1/2 = a_0/2$
Comparison of the coefficients for x^2 yields $3a_3 = a_2$ or $a_3 = a_2/3 = a_0/6$, and so on.

In general, $a_{n+1} = a_n/(n+1)$. This is a recursion relation showing how a coefficient can be obtained from the previous coefficient.

We can thus write for $y(x)$

$$y(x) = a_0\left(1 + x + \frac{x^2}{2!} + \frac{x^3}{3!} + \ldots\right)$$

which can be recognized as the Maclaurin series for $\exp(x)$.

Hence, we can write the solution as $y = a_0 e^x$ in agreement with the solution obtained in the usual way, that is,

$$\frac{dy}{dx} = y \Rightarrow \int \frac{dy}{y} = \int dx \Rightarrow \ln\, y = x + c \Rightarrow y = c\,\exp(x)$$

where c (like a_0) is a constant.

In the next example, the same procedure is applied to a second-order differential equation.

Example 12.2: Series Solution of the Harmonic Oscillator Equation $y'' + y = 0$

Objective: To obtain the series solution of the harmonic oscillator equation $y'' + y = 0$ and show how sine and cosine emerge naturally.
Method Used: Assume a power series for y, differentiate term by term, substitute into the ODE, and derive separate even and odd recursions.
Problem: Solve the second-order differential equation

$$y''(x) + y(x) = 0$$

by the power-series method.

Solution:

We can write $y(x)$ as a power series, as in Example 12.1. The first derivative needs to be differentiated again to give the second derivative, that is

$$y''(x) = 2a_2 + (3)(2)a_3 x + (4)(3)a_4 x^2 + \cdots$$

Collecting powers,

$$2a_2 + a_0 = 0 \Rightarrow a_2 = -a_0/2$$

$$(3)(2)a_3 + a_1 = 0 \Rightarrow a_3 = -a_1/3!$$

$$(4)(3)a_4 + a_2 = 0 \Rightarrow a_4 = -a_2/[(4)(3)] = a_0/4!$$

In general,

$$a_n = -\frac{a_{n-2}}{n(n-1)} \text{ for all } n \geq 2 \text{ or } a_{n+2} = -\frac{a_n}{(n+2)(n+1)} \text{ for all } n \geq 0.$$

This is a **recurrence** formula. Knowing a_0, we can determine all even coefficients, and knowing a_1, all the odd coefficients.

Therefore,

$$y(x) = a_0\left(1 - \frac{x^2}{2!} + \frac{x^4}{4!} - \ldots\right) + a_1\left(x - \frac{x^3}{3!} + \frac{x^5}{5!} - \ldots\right)$$

The even terms can be recognized as the cosine series and the odd terms as the sine series. Therefore, the complete solution to the differential equation is

$$y(x) = a_0\cos(x) + a_1\sin(x)$$

We have already confirmed in Chapter 9 that both series converge for all real x. They are also linearly independent, since one series consists of only even terms and the other only of odd terms. We can also check that the Wronskian is nonzero.

Example 12.3: Using the Wronskian to Find a Second Solution

Consider the simple second-order linear homogeneous ODE of Example 12.2:

$$y''(x) + y(x) = 0$$

Step 1: Find one solution by inspection

We can guess a solution of the form $y_1(x) = \cos(x)$.

Check: $y_1''(x) = -\cos(x)$, so $y_1'' + y_1 = -\cos(x) + \cos(x) = 0$✓

Step 2: Use the Wronskian formula to find a second solution

For a second-order ODE in standard form $y'' + p(x)y' + q(x)y = 0$, if $y_1(x)$ is a known solution, a second linearly independent solution $y_2(x)$ can be found using:

$$y_2(x) = y_1(x) \int \frac{e^{-\int p(x)dx}}{[y_1(x)]^2} dx$$

Here, $p(x) = 0$, so:

$$e^{-\int p(x)dx} = e^0 = 1$$

Thus:

$$y_2(x) = \cos(x) \int \frac{1}{\cos^2(x)} dx$$

Step 3: Evaluate the integral

$$\int \frac{1}{\cos^2(x)} dx = \int \sec^2(x)dx = \tan(x) + C$$

Step 4: Write the second solution

Choosing $C = 0$ for simplicity:

$$y_2(x) = \cos(x) \cdot \tan(x) = \sin(x)$$

Step 5: Verify linear independence

Compute the Wronskian:

$$W(y_1, y_2) = \begin{vmatrix} y_1 & y_2 \\ y_1' & y_2' \end{vmatrix} = \begin{vmatrix} \cos(x) & \sin(x) \\ -\sin(x) & \cos(x) \end{vmatrix} = \cos^2(x) + \sin^2(x) = 1 \neq 0$$

Since $W \neq 0$, $y_1 = \cos(x)$ and $y_2 = \sin(x)$ are linearly independent.

Step 6: General solution

The general solution is:

$$y(x) = C_1 \cos(x) + C_2 \sin(x)$$

as obtained in Example 12.2.

12.3.1 General Procedure for Power-Series Substitution

Although writing an expanded series is straightforward for simple equations, the process becomes cumbersome for more complicated differential equations. We illustrate a systematic approach below by revisiting Example 12.1.

We write the solution in the form

$$y(x) = \sum_{n=0}^{\infty} a_n x^n. \tag{12.9}$$

Differentiating term by term,

$$y'(x) = \sum_{n=1}^{\infty} n\, a_n x^{n-1}, \tag{12.10}$$

Substitute (12.9)–(12.10) into

$$y' - y = 0.$$

Thus,

$$\sum_{n=1}^{\infty} n\, a_n x^{\,n-1} - \sum_{n=0}^{\infty} a_n x^n = 0. \tag{12.11}$$

In order to have the same exponent in the two terms, we set $n' = n - 1$ in the first term, so that $n = n' + 1$. The first term therefore becomes

$$\sum_{n'=0}^{\infty} (n' + 1) a_{n'+1} x^{n'}$$

Since n' is just an index to count the terms, we might as well replace it with n. Equation (12.11) can now be rewritten as

$$\sum_{n=0}^{\infty} [(n + 1)a_{n+1} - a_n]x^n = 0.$$

Equating coefficients yields the recursion

$$a_{n+1} = \frac{a_n}{n + 1}.$$

Recurrence relations allow the construction of the full series from the initial coefficient(s). The same procedure extends to second-order equations, as shown next.

12.3.2 General Method for Second-order Equations

Differentiate (12.10) again:

$$y''(x) = \sum_{n=2}^{\infty} n(n - 1)a_n x^{n-2}.$$

Substitute into the given second-order equation, shift indices to match powers of x, and equate coefficients term by term. This always yields a recurrence relation of the form

$$a_{n+k} = f(n)\, a_n,$$

with $k = 1$ or $k = 2$, depending on the structure of the equation.

We illustrate this in the following examples, using two cases, $y'' - k^2y = 0$ (Example 12.4), and the harmonic oscillator equation (Example 12.5).

Example 12.4: Solve $y'' - k^2y = 0$ by Power Series

Objective: To solve $y'' - k^2y = 0$ by power series and demonstrate how exponential/hyperbolic functions arise from even–odd recurrence.
Method Used: Derive the recurrence relation and separate the even and odd coefficient chains.
Problem: Solve the second-order homogeneous differential equation

$$y''(x) - k^2y(x) = 0,$$

where k is a real constant, by the power-series method.

Solution:
Substitute

$$y(x) = \sum_{n=0}^{\infty} a_n x^n,\ y''(x) = \sum_{n=2}^{\infty} n(n-1)a_n x^{n-2}.$$

Then

$$\sum_{n=2}^{\infty} n(n-1)a_n x^{n-2} - k^2 \sum_{n=0}^{\infty} a_n x^n = 0.$$

To match powers of x, set $n' = n - 2$ in the first sum, so $n = n' + 2$

$$\sum_{n=0}^{\infty} (n+2)(n+1)a_{n+2}x^n - k^2 \sum_{n=0}^{\infty} a_n x^n = 0.$$

Thus,

$$(n+2)(n+1)a_{n+2} - k^2 a_n = 0,$$

or

$$a_{n+2} = \frac{k^2}{(n+2)(n+1)} a_n.$$

This produces:

- an **even-power series** from a_0,
- an **odd-power series** from a_1.

Hence,

$$y(x) = a_0\left(1 + \frac{k^2x^2}{2!} + \frac{k^4x^4}{4!} + \cdots\right) + a_1\left(x + \frac{k^2x^3}{3!} + \frac{k^4x^5}{5!} + \cdots\right),$$

The even series is $\cosh(kx)$ and the odd series is $\sinh(kx)/k$, so the general solution is

$$y(x) = a_0 \cosh(kx) + \frac{a_1}{k} \sinh(kx).$$

Compare with Example 12.2: when $k^2 < 0$ (i.e., $k = i\omega$), we get oscillatory solutions (sine/cosine), while $k^2 > 0$ gives exponential growth/decay.

Concept Insight 12.1: Even–Odd Separation

In many second-order ODEs, the recurrence relation links each coefficient only to the one two places ahead, a_{n+2} in terms of a_n. This automatically divides the solution into two independent series: one built from the even coefficients and the other from the odd ones. The even series typically leads to cosine-like or symmetric functions, while the odd series produces sine-like or antisymmetric solutions. This pattern reappears in Legendre, Hermite, and Laguerre equations and is the key to obtaining two linearly independent solutions.

Derivation Note—Why the Recurrence Relation Jumps by Two Indices

Substituting a power-series expansion into a second-order differential equation produces terms in x^n coming from both y' and y''. Aligning these powers requires shifting indices so that all terms start at $n = 0$. Because differentiation reduces the power by one or two, all terms eventually collect into powers x^n and x^{n+2}, which is why the recurrence relation links a_{n+2} with a_n. This structure is characteristic of second-order linear equations with analytic coefficients.

Example 12.5: Generalized Harmonic Oscillator Equation

Objective: To solve the parametric harmonic oscillator equation $y'' + \omega^2 y = 0$ and show how the frequency parameter ω affects the solution structure, connecting to physical oscillatory systems.
Method Used: Derive the recurrence relation and show how the parameter ω scales the argument of the trigonometric solutions.
Problem: Solve the generalized harmonic oscillator equation

$$y'' + \omega^2 y = 0,$$

where ω is a real constant representing the angular frequency.

Solution:

Assume a power-series solution: $y = \sum_{n=0}^{\infty} a_n x^n$.
Then $y'' = \sum_{n=2}^{\infty} n(n-1)a_n x^{n-2}$.
Substitute into the equation:

$$\sum_{n=2}^{\infty} n(n-1)a_n x^{n-2} + \omega^2 \sum_{n=0}^{\infty} a_n x^n = 0.$$

Shift the index in the first sum (let $n' = n - 2$, so $n = n' + 2$):

$$\sum_{n=0}^{\infty} (n+2)(n+1)a_{n+2} x^n + \omega^2 \sum_{n=0}^{\infty} a_n x^n = 0.$$

Equating coefficients of x^n:

$$\boxed{a_{n+2} = -\frac{\omega^2}{(n+2)(n+1)} a_n.}$$

This recurrence relation generates:
- an **even series** beginning with a_0,
- an **odd series** beginning with a_1.

Even series ($a_0 \neq 0, a_1 = 0$):

$$a_2 = -\frac{\omega^2}{2!}a_0,\ a_4 = -\frac{\omega^2}{4 \cdot 3}a_2 = \frac{\omega^4}{4!}a_0,\ a_6 = -\frac{\omega^6}{6!}a_0, \ldots$$

$$y_{\text{even}}(x) = a_0\left(1 - \frac{(\omega x)^2}{2!} + \frac{(\omega x)^4}{4!} - \frac{(\omega x)^6}{6!} + \cdots\right) = a_0 \cos(\omega x).$$

Odd series ($a_0 = 0, a_1 \neq 0$):

$$a_3 = -\frac{\omega^2}{3!}a_1,\ a_5 = \frac{\omega^4}{5!}a_1,\ a_7 = -\frac{\omega^6}{7!}a_1, \ldots$$

$$y_{\text{odd}}(x) = a_1/\omega\left(x - \frac{(\omega x)^3}{3!} + \frac{(\omega x)^5}{5!} - \cdots\right) = \frac{a_1}{\omega}\sin(\omega x).$$

Thus, the general solution is:

$$\boxed{y(x) = C_1 \cos(\omega x) + C_2 \sin(\omega x),}$$

where $C_1 = a_0$ and $C_2 = a_1/\omega$.

Physical Interpretation:

- The parameter ω determines the **frequency of oscillation**: larger ω means faster oscillation.
- In classical mechanics, for a spring-mass system: $\omega = \sqrt{k/m}$, where k is the spring constant and m is the mass.
- In quantum mechanics, this equation describes the harmonic oscillator potential with characteristic frequency ω.

Connection to Example 12.4: Note that if we replace ω^2 by $-k^2$(i.e., $\omega^2 < 0$), the recurrence relation becomes

$$a_{n+2} = \frac{k^2}{(n+2)(n+1)}a_n,$$

with **positive** coefficients, leading to the exponential/hyperbolic solutions of Example 12.4:

$$y(x) = C_3 \cosh(kx) + C_4 \sinh(kx).$$

This illustrates a fundamental principle: **the sign of the coefficient determines whether solutions are oscillatory (trigonometric) or exponential (hyperbolic).**

The solution to Example 12.5 can also be equivalently written as

$$x(t) = A\exp(i\omega t) + B\exp(-i\omega t)$$

using Euler's relations. Therefore, a second-order linear homogeneous equation of the type $y''(x) + \lambda y(x) = 0$ has oscillatory solutions (in terms of sine and cosine or complex exponential) when $\lambda > 0$ and non-oscillatory solutions if $\lambda < 0$. This result will be important for us in physical problems, where we seek oscillatory solutions. In case $\lambda = 0$, the differential equation reduces to $y'' = 0$, whose solution is a linear function of the type $Ax + B$.

Key Takeaways: Power-Series Method

- Works when $p(x)$ and $q(x)$ are analytic at the expansion point
- Substitute $y = \sum a_n x^n$, differentiate termwise
- Shift indices to align powers of x
- Equate coefficients → recurrence relation
- Even/odd separation often gives two independent solutions

12.4 Classification of Points

The power-series method discussed above seeks solutions by expanding the function about a point, usually $x = 0$. However, this method works only when the coefficients $p(x)$ and $q(x)$ in Equation (12.1) are **analytic** at the expansion point, typically $x = 0$.

Consider the standard second-order linear ODE (12.1),

$$y'' + p(x)y' + q(x)y = 0.$$

A point $x = x_0$ is called an **ordinary point** if both $p(x)$ and $q(x)$ are analytic at x_0. If either coefficient is nonanalytic (diverges or becomes indeterminate), then x_0 is a **singular point**.

12.4.1 Example: Euler–Cauchy Equation

Consider the Euler–Cauchy form:

$$x^2y'' + axy' + by = 0.$$

To compare with (12.1), divide through by x^2:

$$y'' + \frac{a}{x}y' + \frac{b}{x^2}y = 0.$$

Here,

$$p(x) = \frac{a}{x}, \quad q(x) = \frac{b}{x^2}.$$

Both $p(x)$ and $q(x)$ diverge at $x = 0$; therefore $x = 0$ is a **singular point**.

12.4.2 Regular Versus Irregular Singular Points

Even if a point is singular, it may still be possible to obtain a power-series-type solution (generalized series). We examine the behavior of

$$(x - x_0)p(x), \quad (x - x_0)^2 q(x).$$

- If both are **analytic** at x_0, then x_0 is a **regular singular point**.
- If either is **nonanalytic**, the point is an **irregular (essential) singular point**.

At regular singular points, the **Frobenius method** may be applied to find solutions of the form

$$y(x) = \sum_{n=0}^{\infty} a_n (x - x_0)^{n+r},$$

where r is determined from the **indicial equation**.

Note–Classification of Points for ODEs:

Ordinary Point: Both $p(x)$ and $q(x)$ are analytic at x_0.
Regular Singular Point: x_0 is a singular point, but both $(x - x_0)p(x)$ and $(x - x_0)^2 q(x)$ are analytic at x_0.
Irregular Singular Point (Essential Singularity): x_0 is a singular point, and it is not regular.

Concept Insight 12.2–Why We Classify Points

The behavior of a differential equation near a point depends on whether its coefficients remain finite or blow up. At an ordinary point, both coefficients are analytic, and a simple power series always works. At a singular point, however, the function may still admit a controlled series solution, but only if the singularity is mild—this is the idea of a *regular* singular point. The Frobenius method extends ordinary power-series techniques precisely to handle such cases.

Table 12.1 summarizes the classification of points.

12.4.3 Bullet Points for Quick Reference

- **Ordinary Point:**
  Both $p(x)$ and $q(x)$ are finite and differentiable at x_0.
  → Use standard power series $y(x) = \sum_{n=0}^{\infty} a_n (x - x_0)^n$.
- **Regular Singular Point:**
  At least one of $p(x)$ or $q(x)$ diverges, but $(x - x_0)p(x)$ and $(x - x_0)^2 q(x)$ remain finite and analytic.
  → Use Frobenius method $y(x) = \sum_{n=0}^{\infty} a_n (x - x_0)^{n+r}$ with r determined by an indicial equation.
- **Irregular Singular Point:**
  The conditions for a regular singular point fail.
  → Solutions may be more difficult to obtain and often require asymptotic or special-function methods.

In Figure 12.1, we show a decision tree for solving linear second-order differential equations, depending on the nature of the expansion point.

Table 12.1 Classification of points in linear second-order ODEs (use to select power-series vs. Frobenius method).

Consider the standard form of a linear second-order ODE:

$$y''(x) + p(x)y'(x) + q(x)y(x) = 0$$

Point type	Condition(s)	Example ODE
ordinary point	$p(x)$ and $q(x)$ are **analytic** (expressible as convergent power series) at $x = x_0$.	$y'' + y = 0$ at $x = 0$ (since $p = 0, q = 1$ are constants)
Regular singular point	$(x - x_0)p(x)$ and $(x - x_0)^2 q(x)$ are analytic at $x = x_0$.	Legendre's equation $(1 - x^2)y'' - 2xy' + l(l + 1)y = 0$ at $x = \pm 1$ (see Section 12.5.3)
Irregular singular point	At least one of $(x - x_0)p(x)$ or $(x - x_0)^2 q(x)$ is **not analytic** at $x = x_0$.	$x^4 y'' + y = 0$ at $x = 0$ (since $xp(x) = 0$ is analytic, but $x^2 q(x) = x^2/x^4 = 1/x^2$ is **not** analytic at $x = 0$).

Notes:

- An **analytic function** can be expanded in a Taylor series about the point.
- At a **regular singular point**, the Frobenius method can be used to find series solutions.
- At an **irregular singular point**, series solutions may not exist or may require more advanced methods.
- A point that is **not ordinary** is called a **singular point**.

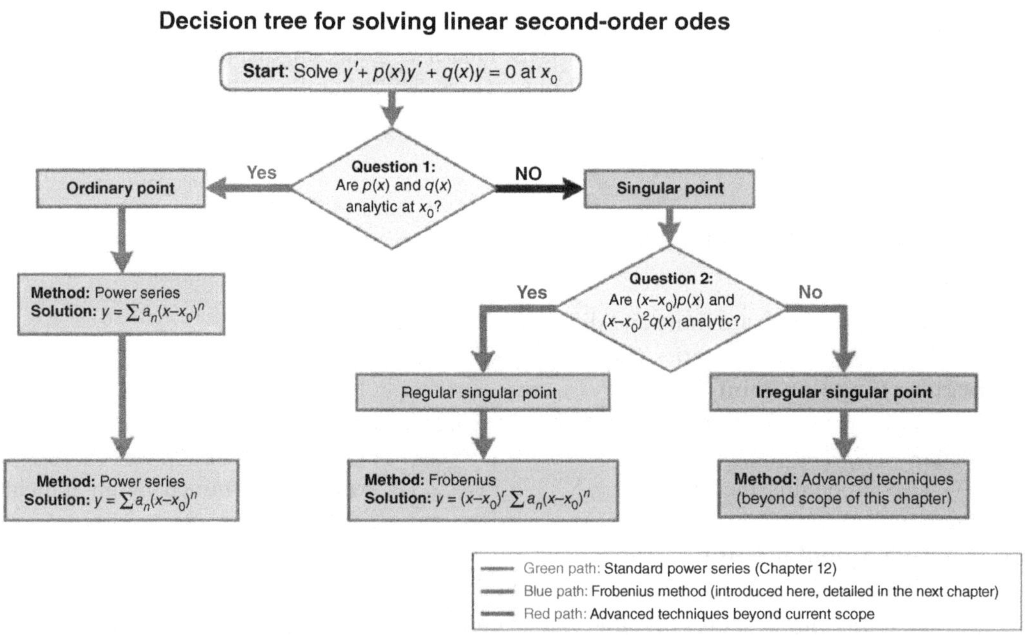

Figure 12.1 Algorithm for solving linear second-order ODEs (follow based on point classification). For a color version of this figure, please see the electronic version of this book.

12.5 Case Study: Legendre's Equation

During the solution of the rigid rotator model or the hydrogen atom, the process of separation of variables leads to the emergence of the angular equation in θ.

$$(1-x^2)y''(x) - 2xy'(x) + \lambda y = 0$$

Here, the substitution $x = \cos\theta$ is made, where λ denotes the separation constant. This is the **Legendre equation**.

Written in the standard form (12.1), this equation becomes

$$y''(x) - \frac{2x}{1-x^2}y'(x) + \frac{\lambda}{1-x^2}y = 0$$

The presence of the term $(1-x^2)$ in the denominator of $p(x)$ and $q(x)$ shows that the equation is divergent at the points $x = \pm 1$. However, $x = 0$ is a regular point since both $p(x)$ and $q(x)$ are analytic at that point. This means that the series solution method is valid for expansion about $x = 0$.

Math Box 12.2: Frobenius-Style Series Expansion About $x = 0$

Although Legendre's equation has singular points at $x = \pm 1$, the point $x = 0$ is **ordinary**, and an ordinary power series is sufficient. Nevertheless, beginning with a Frobenius-type expansion clarifies why no singular exponent appears and connects the discussion with equations that genuinely require the Frobenius method.

We assume a general Frobenius form

$$y = x^r \sum_{n=0}^{\infty} a_n x^n,\ a_0 \neq 0.$$

Substituting into

$$(1-x^2)y'' - 2xy' + \lambda y = 0$$

and collecting the lowest power of x gives the **indicial equation**

$$r(r-1) = 0.$$

Thus, the two exponents are:

$$\boxed{r = 0, 1}.$$

The detailed derivation is provided in Appendix J12. Thus, the two **Frobenius exponents** are $r = 0$ and $r = 1$; since the point is ordinary, no logarithmic term appears.

At an ordinary point, the Frobenius method always reduces to an ordinary power series with $r = 0$; nonzero r and logarithmic behavior arise only at regular singular points.

Teacher's Note/Method Reminder for Students–When to Use Power Series Versus Frobenius

In practice, students often struggle with *when* to use a plain power series and *when* a Frobenius expansion is needed. The decision is entirely dictated by the nature of the expansion point for the standard form

$$y'' + p(x)y' + q(x)y = 0.$$

- **Step 1: Classify the point.**
  - If $p(x)$ and $q(x)$ are analytic at x_0, then x_0 is an ordinary point. A simple Taylor/Maclaurin series $y = \sum a_n(x - x_0)^n$ is guaranteed to work.
  - If at least one of p, q diverges at x_0, it is a singular point. Then check $(x - x_0)p(x)$ and $(x - x_0)^2 q(x)$:
  - If both are analytic at x_0, it is a regular singular point→Frobenius method with $y = (x - x_0)^r \sum a_n(x - x_0)^n$.
  - Otherwise, it is an irregular singular point and neither basic power series nor simple Frobenius is usually sufficient.
- **Step 2: Explain that Frobenius at an ordinary point collapses to Taylor.**
  When this classification is done in front of students, stress that at an ordinary point the Frobenius ansatz *always* yields $r = 0$ (and possibly $r = 1$ for a second solution), so one recovers an ordinary power series. In Math Box 12.2, the Frobenius form at $x = 0$ for Legendre's equation is used deliberately as a bridge to later regular singular problems, not because it is actually required there.
- **Step 3: Tie the choice of method to physics.**
  Emphasize that the *mathematical* method (Taylor vs Frobenius) is chosen from the classification, but the *physical* acceptability of a solution is decided by boundary conditions. For Legendre's equation the radius of convergence is set by the singularities at $x = \pm 1$, and demanding finiteness at these endpoints forces series truncation and hence quantized $\lambda = l(l + 1)$. The same "classification → series → truncation → quantization" pattern reappears later for Hermite and Laguerre equations in the next chapter.

We therefore proceed to solve Legendre's equation using the **series solution method**.

Example 12.6: Legendre Equation–Series Solution

Objective: To derive the recurrence relation for Legendre's equation and identify the two independent series solutions.
Method Used: Apply the power-series substitution and analyze the coefficient recursion to reveal the structure of the Legendre series.
Problem: Solve Legendre's equation

$$(1 - x^2)y''(x) - 2xy'(x) + \lambda y = 0$$

by the power-series method.

Solution:
Substitution of the power series in the given equation yields

$$(1-x^2)\sum_{n=2}^{\infty} n(n-1)a_n x^{n-2} - 2x\sum_{n=1}^{\infty} n a_n x^{n-1} + \lambda \sum_{n=0}^{\infty} a_n x^n = 0$$

which, on simplification gives

$$\sum_{n=2}^{\infty} n(n-1)a_n x^{n-2} - \sum_{n=1}^{\infty} n(n-1)a_n x^n - 2\sum_{n=0}^{\infty} n a_n x^n + \lambda \sum_{n=0}^{\infty} a_n x^n = 0$$

We set $n' = n - 2$ in the first term and then replace n' by n to give

$$\sum_{n=0}^{\infty}(n+2)(n+1)a_{n+2}x^n - \sum_{n=1}^{\infty} n(n-1)a_n x^n - 2\sum_{n=0}^{\infty} n a_n x^n + \lambda \sum_{n=0}^{\infty} a_n x^n = 0$$

resulting in the recurrence relation ($n \geq 0$)

$$a_{n+2} = \frac{n(n-1) + 2n - \lambda}{(n+2)(n+1)} a_n = \frac{n(n+1) - \lambda}{(n+2)(n+1)} a_n$$

This gives two independent series: an even series starting with a_0 and an odd series starting with a_1.

As we found in Example 12.6, the power-series method gives two infinite series solutions, one with even powers and the other with odd. Both series converge for $|x| < 1$, but generally **diverge at the endpoints** $x = \pm 1$. In physical problems where the domain includes these boundary points (such as when $x = \cos\theta$ in spherical coordinates), we require solutions to remain finite everywhere on $[-1, 1]$.

This physical **boundary condition** can only be satisfied if we choose λ such that one of the two series **terminates** in a polynomial. As shown later in Section 12.5.2, this occurs when

$\lambda = l(l+1)$, $l = 0, 1, 2, \ldots$ giving the recurrence relation (12.13). The resulting polynomial solutions are the **Legendre polynomials** $P_l(x)$.

For the even series ($a_1 = 0, a_0 \neq 0$):

From the recurrence relation:

$$a_{n+2} = \frac{n(n+1) - \lambda}{(n+2)(n+1)} a_n$$

$$a_2 = \frac{0(1) - \lambda}{2 \cdot 1} a_0 = -\frac{\lambda}{2} a_0$$

$$a_4 = \frac{2(3) - \lambda}{4 \cdot 3} a_2 = \frac{6 - \lambda}{12} \cdot \left(-\frac{\lambda}{2} a_0\right) = -\frac{\lambda(6-\lambda)}{24} a_0$$

$$a_6 = \frac{4(5) - \lambda}{6 \cdot 5} a_4 = \frac{20 - \lambda}{30} \cdot \left(-\frac{\lambda(6-\lambda)}{24} a_0\right) = -\frac{\lambda(6-\lambda)(20-\lambda)}{720} a_0$$

Therefore, the even series is:

$$y_{\text{even}}(x) = a_0\left[1 - \frac{\lambda}{2}x^2 - \frac{\lambda(6-\lambda)}{24}x^4 - \frac{\lambda(6-\lambda)(20-\lambda)}{720}x^6 - \cdots\right]$$

For the odd series ($a_0 = 0$, $a_1 \neq 0$):

$$a_3 = \frac{1(2)-\lambda}{3\cdot 2}a_1 = \frac{2-\lambda}{6}a_1$$

$$a_5 = \frac{3(4)-\lambda}{5\cdot 4}a_3 = \frac{12-\lambda}{20}\cdot\frac{2-\lambda}{6}a_1 = \frac{(2-\lambda)(12-\lambda)}{120}a_1$$

$$a_7 = \frac{5(6)-\lambda}{7\cdot 6}a_5 = \frac{30-\lambda}{42}\cdot\frac{(2-\lambda)(12-\lambda)}{120}a_1 = \frac{(2-\lambda)(12-\lambda)(30-\lambda)}{5040}a_1$$

Therefore, the odd series is:

$$y_{\text{odd}}(x) = a_1\left[x + \frac{2-\lambda}{6}x^3 + \frac{(2-\lambda)(12-\lambda)}{120}x^5 + \frac{(2-\lambda)(12-\lambda)(30-\lambda)}{5040}x^7 + \cdots\right]$$

Observation: Notice that:

- For the **even series**, the numerator contains factors: λ, $(6-\lambda)$, $(20-\lambda)$, ...
  - These correspond to $0(1), 2(3), 4(5), \ldots = n(n+1)$ for even n
- For the **odd series**, the numerator contains factors: $(2-\lambda)$, $(12-\lambda)$, $(30-\lambda)$, ...
  - These correspond to $1(2), 3(4), 5(6), \ldots = n(n+1)$ for odd n

Key insight: If $\lambda = l(l+1)$ for some non-negative integer l, then:

- If l is **even**, one of the factors in the even series vanishes → series terminates.
- If l is **odd**, one of the factors in the odd series vanishes → series terminates.

This is precisely the mechanism that produces the Legendre polynomials!

Concept Insight 12.3–Even–Odd Structure of Legendre Polynomials

Legendre's equation is invariant under the transformation $x \mapsto -x$, so its solutions naturally split into even and odd functions. This symmetry causes the recurrence relation to separate into two independent sequences of coefficients—one generating even-degree polynomials and the other generating odd-degree polynomials. As a result, $P_0, P_2, P_4, \ldots$ are even functions, while $P_1, P_3, P_5, \ldots$ are odd.

12.5.1 Boundary Behavior and Divergence

In many physical problems (e.g., the rigid rotator and the hydrogen atom), the variable

$$x = \cos\theta,$$

and the points $x = \pm 1$ correspond to the poles $\theta = 0, \pi$. A physically meaningful solution cannot diverge at these points. Thus, a boundary condition is imposed: **The solution must remain finite at $x = \pm 1$.**

For Legendre's equation, the non-terminating series solutions behave like the Legendre functions of the second kind, which diverge at $x = \pm 1$. Specifically, the non-terminating series $Q_l(x)$ exhibits logarithmic divergence $Q_l(x) \sim \ln|1 - x|$ as $x \to 1^-$. Imposing finiteness therefore forces one of the series to terminate in a polynomial. This **truncation** produces the **Legendre polynomials, $P_l(x)$** (Figure 12.2).

12.5.2 Truncation → Quantization

Let us examine the recurrence relation we derived in Example 12.6:

$$a_{n+2} = \frac{n(n+1) - \lambda}{(n+1)(n+2)} a_n$$

This relation shows that the series will terminate if and only if the numerator becomes zero for some non-negative integer n. Let us call this specific integer l.

$$l(l+1) - \lambda = 0$$

This gives us the crucial condition for **physically acceptable** solutions:

$$\lambda = l(l+1), \text{ where } l = 0, 1, 2, \ldots$$

By setting λ equal to this value, the numerator becomes zero when $n = l$. This forces a_{l+2} to be zero, and subsequently all higher coefficients ($a_{l+4}, a_{l+6}, \ldots$) also become zero.

This truncation converts *one* of the two series (either the even or the odd one) into a polynomial of degree l. To ensure the *entire* solution remains finite, we must also discard the *other*, non-terminating series by setting its arbitrary constant to zero (e.g., if l is even, a_1 is set to 0, discarding the odd series). Similarly, if l is odd, a_0 is set to 0, discarding the even series.

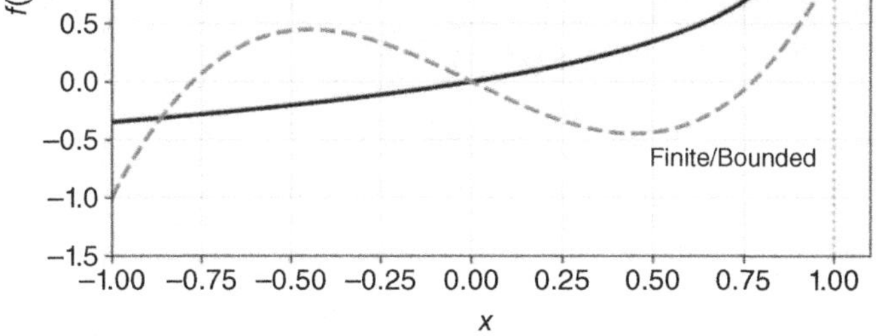

Figure 12.2 Behavior of a non-terminating Legendre series near $x = 1$. The function diverges as $x \to 1$, whereas a truncated polynomial solution remains finite. Physical boundary conditions, therefore, force termination of the series.

Fact Box 12.1: Quantization Condition from Boundary Requirements

Finite and physically meaningful solutions arise only when the series terminates.
If the numerator of the recurrence relation vanishes at $n = l$, that is,

$$\lambda = l(l+1), \quad l = 0, 1, 2, \ldots,$$

all higher coefficients become zero and the solution reduces to a polynomial of degree l, the Legendre polynomial $P_l(x)$.

This truncation condition is the mathematical origin of **quantized values of l** in quantum mechanics.

The same pattern appears in the rigid rotator, electronic angular momentum, and other systems where boundary conditions enforce discrete allowed states.

Why Finiteness Matters in Quantum Mechanics

In the rigid rotator or hydrogen atom, $x = \cos\theta$ represents the angular coordinate. The points $x = \pm 1$ correspond to $\theta = 0, \pi$ (the poles). A **physically acceptable wavefunction must be finite at all points in space**, including these poles. This boundary condition selects only the terminating series solutions, which occur when $\lambda = l(l+1)$ for integer l—mathematically generating the **quantization of angular momentum.**

From Divergence to Quantization: The Mathematical Mechanism

1. **Series solution**→ recurrence $a_{n+2} = \dfrac{n(n+1)-\lambda}{(n+2)(n+1)} a_n$
2. **For large n**: $a_{n+2} \approx a_n$→ series behaves roughly like a series with coefficients of comparable magnitude, which typically diverges at $|x| = 1$.
3. **To avoid divergence**: numerator must vanish for some $n = l$.
4. **Condition**: $\lambda = l(l+1)$ where $l = 0, 1, 2, \ldots$.
5. **Result**: Series truncates to polynomial → **quantized l values.**

This pattern—boundary conditions → series truncation → discrete parameters—reappears throughout quantum mechanics (harmonic oscillator, hydrogen atom, etc.).

12.5.3 Series Solution

With this physically required condition, Legendre's equation is now written in its familiar form:

$$(1-x^2)y'' - 2xy' + l(l+1)y = 0. \tag{12.12}$$

To write it in the standard form, divide by $1 - x^2$:

$$y'' - \frac{2x}{1-x^2} y' + \frac{l(l+1)}{1-x^2} y = 0.$$

Thus,

$$p(x) = -\frac{2x}{1-x^2}, \quad q(x) = \frac{l(l+1)}{1-x^2}.$$

Inspection shows:

- At $x = 0$: both $p(x)$ and $q(x)$ are analytic → **ordinary point**.
- At $x = \pm 1$: both coefficients diverge → **singular points**.

Check regularity at $x = 1$:

$$(x-1)p(x) = -(x-1)\frac{2x}{1-x^2} = -\frac{2x(x-1)}{(1-x)(1+x)} = \frac{2x}{1+x},$$

which is analytic near $x = 1$.
Similarly,

$$(x-1)^2 q(x) = (x-1)^2 \frac{l(l+1)}{(1-x)(1+x)} = -\frac{l(l+1)(x-1)}{1+x},$$

also analytic.
For $x = -1$,

$$(x+1)p(x) = -(x+1)\frac{2x}{1-x^2} = -\frac{2x(x+1)}{(1-x)(1+x)} = -\frac{2x}{1-x},$$

and

$$(x+1)^2 q(x) = (x+1)^2 \frac{l(l+1)}{(1-x)(1+x)} = \frac{l(l+1)(x+1)}{1-x},$$

are analytic near $x = -1$.
Thus, **both** $x = 1$ **and** $x = -1$ **are regular singular points**.

It can also be shown (using variable transformations) that the equation has a regular singular point at $x = \infty$.

Since $x = 0$ is an ordinary point, we may use the **power-series method** to obtain two independent solutions and then impose physical conditions (finiteness at $x = \pm 1$) to obtain the **Legendre polynomials**.

Concept Insight 12.4—Power-Series Expansion

Since one of the exponents is $r = 0$, the Frobenius method reduces to an **ordinary power-series expansion** about the origin.

No singular behavior arises at $x = 0$. This explains why Legendre polynomials are usually developed from a simple series and why the exponent r disappears from subsequent work.

We therefore now set $r = 0$ and proceed with a standard power-series derivation.

Since $x = 0$ is an ordinary point, we look for solutions of the form

$$y(x) = \sum_{n=0}^{\infty} a_n x^n.$$

Substitution of the power series into (12.12) yields

$$(1-x^2)\sum_{n=2}^{\infty} n(n-1)a_n x^{n-2} - 2x\sum_{n=1}^{\infty} n a_n x^{n-1} + l(l+1)\sum_{n=0}^{\infty} a_n x^n = 0$$

which, on simplification, gives

$$\sum_{n=2}^{\infty} n(n-1)a_n x^{n-2} - \sum_{n=1}^{\infty} n(n-1)a_n x^n - 2\sum_{n=0}^{\infty} n a_n x^n + l(l+1)\sum_{n=0}^{\infty} a_n x^n = 0.$$

We set $n' = n - 2$ in the first term and then replace n' by n to give

$$\sum_{n=0}^{\infty}(n+2)(n+1)a_{n+2}x^n - \sum_{n=1}^{\infty} n(n-1)a_n x^n - 2\sum_{n=0}^{\infty} n a_n x^n + l(l+1)\sum_{n=0}^{\infty} a_n x^n = 0,$$

resulting in the recurrence relation ($n \geq 0$)

$$a_{n+2} = \frac{n(n-1) + 2n - l(l+1)}{(n+2)(n+1)} a_n = \frac{n(n+1) - l(l+1)}{(n+2)(n+1)} a_n.$$

Hence, the recurrence relation is

$$a_{n+2} = \frac{n(n+1) - l(l+1)}{(n+2)(n+1)} a_n, n \geq 0. \tag{12.13}$$

This relation gives **two independent series**:

- one with **even** powers of x, from a_0
- one with **odd** powers of x, from a_1

Both series converge for $|x| < 1$ but generally diverge at $x = \pm 1$, as discussed in Section 12.5.1.

12.5.3.1 Radius of Convergence

We can determine the radius of convergence of the Legendre series using the **ratio test**. For large n, the recurrence relation (12.13) gives:

$$\frac{a_{n+2}}{a_n} = \frac{n(n+1) - l(l+1)}{(n+2)(n+1)} \approx \frac{n^2}{n^2} = 1 \text{ as } n \to \infty.$$

For a power series $\sum a_n x^n$, the ratio test examines:

$$\lim_{n\to\infty}\left|\frac{a_{n+2}x^{n+2}}{a_n x^n}\right| = \lim_{n\to\infty}\left|\frac{a_{n+2}}{a_n}\right| \cdot |x|^2 = |x|^2.$$

The series converges when this limit is less than 1:

$$|x|^2 < 1 \Rightarrow |x| < 1.$$

Thus, the **radius of convergence is** $R = 1$.

This result is not accidental: it reflects the location of the **singular points** of Legendre's equation at $x = \pm 1$. A power series about $x = 0$ can converge at most up to the nearest singularity, which in this case is at distance 1 from the origin.

12.5.3.2 Condition for Polynomial Solutions

The series will **terminate** (i.e., become a polynomial) if the numerator in Equation (12.13) becomes zero:

$$n(n+1) - l(l+1) = 0.$$

This occurs when

$$l = 0, 1, 2, \ldots$$

and specifically, when the index $n = l$.

Then $a_{l+2} = 0$, and all higher coefficients also vanish.
Thus, **Legendre polynomials arise when l is a non-negative integer**.
These polynomials are finite everywhere on $[-1, 1]$ and hence physically acceptable.
If we choose

$$a_0 = 1, \quad a_1 = 0,$$

we obtain the **even solution**,

$$y_1(x) = 1 - \frac{l(l+1)}{2!}x^2 + \frac{l(l+1)(l-2)(l+3)}{4!}x^4 - \cdots, \tag{12.14}$$

while choosing

$$a_0 = 0, \quad a_1 = 1,$$

gives the **odd solution**,

$$y_2(x) = x - \frac{l(l+1)-2}{3!}x^3 + \frac{(l-1)(l+2)(l(l+1)-12)}{5!}x^5 - \cdots. \tag{12.15}$$

Both series converge for $|x| < 1$; the radius of convergence is unity.
Thus, the general solution is

$$y(x) = C_1 y_1(x) + C_2 y_2(x), \; |x| < 1. \tag{12.16}$$

The series solution is only valid within the radius of convergence, which is determined by the nearest singularity of $p(x)$ or $q(x)$.

In Example 12.7, the expression for $P_2(x)$ is derived from the recurrence relations.

Example 12.7: Constructing $P_2(x)$ from the Recurrence Relation

Objective: To explicitly demonstrate how setting $l = 2$ in the Legendre recurrence relation produces the polynomial $P_2(x) = \frac{1}{2}(3x^2 - 1)$.
Method Used: Apply the recurrence relation (12.13) with $l = 2$, show where truncation occurs, and normalize the result.
Problem: Use the recurrence relation for Legendre's equation to construct $P_2(x)$.

Solution:

From Equation (12.13), the recurrence relation is:

$$a_{n+2} = \frac{n(n+1) - l(l+1)}{(n+2)(n+1)} a_n.$$

For $l = 2$, we have $l(l+1) = 2(3) = 6$.

Step 1: Even series (set $a_1 = 0$, start with a_0)

For $n = 0$:

$$a_2 = \frac{0(1) - 6}{2 \cdot 1} a_0 = \frac{-6}{2} a_0 = -3a_0.$$

For $n = 2$:

$$a_4 = \frac{2(3) - 6}{4 \cdot 3} a_2 = \frac{6 - 6}{12} a_2 = 0.$$

The series terminates! Since $a_4 = 0$, all subsequent even coefficients vanish: $a_6 = a_8 = \cdots = 0$.

The even solution is:

$$y_{\text{even}}(x) = a_0(1 - 3x^2).$$

Step 2: Odd series (set $a_0 = 0$, start with a_1)

For $n = 1$:

$$a_3 = \frac{1(2) - 6}{3 \cdot 2} a_1 = \frac{-4}{6} a_1 = -\frac{2}{3} a_1.$$

For $n = 3$:

$$a_5 = \frac{3(4) - 6}{5 \cdot 4} a_3 = \frac{6}{20} a_3 = \frac{3}{10} \cdot \left(-\frac{2}{3} a_1\right) = -\frac{1}{5} a_1.$$

The odd series **does not terminate** and continues: $a_5 \neq 0$, $a_7 \neq 0$, and so on. This series diverges at $x = \pm 1$, so we **discard it** by setting $a_1 = 0$.

Step 3: Normalize

The physically acceptable solution is the even polynomial:

$$P_2(x) = A(1 - 3x^2),$$

where A is chosen so that $P_2(1) = 1$ (standard normalization):

$$P_2(1) = A(1 - 3) = -2A = 1 \Rightarrow A = -\frac{1}{2}.$$

Therefore:

$$\boxed{P_2(x) = -\frac{1}{2}(1 - 3x^2) = \frac{1}{2}(3x^2 - 1),}$$

Key Observations:

1. **Truncation occurs when the numerator vanishes**: For $n = 2$, we have $n(n+1) = 6 = l(l+1)$, so $a_4 = 0$.
2. **Only one series terminates**: The odd series continues indefinitely and must be discarded for physical acceptability.
3. **The degree matches l**: The polynomial is degree 2, corresponding to $l = 2$.
4. **Normalization is conventional**: The choice $P_l(1) = 1$ is standard but arbitrary.

Concept Insight 12.5 – Radius of Convergence

The power series obtained from a differential equation is valid only up to the nearest singularity of the equation, not necessarily the nearest singularity of the solution. If the coefficients $p(x)$ and $q(x)$ contain a singularity at a point $x = x_s$, then a power series about x_0 converges at most for $|x - x_0| < |x_s - x_0|$. This is why series expansions for Legendre's equation about $x = 0$ converge only on $[-1, 1]$: the singularities at $x = \pm 1$ lie exactly at the boundary of the convergence region.

12.5.4 Legendre Polynomials

Legendre's equation plays a central role in physics and quantum mechanics. It arises whenever a problem possesses spherical symmetry, most notably in the angular part of the Schrödinger equation and in classical potential theory. Its solutions form a sequence of orthogonal polynomials that play a central role in molecular structure, atomic spectroscopy, electrostatics, and quantum mechanics. These functions emerge naturally when the power-series solution of the differential equation terminates, producing finite solutions on the interval $[-1, 1]$. We now examine how these polynomials arise and explore their main properties.

As shown in Example 12.6, the power-series method yields **two solutions**, one containing only even powers of x and the other only odd powers. Although these series converge for $|x| < 1$, both generally **diverge at** $x = \pm 1$.

12.5.5 General Solution for Integer l

When l is a **non-negative integer**, the recurrence relation (12.13) gives

$$a_{l+2} = 0, \tag{12.17}$$

and all higher coefficients vanish, producing a **polynomial** of degree l.

These polynomials are the **Legendre polynomials,** $P_l(x)$.

Legendre's equation can be written concisely as

$$\frac{d}{dx}\left[(1 - x^2)\frac{dP_l}{dx}\right] + l(l + 1)P_l(x) = 0, \tag{12.18}$$

or, equivalently,

$$(1 - x^2)P_l''(x) - 2xP_l'(x) + l(l + 1)P_l(x) = 0. \tag{12.19}$$

In physical applications (e.g., spherical harmonics), the domain is

$$-1 \leq x \leq 1, \quad x = \cos\theta.$$

Legendre polynomials are standardized so that

$$P_l(1) = 1, \quad P_l(-1) = (-1)^l.$$

They satisfy the important **orthogonality relation**:

$$\int_{-1}^{1} P_l(x)P_m(x)dx = \frac{2}{2l + 1}\delta_{lm}.$$

Orthogonality arises because Legendre's equation is a Sturm–Liouville problem (see next chapter) on the interval $[-1, 1]$ with a unit weight function. Here $w(x) = 1$ is the weight function on $[-1, 1]$, so the eigenfunctions $P_l(x)$ with different eigenvalues $l(l + 1)$ are orthogonal under this weight. Solutions of such problems automatically form sets of orthogonal functions when the eigenvalues are different. Since each Legendre polynomial corresponds to a distinct eigenvalue $l(l + 1)$, the polynomials are mutually orthogonal on $[-1, 1]$. This property makes them ideal basis functions for expanding angular distributions and potentials in physical chemistry.

These match the patterns: parity, increasing complexity, alternation of even/odd powers.

Figure 12.3 plots these polynomials.

Table 12.2 First few $P_l(x)$ used repeatedly in later chapters.

$$P_0(x) = 1,$$
$$P_1(x) = x,$$
$$P_2(x) = \frac{1}{2}(3x^2 - 1),$$
$$P_3(x) = \frac{1}{2}(5x^3 - 3x),$$
$$P_4(x) = \frac{1}{8}(35x^4 - 30x^2 + 3).$$

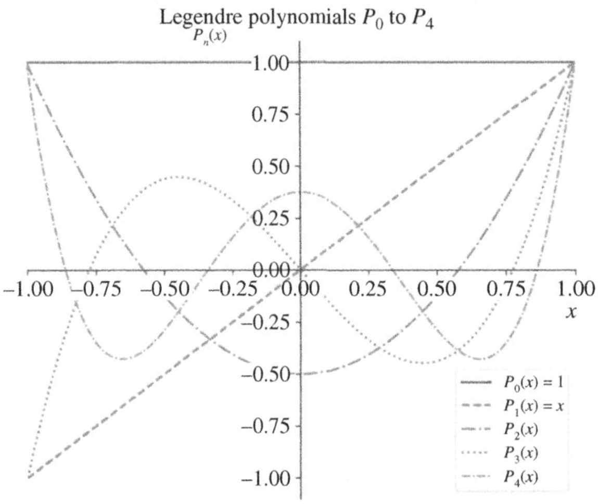

Figure 12.3 Plots of the first few Legendre polynomials (Table 12.2).

According to whether l is an even or odd integer, either $y_1(x)$ or $y_2(x)$ in Equation (12.16) terminates to give a multiple of the corresponding Legendre polynomial $P_l(x)$. In either case, however, the other series does not terminate and therefore converges only for $|x| < 1$. Whereas the above polynomials $P_l(x)$ are called **Legendre functions of the first kind**, the infinite series are called **Legendre functions of the second kind** and are designated as $Q_l(x)$.

The general solution of Legendre's equation for integer l is therefore

$$y_l(x) = c_1 P_l(x) + c_2 Q_l(x)$$

where $P_l(x)$ is a polynomial of order l and so converges for all x and $Q_l(x)$ is an infinite series that converges only for $|x| < 1$. Note that, for odd l, $P_l(x)$ is a polynomial of only odd terms, and $Q_l(x)$ an infinite series containing even terms only and vice versa. Since both solutions are linearly independent, we may use the Wronskian method to obtain one from the other. The Legendre functions of the second kind also have physical applications, but, in this course, we shall be dealing only with those of the first kind, and, therefore, we will study these in some detail in the next chapter.

12.5.5.1 Legendre Functions of the First and Second Kinds

For integer l:

- The truncated polynomial solution is $P_l(x)$: **first kind**.
- The non-terminating series is $Q_l(x)$: **second kind**.

The general solution is

$$y(x) = C_1 P_l(x) + C_2 Q_l(x), \; |x| < 1. \tag{12.20}$$

The $Q_l(x)$ diverge at $x = \pm 1$ and are therefore omitted in many physical problems. This divergence is logarithmic.

12.5.5.2 Obtaining the Second Solution

The functions $Q_l(x)$ arise from the non-terminating series and can also be obtained using the Wronskian method (Equation 12.8). Unlike the polynomials, $Q_l(x)$ diverge logarithmically at $x = \pm 1$:

$$Q_l(x) \sim \ln\left|\frac{1+x}{1-x}\right| \text{ as } x \to \pm 1.$$

In most physical problems involving spherical coordinates (where $x = \cos\theta$ ranges over $[-1, 1]$), boundary conditions require solutions to remain finite at the poles ($\theta = 0, \pi$). Therefore, we set $C_2 = 0$ and retain only the polynomial solutions $P_l(x)$.

The Legendre functions of the second kind have specialized applications in potential theory and scattering problems where the domain excludes the endpoints $x = \pm 1$, but in most quantum mechanical problems, only the polynomials $P_l(x)$ are needed.

⚠ Common Pitfalls

- **Confusing ordinary and singular points.**
  A point where coefficients diverge may still be a *regular* singular point if the Frobenius conditions are satisfied.
- **Assuming a power series always converges everywhere.**
  Its radius of convergence is limited by the nearest singularity.
- **Mixing up the even and odd recursions.**
  Each sequence forms an independent solution; combining them incorrectly leads to errors.
- **Ignoring boundary behavior.**
  A mathematically correct infinite series may diverge at $x = \pm 1$, making it physically unacceptable.
- **Misinterpreting the Wronskian.**
  $W(x_0) = 0$ at a single point does *not* imply linear dependence; it must vanish on an entire interval.
- **Confusing notation.**
  The index l in Legendre polynomials is not a quantum number by itself; it becomes meaningful only in a physical context.

12.5.5.3 Method Reminder: Verifying a Solution

Whenever a function is proposed as a solution of a differential equation, it should be substituted back into the equation to check that all terms cancel. For polynomial solutions such as the Legendre polynomials, this verification is straightforward and confirms both the correctness of the recurrence relation and the validity of the truncation. Cultivating this habit helps detect algebraic errors and ensures that each derived series or polynomial genuinely satisfies the original equation.

The ideas developed in this chapter extend far beyond Legendre's equation. Many of the most important differential equations in quantum mechanics, spectroscopy, and mathematical physics lead to similar power-series solutions and to families of special functions such as Hermite and Laguerre polynomials. In the next chapter, we apply the same methods to these equations and explore their role in vibrational, rotational, and electronic structure problems.

Concept Insight 12.6–Associated Legendre Functions (Preview)

In many physical applications, especially spherical harmonics, Legendre polynomials appear in a generalized form called *associated Legendre functions*, $P_l^m(x)$, where m arises from angular dependence. These functions are generated by differentiating the ordinary Legendre polynomials and therefore share the same mathematical origin, though their detailed properties will be discussed in the next chapter.

Key Takeaways: Truncation Mechanism

1. Series solutions may diverge at domain boundaries.
2. Physical solutions must remain finite everywhere.
3. Finiteness forces the recurrence numerator to vanish.
4. This selects discrete parameter values (quantization).
5. Result: polynomial solutions instead of infinite series.

12.6 Connections to Chemistry and Physics

The methods developed in this chapter appear repeatedly across physical chemistry and quantum mechanics. Legendre polynomials arise in rotational spectroscopy, the angular part of the hydrogen atom, and multipole expansions in electrostatics. Power-series and Frobenius methods underlie the solutions of the harmonic oscillator, Morse potential, and radial Schrödinger equations. Recurrence relations and orthogonality properties form the basis of expansions used in scattering theory, molecular orbital calculations, and vibrational analysis. These mathematical tools therefore provide a unified framework for understanding a wide range of molecular and spectroscopic phenomena.

The methods introduced in this chapter—power series, classification of singular points, and the Wronskian—form the foundation for solving the differential equations of quantum mechanics. We have seen how **boundary conditions force series truncation**, leading to discrete solutions in Legendre's equation. In the next chapter, we will apply this same mathematical toolkit to other key equations:

- **Hermite's equation** $y'' - 2xy' + 2ny = 0$ (quantum harmonic oscillator)
- **Laguerre's equation** $xy'' + (1 - x)y' + ny = 0$ (hydrogen atom radial part)
- **Associated Legendre equation** (angular momentum with magnetic quantum number)

Each will yield families of orthogonal polynomials with recurrence relations, generating functions, and physical interpretations—extending the pattern established here with Legendre's equation.

12.7 Summary

Linear second-order differential equations play a central role in mathematical physics and chemistry. In this chapter, we examined their structure, methods of solution, and their importance in physical applications.

A linear homogeneous second-order ODE of the form

$$y'' + p(x)y' + q(x)y = 0$$

possesses **two linearly independent solutions**, and any linear combination of these forms the **general solution**.

The **Wronskian**,

$$W = y_1y_2' - y_2y_1',$$

provides a convenient test for determining linear independence. It satisfies

$$W(x) = C\,e^{-\int p(x)dx},$$

and is therefore nonzero throughout an interval if it is nonzero at any single point in that interval.

The **power-series method** is a systematic approach to finding solutions when the coefficients of the differential equation are analytic. By substituting

$$y(x) = \sum_{n=0}^{\infty} a_n x^n,$$

and equating coefficients of like powers, one obtains **recurrence relations** that determine the series coefficients.

The classification of points is crucial:

- **Ordinary point:** $p(x)$ and $q(x)$ analytic
- **Regular singular point:** $(x - x_0)p(x)$ and $(x - x_0)^2 q(x)$ analytic
- **Irregular singular point:** at least one of these expressions nonanalytic

At regular singular points, solutions can be obtained using the **Frobenius method**.

Legendre's differential equation,

$$(1 - x^2)y'' - 2xy' + l(l + 1)y = 0,$$

arises in problems possessing spherical symmetry, including the angular part of the Schrödinger equation. For integer values of l, one of the two series solutions terminates, giving the **Legendre polynomials** $P_l(x)$, which are finite on the interval $[-1, 1]$ and satisfy the orthogonality relation:

$$\int_{-1}^{1} P_l(x)P_m(x)dx = \frac{2}{2l + 1}\delta_{lm}.$$

The non-terminating counterpart, $Q_l(x)$, is the **Legendre function of the second kind** and diverges at the endpoints. Thus, in physical applications, the polynomial solutions $P_l(x)$ are typically chosen.

Further Reading

M. L. Boas, *Mathematical Methods in the Physical Sciences*, 3rd ed., Wiley, 2006. ISBN: 978-0471198260

G. B. Arfken, H. J. Weber, and F. E. Harris, *Mathematical Methods for Physicists*, 7th ed., Academic Press (Elsevier), 2012. ISBN: 978-0123846549 DOI: https://doi.org/10.1016/C2009-0-30629-7

D. J. Griffiths, *Introduction to Quantum Mechanics*, 2nd ed., Pearson Prentice Hall, 2015. ISBN: 978-0131118928

M. Abramowitz and I. A. Stegun (eds.), *Handbook of Mathematical Functions with Formulas, Graphs, and Mathematical Tables*, Dover Publications, 1965. ISBN: 978-0486612720 DOI (NIST Digital Library): https://doi.org/10.6028/NBS.MONO.55

Exercises

A. Conceptual Questions

1. Explain why Legendre polynomials of different orders are orthogonal.

2. Distinguish between ordinary and regular singular points of a second-order linear differential equation.

3. State the physical systems modeled respectively by the
   - Hermite equation,
   - Laguerre equation, and
   - Legendre equation.

B. Problems: Warm-up

1. **Identifying Ordinary and Singular Points**
   Determine which of the following points are ordinary, singular, or regular singular for the differential equation:

   $$(1-x^2)y'' - 2xy' + y = 0. \text{ Consider } x = 0,\ x = 1, \text{ and } x = 2.$$

2. **Wronskian Check**
   Compute the Wronskian of the pair $y_1 = e^x$, $y_2 = xe^x$.
   State whether they are linearly independent on $(-\infty, \infty)$.

3. **Simple Power-Series Expansion**
   Find the first four nonzero terms of the power-series solution of

   $$y' - y = 0$$

   about $x = 0$, assuming $y(0) = 1$.

4. **Testing for Recurrence Structure**
   For the series

$$y = \sum_{n=0}^{\infty} a_n x^n,$$

differentiate twice and write

$$y'', y', y$$

all in the form $\sum(\text{coefficient})x^n$.

5. **Even/Odd Separation**
   Show that the general solution of

$$y'' - y = 0$$

can be written as the sum of an even function and an odd function.

C. Problems

1. **Consider the differential equation**

$$(1 - x^2)y'' - 2xy' + \lambda y = 0.$$

   a. What are the singular points of the equation?
   b. Solve the equation by the power-series method.
   c. What is the radius of convergence of the power-series solutions?
   d. Deduce the values of λ for which one of the solutions becomes analytic at the singular points and reduces to a polynomial.
   e. Develop explicitly several such polynomials, standardizing them by the condition

$$P_l(1) = 1.$$

2. **Consider the Legendre differential equation**

$$(1 - x^2)y'' - 2xy' + \lambda y = 0 \ (\lambda \text{ real constant}).$$

   a. **Show that the Wronskian** of any two solutions is

$$W(x) = \frac{C}{1 - x^2},$$

   where C is a constant depending on the particular pair of solutions.
   b. For the specific pair of solutions $P_0(x)$ and $Q_0(x)$, the constant is $C = 1$. Show that the Wronskian of $P_0(x)$ and $Q_0(x)$ is

$$W[P_0, Q_0](x) = \frac{1}{1 - x^2}.$$

   c. What is the Wronskian of the two solutions of the Laguerre equation

$$xy'' + (1 - x)y' + ny = 0 \ (n \text{ constant})?$$

d. What is the Wronskian of two solutions of the Hermite equation

$$y'' - 2xy' + 2ny = 0\,(n\,\text{constant})?$$

3. Apply the power-series method to $y'' - 2xy' + 2ny = 0$ and determine the condition on n for polynomial solutions.

D. Advanced Problems

1. **Schrödinger Equation in One Dimension**
   The one-dimensional Schrödinger equation for a particle in a potential $V(x)$ is

$$\frac{d^2\psi}{dx^2} + 2m[E - V(x)]\psi = 0.$$

a. Using

$$\psi = e^{-x^2/2}u(x)\,\text{with}\ \epsilon = E/\hbar\omega\ \text{in the usual dimensionless convention,}$$

we obtain

$$u'' - 2xu' + (2\epsilon - 1)u = 0.$$

where $2\epsilon - 1 = 2n$
For the case $2\epsilon = 1$, show that the equation reduces to

$$u'' - 2xu' = 0.$$

b. **Solve this equation** by the power-series method.
Show that the odd series can be written in closed form as

$$u_{\text{even}}(x) = A\int e^{x^2}\,dx,$$

and that this solution **diverges** as $|x| \to \infty$, violating normalizability.
Show that the **even series** is trivial, reducing to the constant $a - 0$, and remains finite for all x.
Hence only the **Gaussian factor** survives for this case.

2. **Now consider the case $2\epsilon \neq 1$**
   Write

$$\psi(x) = e^{-x^2/2}u(x)$$

and substitute into the Schrödinger equation to obtain

$$u'' - 2xu' + (2\varepsilon - 1)u - 0.$$

This has the same form as the **Hermite equation**.
Use the power-series method to show:

- Both series solutions diverge unless the series **truncates**,
- This occurs only for integer values of n,

- The resulting functions are the **Hermite polynomials** $H_n(x)$.
  Find the corresponding quantized energies:

$$E_n = \left(n + \frac{1}{2}\right)\hbar\omega.$$

Show that the same truncation mechanism leads to quantized energies E_n.

3. Solve the Legendre differential equation

$$(1 - x^2)y'' - 2xy' + l(l + 1)y = 0$$

by the power-series method and deduce the values of l for which the solutions become polynomials.

4. **Repeat Problem** 3 for **the Laguerre differential equation**

$$xy'' + (1 - x)y' + ny = 0,$$

and deduce the values of n for which the solutions become **Laguerre polynomials**. Find the first few such polynomials.

5. Show that the Schrödinger equation for the hydrogen atom leads to the associated Legendre differential equation for the angular part of the wavefunction.

6. Demonstrate how the Wronskian formula can be used to derive the closed-form second independent solution $Q_0(x)$.

13

Special Functions

The orthogonal features, when combined, can explode into complexity.
—Yukihiro Matsumoto

Learning Outcomes

After studying this chapter, students should be able to:

1. **Understand** the definition and properties of Legendre, Laguerre, and Hermite polynomials.
2. **Derive** Rodrigues's and generating function formulas for special functions.
3. **Apply** orthogonality and recurrence relations in quantum mechanics.
4. **Recognize** the role of associated Legendre and Laguerre functions in hydrogenic systems.
5. **Expand** functions using orthogonal polynomial series.

13.1 Introduction

In the previous chapter, we had seen that Legendre's differential equation has two solutions, which may diverge at the regular singular points $x = \pm 1$. One of the solutions can be made analytic by setting λ in the equation

$$(1 - x^2)y''(x) - 2xy'(x) + \lambda y = 0 \tag{13.1}$$

equal to $l(l + 1)$, where λ is the separation constant arising in the Legendre differential equation and l is an integer, truncating the series to a polynomial of degree l, which is then analytic at the two points. These polynomials are known as **Legendre polynomials** and have high significance in quantum chemistry. They are involved in quantum-mechanical problems involving spherical symmetry, such as the rigid rotator and the hydrogen atom.

Besides Legendre polynomials, there are other named polynomials, such as the Hermite and Laguerre polynomials, that we shall encounter in our study of quantum mechanics. Hermite polynomials are involved in quantum-mechanical solutions to the harmonic oscillator problem, and Laguerre polynomials form a solution to the radial part of the hydrogen atom. These are so important that we devote an entire chapter to these polynomials, beginning with the Legendre polynomials that we have already obtained in the previous chapter.

Historical Note 13.1: Adrien-Marie Legendre (1752–1833)

Adrien-Marie Legendre was a French mathematician whose work laid essential foundations for analysis, number theory, and mathematical physics. The **Legendre polynomials** arose in his study of Newtonian gravitational potentials, particularly in the solution of **Laplace's equation** in spherical coordinates. These polynomials naturally appear when separating variables in problems possessing spherical symmetry.

Legendre introduced these functions in the context of celestial mechanics, but their importance soon extended far beyond astronomy. In physics and chemistry, Legendre polynomials play a central role in **electrostatics, quantum mechanics, angular momentum theory, and molecular orbital symmetry**, where they underpin the angular dependence of wavefunctions and potential expansions.

Legendre's work exemplifies how abstract mathematical constructions, motivated by physical problems, can acquire lasting and universal significance across the sciences.

13.2 Properties of Legendre Polynomials

As we shall see, all special functions are characterized by a Rodrigues's formula, a generating function, and a few recurrence relations. Some of these properties will be useful in our study of quantum mechanics, and so we discuss some of the important ones. We begin with Legendre polynomials, which we have already introduced in the previous chapter.

13.2.1 Orthogonality

One important property of Legendre and other special functions is their mutual orthogonality over an interval range. For Legendre polynomials, the interval range is $-1 \leq x \leq 1$,

$$\int_{-1}^{1} P_l(x) P_k(x)\, dx = 0, l \neq k. \tag{13.2}$$

This orthogonality relation means the Legendre polynomials are orthogonal on $[-1, 1]$ with a weight function $w(x) = 1$.

The proof of Equation (13.2) using the Legendre equation, and other proofs are given in Appendix J.6 for the interested reader. Here we prove the orthogonality of two Legendre polynomials ($P_1(x)$ and $P_3(x)$) in Example 13.1.

For ready reference, some Legendre polynomials are given below (see also Chapter 12):

$$\begin{aligned}
P_0(x) &= 1,\\
P_1(x) &= x,\\
P_2(x) &= \frac{1}{2}(3x^2 - 1),\\
P_3(x) &= \frac{1}{2}(5x^3 - 3x),\\
P_4(x) &= \frac{1}{8}(35x^4 - 30x^2 + 3),\\
P_5(x) &= \frac{1}{8}(63x^5 - 70x^3 + 15x).
\end{aligned} \tag{13.3}$$

Example 13.1: Orthogonality of $P_1(x)$ and $P_3(x)$

Show that $P_1(x)$ and $P_3(x)$ are mutually orthogonal.

Solution:

Inserting the respective polynomials from Equation (13.3) in Equation (13.2), we obtain

$$\frac{1}{2}\int_{-1}^{1} x(5x^3 - 3x)dx = \left[\frac{5x^5}{5} - \frac{3x^3}{3}\right]_{-1}^{1} = 0.$$

Figure 13.1 illustrates the orthogonality of two polynomials, $P_2(x)$ and $P_3(x)$.

Interpretation: The overlap integral between two different Legendre polynomials vanishes, confirming their **orthogonality** under the unit weight function $w(x) = 1$. This property forms the mathematical foundation for expanding angular parts of wavefunctions (spherical harmonics) in quantum mechanics.

13.2.2 Rodrigues's Formula

Legendre polynomials can also be expressed using Rodrigues's formula

$$P_l(x) = \frac{1}{2^l l!}\frac{d^l}{dx^l}(x^2 - 1)^l. \tag{13.4}$$

By inspection, $P_l(x)$ has degree l and we have $P_l(-x) = (-1)^l P_l(x)$, since differentiating an even function l times yields **parity** $(-1)^l$.

Example 13.2 illustrates the use of Rodrigues's formula to obtain Legendre's polynomials.

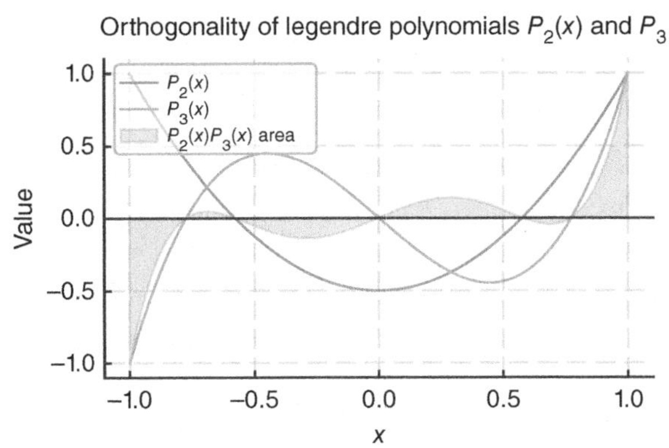

Figure 13.1 Orthogonality of Legendre polynomials $P_2(x)$ and $P_3(x)$.
Key features:
The shaded region represents the integral $\int_{-1}^{1} P_2(x)P_3(x)\,dx = 0$, illustrating that the positive and negative areas cancel exactly. The product $P_2(x)P_3(x)$ is an odd function (even × odd = odd). The integral of any odd function over a symmetric interval [−1, 1] is exactly zero.

Example 13.2: Rodrigues' Formula for Legendre Polynomials

Use Rodrigues's formula to verify the first five Legendre polynomials.

Solution:

According to Rodrigues's formula (13.4),

$$P_l(x) = \frac{1}{2^l l!}\frac{d^l}{dx^l}(x^2-1)^l.$$

For $l = 0$,

$$P_0(x) = \frac{1}{2^0 0!}\frac{d^0}{dx^0}(x^2-1)^0 = 1.$$

For $l = 1$,

$$P_1(x) = \frac{1}{2^1 1!}\frac{d}{dx}(x^2-1) = \frac{1}{2}\cdot 2x = x.$$

For $l = 2$,

$$P_2(x) = \frac{1}{2^2 2!}\frac{d^2}{dx^2}(x^2-1)^2.$$

First derivative:

$$\frac{d}{dx}(x^2-1)^2 = 2(x^2-1)\cdot 2x = 4x(x^2-1).$$

Second derivative:

$$\begin{aligned}\frac{d^2}{dx^2}(x^2-1)^2 &= \frac{d}{dx}[4x(x^2-1)] = 4(x^2-1) + 4x\cdot 2x = 4(x^2-1+2x^2)\\ &= 4(3x^2-1).\end{aligned}$$

Thus:

$$P_2(x) = \frac{1}{4\cdot 2}\cdot 4(3x^2-1) = \frac{1}{2}(3x^2-1).$$

For $l = 3$,

$$P_3(x) = \frac{1}{2^3 3!}\frac{d^3}{dx^3}(x^2-1)^3.$$

First derivative:

$$\frac{d}{dx}(x^2-1)^3 = 3(x^2-1)^2\cdot 2x = 6x(x^2-1)^2.$$

Second derivative:

$$\begin{aligned}\frac{d^2}{dx^2}(x^2-1)^3 &= \frac{d}{dx}\left[6x(x^2-1)^2\right] = 6(x^2-1)^2 + 6x\cdot 2(x^2-1)2x\\ &= 6(x^2-1)^2 + 24x^2(x^2-1).\end{aligned}$$

Third derivative (simplifying via binomial expansion or direct differentiation):

$$(x^2-1)^3 = x^6 - 3x^4 + 3x^2 - 1,$$

$$\frac{d^3}{dx^3}(x^2-1)^3 = \frac{d^3}{dx^3}(x^6 - 3x^4 + 3x^2 - 1) = 120x^3 - 72x = 24(5x^3 - 3x).$$

Thus,

$$P_3(x) = \frac{1}{48} \cdot 24(5x^3 - 3x) = \frac{1}{2}(5x^3 - 3x).$$

For $l = 4$,

$$P_4(x) = \frac{1}{2^4 4!} \frac{d^4}{dx^4}(x^2-1)^4.$$

Let $f(x) = (x^2-1)^4$. Expanding:

$$f(x) = x^8 - 4x^6 + 6x^4 - 4x^2 + 1.$$

Now differentiate four times:

$$f'(x) = 8x^7 - 24x^5 + 24x^3 - 8x,$$
$$f''(x) = 56x^6 - 120x^4 + 72x^2 - 8,$$
$$f'''(x) = 336x^5 - 480x^3 + 144x,$$
$$f^{(4)}(x) = 1680x^4 - 1440x^2 + 144 = 48(35x^4 - 30x^2 + 3).$$

Therefore,

$$P_4(x) = \frac{1}{16 \cdot 24} \cdot 48(35x^4 - 30x^2 + 3) = \frac{48}{384}(35x^4 - 30x^2 + 3)$$
$$= \frac{1}{8}(35x^4 - 30x^2 + 3).$$

All results agree with Equation (13.3).

Rodrigues's formula can be used to derive certain important properties of Legendre polynomials.

13.2.3 Normalization of Legendre Polynomials

We can use Rodrigues's formula to show that (see Appendix J.6 for proof)

$$I_l = \int_{-1}^{1} [P_l(x)]^2 dx = \frac{2}{2l+1}. \tag{13.5}$$

Example 13.3 verifies Equation (13.5) for the first three Legendre polynomials.

Example 13.3: Normalization of Legendre Polynomials

Show that the first three Legendre polynomials obey Equation (13.5).

Solution:

$$\int_{-1}^{1} P_0(x)^2 dx = \int_{-1}^{1} dx = 2,$$

$$\int_{-1}^{1} P_1(x)^2 dx = \int_{-1}^{1} x^2 dx = \left[\frac{x^3}{3}\right]_{-1}^{1} = \frac{2}{3},$$

$$\int_{-1}^{1} P_2(x)^2 dx = \frac{1}{4}\int_{-1}^{1} (3x^2-1)^2 dx = \frac{1}{4}\left[\frac{9x^5}{5} - \frac{6x^3}{3} + x\right]_{-1}^{1} = \frac{2}{5}.$$

Both results (13.2) and (13.5) can be combined into a single equation

$$\int_{-1}^{1} P_k(x)P_l(x)dx = \frac{2}{2l+1}\delta_{kl}, \tag{13.6}$$

where δ_{kl} is the Kronecker delta, which is equal to 1 when $k = l$, and is equal to zero when $k \neq l$.

Thus, the Legendre polynomials constitute a **complete orthogonal set** on the interval $[-1, 1]$, allowing functions defined on this interval to be represented through Fourier–Legendre expansions.

Concept Insight 13.1: Completeness

Legendre polynomials form a complete set of orthogonal functions on the interval $[-1, 1]$. This means that any reasonably well-behaved function on this interval can be expanded as a series of Legendre polynomials, just as Fourier series expand periodic functions. Completeness is a consequence of Legendre's equation being a Sturm–Liouville problem (see Section 13.4) with a discrete spectrum. In physical chemistry, this property enables expansions of angular distributions, potential functions, and scattering amplitudes.

13.2.4 Generalized Fourier–Legendre Series

The mutual orthogonality and completeness of the $P_l(x)$ means that any reasonable $f(x)$ (i.e., one obeying the Dirichlet conditions discussed in Chapter 9) can be expressed in the interval $|x| < 1$ as an infinite sum of Legendre polynomials,

$$f(x) = \sum_{l=0}^{\infty} c_l P_l(x). \tag{13.7}$$

Multiplying this equation by $P_k(x)$ and integrating from $x = -1$ to $x = 1$, we obtain

$$\int_{-1}^{1} P_k(x)f(x)dx = \sum_{l=0}^{\infty} c_l \int P_k(x)P_l(x)dx.$$

Because of their mutual orthogonality, the only term that survives in the summation is when $l = k$, and then the integral equals $2/(2k+1)$. Therefore,

$$\int_{-1}^{1} P_k(x)f(x)dx = c_k\frac{2}{2k+1} \Longrightarrow c_k = \frac{2k+1}{2}\int_{-1}^{1} P_k(x)f(x)dx.$$

Therefore, the coefficients c_l in Equation (13.7) are given by

$$c_l = \frac{2l+1}{2} \int_{-1}^{1} f(x) P_l(x)\, dx. \tag{13.8}$$

For polynomial functions $f(x)$, this sum has only a finite number of terms, the highest power of l needed being equal to the degree of the polynomial (Example 13.4).

Example 13.4: Legendre Series Expansion of a Polynomial

Express the function $f(x) = 1 + x + x^2$ as a sum of Legendre polynomials.

Solution:

From Equation (13.8),

$$c_0 = \frac{1}{2} \int_{-1}^{+1} (1 + x + x^2) dx = \frac{1}{2} \left[x + \frac{x^2}{2} + \frac{x^3}{3} \right]_{-1}^{1} = \frac{4}{3},$$

$$c_1 = \frac{3}{2} \int_{-1}^{+1} x(1 + x + x^2) dx = \frac{3}{2} \left[\frac{x^2}{2} + \frac{x^3}{3} + \frac{x^4}{4} \right]_{-1}^{1} = 1,$$

$$c_2 = \frac{5}{4} \int_{-1}^{+1} (3x^2 - 1)(1 + x + x^2) dx = \frac{5}{4} \left[-x - \frac{x^2}{2} + \frac{2x^3}{3} + \frac{3x^4}{4} + \frac{3x^5}{5} \right]_{-1}^{1} = \frac{2}{3}.$$

Therefore,

$$1 + x + x^2 = \frac{4}{3} P_0(x) + P_1(x) + \frac{2}{3} P_2(x),$$

which can be easily verified by substitution:

$$\frac{4}{3} P_0(x) + P_1(x) + \frac{2}{3} P_2(x) = \frac{4}{3}(1) + x + \frac{2}{3}\left(\frac{1}{2}(3x^2 - 1) \right) = 1 + x + x^2 = f(x).$$

13.2.5 Generating Function

Legendre functions can also be described in terms of their generating function. The generating function for a series of functions $f_n(x)$ for $n = 0, 1, 2, \ldots$ is a function $G(x, t)$, containing as well as x, a dummy variable t, such that

$$G(x, t) = \sum_{n=0}^{\infty} f_n(x) t^n,$$

i.e., $f_n(x)$ is the coefficient of t^n in the expansion of G in powers of t.

The generating function of Legendre polynomials is

$$G(x,t) = \frac{1}{\sqrt{1 - 2xt + t^2}} = \sum_{n=0}^{\infty} P_n(x)t^n. \tag{13.9}$$

By expanding the left-hand side in a binomial expansion, the respective Legendre polynomials can be generated (see also Example 9.2).

The **standardization condition**, i.e., $P_n(x) = 1$ at $x = 1$ is easily verified, since, at $x = 1$, Equation (13.9) becomes

$$G(1,t) = \left(1 - 2t + t^2\right)^{-1/2} = \left((1-t)^2\right)^{-1/2} = \frac{1}{1-t} = 1 + t + t^2 + \ldots = \sum_{n=0}^{\infty} P_n(1)t^n.$$

Since the coefficients of all t^n are unity, all the $P_n(1) = 1$ as defined by the generating function.

Having established that the generating function (13.9) is a faithful representation of the Legendre polynomials, we examine whether it can be used to derive the other relations with respect to Legendre polynomials (Examples 13.5 and 13.6).

Example 13.5: Legendre Normalization via Generating Function

Use the generating function (13.9) to show that the Legendre polynomials obey Equation (13.5).

Solution:

Multiplying the generating function from Equation (13.9) by itself and integrating between −1 and 1, making use of the orthogonality condition (13.2) to simplify the left-hand side, we get

$$\int_{-1}^{1} \sum_{k=0}^{\infty} P_k(x)t^k \sum_{l=0}^{\infty} P_l(x)t^l dx = \int_{-1}^{1} \sum_{k=0}^{\infty}\sum_{l=0}^{\infty} P_l(x)P_k(x)t^{l+k}dx = \int_{-1}^{1} \sum_{l=0}^{\infty} P_l^2(x)t^{2l}dx.$$

This is equal to

$$\int_{-1}^{1} \left(1 - 2xt + t^2\right)^{-1} dx = -\frac{1}{2t}\left[\ln(1 - 2xt + t^2)\right]_{-1}^{1} = \frac{1}{t}\ln\frac{1+t}{1-t}.$$

Expanding the two logarithms as Taylor series in t about $t = 0$, we have

$$\ln(1+t) = t - \frac{t^2}{2} + \frac{t^3}{3} - \frac{t^4}{4} + \frac{t^5}{5} - \ldots,$$

$$\ln(1-t) = -t - \frac{t^2}{2} - \frac{t^3}{3} - \frac{t^4}{4} - \frac{t^5}{5} - \ldots,$$

$$\therefore \ln(1+t) - \ln(1-t) = 2\left(t + \frac{t^3}{3} + \frac{t^5}{5} + \cdots\right) = 2\sum_{l=0}^{\infty} \frac{t^{2l+1}}{2l+1}.$$

Hence,

$$\int_{-1}^{1} \sum_{l=0}^{\infty} P_l^2(x)t^{2l}dx = 2\sum_{l=0}^{\infty} \frac{t^{2l}}{2l+1}.$$

Equating powers of t^l on both sides of the equation, Equation (13.5) results.

Example 13.6: Legendre Parity via Generating Function

Use the generating function (13.9) to show that the Legendre polynomials obey the parity relation $P_n(-x) = (-1)^n P_n(x)$.

Solution:

Using the generating function (13.9),

$$G(x,t) = (1 - 2xt + t^2)^{-1/2} = \sum_{n=0}^{\infty} P_n(x)t^n,$$

$$G(-x,-t) = \left(1 - 2(-x)(-t) + (-t)^2\right)^{-1/2} = (1 - 2xt + t^2)^{-1/2} = \sum_{n=0}^{\infty} P_n(-x)(-t)^n.$$

Therefore,

$$\sum_{n=0}^{\infty} P_n(x)(t)^n = \sum_{n=0}^{\infty} P_n(-x)(-t)^n.$$

Equating powers of t^n in both equations results in the above parity relation, since for odd n, $P_n(-x) = -P_n(x)$, whereas for even n, $P_n(-x) = P_n(x)$.

13.2.5.1 Recurrence Relations

Though we had verified the first few Legendre polynomials using the generating function, both this and Rodrigues's formula (Example 13.2) are cumbersome to use. We can derive certain recurrence formulae to make the task easier. Some of these recurrence formulae are useful in other contexts as well, such as in deriving the selection rules for rotational spectroscopy and electronic transitions of the hydrogen atom. Many more recurrence formulae can be derived by combining these, but we focus on a few of the most useful ones.

We differentiate $G(x,t)$ from Equation (13.9) with respect to t to give

$$\frac{\partial G(x,t)}{\partial t} = -\frac{1}{2}(-2x + 2t)(1 - 2xt + t^2)^{-3/2} = \frac{x - t}{(1 - 2xt + t^2)^{3/2}} = \frac{(x - t)G(x,t)}{(1 - 2xt + t^2)},$$

$$\therefore (1 - 2xt + t^2)\frac{\partial G(x,t)}{\partial t} + (t - x)G(x,t) = 0.$$

Using Equation (13.9), we obtain

$$(1 - 2xt + t^2)\sum_{l=1}^{\infty} P_l(x) l t^{l-1} + (t - x)\sum_{l=0}^{\infty} P_l(x)t^l = 0$$

$$\Rightarrow \sum_{l=1}^{\infty} P_l(x) l t^{l-1} - 2\sum_{l=1}^{\infty} xP_l(x) l t^l + \sum_{l=1}^{\infty} P_l(x) l t^{l+1} + \sum_{l=0}^{\infty} P_l(x)t^{l+1} - \sum_{l=0}^{\infty} xP_l(x)t^l = 0.$$

As usual, we change the exponent of t uniformly to l,

$$\sum_{l=0}^{\infty} P_{l+1}(x)(l+1)t^l - 2\sum_{l=1}^{\infty} xP_l(x) l t^l + \sum_{l=2}^{\infty} P_{l-1}(x)(l-1)t^l$$
$$+ \sum_{l=1}^{\infty} P_{l-1}(x)t^l - \sum_{l=0}^{\infty} xP_l(x)t^l = 0.$$

Thus, the recurrence relation is

$$(l+1)P_{l+1}(x) = (2l+1)xP_l(x) - lP_{l-1}(x). \tag{13.10}$$

This three-term recurrence greatly simplifies polynomial generation (Example 13.7). This is a very useful result that allows us to derive selection rules for rotational transitions of a rigid rotator and electronic transitions of a hydrogen atom.

Example 13.7: Iterative Generation of Legendre Polynomials

Use Equation (13.10) to find Legendre polynomials up to $P_3(x)$, given that $P_0(x) = 1$.

Solution:

For $l = 0$, Equation (13.10) becomes

$$P_1(x) = xP_0(x) \Rightarrow P_1(x) = x.$$

Setting $l = 1$,

$$2P_2(x) = 3xP_1(x) - P_0(x) \Rightarrow P_2(x) = \frac{1}{2}(3x^2 - 1).$$

For $l = 2$,

$$3P_3(x) = 5xP_2(x) - 2P_1(x) \Rightarrow P_3(x) = \frac{1}{2}(5x^3 - 3x).$$

Figure 13.2 plots the first four Legendre polynomials.

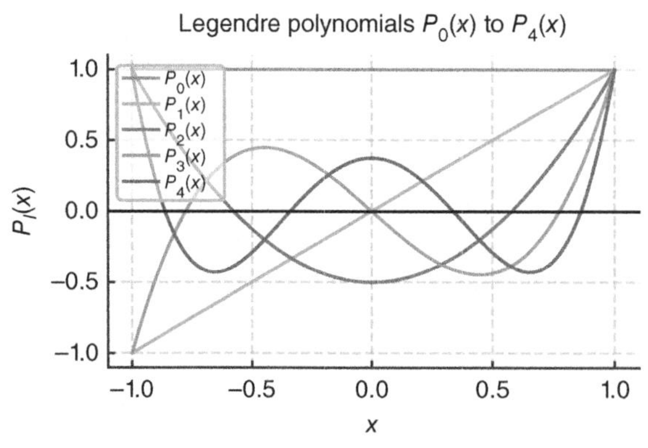

Figure 13.2 Plot of Legendre polynomials $P_0(x)$ to $P_4(x)$.
Key features:

- $P_0(x) = 1$: constant function;
- $P_1(x) = x$: straight line through the origin;
- $P_2(x) = \frac{1}{2}(3x^2 - 1)$: parabola symmetric about the y-axis;
- all satisfy $P_n(1) = 1$;
- even $n \rightarrow$ even functions; odd $n \rightarrow$ odd functions (alternating symmetry).

Interpretation: The figure demonstrates how the complexity of $P_n(x)$ increases with order n. Even-order functions are symmetric and odd-order functions antisymmetric about the origin, emphasizing their role as orthogonal basis functions on the interval $[-1,1]$.

Another recurrence relation can be obtained from the generating function by differentiating both sides of Equation (13.9) with respect to x,

$$t(1-2xt+t^2)^{-3/2} = \sum_{l=0}^{\infty} P_l'(x)t^l$$

$$\Rightarrow tG(x,t) = (1-2xt+t^2)\sum_{l=0}^{\infty} P_l'(x)t^l$$

$$\Rightarrow \sum_{l=0}^{\infty} P_l(x)t^{l+1} = \sum_{l=0}^{\infty} P_l'(x)t^l - 2x\sum_{l=0}^{\infty} P_l'(x)t^{l+1} + \sum_{l=0}^{\infty} P_l'(x)t^{l+2}$$

$$\Rightarrow \sum_{l=0}^{\infty} P_{l-1}(x)t^l = \sum_{l=0}^{\infty} P_l'(x)t^l - 2x\sum_{l=1}^{\infty} P_{l-1}'(x)t^l + \sum_{l=2}^{\infty} P_{l-2}'(x)t^l$$

Collecting terms, we can write the recurrence relation as

$$\sum_{l=1}^{\infty} P_{l-1}(x)t^l = \sum_{l=0}^{\infty} P_l'(x)t^l - 2x\sum_{l=1}^{\infty} P_{l-1}'(x)t^l + \sum_{l=2}^{\infty} P_{l-2}'(x)t^l$$

For $l=0$, the only two terms imply that $P_0'(x)=0$ or $P_0(x)$ is a constant.

For $l=1$, $P_0(x) = P_1'(x) - 2xP_0'(x) = P_1'(x)$.

For $l \geq 2$, the recurrence relation is $P_{l-1}(x) = P_l'(x) - 2xP_{l-1}'(x) + P_{l-2}'(x)$, which can be rewritten as

$$P_{l+1}(x) = P_{l+2}'(x) - 2xP_{l+1}'(x) + P_l'(x) \quad (l \geq 0). \tag{13.11}$$

Several other recurrence relations can be derived by combining these two recurrence relations. For example,

$$(2l+1)P_l(x) = P_{l+1}'(x) - P_{l-1}'(x).$$

Flowchart 13.1 summarizes the steps from the Legendre differential equation to quantum mechanics.

13.3 Associated Legendre Polynomials

The polynomials that are actually involved in the hydrogen atom solutions are the associated Legendre polynomials, P_l^m. We shall only consider the case of l and m both integers, which is the case for most physical applications, and only Legendre polynomials of the first kind, though it must be remembered that the general solution to the associated Legendre equation

$$(1-x^2)y''(x) - 2xy'(x) + \left[l(l+1) - \frac{m^2}{1-x^2}\right]y(x) = 0 \tag{13.12}$$

is

$$y(x) = c_1P_l^m(x) + c_2Q_l^m(x). \tag{13.13}$$

However, in physical applications, $x = \cos\theta$, where θ is the polar angle, which varies between 0 and π. Hence, x varies between -1 and $+1$. We, therefore, require $y(x)$ to remain finite at both

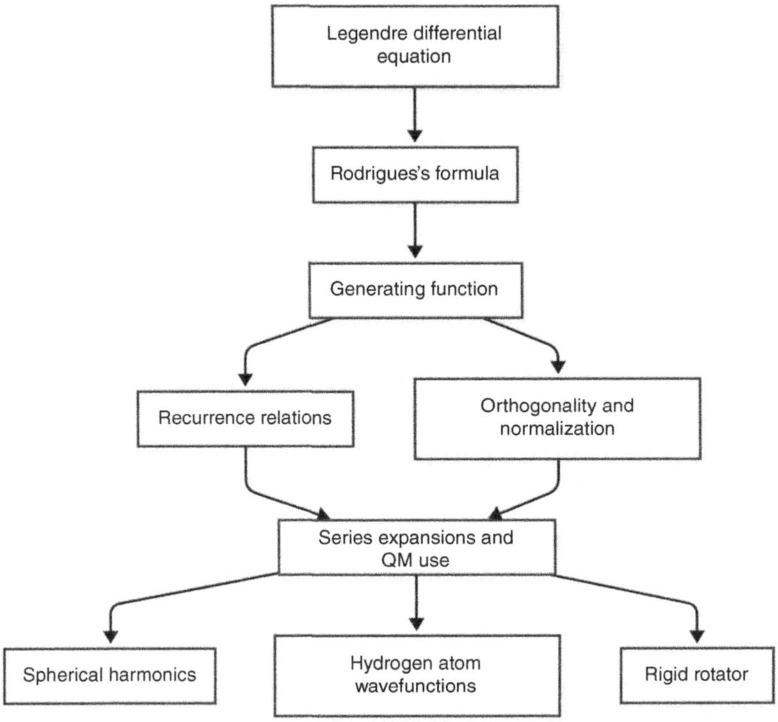

Flowchart 13.1 Conceptual links of Legendre polynomials to quantum mechanics.

these end points. As for the regular Legendre solutions, the functions of the first kind $P_l^m(x)$ are polynomials (hence finite at $x = \pm 1$), whereas $Q_l^m(x)$ behave logarithmically near the end points and diverge at $x = \pm 1$. As in the case of Legendre polynomials, we set c_2 in Equation (13.13) equal to zero. Henceforth, we shall only deal with functions of the first kind.

In terms of $P_l^m(x)$, the associated Legendre equation has the form

$$(1 - x^2)\frac{d^2}{dx^2}P_l^m(x) - 2x\frac{d}{dx}P_l^m(x) + \left[l(l+1) - \frac{m^2}{1-x^2}\right]P_l^m(x) = 0. \tag{13.14}$$

It can be observed that the Legendre equation (13.1) is just the associated Legendre equation with $m = 0$. Interestingly, this equation can be obtained from the regular Legendre equation (13.1) by differentiating m times with respect to x. For positive values of m, the associated Legendre polynomials can be obtained from Legendre polynomials by using the relation

$$P_l^m(x) = (1 - x^2)^{m/2}\frac{d^m}{dx^m}P_l(x). \tag{13.15}$$

These are polynomials of degree l for m even, and $\sqrt{1 - x^2}$ times a polynomial of degree $l - 1$ for m odd.

The **parity relation** is

$$P_l^m(-x) = (-1)^{l+m}P_l^m(x). \tag{13.16}$$

Also, since m appears as its square, both positive and negative values of m obey the same Equation (13.14) and are hence the solutions for positive and negative m are proportional to each other. Conventionally, the associated Legendre equations for negative m are expressed as

$$P_l^{-m}(x) = (-1)^m\frac{(l - |m|)!}{(l + |m|)!}P_l^m(x). \tag{13.17}$$

Moreover, m is an integer since one can take only an integer order derivative. According to Equation (13.15), since $P_l(x)$ is a polynomial of order l, it can be differentiated a maximum number of l times. Therefore, m is restricted to $-l \leq m \leq l$, or, in other words, m can take on $2l + 1$ values ranging from $-l$ to $+l$ through zero. Readers can relate m to the magnetic quantum number and l to the azimuthal quantum number.

For example, let $l = 2$. Then, the maximum number of derivatives that can be taken is two. We have

$$P_2^0(x) = P_2(x) = \frac{1}{2}(3x^2 - 1),$$

$$P_2^1(x) = (1 - x^2)^{1/2} \frac{d}{dx} P_2(x) = 3x\sqrt{1 - x^2},$$

$$P_2^2(x) = (1 - x^2) \frac{d^2}{dx^2} P_2(x) = 3(1 - x^2).$$

From Equation (13.17),

$$P_2^{-1}(x) = -\frac{1}{6} P_2^1(x) = -\frac{1}{2} x\sqrt{1 - x^2},$$

$$P_2^{-2}(x) = \frac{1}{24} P_2^2(x) = \frac{1}{8}(1 - x^2).$$

In problems with spherical symmetry, the solutions of the associated Legendre equation, together with the azimuthal factor $e^{im\phi}$, lead naturally to the **spherical harmonics**.

These spherical harmonics are listed in Table 13.1 and plotted in Figure 13.3. **Real forms** or **complex forms** may be used; Table 13.1 gives the **complex, normalized** spherical harmonics used in chemistry and quantum mechanics:

$$Y_l^m(\theta, \phi) = (-1)^m \sqrt{\frac{2l + 1}{4\pi} \frac{(l - m)!}{(l + m)!}} P_l^m(\cos\theta)\, e^{im\phi}.$$

Table 13.1 The first few spherical harmonics $Y_l^m(\theta, \phi)$ for $l = 0, 1, 2$ with the Condon–Shortley phase carried in Y_l^m (Equation above), consistent with $P_l^m(x)$ as defined in Equation (13.15).

Orbital	l	m	$Y_l^m(\theta, \phi)$
s	0	0	$Y_0^0 = \frac{1}{\sqrt{4\pi}}$
p	1	0	$Y_1^0 = \sqrt{\frac{3}{4\pi}} \cos\theta$
	1	+1	$Y_1^{+1} = -\sqrt{\frac{3}{8\pi}} \sin\theta\, e^{+i\phi}$
	1	−1	$Y_1^{-1} = +\sqrt{\frac{3}{8\pi}} \sin\theta\, e^{-i\phi}$
d	2	0	$Y_2^0 = \sqrt{\frac{5}{16\pi}}(3\cos^2\theta - 1)$
	2	+1	$Y_2^{+1} = -\sqrt{\frac{15}{8\pi}} \cos\theta \sin\theta\, e^{+i\phi}$

(Continued)

Table 13.1 (Continued)

Orbital	l	m	$Y_l^m(\theta, \phi)$
	2	−1	$Y_2^{-1} = +\sqrt{\frac{15}{8\pi}} \cos\theta \sin\theta\, e^{-i\phi}$
	2	+2	$Y_2^{+2} = \sqrt{\frac{15}{32\pi}} \sin^2\theta\, e^{+2i\phi}$
	2	−2	$Y_2^{-2} = \sqrt{\frac{15}{32\pi}} \sin^2\theta\, e^{-2i\phi}$

Notes for Chemistry Students:

- $l = 0, 1, 2$ correspond to **s-, p-, d-orbitals.**
- m controls the orientation of the orbital in space.
- Real forms of spherical harmonics (e.g., p_x, p_y, p_z) are linear combinations of the complex forms above.

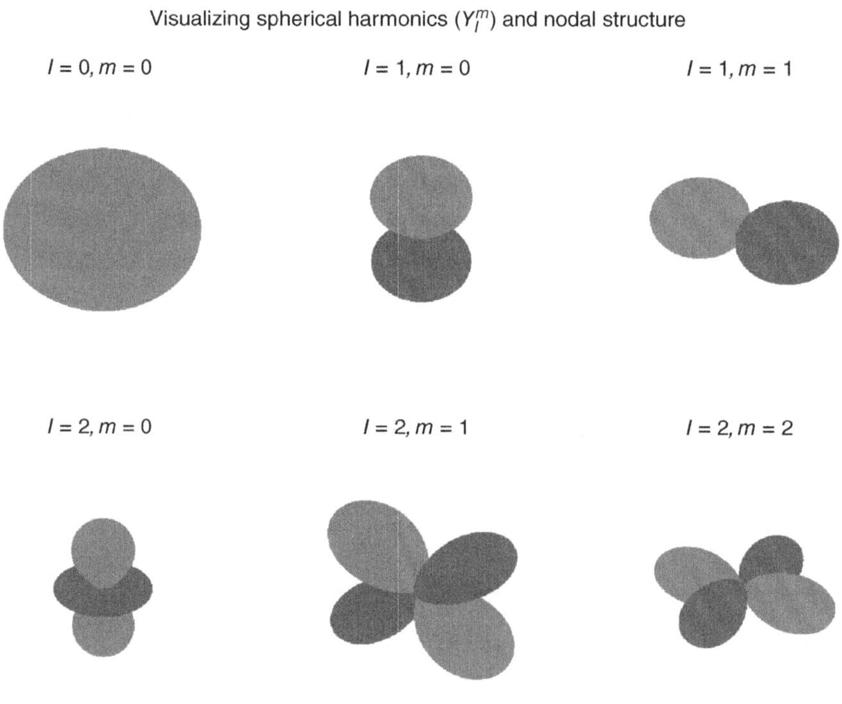

Figure 13.3 Visualization of the first few spherical harmonics, $Y_l^m(\theta, \phi)$. These functions are the product of the associated Legendre polynomials $P_l^m(\cos\theta)$ and the exponential $e^{im\phi}$. The color change (red to blue, online version) indicates a change in sign, representing a nodal plane where the probability of finding an electron is zero. This angular structure forms the familiar shapes of s, p, and d atomic orbitals.

Textbox 13.1: Real Spherical Harmonics as Linear Combinations of Complex Y_l^m

Chemistry uses **real** spherical harmonics because they correspond to the familiar orbital shapes: p_x, p_y, d_{xy}, $d_{x^2-y^2}$, and so on. These are obtained from the **complex** spherical harmonics by taking symmetric or antisymmetric linear combinations.

The general rules are:
For $m > 0$:

$$Y^{\text{real}}_{l,m} = \frac{1}{\sqrt{2}}(Y_l^{-m} + (-1)^m Y_l^m).$$

For $m < 0$:

$$Y^{\text{real}}_{l,m} = \frac{i}{\sqrt{2}}(Y_l^{-m} - (-1)^m Y_l^m).$$

For $m = 0$:

$$Y^{\text{real}}_{l,0} = Y_l^0.$$

Applying these rules to the complex Y_l^m yields the real orbital forms tabulated in the Expanded Table 13.1 below.

Expanded Table 13.1: Real Spherical Harmonics as Linear Combinations

(For $l = 1, 2$; suitable for orbital plots)

p-orbitals $l = 1$

Orbital	**Linear combination (from complex Y_1^m)**	**Simplified real form**
p_z	Y_1^0	$\sqrt{\frac{3}{4\pi}}\cos\theta$
p_x	$\frac{1}{\sqrt{2}}(Y_1^{-1} - Y_1^{+1})$	$\sqrt{\frac{3}{4\pi}}\sin\theta\cos\phi$
p_y	$\frac{i}{\sqrt{2}}(Y_1^{-1} + Y_1^{+1})$	$\sqrt{\frac{3}{4\pi}}\sin\theta\sin\phi$

These match the usual Cartesian shapes (Figure 13.3).

d-orbitals $l = 2$

Orbital	**Linear combination (from complex Y_2^m)**	**Simplified real form**
d_{z^2}	Y_2^0	$\sqrt{\frac{5}{16\pi}}(3\cos^2\theta - 1)$
d_{xz}	$\frac{1}{\sqrt{2}}(Y_2^{-1} - Y_2^{+1})$	$\sqrt{\frac{15}{4\pi}}\sin\theta\cos\theta\cos\phi$
d_{yz}	$\frac{i}{\sqrt{2}}(Y_2^{-1} + Y_2^{+1})$	$\sqrt{\frac{15}{4\pi}}\sin\theta\cos\theta\sin\phi$
d_{xy}	$\frac{i}{\sqrt{2}}(Y_2^{-2} - Y_2^{+2})$	$\sqrt{\frac{15}{4\pi}}\sin^2\theta\sin\phi\cos\phi$
$d_{x^2-y^2}$	$\frac{1}{\sqrt{2}}(Y_2^{-2} + Y_2^{+2})$	$\sqrt{\frac{15}{16\pi}}\sin^2\theta\left(\cos^2\phi - \sin^2\phi\right)$

These produce the standard five d-orbital shapes (Figure 13.3) used throughout inorganic chemistry, ligand field theory, and computational chemistry.

13.3.1 Properties of Associated Legendre Functions $P_l^m(x)$

13.3.1.1 Mutual Orthogonality and Normalization

The associated Legendre functions obey the relation

$$\int_{-1}^{1} P_l^m(x)P_k^m(x)dx = \frac{2}{2l+1}\frac{(l+|m|)!}{(l-|m|)!}\delta_{lk}. \tag{13.18}$$

Note that the m value should be the same for the two associated Legendre functions for this relation to hold. This condition can be proved using Rodrigues's formula, as in the case of regular Legendre polynomials.

This condition means that, like the regular Legendre polynomials, the associated Legendre polynomials can be used to expand a function $f(x)$ in the interval $[-1,1]$ in a series of the form

$$f(x) = \sum_{k=0}^{\infty} c_{m+k}P_{m+k}^m(x), \tag{13.19}$$

where the coefficient c_l is given by

$$c_l = \frac{2l+1}{2}\frac{(l-|m|)!}{(l+|m|)!}\int_{-1}^{1} f(x)P_l^m(x)dx.$$

We note thát the series takes the form (13.19) because $P_l^m(x) = 0$ for $m > l$; equivalently, the sum in Equation (13.19) runs over degrees $l = m,\ m+1,\ m+2, \ldots$, indexed by $k = l - m$.

13.4 Other Orthogonal Polynomials

Legendre polynomials belong to a class of polynomials called orthogonal polynomials. Other such polynomials that we shall use are Hermite and Laguerre polynomials. We shall first develop a general theory of these polynomials.

Families of orthogonal polynomials arise as solutions of distinct **Sturm–Liouville problems**, with different **boundary conditions** leading naturally to different **weight functions** and **intervals of orthogonality** (Table 13.1) .

A **Sturm–Liouville problem** consists of a second-order linear differential equation of the form

$$\frac{d}{dx}\left[p(x)\frac{dy}{dx}\right] + [\lambda w(x) - q(x)]y = 0,$$

together with appropriate boundary conditions. Such problems possess a discrete set of eigenvalues λ, and the corresponding eigenfunctions are orthogonal with respect to the **weight function** $w(x)$ over a specified interval. Many of the special functions encountered in mathematical physics—such as Legendre, Hermite, and Laguerre polynomials—arise naturally as solutions of Sturm–Liouville problems with different choices of $p(x)$, $q(x)$, $w(x)$, and boundary conditions (Table 13.2).

Table 13.2 Sturm–Liouville forms of commonly encountered orthogonal polynomials, showing the associated interval and weight function.

Polynomial family	Symbol	Differential equation (Sturm–Liouville form)	Interval	Weighting function
Legendre	$P_l(x)$	$\frac{d}{dx}\left[(1-x^2)\frac{dy}{dx}\right]+l(l+1)y=0$	$-1\leq x\leq 1$	1
Associated Legendre	$P_l^m(x)$	$\frac{d}{dx}\left[(1-x^2)\frac{dy}{dx}\right]+\left[l(l+1)-\frac{m^2}{1-x^2}\right]y=0$	$-1\leq x\leq 1$	1
Hermite	$H_n(x)$	$\frac{d}{dx}\left(e^{-x^2}\frac{dy}{dx}\right)+2n\,e^{-x^2}y=0$	$-\infty<x<\infty$	e^{-x^2}
Laguerre	$L_n(x)$	$\frac{d}{dx}\left(e^{-x}\frac{dy}{dx}\right)+n\,e^{-x}y=0$	$0\leq x<\infty$	e^{-x}
Associated Laguerre	$L_n^\alpha(x)$	$\frac{d}{dx}\left(x^\alpha e^{-x}\frac{dy}{dx}\right)+n\,x^\alpha e^{-x}y=0$	$0\leq x<\infty$	$x^\alpha e^{-x}$

13.4.1 Legendre Polynomials

Thus, the Legendre polynomial is a special case of a **Sturm–Liouville problem**, whose eigenfunctions form orthogonal and complete sets under appropriate boundary conditions.

Math Box 13.1: Derivation of Legendre's Equation in Compact Form

Legendre's equation arises from solving Laplace's equation in spherical coordinates under azimuthal symmetry. The standard form is:

$$(1-x^2)\frac{d^2y}{dx^2}-2x\frac{dy}{dx}+l(l+1)y=0, \tag{1}$$

where $x=\cos\theta$ and $y=P_l(x)$ are the Legendre polynomials.

Step 1: Start with the expanded form

Equation (1) can be rewritten in **divergence (Sturm–Liouville) form**:

$$\frac{d}{dx}\left[(1-x^2)\frac{dy}{dx}\right]+l(l+1)y=0. \tag{2}$$

Step 2: Expand the derivative in (2)

Apply the product rule:

$$\frac{d}{dx}\left[(1-x^2)\frac{dy}{dx}\right]=\frac{d(1-x^2)}{dx}\cdot\frac{dy}{dx}+(1-x^2)\frac{d^2y}{dx^2}.$$

Since

$$\frac{d}{dx}(1-x^2)=-2x,$$

we obtain

$$\frac{d}{dx}\left[(1-x^2)\frac{dy}{dx}\right]=-2x\frac{dy}{dx}+(1-x^2)\frac{d^2y}{dx^2}.$$

Step 3: Substitute back into (2)

$$-2x\frac{dy}{dx} + (1-x^2)\frac{d^2y}{dx^2} + l(l+1)y = 0.$$

Rearranging terms:

$$(1-x^2)\frac{d^2y}{dx^2} - 2x\frac{dy}{dx} + l(l+1)y = 0.$$

This is exactly Equation (1).

Step 4: Why the compact form (2) is useful

The form:

$$\frac{d}{dx}\left[(1-x^2)\frac{dP_l}{dx}\right] + l(l+1)P_l = 0$$

is the **Sturm–Liouville form** of Legendre's equation, with:

- **weight function** $w(x) = 1$;
- **eigenvalue** $\lambda = l(l+1)$;
- **interval** $x \in [-1, 1]$.

This form is convenient for proving **orthogonality** via integration by parts, since boundary terms vanish at $x = \pm 1$.

Summary of equivalence:

$$\boxed{\frac{d}{dx}\left[(1-x^2)\frac{dP_l}{dx}\right] + l(l+1)P_l = 0} \Longleftrightarrow \boxed{(1-x^2)P_l'' - 2xP_l' + l(l+1)P_l = 0.}$$

Both represent the same differential equation satisfied by Legendre polynomials $P_l(x)$.

It may be pointed out that, in quantum-mechanical applications, $x = \cos\theta$ is used for Legendre polynomials. Since θ is the polar angle in spherical polar coordinates, its interval is $[0, \pi]$. This substitution correctly maps the domain of x to $[-1,1]$, as required for the polynomials.

13.4.2 Orthogonality

A set of functions $\varphi_0(x), \varphi_1(x), \ldots$ is said to form an orthonormal set over the interval with respect to a **weight function** $w(x) \geq 0$ if it satisfies the condition

$$\int_a^b \varphi_m(x)\varphi_n(x)w(x)\,dx = \delta_{mn}. \tag{13.20}$$

The interval and weight function completely define the polynomial. To see this, let $\varphi_0(x)$ be 1. Then, assuming that $\varphi_1(x)$ is of the form $c_1 + c_2x$, and using condition (13.20), we obtain

$$\int_a^b w(x)\varphi_0(x)\varphi_1(x)dx = 0 \Rightarrow c_1\int_a^b w(x)dx + c_2\int_a^b w(x)xdx. \tag{13.21}$$

This gives c_2 in terms of c_1 and hence a constant times φ_1 is obtained. The constant can be found by using the standardizing convention for that particular integral. For example, for Legendre polynomials, the condition is $P_n(1) = 1$. Reference to Equation (13.7) shows that $w(x) = 1$ for the Legendre polynomials. This process can be repeated by assuming that $\varphi_2(x) = c_3 + c_4x + c_5x^2$ and requiring that it is orthogonal to both $\varphi_0(x)$ and $\varphi_1(x)$, giving two equations. Thus, $\varphi_2(x)$ is obtained to within a multiplicative constant that can be determined by the normalization condition. This process is called the *Gram–Schmidt orthogonalization* procedure, which we have already introduced in Chapter 5.

The Gram–Schmidt procedure is a systematic method for generating an orthogonal set of polynomials from the simple set $1, x, x^2,...,$ given a specific interval and weight function (Examples 13.8 and 13.9).

Example 13.8: Gram–Schmidt Orthogonalization for Legendre Polynomials

Use the Gram–Schmidt orthogonalization procedure to obtain the first two Legendre polynomials.

Solution:

Let $\varphi_0(x) = 1$ and $\varphi_1(x) = c_1 + c_2x$. Then, from Equation (13.21) with $w(x) = 1$, $a = -1$, $b = 1$, we obtain

$$\int_a^b w(x)\varphi_0(x)\varphi_1(x)dx = 0 \Rightarrow c_1\int_{-1}^{1} dx + c_2\int_{-1}^{1} xdx = 2c_1 \Rightarrow c_1 = 0,$$

since the integrand in the second term is an odd function of x and thus vanishes. Therefore, $\varphi_1(x) = c_2x$. Legendre polynomials are standardized by the condition $P_n(1) = 1$. Hence, $c_2 = 1$ and $\varphi_1(x) = x$.

Now, let $\varphi_2(x) = c_3 + c_4x + c_5x^2$. By requiring that it be orthogonal to both $\varphi_0(x)$ and $\varphi_1(x)$, we have the two conditions

$$\int_a^b w(x)\varphi_0(x)\varphi_2(x)dx = 0 \Rightarrow c_3\int_{-1}^{1} dx + c_4\int_{-1}^{1} xdx + c_5\int_{-1}^{1} x^2dx = 2c_3 + \frac{2}{3}c_5 \Rightarrow c_5 = -3c_3$$

and

$$\int_a^b w(x)\varphi_1(x)\varphi_2(x)dx = 0 \Rightarrow c_3\int_{-1}^{1} xdx + c_4\int_{-1}^{1} x^2dx + c_5\int_{-1}^{1} x^3dx = \frac{2}{3}c_4 \Rightarrow c_4 = 0.$$

Therefore,

$$\varphi_2(x) = c_3(3x^2 - 1)/2.$$

The above example shows that it is enough to know the interval and the weighting function in order to generate the polynomials.

We illustrate this with another example.

Example 13.9: Gram–Schmidt Generation of Hermite Polynomials

Use Gram–Schmidt orthogonalization to generate the first few polynomials that are orthogonal over the interval $-\infty < x < \infty$ with weighting function e^{-x^2}. Use the convention that the coefficient of x^n is 2^n.

Solution:

Let $\varphi_0(x) = 1$ and $\varphi_1(x) = c_1 + c_2 x$. Then, from Equation (13.21) with $w(x) = e^{-x^2}$, $a = -\infty$, $b = \infty$, we obtain

$$\int_a^b w(x)\varphi_0(x)\varphi_1(x)dx = 0 \Rightarrow c_1 \int_{-\infty}^{\infty} e^{-x^2}dx + c_2 \int_{-\infty}^{\infty} xe^{-x^2}dx = \sqrt{\pi}c_1 \Rightarrow c_1 = 0,$$

where we have used the standard integral and the second term vanishes because it is the integral of an odd function. Therefore, $\varphi_1(x) = c_2 x$. We require the coefficient of x^n to be 2^n and hence $c_2 = 2$ or $\varphi_1(x) = 2x$.

Now, let $\varphi_2(x) = c_3 + c_4 x + c_5 x^2$. By requiring that it be orthogonal to both $\varphi_0(x)$ and $\varphi_1(x)$, we have the two conditions

$$\int_a^b w(x)\varphi_0(x)\varphi_2(x)dx = 0$$

$$\Rightarrow c_3 \int_{-\infty}^{\infty} e^{-x^2}dx + c_4 \int_{-\infty}^{\infty} xe^{-x^2}dx + c_5 \int_{-\infty}^{\infty} x^2 e^{-x^2}dx = \sqrt{\pi}c_3 + \frac{\sqrt{\pi}}{2}c_5$$

$$\Rightarrow c_5 = -2c_3$$

and

$$\int_a^b w(x)\varphi_1(x)\varphi_2(x)dx = 0$$

$$\Rightarrow 2c_3 \int_{-\infty}^{\infty} xe^{-x^2}dx + 2c_4 \int_{-\infty}^{\infty} x^2 e^{-x^2}dx + 2c_5 \int_{-\infty}^{\infty} x^3 e^{-x^2}dx = \sqrt{\pi}c_4$$

$$\Rightarrow c_4 = 0.$$

Therefore,

$$\varphi_2(x) = c_3 - 2c_3 x^2.$$

Applying the condition, we obtain $-2c_3 = 2^2$, or $c_3 = -2$, and $\varphi_2(x) = 4x^2 - 2$.
Continuing, we find that $\varphi_3(x) = 8x^3 - 12x$ and $\varphi_4(x) = 16x^4 - 48x^2 + 12$.

13.4.3 Hermite Polynomials

The polynomials that we have just obtained in Example 13.9 are *Hermite polynomials*, $H_n(x)$, which occur in the quantum-mechanical treatment of the harmonic oscillator problem. Note that

$H_n(-x) = (-1)^n H_n(x)$, i.e., alternate polynomials have even and odd **parity**, just like the Legendre polynomials.

Historical Note 13.2: Charles Hermite (1822–1901)

Charles Hermite was a French mathematician renowned for his contributions to algebra, analysis, and number theory. The **Hermite polynomials** emerged from his investigations into orthogonal polynomials and Gaussian-weighted function spaces. Their most profound physical significance appeared later, with the development of quantum mechanics.

Hermite polynomials form the natural basis functions for the **quantum harmonic oscillator**, one of the most fundamental exactly solvable systems in physics and chemistry. The Gaussian factor and associated Hermite polynomials arise directly from the structure of the Schrödinger equation for quadratic potentials.

Hermite's work illustrates the deep connection between orthogonal polynomials and physical observables, particularly in systems governed by quadratic Hamiltonians, vibrational spectroscopy, and ladder-operator formalisms.

Figure 13.4 plots the first few Hermite polynomials.

Interpretation: Hermite polynomials exhibit **alternating parity** and increasing oscillations with higher order n. Their orthogonality with respect to the weight function e^{-x^2} and nodal structure make them fundamental in describing **harmonic oscillator eigenfunctions** in quantum mechanics.

The **orthogonality** of Hermite polynomials $H_2(x)$ and $H_3(x)$ is illustrated in Figure 13.5, where we have scaled the areas to emphasize them,

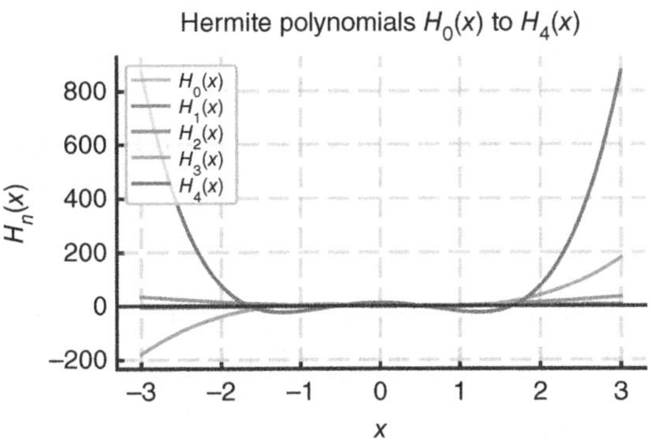

Figure 13.4 Plot of Hermite polynomials $H_0(x)$ to $H_4(x)$.
Key features:
- $H_0(x) = 1$−constant function;
- $H_1(x) = 2x$−linear, odd symmetry;
- $H_2(x) = 4x^2 - 2$−even, quadratic shape;
- $H_3(x) = 8x^3 - 12x$−cubic, odd symmetry;
- $H_4(x) = 16x^4 - 48x^2 + 12$−even, quartic shape;
- even n: even functions | odd n: odd functions.

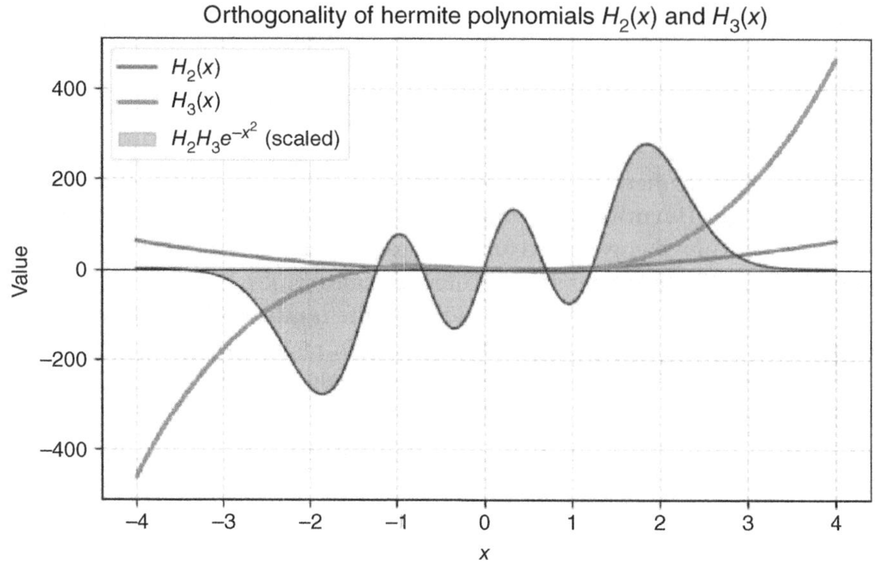

Figure 13.5 Orthogonality of Hermite polynomials $H_2(x)$ and $H_3(x)$. The shaded region represents the integral $\int_{-\infty}^{\infty} H_2(x)H_3(x)e^{-x^2}\,dx = 0$, where alternating positive and negative lobes cancel symmetrically.

Interpretation: Hermite polynomials are **orthogonal** on $(-\infty, \infty)$ with respect to the weight e^{-x^2}, forming the basis of harmonic oscillator eigenfunctions in quantum mechanics. The visualization emphasizes how their product integrates to zero due to alternating sign regions.

Textbox 13.2: Hermite Orthogonality

Observation

The shaded region shows the weighted product $H_2(x)H_3(x)e^{-x^2}$. Positive and negative lobes cancel symmetrically, demonstrating that Hermite polynomials of different orders are orthogonal on $(-\infty, \infty)$ with respect to the weight e^{-x^2}.

We have demonstrated in Example 13.9 that it is enough to know the interval and weighting function, along with their defining convention, to generate orthogonal polynomials. In Table 13.2, some orthogonal polynomials and their defining intervals and weighting functions are listed. These are the ones that occur commonly in quantum mechanics.

Associated Legendre functions: A clarification on the **weighting function** for the associated Legendre functions is important. The orthogonality integral in x (Equation 13.18) has an implicit weight function of $w(x) = 1$. However, when changing the variable of integration from x to θ, the differential $dx = -\sin\theta\, d\theta$ must also be substituted.

This transforms the integral:

$$\int_{-1}^{1} P_l^m(x)P_k^m(x)dx = \int_{\pi}^{0} P_l^m(\cos\theta)P_k^m(\cos\theta)(-\sin\theta\, d\theta)$$

$$= \int_{0}^{\pi} P_l^m(\cos\theta)P_k^m(\cos\theta)\sin\theta\, d\theta.$$

Therefore, the correct weighting function for the associated Legendre polynomials when expressed as functions of θ is $w(\theta) = \sin\theta$.

13.4.4 Laguerre Polynomials

The associated Legendre polynomials reduce to the regular Legendre polynomials when $m = 0$, and the **associated Laguerre polynomials** ($L_n^\alpha(x)$) to the unassociated ones when $\alpha = 0$ (Table 13.2). The latter are involved in the quantum-mechanical solution to the radial part of the hydrogen atom. A few Laguerre polynomials are listed below:

$$\begin{aligned}
L_0(x) &= 1,\\
L_1(x) &= 1 - x,\\
L_2(x) &= \frac{1}{2}(x^2 - 4x + 2),\\
L_3(x) &= \frac{1}{6}(-x^3 + 9x^2 - 18x + 6).
\end{aligned} \tag{13.22}$$

Math Box 13.2: A Note on Laguerre Polynomial Conventions

In the scientific literature, there are two common conventions for normalization of Laguerre polynomials ($L_n(x)$):

1. **The Standard (Modern) Convention:** $L_n(0) = 1$. This is used by most modern physics textbooks (e.g., Arfken & Weber), quantum chemistry software (Gaussian, ORCA), and computer algebra systems (Python/SciPy, Mathematica).
2. **The Historical Convention:** $L_n(0) = n!$. This is found in some older quantum mechanics texts.

This book utilizes the Standard Convention ($L_n(0) = 1$) to ensure consistency with modern computational tools. If you encounter formulas in older literature containing factors of $n!$ or $(n!)^2$ in Laguerre expressions, it is likely due to this difference in normalization.

The **defining condition** for the Laguerre polynomials is that $L_n(0) = 1$. The first few Laguerre polynomials are plotted in Figure 13.6.

Interpretation: Laguerre polynomials form an orthogonal set over $[0, \infty]$ with weight e^{-x}. They describe the **radial part of the hydrogen atom wavefunctions**, exhibiting more oscillations as the quantum number increases.

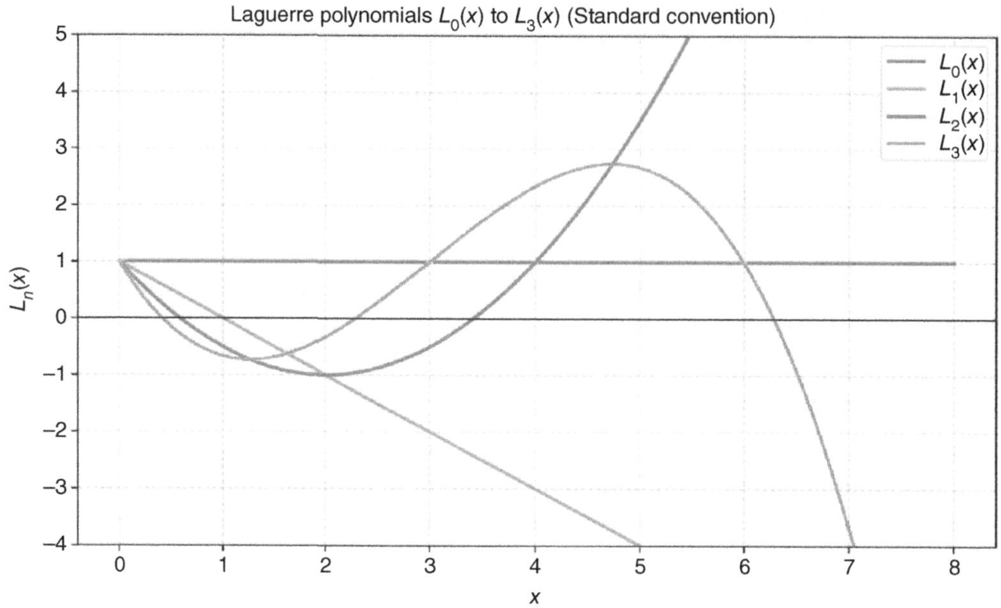

Figure 13.6 Plot of Laguerre polynomials $L_0(x)$ to $L_3(x)$.
Key features:
- $L_1(x) = 1 - x$.
- $L_2(x) = \frac{1}{2}(x^2 - 4x + 2)$.
- $L_3(x) = \frac{1}{6}(-x^3 + 9x^2 - 18x + 6)$.
- Increasing oscillatory behavior for higher n.

Historical Note 13.3: Edmond Laguerre (1834–1886)

Edmond Laguerre was a French mathematician whose work focused on geometry, analysis, and special functions. The **Laguerre polynomials** were introduced in the study of differential equations and orthogonal expansions with exponential weight functions.

Their most important physical application arose with the formulation of quantum mechanics, where **associated Laguerre polynomials** appear in the radial solutions of the **hydrogen atom Schrödinger equation**. They describe the radial dependence of atomic orbitals and are fundamental to understanding atomic structure, spectroscopy, and selection rules.

Laguerre polynomials thus provide a striking example of how mathematical tools developed in the 19th century became indispensable in 20th-century quantum theory, especially in problems involving Coulombic potentials and radial symmetry.

We now show that the first few Laguerre polynomials are **orthogonal** with respect to their weighting function (Figure 13.7 and Example 13.10).

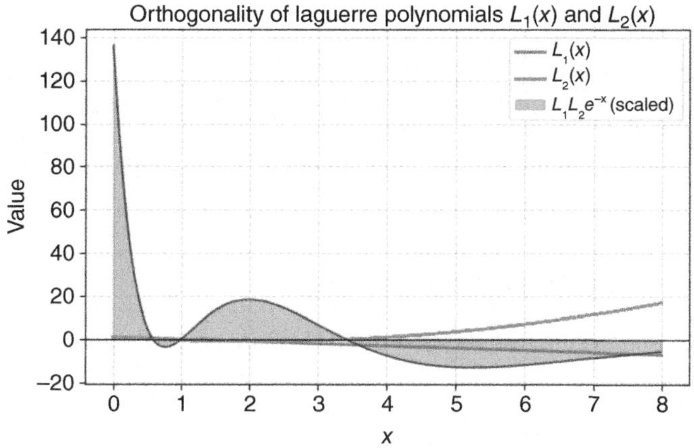

Figure 13.7 Orthogonality of Laguerre polynomials $L_1(x)$ and $L_2(x)$.

Example 13.10: Orthogonality of Laguerre Polynomials

Show that $L_0(x)$ is orthogonal to $L_1(x)$ and $L_2(x)$ with respect to the weighting function e^{-x} over the interval range $(0,\infty)$.

Solution:

$$\int_0^\infty L_0(x)L_1(x)e^{-x}dx = \int_0^\infty (1-x)e^{-x}dx = \int_0^\infty e^{-x}dx - \int_0^\infty xe^{-x}dx = 1-1=0,$$

$$\begin{aligned} 2\int_0^\infty L_0(x)L_2(x)e^{-x}dx &= \int_0^\infty (x^2-4x+2)e^{-x}dx \\ &= \int_0^\infty x^2e^{-x}dx - 4\int_0^\infty xe^{-x}dx + 2\int_0^\infty e^{-x}dx \\ &= 2-4+2=0. \end{aligned}$$

Interpretation: The positive and negative shaded areas cancel, demonstrating orthogonality. This property underlies the radial part of hydrogenic wavefunctions.

Textbox 13.3: Laguerre Orthogonality

Observation

The shaded area represents the weighted product $L_1(x)L_2(x)e^{-x}$. Small positive and negative regions cancel when integrated over $[0,\infty]$, confirming the orthogonality of Laguerre polynomials under the exponential weight e^{-x}.

13.4.5 Generating Functions

Similarly, the other polynomials listed in Table 13.2 form orthogonal sets. This can be proved using the **generating functions**, which are the key to the properties of these polynomials. The generating functions of these polynomials are listed in Table 13.3.

Example 13.11 illustrates the use of the generating function in Table 13.3 for generating the first three Hermite polynomials.

Example 13.11: Hermite Polynomials from Generating Function

Use the generating function in Table 13.3 to generate the first three Hermite polynomials.

Solution:

Expanding the exponential:

$$e^y = 1 + y + \frac{y^2}{2!} + \frac{y^3}{3!} + \cdots$$

and setting $y = 2xt - t^2$, we can write

$$\begin{aligned} e^{2xt-t^2} &= 1 + (2xt - t^2) + \frac{(2xt - t^2)^2}{2!} + O(t^3) \\ &= 1 + (2x)t + (2x^2 - 1)t^2 + O(t^3) \\ &= H_0(x) + H_1(x)t + \frac{H_2(x)}{2!}t^2 + O(t^3). \end{aligned}$$

Hence, $H_0(x) = 1$, $H_1(x) = 2x$, and $H_2(x) = 4x^2 - 2$.

Table 13.3 The generating functions for the orthogonal polynomials listed in Table 13.2.

Polynomial	Generating function	Condition
$P_l(x)$	$(1 - 2xt + t^2)^{-1/2} = \sum_{l=0}^{\infty} P_l(x)t^l$	$\lvert x\rvert \le 1,\ \lvert t\rvert < 1$
$P_l^m(x)$	$\frac{(2m)!}{2^m m!}\frac{(1 - x^2)^{m/2}t^m}{(1 - 2xt + t^2)^{m+1/2}} = \sum_{l=m}^{\infty} P_l^m(x)t^l$	$\lvert x\rvert < 1,\ \lvert t\rvert < 1$
$H_n(x)$	$e^{2xt-t^2} = \sum_{n=0}^{\infty} \frac{H_n(x)}{n!}t^n$	$x \in (-\infty, \infty),\ \lvert t\rvert < \infty$
$L_n(x)$	$\frac{e^{-xt/(1-t)}}{1 - t} = \sum_{n=0}^{\infty} L_n(x)t^n$	$x \ge 0,\ \lvert t\rvert < 1$
$L_n^{\alpha}(x)$	$\frac{e^{-xt/(1-t)}}{(1 - t)^{\alpha+1}} = \sum_{n=0}^{\infty} L_n^{(\alpha)}(x)t^n$	$x \ge 0,\ \lvert t\rvert < 1,\ \alpha > -1$

13.4.6 Recurrence Relations

The generating functions can be used to derive the **recurrence relations** (Table 13.4), which are very useful in many applications. For example, for the Hermite polynomials, we first take the derivatives with respect to t of both sides of the generating function (Table 13.3)

$$(2x - 2t)e^{2xt-t^2} = \sum_{n=0}^{\infty} \frac{nH_n(x)}{n!} t^{n-1}$$

$$\therefore 2(x - t) \sum_{n=0}^{\infty} \frac{H_n(x)}{n!} t^n = \sum_{n=0}^{\infty} \frac{nH_n(x)}{n!} t^{n-1}$$

$$\Rightarrow 2 \sum_{n=0}^{\infty} x \frac{H_n(x)}{n!} t^n - 2 \sum_{n=0}^{\infty} \frac{H_n(x)}{n!} t^{n+1} = \sum_{n=0}^{\infty} \frac{nH_n(x)}{n!} t^{n-1}.$$

Equating all powers to t^n, we obtain

$$2x \sum_{n=0}^{\infty} \frac{H_n(x)}{n!} t^n - 2 \sum_{n=1}^{\infty} \frac{H_{n-1}(x)}{(n-1)!} t^n = \sum_{n=0}^{\infty} \frac{(n+1)H_{n+1}(x)}{(n+1)!} t^n.$$

Multiplying throughout by $n!$, we obtain

$$2xH_n(x) - 2nH_{n-1}(x) = H_{n+1}(x), \quad n \geq 1,$$

or

$$xH_n(x) = nH_{n-1}(x) + \frac{1}{2} H_{n+1}(x), \quad n \geq 1.$$

Another recurrence relation for Hermite polynomials can be obtained by differentiating the generating function, this time with respect to x,

$$(2t)e^{2xt-t^2} = \sum_{n=0}^{\infty} \frac{H_n'(x)}{n!} t^n \Rightarrow 2 \sum_{n=0}^{\infty} \frac{H_n(x)}{n!} t^{n+1} = \sum_{n=1}^{\infty} \frac{H_n'(x)}{n!} t^n \Rightarrow 2 \sum_{n=1}^{\infty} \frac{H_{n-1}(x)}{(n-1)!} t^n = \sum_{n=1}^{\infty} \frac{H_n'(x)}{n!} t^n.$$

Hence, the recurrence relation is

$$2 \frac{H_{n-1}(x)}{(n-1)!} = \frac{H_n'(x)}{n!} \Rightarrow H_n'(x) = 2nH_{n-1}(x), \quad n \geq 1. \tag{13.23}$$

Table 13.4 Some recurrence formulae for the orthogonal polynomials listed in Table 13.2.

Polynomial	Recurrence formula	Condition
$P_l(x)$	$(l+1)P_{l+1}(x) - (2l+1)xP_l(x) + lP_{l-1}(x) = 0$	$l \geq 1$
$P_l^m(x)$	$(l-m+1)P_{l+1}^m(x) - (2l+1)xP_l^m(x) + (l+m)P_{l-1}^m(x) = 0$	$l \geq m+1, \quad \|m\| \leq l-1$
$H_n(x)$	$H_{n+1}(x) - 2xH_n(x) + 2nH_{n-1}(x) = 0$	$n \geq 1$
$L_n(x)$	$(n+1)L_{n+1}(x) + (x-1-2n)L_n(x) + nL_{n-1}(x) = 0$	$n \geq 1$
$L_n^\alpha(x)$	$(n+1)L_{n+1}^\alpha(x) - (2n+\alpha+1-x)L_n^\alpha(x) + (n+\alpha)L_{n-1}^\alpha(x) = 0$	$n \geq 1$

Example 13.12 illustrates the use of the recurrence formula (Table 13.4) for Laguerre polynomials to obtain the third polynomial given the first two.

Example 13.12: Generating $L_3(x)$ from Lower-Order Polynomials

Use the recurrence formula in Table 13.4 and the fact that $L_0(x) = 1$, $L_1(x) = 1 - x$, and $L_2(x) = \frac{1}{2}(x^2 - 4x + 2)$ to derive $L_3(x)$.

Solution:

The standard recurrence relation for Laguerre polynomials is given by (Table 13.4):

$$(n+1)L_{n+1}(x) + (x - 2n - 1)L_n(x) + nL_{n-1}(x) = 0$$

Setting $n = 2$ yields:

$$3L_3(x) + (x - 5)L_2(x) + 2L_1(x) = 0$$

Substituting the expressions for $L_1(x)$ and $L_2(x)$:

$$3L_3(x) + \frac{1}{2}(x-5)(x^2 - 4x + 2) + 2(1 - x) = 0$$

Solving for $L_3(x)$ gives:

$$3L_3(x) = -\frac{1}{2}(x^3 - 9x^2 + 22x - 10) - 2(1 - x)$$

$$3L_3(x) = \frac{1}{2}(-x^3 + 9x^2 - 18x + 6)$$

$$\boxed{L_3(x) = \frac{1}{6}(-x^3 + 9x^2 - 18x + 6)}$$

The generating function can be further used to derive the differential equations satisfied by the orthogonal polynomials. For example, for Hermite polynomials, the combination of the two recurrence relations (13.23) and the one in Table 13.3 gives

$$H_n'(x) = 2nH_{n-1}(x) = 2xH_n(x) - H_{n+1}(x). \tag{13.24}$$

Differentiation of this equation with respect to x gives

$$H_n''(x) = 2H_n(x) + 2xH_n'(x) - H_{n+1}'(x). \tag{13.25}$$

Using Equation (13.23) for the last term,

$$H_{n+1}'(x) = 2(n+1)H_n(x),$$

and substitution in (13.25) gives

$$H_n''(x) = 2H_n(x) + 2xH_n'(x) - 2(n+1)H_n(x).$$

Rearrangement gives

$$H_n''(x) - 2xH_n'(x) + 2nH_n(x) = 0, \tag{13.26}$$

which is Hermite's equation (Table 13.5).

13.4.7 Differential Equations

In quantum mechanics, we shall come across these polynomials as solutions to differential equations. The orthogonal polynomials listed in Table 13.2 satisfy the differential equations given in Table 13.5.

13.4.8 Orthogonality and Completeness Conditions

Generating functions can also be useful for obtaining the orthogonality and completeness conditions such as

$$\int_a^b w(x)\varphi_n(x)\varphi_m(x)dx = h_n\delta_{mn}. \tag{13.27}$$

These are given in Table 13.6 for the various polynomials.

For example, for the Hermite polynomials, using the generating function and the weighting function, we can write

$$\int_{-\infty}^{\infty} e^{-x^2}G(x,s)G(x,t)dx = \int_{-\infty}^{\infty} e^{-x^2}e^{2xs-s^2}e^{2xt-t^2}dx. \tag{13.28}$$

Table 13.5 The differential equations for the orthogonal polynomials.

Polynomial	Differential equation	Application
$P_l^m(x)$	$(1-x^2)y''(x) - 2xy'(x) + \left(l(l+1) - \frac{m^2}{1-x^2}\right)y(x) = 0$	Spherical harmonics
$H_n(x)$	$y''(x) - 2xy'(x) + 2ny(x) = 0$	Quantum harmonic oscillator
$L_n^\alpha(x)$	$xy''(x) + (\alpha+1-x)y'(x) + ny(x) = 0$	Radial part of hydrogen atom

Table 13.6 The integral conditions (13.30) for the orthogonal polynomials listed in Table 13.2.

Polynomial	Integral condition
$P_l(x)$	$\int_{-1}^{1} P_l(x)P_k(x)dx = \frac{2}{2l+1}\delta_{lk}$
$P_l^m(x)$	$\int_{-1}^{1} P_l^m(x)P_k^m(x)dx = \frac{2}{2l+1}\frac{(l+\lvert m\rvert)!}{(l-\lvert m\rvert)!}\delta_{lk}$
$H_n(x)$	$\int_{-\infty}^{\infty} e^{-x^2}H_n(x)H_m(x)dx = 2^n\sqrt{\pi}n!\delta_{mn}$
$L_n(x)$	$\int_0^{\infty} e^{-x}L_n(x)L_m(x)dx = \delta_{nm}$
$L_n^\alpha(x)$	$\int_0^{\infty} e^{-x}x^\alpha L_n^\alpha(x)L_m^\alpha(x)dx = \frac{(n+\alpha)!}{n!}\delta_{nm}$

The exponent can be written as a sum of squares by writing

$$-(x^2 - 2xs + s^2 - 2xt + t^2) = -(x - (s + t))^2 + 2st.$$

Therefore, Equation (13.28) becomes

$$\int_{-\infty}^{\infty} e^{-x^2} G(x, s)G(x, t)dx = e^{2st} \int_{-\infty}^{\infty} e^{-(x-s-t)^2} dx = \sqrt{\pi} e^{2st} = \sqrt{\pi} \sum_{n=0}^{\infty} \frac{(2st)^n}{n!}. \tag{13.29}$$

Now, the left-hand side of (13.29) can also be expressed as a series expansion (Table 13.3) as

$$\int_{-\infty}^{\infty} e^{-x^2} G(x, s)G(x, t)dx = \sum_{n=0}^{\infty} \sum_{m=0}^{\infty} \left(\int_{-\infty}^{\infty} e^{-x^2} H_m(x)H_n(x) \right) \frac{s^m t^n}{m!n!}. \tag{13.30}$$

Comparison with Equation (13.29) reveals that $m = n$. Equating the same powers of s and t in Equations (13.29) and (13.30) gives

$$\int_{-\infty}^{\infty} e^{-x^2} [H_n(x)]^2 dx \frac{(st)^n}{(n!)^2} = \sqrt{\pi} \frac{2^n (st)^n}{n!} \Rightarrow \int_{-\infty}^{\infty} e^{-x^2} [H_n(x)]^2 dx = \sqrt{\pi} 2^n n! \tag{13.31}$$

as given in Table 13.6.

The completeness of the orthogonal polynomials implies that any function obeying the Dirichlet conditions and satisfying the boundary conditions of the polynomial can be expanded in terms of the polynomial.

13.4.9 Rodrigues's Formulae

Finally, series expansion and differentiation of the generating functions yield the Rodrigues's formulae, which may also be used to derive the orthogonality and completion relations. For example, differentiation of the generating function n times with respect to t and then setting t equal to zero yields the Rodrigues's formula for Hermite polynomials.

The Rodrigues's formulae for the polynomials are listed in Table 13.7.

Table 13.7 The Rodrigues's formulae for the orthogonal polynomials listed in Table 13.2.

Polynomial	Rodrigues's formula
$P_l(x)$	$\frac{1}{2^l l!} \left(\frac{d^l}{dx^l} \right) (x^2 - 1)^l$
$P_l^m(x)$	$(1 - x^2)^{m/2} \frac{d^m}{dx^m} P_l(x) m > 0$
	$P_l^{-m}(x) = (-1)^m \frac{(l - \lvert m \rvert)!}{(l + \lvert m \rvert)!} P_l^m(x)$
$H_n(x)$	$(-1)^n e^{x^2} \frac{d^n}{dx^n} e^{-x^2}$
$L_n(x)$	$\frac{e^x}{n!} \frac{d^n}{dx^n} (x^n e^{-x})$
$L_n^\alpha(x)$	$L_n^{(\alpha)}(x) = \frac{x^{-\alpha} e^x}{n!} \frac{d^n}{dx^n} (e^{-x} x^{n+\alpha})$

Teacher's Note: Which Special Function Appears Where?

Students often find it hard to remember *which* orthogonal polynomials arise in *which* quantum problems. A simple way to organize this is to look at three features of the differential equation:

- the coordinate/domain,
- the weight factor in the scalar product,
- and the boundary conditions.

1. **Legendre and associated Legendre (P_l, P_l^m)**
   - *Coordinate/domain:* Angular part in spherical coordinates; $x = \cos\theta \in [-1, 1]$.
   - *Weight:* $w(x) = 1$ in x; in θ, the measure carries a factor $\sin\theta$.
   - *Boundary condition:* Wavefunction must be finite at the poles $\theta = 0, \pi$ (i.e., at $x = \pm 1$).
   - *Typical appearance:* Rigid rotator, angular part of the hydrogenic Schrödinger equation, multipole expansions in electrostatics.
2. **Laguerre and associated Laguerre $(L_n, L_n^{(\alpha)})$**
   - *Coordinate/domain:* Radial coordinate after suitable scaling; $r \in [0, \infty]$.
   - *Weight:* $w(x) = e^{-x}$ for $L_n(x)$; $w(x) = x^{\alpha}e^{-x}$ for $L_n^{(\alpha)}(x)$, with $\alpha > -1$.
   - *Boundary conditions:* Regular at the origin and square-integrable as $r \to \infty$.
   - *Typical appearance:* Radial part of hydrogen-like atoms; some radial diffusion and bound-state problems.
3. **Hermite (H_n)**
   - *Coordinate/domain:* Cartesian coordinate for the oscillator; $x \in (-\infty, \infty)$.
   - *Weight:* $w(x) = e^{-x^2}$.
   - *Boundary condition:* Square-integrable on the whole line (wavefunction $\to 0$ as $|x| \to \infty$).
   - *Typical appearance:* One-dimensional quantum harmonic oscillator; also in Gaussian integrals and normal-mode analysis.

How to present this to students:
When introducing a new Schrödinger equation:

1. Put it into Sturm–Liouville form and identify the interval and weight.
2. Match "interval + weight" with Table 13.1–13.8 to see whether the natural basis is Legendre, associated Legendre, Laguerre, associated Laguerre, or Hermite.
3. Use the boundary conditions (finiteness, normalizability) to explain why only certain discrete parameter values (e.g., l, n) are allowed, linking back to the truncation and quantization mechanisms developed in Chapters 12 and 13.

13.5 Summary

In this chapter, we have summarized the properties of special functions, which are useful for the quantum-mechanical solution to many problems. Detailed derivations are not required for a student of chemistry, but a few representative examples have been provided for the curious reader in Appendix J.6. Detailed properties are provided in the previously given tables for easy reference. The **quantum-mechanical applications** are summarized in Table 13.8.

Table 13.8 Orthogonal polynomials and their quantum-mechanical applications.

Polynomial/function	Differential equation	Weight function	Domain	Standardization	Quantum-mechanical application
Legendre $P_l(x)$	$(1-x^2)y'' - 2xy' + l(l+1)y = 0$	1	$-1 \leq x \leq 1$	$P_l(1) = 1$	Angular part of wavefunction in central potentials (spherical harmonics, hydrogen atom)
Associated Legendre $P_l^m(x)$	$(1-x^2)y'' - 2xy' + [l(l+1) - m^2/(1-x^2)]\ y = 0$	1	$-1 \leq x \leq 1$	Condon–Shortley phase: $P_l^m(x) \sim (-1)^m$ (normalized via spherical-harmonic convention)	Spherical harmonics $Y_l^m(\theta, \varphi)$; angular momentum eigenfunctions
Laguerre $L_n(x)$	$xy'' + (1-x)y' + ny = 0$	e^{-x}	$0 \leq x < \infty$	$L_n(0) = 1$	Radial wavefunctions of hydrogenic atoms ($R_{nl}(r)$)
Associated Laguerre $L_n^{(\alpha)}(x)$	$xy'' + (\alpha + 1 - x)y' + ny = 0$	$x^{\alpha}e^{-x}$	$0 \leq x < \infty$	$L_n^{(\alpha)}(0) = \left(\frac{n+\alpha}{n}\right)(n+\alpha)$ (so for $\alpha = 0$, $L_n(0) = 1$)	Hydrogen atom radial part with $\alpha = 2l + 1$
Hermite $H_n(x)$	$y'' - 2xy' + 2ny = 0$	e^{-x^2}	$-\infty < x < \infty$	H_n leading coefficient is 2^n	One-dimensional harmonic oscillator eigenfunctions
Chebyshev (first kind) $T_n(x)$	$(1-x^2)y'' - xy' + n^2y = 0$	$(1-x^2)^{-1/2}$	$-1 \leq x \leq 1$		Angular motion of rigid rotator, molecular vibrations (approximate)
Chebyshev (second kind) $U_n(x)$	$(1-x^2)y'' - 3xy' + n(n+2)y = 0$	$(1-x^2)^{1/2}$	$-1 \leq x \leq 1$		Quantum wells and vibrational models (approximate orthogonal basis)

Notes:

1. All these functions are **orthogonal** under their respective weight functions over the stated domains.
2. Their occurrence in quantum mechanics arises naturally from the **separation of variables** in the Schrödinger equation.
3. Legendre and Laguerre families dominate atomic and molecular models, while Hermite functions appear in oscillator and Gaussian wave packet analyses.

In summary, the Schrödinger equation for key model systems separates into angular and radial parts. The angular solutions are spherical harmonics, built from associated Legendre polynomials. The radial solution for the hydrogen atom involves associated Laguerre polynomials, while the harmonic oscillator yields wavefunctions based on Hermite polynomials. The orthogonality of these polynomials directly reflects the orthogonality of the corresponding quantum-mechanical wavefunctions.

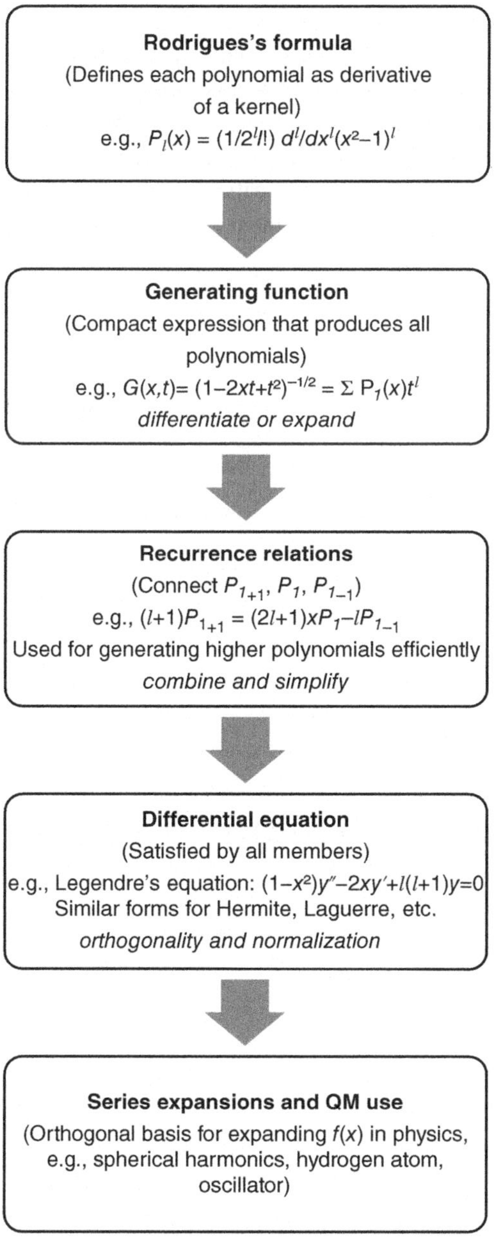

Figure 13.8 Algorithmic flow for manipulating orthogonal polynomials.

Quantum Systems → Separated Equations → Special Functions

Hydrogen Atom:
Angular Part → Associated Legendre Equation → $P_l^m(x)$ → Spherical Harmonics;
Radial Part → Associated Laguerre Equation → $L_{n-l-1}^{2l+1}(x)$ → Radial Wavefunctions.
Harmonic Oscillator:
1D Schrödinger Equation → Hermite Equation → $H_n(x)$ → Wavefunctions $\psi_n(x)$.

Figure 13.8 shows an algorithmic flow while manipulating orthogonal polynomials.

This schematic summarizes the practical steps used when working with special functions such as Legendre, Laguerre, and Hermite polynomials. Starting from the generating function or Rodrigues's form, one may **differentiate or expand** to obtain explicit polynomial expressions. Subsequent steps **combine and simplify** these expressions to derive recurrence relations or closed forms. Finally, the **orthogonality and normalization** conditions are verified to ensure that the functions form complete, physically meaningful bases—applicable in problems such as the hydrogen atom, harmonic oscillator, and spherical harmonics.

Further Reading

G. B. Arfken, H. J. Weber and F. E. Harris, *Mathematical Methods for Physicists*, 7th ed., Academic Press (Elsevier), 2012.
J. Mathews and R. L. Walker, *Mathematical Methods of Physics*, W. A. Benjamin, 1964.
M. Abramowitz and I. A. Stegun (eds.), *Handbook of Mathematical Functions*, Dover Publications, 1965.
D. A. McQuarrie, *Quantum Chemistry*, 2nd ed., University Science Books, 2007.

Exercises

A. Conceptual Questions

1. **Representation of Polynomials**
   - **a.** Using the principle of mathematical induction, show that any polynomial $f(x)$ of degree n can be represented in the form

     $$f(x) = \sum_{l=0}^{n} a_l P_l(x),$$

     where $P_l(x)$ are Legendre polynomials.
   - **b.** Show that

     $$\int_{-1}^{1} P_l(x) P_k(x)\, dx = 0 \text{ for } k \neq m.$$

   - **c.** Represent the following polynomials in the form (a):

     $$1,\ x,\ x^2,\ x^3,\ 1 - 3x + x^2.$$

2. **Expansion in Legendre Polynomials**
For the polynomials given below, expand each in terms of Legendre polynomials $P_l(x)$:
a. $f(x) = x^2$;
b. $f(x) = x^3$;
c. $f(x) = 1 - 3x + x^2$;
d. $f(x) = 3x^4 - 5x^2 + 2$.

B. Problems

1. **Rodrigues's Formula (Laguerre Polynomials)**
Using Rodrigues's formula for Laguerre polynomials given in Table 13.7, show that

$$L_n(x) = \frac{e^x}{n!}\frac{d^n}{dx^n}(x^n e^{-x}).$$

2. **Laguerre Differential Equation**
Verify that $L_n(x)$ satisfies Laguerre's differential equation (with $\alpha = 0$) given in Table 13.5 for $n = 0, 1, 2, 3$:

$$xy'' + (1 - x)y' + ny = 0.$$

3. **Orthogonality of Laguerre Polynomials**
Verify by direct integration that $L_0(x), L_1(x), L_2(x)$ are orthogonal on the positive x-axis $0 < x < \infty$ with respect to the weight function e^{-x}:

$$\int_0^\infty L_m(x)L_n(x)e^{-x}\,dx = 0\,(m \neq n).$$

C. Advanced/Applied Problems

1. **Properties of Hermite Polynomials**
a. By differentiating the generating function (Table 13.3) with respect to x, show that the Hermite polynomials satisfy the recurrence relation

$$\frac{dH_n(x)}{dx} = 2nH_{n-1}(x)(n \geq 1).$$

b. By differentiating the generating function (Table 13.3) with respect to t, show that the Hermite polynomials satisfy the relation

$$H_{n+1}(x) = 2xH_n(x) - 2nH_{n-1}(x)(n \geq 1).$$

c. Using these two recurrence relations, show that the Hermite polynomials obey Hermite's differential equation given in Table 13.5:

$$y'' - 2xy' + 2ny = 0.$$

14

Partial Differential Equations

The next great era of awakening of human intellect may well produce a method of understanding the qualitative content of equations. Today we cannot see that the water flow equations contain such things as the barber pole structure of turbulence that one sees between rotating cylinders. Today we cannot see whether Schrödinger's equation contains frogs, musical composers, or morality—or whether it does not.

—Richard Feynman (The Feynman Lectures on Physics)

Learning Outcomes

After studying this chapter, the reader will be able to:

1. **Define** partial differential equations and **classify** their types.
2. **Derive** and **solve** the one-dimensional wave equation using separation of variables.
3. **Interpret** eigenvalues and eigenfunctions in physical systems.
4. **Apply** the Fourier series to solve PDEs with given initial and boundary conditions.
5. **Understand** the emergence of quantum-mechanical operators from the classical wave equation.

14.1 Introduction

Ordinary differential equations (ODEs) involve one variable. However, many physical problems can be formulated in terms of differential equations involving functions of more than one variable. These are known as **partial differential equations (PDEs)**. The example of a stretched string will serve as an introduction to the formulation of such problems and methods of their solution.

14.2 The "Big Three" Partial Differential Equations

The **one-dimensional wave equation** is a cornerstone of mathematical physics, but it is one of three fundamental linear PDEs that describe a vast range of phenomena. These "big three" second-order equations share a common structure but exhibit profoundly different behaviors. Understanding them together provides a powerful toolkit for modeling the physical world.

Table 14.1 summarizes their key characteristics.

A deep understanding of the one-dimensional wave equation and the method of separation of variables provides the essential toolkit for tackling all of them. The procedure is analogous, but the

Table 14.1 The "big three" partial differential equations.

PDE name	Standard form (1D)	Type of problem	Physical and chemical applications	Time behavior
Wave equation	$\partial^2 u/\partial t^2 = c^2(\partial^2 u/\partial x^2)$	Initial-boundary value problem	Vibrating string, sound waves, electromagnetic radiation	Oscillatory, reversible
Heat equation	$\partial u/\partial t = \alpha(\partial^2 u/\partial x^2)$	Initial-boundary value problem	Heat flow, chemical diffusion, dopant implantation, drug release	Decaying toward equilibrium, irreversible
Laplace's equation	$\partial^2 u/\partial x^2 + \partial^2 u/\partial y^2 = 0$	Boundary value problem	Steady-state temperature, electrostatics, potential fluid flow	Static, no time dependence

character of the resulting time equation—oscillatory for waves, decaying for diffusion, or nonexistent for Laplace—is what defines the unique physics of each system. We will now develop this method in detail.

14.3 The Stretched String: Wave Equation

14.3.1 Derivation and Boundary Conditions

Consider a thin wire or a string stretched horizontally between two ends at $x = 0$ and $x = L$ (Figure 14.1). Suppose it is displaced from its original straight shape by a small force, producing a transverse motion of the wire.

The motion is described by the one-dimensional wave equation:

$$\frac{\partial^2 u(x,t)}{\partial t^2} = c^2 \frac{\partial^2 u(x,t)}{\partial x^2}, \tag{14.1}$$

where $u(x, t)$ is the deflection of the string at point x along the string, and c is the velocity of propagation along the string. This is a **linear partial differential equation** of the second order in two variables (x and t) with constant coefficients.

Since a typical string of length L is fixed at its ends ($x = 0$ and $x = L$), any valid solution should satisfy the **boundary conditions**

$$u(0,t) = 0, \quad u(L,t) = 0 \tag{14.2}$$

for all values of t.

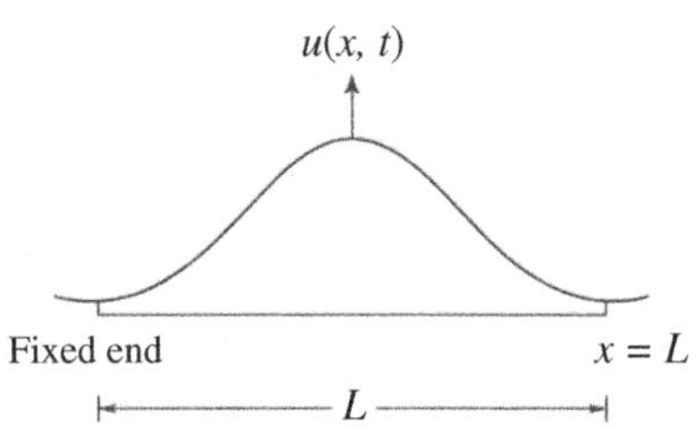

Figure 14.1 The stretched string at time t.

At the instant the string is released ($t = 0$), it has a certain deflection function $u_0(x)$ and a distribution of velocities $v_0(x)$. These are called the **initial conditions** and are expressed as

$$u(x, 0) = u_0(x), \quad \left(\frac{\partial u}{\partial t}\right)_{t=0} = v_0(x), \tag{14.3}$$

where $u_0(x)$ and $v_0(x)$ represent the displacements and velocities of all points of the string at time $t = 0$.

We now illustrate the solution of the wave equation by the **method of separation of variables**.

14.3.2 Method of Separation of Variables

Since the additional conditions imposed on $u(x, t)$ are of two types—(a) those imposed on x through the boundary conditions (14.2), and (b) those involving t through the initial conditions (14.3)—it is reasonable to expect that $u(x, t)$ may be written as a product of a **spatial function** $X(x)$ and a **temporal function** $T(t)$:

$$u(x, t) = X(x)\,T(t). \tag{14.4}$$

If we can solve for $X(x)$, imposing the boundary conditions (14.2) on it will ensure that they hold for $u(x, t)$ as well. Similarly, $T(t)$ may be chosen to satisfy the initial conditions (14.3).

Substituting Equation (14.4) into Equation (14.1) gives

$$X(x)\,\frac{d^2T}{dt^2} = c^2\,T(t)\,\frac{d^2X}{dx^2}. \tag{14.5}$$

Note that partial derivatives become ordinary derivatives after separation: $\frac{d^2T}{dt^2}$ not $\frac{\partial^2 T}{\partial t^2}$.

Dividing both sides by $X(x)T(t)$, we obtain

$$\frac{1}{c^2T}\,\frac{d^2T}{dt^2} = \frac{1}{X}\,\frac{d^2X}{dx^2}.$$

The left-hand side of this equation is a function of t alone, while the right-hand side depends only on x. The two sides can be equal only if both are independently equal to the same constant, denoted by λ, called the **separation constant**,

$$\frac{1}{c^2T}\,\frac{d^2T}{dt^2} = \frac{1}{X}\,\frac{d^2X}{dx^2} = \lambda. \tag{14.6}$$

14.3.2.1 Equation for $X(x)$

The equation for $X(x)$ is therefore

$$\frac{d^2X}{dx^2} - \lambda X = 0. \tag{14.7}$$

This is a linear, second-order, homogeneous differential equation of the type

$$\frac{d^2y}{dx^2} - \lambda y = 0.$$

Equation (14.7) is a classic **eigenvalue problem** of the form $\hat{L}X = \lambda X$, where $\hat{L}$ is the linear differential operator $\frac{d^2}{dx^2}$ subject to the fixed-end boundary conditions $X(0) = X(L) = 0$.

For different signs of λ:

Case 1: $\lambda = 0$

From (14.7), $\dfrac{d^2X}{dx^2} = 0$.

Since the second derivative is zero, $X(x)$ must be a linear function.

Thus,

$$X(x) = Cx + D.$$

Applying $u(0, t) = 0$ gives $X(0) = 0$, which implies $D = 0$. Applying $u(L, t) = 0$ gives $X(L) = 0$.

Applying $X(L) = 0$, $CL = 0 \Rightarrow C = 0$.

This gives only the trivial solution $X(x) = 0$, which is physically meaningless.

Case 2: $\lambda = k^2 > 0$, where k is a real constant

From (14.7),

$$\frac{d^2X}{dx^2} = k^2X.$$

Let $X = e^{\alpha x}$.

Then

$$\frac{d^2X}{dx^2} = \alpha^2 e^{\alpha x} = \alpha^2 X.$$

Therefore, $\alpha^2 = k^2$ or $\alpha = \pm k$.

Thus, the general solution is

$$X(x) = Ee^{kx} + Fe^{-kx}.$$

Applying $X(0) = 0$, $E + F = 0 \Rightarrow F = -E$.

So $X(x) = E\left(e^{kx} - e^{-kx}\right) = 2E \sinh(kx)$.

Applying $X(L) = 0$, $2E \sinh(kL) = 0$.

Since $\sinh(kL) \neq 0$ for $k, L > 0$, we must have $E = 0$, giving again only the trivial solution.

Case 3: $\lambda = -k^2 < 0$

From (14.7),

$$\frac{d^2X}{dx^2} + k^2X = 0.$$

Let $X = e^{\beta x}$. Then

$$\frac{d^2X}{dx^2} = \beta^2 e^{\beta x} = \beta^2 X.$$

Thus, $\beta^2 = -k^2 \Longrightarrow \beta = \pm ik$.

Therefore, the general solution is

$$X(x) = Ae^{ikx} + Be^{-ikx}. \tag{14.8}$$

Imposing the first boundary condition, $X(0) = 0$, implies $A + B = 0$.

Thus, Equation (14.8) becomes

$$X(x) = A\left(e^{ikx} - e^{-ikx}\right) = 2A \sin(kx).$$

Applying the second boundary condition, $X(L) = 0$,

$$2A \sin(k\,L) = 0. \tag{14.9}$$

Since $A \neq 0$, this is possible only if the argument is an integral multiple of π:

$$k\,L = n\pi, \quad n = 1, 2, 3, \dots .$$

Thus, the **allowed** values of the separation constant are

$$\lambda_n = -\left(\frac{n\pi}{L}\right)^2. \tag{14.10}$$

These are the **eigenvalues** of the operator d^2/dx^2. The corresponding **eigenfunctions** are obtained by substituting (14.10) into (14.9):

$$X_n(x) = c_n \sin\left(\frac{n\pi x}{L}\right). \tag{14.11}$$

Here c_n is an arbitrary nonzero constant. Equation (14.11) shows that there are infinitely many such eigenfunctions. Note, however, that $n = 0$ is excluded since it would give a trivial solution $X = 0$.

Conclusion: Only $\lambda < 0$ yields nontrivial solutions satisfying both boundary conditions.

14.3.2.2 Equation for $T(t)$

Substituting the separation constant $\lambda_n = -\left(\frac{n\pi}{L}\right)^2$ into the time-dependent equation obtained from (14.6), we get

$$\frac{d^2T_n}{dt^2} + c^2\left(\frac{n\pi}{L}\right)^2 T_n = 0. \tag{14.12}$$

The general solution of this equation is

$$T_n(t) = a_n \cos(\omega_n t) + b_n \sin(\omega_n t), \tag{14.13}$$

where

$$\omega_n = c\,\frac{n\pi}{L}. \tag{14.14}$$

Substituting Equations (14.11) and (14.13) into Equation (14.4), we obtain the complete solution

$$u_n(x,t) = [A_n \cos(\omega_n t) + B_n \sin(\omega_n t)] \sin\left(\frac{n\pi x}{L}\right), \tag{14.15}$$

where $A_n = a_n c_n$ and $B_n = b_n c_n$ are arbitrary constants.

14.3.3 Normal Modes and Quantization

Equation (14.15) is the **general solution** of the wave equation. It remains to apply the initial conditions to obtain a particular solution. Each individual solution $u_n(x, t)$ represents a possible motion of the stretched string corresponding to specific initial conditions. These are the **characteristic modes (normal modes)** of vibration of the string. Each represents a harmonic vibration with frequency ω_n given by (14.14).

Hence, the vibrations of the string are **quantized**, even though we are dealing with a classical system. Quantization arises in both classical and quantum-mechanical systems whenever boundary conditions impose discrete solutions. The integers n determine the characteristic values or **eigenvalues**.

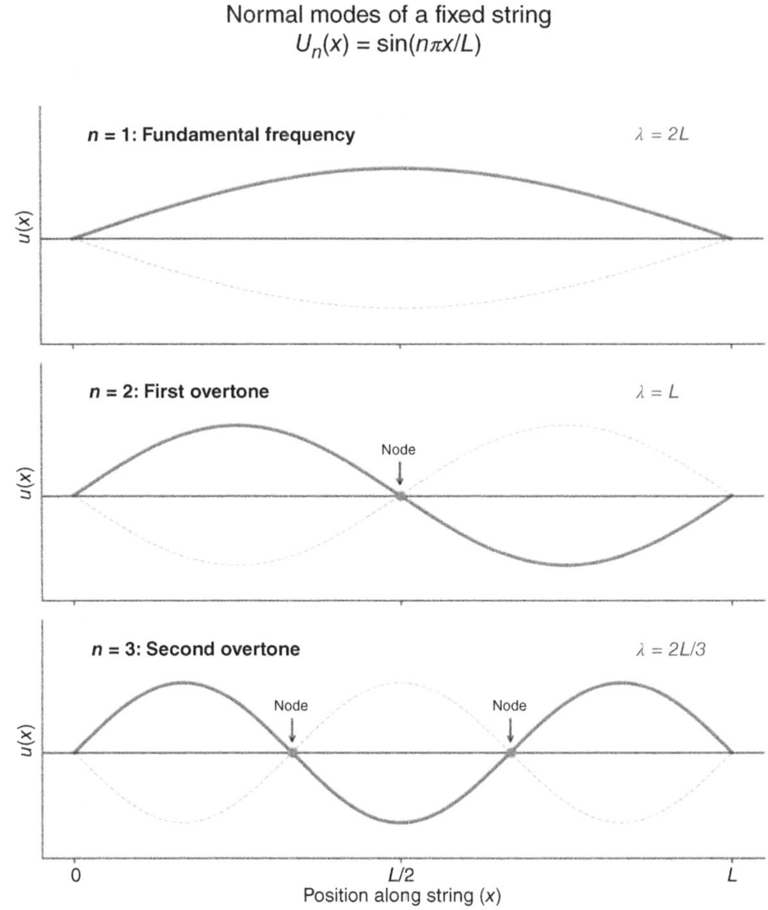

Figure 14.2 The first three normal modes of a vibrating string fixed at both ends ($x = 0$ and $x = L$).

The first mode ($n = 1$) is called the **fundamental**, while those corresponding to $n = 2, 3, \ldots$ are called the first, second, and so on **overtones**, respectively. This terminology is also used in the description of molecular vibrations. Figure 14.2 shows the first three normal modes:

- **(Top) Fundamental Mode ($n = 1$):** The string vibrates as a single arc. Wavelength $\lambda = 2L$.
- **(Middle) First Overtone ($n = 2$):** A node exists at the center $x = L/2$. Wavelength $\lambda = L$.
- **(Bottom) Second Overtone ($n = 3$):** Two nodes exist at $x = L/3$ and $x = 2L/3$. Wavelength $\lambda = 2L/3$.

14.3.4 Applying Initial Conditions

Applying the first initial condition $u(x, 0) = u_0(x)$ to Equation (14.15), we obtain

$$u_0(x) = \sum_{n=1}^{\infty} A_n \sin\left(\frac{n\pi x}{L}\right). \tag{14.16}$$

Similarly, applying the second initial condition $\left(\frac{\partial u}{\partial t}\right)_{t=0} = v_0(x)$ yields

$$v_0(x) = \sum_{n=1}^{\infty} \omega_n B_n \sin\left(\frac{n\pi x}{L}\right). \tag{14.17}$$

Any arbitrary function $u(x, t)$ will satisfy initial conditions of the form (14.16) and (14.17), and no other. Therefore, a solution based on a **single normal mode**, $u_n(x, t)$, is generally insufficient. To satisfy an *arbitrary* initial shape, we must use the **principle of superposition** and construct a general solution from a linear combination (an infinite series) of *all* possible modes.

Example 14.1: Plucked String With a Single Mode

Consider a string of length L fixed at both ends, released from rest. Its initial displacement is given by $u(x, 0) = u_0(x) = A \sin\left(\frac{\pi x}{L}\right)$, and its initial velocity is zero, $\frac{\partial u}{\partial t}(x, 0) = v_0(x) = 0$. Determine the subsequent motion $u(x, t)$.

Solution:

The general solution for the wave equation with these boundary conditions was found to be

$$u(x, t) = \sum_{n=1}^{\infty} [A_n \cos(\omega_n t) + B_n \sin(\omega_n t)] \sin\left(\frac{n\pi x}{L}\right),$$

where the normal mode frequencies are $\omega_n = \frac{cn\pi}{L}$.

The coefficients A_n and B_n are determined by the initial conditions.

Step 1: Apply the initial velocity condition.

The general solution's time derivative is

$$\frac{\partial u}{\partial t}(x, t) = \sum_{n=1}^{\infty} [-\omega_n A_n \sin(\omega_n t) + \omega_n B_n \cos(\omega_n t)] \sin\left(\frac{n\pi x}{L}\right).$$

Applying the condition $\frac{\partial u}{\partial t}(x, 0) = 0$,

$$\frac{\partial u}{\partial t}(x, 0) = \sum_{n=1}^{\infty} \omega_n B_n \sin\left(\frac{n\pi x}{L}\right) = 0.$$

For this Fourier sine series to be zero for all $x \in [0, L]$, each coefficient must be zero,

$$\omega_n B_n = 0 \text{ for all } n.$$

Since $\omega_n \neq 0$, this implies

$$B_n = 0 \text{ for all } n = 1, 2, 3.$$

Thus, the solution simplifies to a sum of cosines in time:

$$u(x, t) = \sum_{n=1}^{\infty} A_n \cos(\omega_n t) \sin\left(\frac{n\pi x}{L}\right).$$

Step 2: Apply the initial displacement condition.

We now apply $u(x, 0) = u_0(x) = A \sin\left(\frac{\pi x}{L}\right)$,

$$u(x, 0) = \sum_{n=1}^{\infty} A_n \sin\left(\frac{n\pi x}{L}\right) = A \sin\left(\frac{\pi x}{L}\right).$$

This is an equation in the form of a Fourier sine series. The coefficients A_n are given by the orthogonality of the sine functions,

$$A_n = \frac{2}{L} \int_0^L u_0(x) \sin\left(\frac{n\pi x}{L}\right) dx.$$

Substituting $u_0(x) = A \sin\left(\frac{\pi x}{L}\right)$,

$$A_n = \frac{2}{L} \int_0^L A \sin\left(\frac{\pi x}{L}\right) \sin\left(\frac{n\pi x}{L}\right) dx.$$

Using the orthogonality relation,

$$\int_0^L \sin\left(\frac{m\pi x}{L}\right) \sin\left(\frac{n\pi x}{L}\right) dx = \frac{L}{2}\delta_{mn},$$

where δ_{mn} is the Kronecker delta ($\delta_{mn} = 1$ if $m = n$, 0 otherwise), we find:
for $n = 1$,

$$A_1 = \frac{2}{L} \cdot A \cdot \frac{L}{2} = A;$$

for $n \neq 1$,

$$A_n = 0.$$

Step 3: Write the final solution.

Substituting the coefficients $A_1 = A$, $A_n = 0$ for $n \neq 1$, and $B_n = 0$ for all n into the general solution, we obtain the final result:

$$u(x, t) = A \cos\left(\frac{\pi ct}{L}\right) \sin\left(\frac{\pi x}{L}\right).$$

Interpretation:

This result has a clear physical meaning:

- The string vibrates purely in its fundamental mode (the first normal mode, $n = 1$).
- The spatial part, $\sin(\pi x/L)$, is the fixed shape of this mode—a single arch with maximum displacement at the center, $x = L/2$ (Figure 14.1).
- The temporal part, $\cos(\pi ct/L)$, describes the harmonic oscillation in time of every point on the string. All points move in phase with each other, reaching their maximum displacement simultaneously.
- The angular frequency of this oscillation is $\omega_1 = \frac{\pi c}{L}$.

This example beautifully illustrates that when the initial shape of the string matches exactly one of the eigenfunctions (normal modes), the resulting motion is purely that single mode. For more complex initial shapes (like the triangular pulse in Problem 3), multiple modes are excited, and the solution is a superposition of all of them.

14.3.5 Principle of Superposition and Fourier Series Solution

To overcome this difficulty, we use the **principle of superposition**. Since our PDE is linear and homogeneous, and so are its boundary conditions, it can be shown that any linear combination of two of its solutions, $u_n(x, t)$ and $u_m(x, t)$, is also a solution of the PDE and its boundary conditions. That is, if

$$u_1(x, t) = X_1(x)T_1(t), \quad u_2(x, t) = X_2(x)T_2(t),$$

then

$$u(x, t) = a_1u_1(x, t) + a_2u_2(x, t)$$

is also a solution. This follows directly from the linearity of the wave operator: since each u_i satisfies $\frac{\partial^2 u_i}{\partial t^2} = c^2\frac{\partial^2 u_i}{\partial x^2}$,

$$\frac{\partial^2 u}{\partial t^2} = a_1\frac{\partial^2 u_1}{\partial t^2} + a_2\frac{\partial^2 u_2}{\partial t^2} = a_1c^2\frac{\partial^2 u_1}{\partial x^2} + a_2c^2\frac{\partial^2 u_2}{\partial x^2} = c^2\frac{\partial^2 u}{\partial x^2}$$

so u satisfies the wave equation. The boundary conditions are satisfied likewise, since u_1 and u_2 each vanish at $x = 0$ and $x = L$, and so does any linear combination of them. By induction, a linear combination of any number of terms—even infinite—will also be a valid solution.

For a single mode $u_n(x, t) = X_n(x)T_n(t)$, the time factor satisfies $\frac{d^2T_n}{dt^2} + \omega_n^2 T_n = 0$, with $\omega_n = cn\pi/L$. Hence, the most general solution can be expressed as the **superposition of normal modes**:

$$u(x, t) = \sum_{n=1}^{\infty}[A_n \cos(\omega_n t) + B_n \sin(\omega_n t)] \sin\left(\frac{n\pi x}{L}\right). \tag{14.18}$$

Equation (14.18) satisfies both the PDE and the boundary conditions provided the series converges.

Setting $t = 0$ and using Equation (14.16), we can express $u_0(x)$ as a **Fourier sine series**:

$$u_0(x) = \sum_{n=1}^{\infty} A_n \sin\left(\frac{n\pi x}{L}\right). \tag{14.19}$$

The solution is a superposition of normal modes, each with a spatially fixed shape (the eigenfunction $\sin(n\pi x/L)$) and a harmonic oscillation in time. The initial conditions $u_0(x)$ and $v_0(x)$ determine how much of each mode is excited (coefficients A_n and B_n).

14.3.6 Determination of Coefficients

Multiplying both sides of Equation (14.19) by $\sin\left(\frac{m\pi x}{L}\right)$ and integrating from 0 to L,

$$\int_0^L u_0(x) \sin\left(\frac{m\pi x}{L}\right)dx = \sum_{n=1}^{\infty} A_n \int_0^L \sin\left(\frac{n\pi x}{L}\right) \sin\left(\frac{m\pi x}{L}\right)dx.$$

Due to the orthogonality condition

$$\int_0^L \sin\left(\frac{n\pi x}{L}\right)\sin\left(\frac{m\pi x}{L}\right)dx = \frac{L}{2}\delta_{mn}$$

all terms in the sum on the right-hand side vanish except when $n = m$:

$$\int_0^L u_0(x)\sin\left(\frac{m\pi x}{L}\right)dx = A_m\int_0^L \sin^2\left(\frac{m\pi x}{L}\right)dx = A_m\frac{L}{2}.$$

Therefore, we obtain

$$A_n = \frac{2}{L}\int_0^L u_0(x)\sin\left(\frac{n\pi x}{L}\right)dx. \tag{14.20}$$

Similarly, differentiating Equation (14.18) with respect to t, setting $t = 0$, and using Equation (14.17), we obtain

$$B_n = \frac{2}{L\omega_n}\int_0^L v_0(x)\sin\left(\frac{n\pi x}{L}\right)dx. \tag{14.21}$$

If the string is released from rest ($v_0(x) = 0$), then $B_n = 0$. If the initial displacement is zero ($u_0(x) = 0$), then $A_n = 0$.

The procedure can be extended to investigate vibrations in two dimensions, such as in stretched membranes. We shall, however, develop the method in three dimensions.

14.4 The Unified View: Separation of Variables for the "Big Three"

The power of the separation of variables method is its generality. It can be applied to the heat and Laplace equations using the same logic we applied to the wave equation. The key difference lies in the ordinary differential equation for time, $T(t)$.

Table 14.2 summarizes the solutions for the three fundamental PDEs.

Table 14.2 Physical behavior of separated solutions (for $\lambda < 0$).[a]

PDE type	Governing equation	Spatial solution $X(x)$	Time/other solution $T(t)$ or $Y(y)$	Physical interpretation
Wave equation	$\frac{\partial^2 u}{\partial t^2} = c^2\frac{\partial^2 u}{\partial x^2}$	**Oscillatory** (sin kx, cos kx)	**Oscillatory** (sin ωt, cos ωt)	**Propagating wave:** Energy moves reversibly; the system vibrates indefinitely without damping
Heat equation	$\frac{\partial u}{\partial t} = \alpha\frac{\partial^2 u}{\partial x^2}$	Oscillatory (sin kx, cos kx)	**Exponential decay** $\left(e^{-\alpha k^2 t}\right)$	**Diffusion:** Irreversible process; initial profiles smooth out and decay to equilibrium

PDE type	Governing equation	Spatial solution $X(x)$	Time/other solution $T(t)$ or $Y(y)$	Physical interpretation
Laplace equation	$\frac{\partial^2 u}{\partial x^2} + \frac{\partial^2 u}{\partial y^2} = 0$	Oscillatory ($\sin kx$, $\cos kx$)	**Hyperbolic/ exponential** ($e^{\pm ky}$, $\sinh ky$)	**Steady state:** No time dependence. Values decay or grow spatially away from boundaries

[a]Recall from Section 14.3 that only $\lambda < 0$ gives nontrivial fixed-end solutions for the string.

Table 14.3 Analogy between a stretched string and a molecule.

System	Stretched string	Polyatomic molecule (e.g., CO_2)
Governed by	One-dimensional wave equation	Newton's laws/Schrödinger equation
Constraint	Fixed ends: ($u(0, t) = u(L, t) = 0$)	Molecular geometry and force field
Mathematical consequence	Boundary value problem	Boundary value problem
Solutions	Eigenfunctions $\left(X_n(x) = \sin\left(\frac{n\pi x}{L}\right)\right)$	Normal modes (eigenvectors of the dynamical matrix)
Frequencies	Eigenvalues $\left(\omega_n = \frac{cn\pi}{L}\right)$	Vibrational frequencies ($\tilde{\nu}_j$) (cm^{-1})

Thus, the separation of variables reveals the core nature of each PDE: the wave equation propagates, the heat equation dissipates, and Laplace's equation equilibrates.

14.5 Molecular Vibrations: From Strings to Spectra

In earlier chapters, you were introduced to the concept of **normal modes of vibration** for molecules like CO_2 and H_2O. We discovered that these molecules have specific, discrete patterns of vibration—symmetric stretches, bends, and asymmetric stretches—each with its own characteristic frequency. At the time, we treated these as results of solving the secular equation. Now, with the solution to the wave equation for a stretched string, we can appreciate the profound *mathematical unity* underlying these physical phenomena.

Let us draw a direct analogy (Table 14.3):

The key insight is identical: **the imposition of physical constraints (boundary conditions) leads to a quantized set of solutions**. For the string, the fixed ends quantize the wavelength, allowing only standing waves where the length (L) is a half-integer multiple of the wavelength ($L = n(\lambda/2)$). This directly gives the discrete set of frequencies (ω_n).

In a molecule, the "boundary conditions" are more complex but conceptually similar. They are the specific equilibrium positions of the atoms and the restorative forces of the chemical bonds. Just as the string cannot vibrate with an arbitrary frequency, the molecule cannot vibrate with an arbitrary pattern or frequency. It is constrained to its **normal modes**—the eigenfunctions of the system—each with a discrete **vibrational frequency**—the eigenvalues.

14.6 Fourier Superposition and Physical Interpretation

Recall our general solution for the string:

$$u(x,t) = \sum_{n=1}^{\infty} [A_n \cos(\omega_n t) + B_n \sin(\omega_n t)] \sin\left(\frac{n\pi x}{L}\right).$$

An *arbitrary* vibration of the string is a **superposition of its normal modes**. The initial pluck ($u_0(x)$) determines how much of each mode is excited (the coefficients (A_n)).

This is precisely analogous to a real molecular vibration. When you shine IR light on a molecule, you are not seeing a single, pure normal mode in isolation. A molecule's instantaneous motion may involve a superposition of modes, but IR spectroscopy probes individual normal mode transitions. The initial "pluck" (the interaction with the electromagnetic field) excites a superposition of modes allowed by the selection rules. However, the *fundamental, quantized building blocks* of any complex molecular motion are these normal modes, just as the sine waves are the fundamental building blocks for any shape of the vibrating string.

14.7 Fundamentals and Overtones

In our string solution, the mode with ($n = 1$) is the **fundamental** frequency. The modes with ($n = 2, 3, \ldots$) are the **overtones**. In molecular spectroscopy, we see the same phenomenon. The strongest IR absorption often corresponds to the excitation from the ground vibrational state to the first excited state of a normal mode (the fundamental transition). However, we can also observe weaker **overtone** bands ($\Delta v = 2, 3, \ldots$), which correspond to exciting higher vibrational quanta, much like exciting the ($n = 2, 3, \ldots$) modes of the string.

Conclusions

The mathematics of the humble stretched string—eigenfunctions, eigenvalues, boundary conditions, and superposition—provides the essential conceptual framework for understanding the vibrational spectra of complex molecules. The discrete frequencies in an infrared spectrum are not a mysterious quantum effect but a direct consequence of the wave nature of matter and energy, constrained by the molecular architecture. The transition from the continuous spectrum of possible waves on an infinite string to the discrete spectrum of a finite string is one of the most important and general concepts in mathematical physics, with profound implications from music to quantum chemistry.

14.8 Three-dimensional Wave Equation and Helmholtz Form

The one-dimensional wave Equation (14.1) can easily be extended to three dimensions as

$$\frac{\partial^2 u}{\partial t^2} = c^2\left(\frac{\partial^2 u}{\partial x^2} + \frac{\partial^2 u}{\partial y^2} + \frac{\partial^2 u}{\partial z^2}\right). \tag{14.22}$$

When we represent the three-dimensional wave function as the product of spatial and temporal parts,

$$u(x,y,z,t) = X(x)\,Y(y)\,Z(z)\,T(t), \tag{14.23}$$

substitution into Equation (14.22) gives, after dividing by $XYZT$,

$$\frac{1}{c^2T}\frac{d^2T}{dt^2} = \frac{1}{X}\frac{d^2X}{dx^2} + \frac{1}{Y}\frac{d^2Y}{dy^2} + \frac{1}{Z}\frac{d^2Z}{dz^2}. \tag{14.24}$$

Each term depends on a different coordinate, so both sides must be equal to a constant, say $-k^2$. The solutions are sinusoidal in each coordinate, as in the one-dimensional case.

Thus, the spatial part of the three-dimensional wave equation may be written as

$$\nabla^2 u + k^2 u = 0, \tag{14.25}$$

where ∇^2 is the Laplacian operator defined by

$$\nabla^2 = \frac{\partial^2}{\partial x^2} + \frac{\partial^2}{\partial y^2} + \frac{\partial^2}{\partial z^2}.$$

Equation (14.25) is known as the **Helmholtz equation**, and its solutions correspond to standing waves in three dimensions.

14.9 From Classical Waves to Quantum Operators

14.9.1 de Broglie Relations and the Wave Equation

In electromagnetism, the electric and magnetic fields also obey a wave equation. This universal nature of wave phenomena provides a conceptual bridge to quantum mechanics, where particles are described by wavefunctions. For electromagnetic radiation, the velocity c is related to the frequency ν and wavelength λ by $c = \nu\lambda$. Also, according to the de Broglie relations,

$$E = h\nu, \quad p = \frac{h}{\lambda}. \tag{14.26}$$

Substituting these into the wave equation, we can replace ν and λ by E and p:

$$\nabla^2 u + \left(\frac{2\pi p}{h}\right)^2 u = 0. \tag{14.27}$$

The k in the classical wave equation is identified with $2\pi/\lambda$, so that $p = \hbar k$.

Expressing this in terms of the **reduced Planck's constant** $\hbar = h/2\pi$, we have

$$\nabla^2 u + \frac{p^2}{\hbar^2} u = 0. \tag{14.28}$$

Multiplying both sides by $-\hbar^2/2m$ gives

$$-\frac{\hbar^2}{2m}\nabla^2 u = \frac{p^2}{2m} u. \tag{14.29}$$

14.9.2 Momentum and Kinetic Energy Operators

This is an **eigenvalue equation**. When the operator $-(\hbar^2/2m)\nabla^2$ acts on its eigenfunction u, it returns the kinetic energy $(p^2/2m)$ multiplied by the same function. In quantum mechanics contexts, the eigenfunction is usually represented as ψ. Hence, the operator

$$\hat{T} = -\frac{\hbar^2}{2m}\nabla^2 \tag{14.30}$$

is identified as the **kinetic energy operator**.

Since, classically, $T = p^2/2m$, a similar relation must hold for the corresponding operators, i.e.,

$$\hat{T} = \frac{\hat{p}^2}{2m}. \tag{14.31}$$

Comparing Equations (14.30) and (14.31), we deduce that the **momentum operator** in coordinate space must be

$$\hat{p} = -i\hbar\frac{\partial}{\partial x}. \tag{14.32}$$

The negative sign arises from the Fourier-transform convention (see Equation 10.6).

In one dimension, operating on a function e^{ikx} gives

$$\hat{p}\, e^{ikx} = -i\hbar\frac{\partial}{\partial x}e^{ikx} = k\hbar\, e^{ikx}. \tag{14.33}$$

Thus, e^{ikx} is a **momentum eigenfunction** with eigenvalue $k\hbar$. The unit of momentum is therefore $\hbar$.

The operators for **momentum** and **position** are the basic building blocks for constructing any operator in quantum mechanics. The classical quantity associated with a given operator is expressed in terms of position and momentum, and then these variables are replaced by their corresponding operators.

In coordinate space, the operators are as follows:

- Position operator: $\hat{x} = x$ (multiplication by x).
- Momentum operator: $\hat{p}_x = -i\hbar\dfrac{\partial}{\partial x}$.

In momentum space, these become:

- Momentum operator: $\hat{p}_x = p_x$ (multiplication by p_x).
- Position operator: $\hat{x} = i\hbar\dfrac{\partial}{\partial p_x}$.

Note that the sign difference arises again from the Fourier-transform convention.

These expressions for the kinetic energy, momentum, and position operators are **universal**, valid for all systems, since no assumption about the specific system was made in deriving them.

14.9.3 Hamiltonian and the Schrödinger Equation

Finally, to obtain the total energy, we add the potential energy. The operator for total energy, known as the **Hamiltonian operator**, is given by

$$\hat{H} = -\frac{\hbar^2}{2m}\nabla^2 + \hat{V}. \tag{14.34}$$

Here $\hat{V}$ is the **potential energy operator**, which differs from system to system and thus gives rise to different solutions.

This is a **heuristic argument** showing how quantum operators emerge from combining wave properties with particle properties and is not rigorous. This is not a derivation of quantum mechanics but a plausibility argument showing how the **Schrödinger equation** emerges naturally from wave-particle duality.

The most direct quantum-mechanical analog of the stretched string is a particle confined to a one-dimensional "box" of length L, where the potential energy V is zero inside and infinite outside. Solving the time-independent Schrödinger equation, $\hat{H}\psi = E\psi$, with the boundary conditions

$\psi(0) = \psi(L) = 0$, yields eigenfunctions $\psi_n(x) = \sqrt{\frac{2}{L}}\sin\left(\frac{n\pi x}{L}\right)$ and energy eigenvalues $E_n = \frac{n^2h^2}{8mL^2}$. These are identical in form to the spatial eigenfunctions of the string, demonstrating how boundary conditions quantize energy levels in quantum systems, just as they quantize vibrational frequencies in classical systems.

It should be emphasized again that what we have done above is **not a derivation** of the Schrödinger equation but rather a logical progression combining the wave and particle aspects of matter and radiation. These developments were the crucial steps leading to the foundation of quantum mechanics.

Quantum mechanics itself is built upon a set of **postulates** (from Latin *postulatum*, "that which is assumed"), since no experiment has ever contradicted them. These postulates will be discussed in detail in my forthcoming book on quantum chemistry.

14.10 In-depth Applications and Examples: The Heat and Laplace Equations

This section is intended as an applications-oriented epilog: it assumes the techniques developed above and shows, without exhaustive derivations, how the same ideas apply to diffusion and potential problems central to chemistry and materials science.

The wave equation is not confined to mechanical systems. This solution finds application in solving the Laplace and Poisson equations, the diffusion equation, vibrations in a membrane, and the Helmholtz equation (Table 14.4). We expand on some of the applications in the next section.

Table 14.4 Comparison of solutions for the three fundamental PDEs.

Property	Wave equation	Heat/diffusion equation	Laplace's equation
PDE	$\partial^2u/\partial t^2 = c^2\partial^2u/\partial x^2$	$\partial u/\partial t = \alpha\partial^2u/\partial x^2$	$\partial^2u/\partial x^2 + \partial^2u/\partial y^2 = 0$
Separation form	$u(x, t) = X(x)T(t)$	$u(x, t) = X(x)T(t)$	$u(x, y) = X(x)Y(y)$
Spatial ODE	$d^2X/dx^2 - \lambda X = 0$	$d^2X/dx^2 - \lambda X = 0$	$d^2X/dx^2 - \lambda X = 0$
Temporal/second spatial ODE	$d^2T/dt^2 - c^2\lambda T = 0$	$dT/dt - \alpha\lambda T = 0$	$d^2Y/dy^2 + \lambda Y = 0$
Sign of λ[a]	$\lambda < 0$ (write as $\lambda = -k^2$)	$\lambda < 0$ (write as $\lambda = -k^2$)	$\lambda < 0$ (write as $\lambda = -k^2$)
Spatial solution	$X_n(x) = \sin(n\pi x/L)$ **Oscillatory**	$X_n(x) = \sin(n\pi x/L)$ **Oscillatory**	$X_n(x) = \cos(n\pi x/a)$, $n = 0, 1, 2, \ldots$ **Oscillatory**
Temporal/second spatial solution	$T_n(t) = A_n\cos(\omega_n t) + B_n\sin(\omega_n t)$ where $\omega_n = cn\pi/L$ **Oscillatory in time**	$T_n(t) = C_ne^{(-\alpha k_n^2 t)}$ where $k_n = n\pi/L$ **Exponential decay**	$Y_n(y) = A_n\sinh(k_ny) + B_n\cosh(k_ny)$ where $k_n = n\pi/a$ **Hyperbolic**
Time behavior	• Periodic oscillation • Energy conserved • Reversible	• Monotonic decay • Energy dissipated • Irreversible	• No time dependence • Steady state • Static equilibrium

(Continued)

Table 14.4 (Continued)

Property	Wave equation	Heat/diffusion equation	Laplace's equation
Physical character	• Wave propagation • Standing waves • Resonance	• Diffusion/ smoothing • Approach to equilibrium • Information loss	• Equilibrium distribution • Potential fields • No dynamics
Problem type	Initial-boundary value Requires $u(x, 0)$ and $\partial u/\partial t(x, 0)$	Initial-boundary value Requires only $u(x, 0)$	Pure boundary value No initial conditions
Chemistry and physics examples	• Vibrating molecules • Sound waves • EM radiation • Quantum wavefunctions	• Heat conduction • Drug diffusion • Chemical diffusion • Chromatography	• Steady-state T • Electrostatic potential • Equilibrium concentration
Classification	**Hyperbolic PDE**	**Parabolic PDE**	**Elliptic PDE**

[a] Recall from Section 14.3 that only $\lambda < 0$ gives nontrivial fixed-end (Dirichlet) solutions for the string; for Laplace's equation with insulating (Neumann) sides, $\lambda = 0$ also yields a nontrivial (constant) mode.

14.10.1 The Heat Equation (or Diffusion Equation)

The heat equation describes how a quantity such as temperature or concentration distributes itself and evolves over time as it diffuses from regions of high density to low density. Its one-dimensional form is

$$\frac{\partial u}{\partial t} = \alpha \frac{\partial^2 u}{\partial x^2}, \tag{14.35}$$

where:

- $u(x, t)$ represents the temperature at a point x and time t,
- α is the thermal diffusivity, a positive constant.

In what follows, the terms "heat equation" and "diffusion equation" will be used interchangeably for Equation (14.35).

The key distinction from the wave equation is the first-order time derivative. This single change has a profound physical consequence: the heat equation is irreversible in time. It describes a process of smoothing and equilibration, where initial irregularities are rapidly damped out, and the system evolves toward a uniform steady state. Solutions typically involve decaying exponential functions in time, multiplied by spatial sine or cosine functions.

14.10.1.1 Chemical and Physical Applications of the Diffusion Equation

While the name "heat equation" originates from thermodynamics, its form is universal for any diffusive process, making it paramount in chemistry and materials science. In these contexts, $u(x, t)$ represents concentration and α is replaced by the diffusion coefficient, D.

Chemical Kinetics and Reactor Design: In diffusion-limited reactions, the rate at which reactants encounter each other is controlled by their diffusion through the solvent. Solving the diffusion equation is essential for modeling reaction rates in heterogeneous systems and designing efficient chemical reactors.

Electrochemistry: The flux of ions to an electrode surface is governed by Fick's laws of diffusion, which are direct analogs of the heat equation. The current in an electrochemical cell can often be derived by solving Equation (14.35) with boundary conditions representing the electrode surface.

Materials Science and Engineering: The deliberate diffusion of dopants (e.g., boron, phosphorus) into a silicon wafer is a critical step in semiconductor manufacturing. The depth and profile of the doped region are calculated using the diffusion equation. Similarly, the annealing of metals to relieve stress relies on diffusive processes modeled by this PDE.

Pharmaceuticals and Drug Delivery: The release of a drug from a polymer matrix or its transport across a cellular membrane is a diffusive process. Modeling this with the diffusion equation allows chemists to design controlled-release drug delivery systems (Example 14.2).

Example 14.2: Drug Diffusion Through a Membrane

Objective: To model the diffusion of a drug through a biological membrane using the one-dimensional diffusion equation, demonstrating an exponential approach to equilibrium.

Method Used: Solve the diffusion equation with fixed concentration boundary conditions relevant to pharmaceutical chemistry.

Problem: A drug diffuses through a membrane of thickness L. One side of the membrane (at $x = 0$) is maintained at constant drug concentration C_0 (the donor compartment), while the other side (at $x = L$) is maintained at zero concentration (the receptor compartment with perfect sink conditions).

Initially, the membrane contains no drug: $C(x, 0) = 0$. Find the concentration profile $C(x, t)$ as a function of time.

Solution:

The diffusion equation is

$$\frac{\partial C}{\partial t} = D\frac{\partial^2 C}{\partial x^2},$$

where D is the diffusion coefficient of the drug in the membrane.

Boundary conditions:

- $C(0, t) = C_0$ for all $t > 0$;
- $C(L, t) = 0$ for all $t > 0$.

Initial condition:

- $C(x, 0) = 0$ for $0 < x < L$.

Step 1: Find the steady-state solution

As $t \to \infty$, the system reaches a steady state where $\frac{\partial C}{\partial t} = 0$. The diffusion equation becomes

$$\frac{d^2 C_{ss}}{dx^2} = 0.$$

Integrating twice:

$$C_{ss}(x) = Ax + B.$$

Applying boundary conditions:

- $C_{ss}(0) = C_0 \Rightarrow B = C_0$;
- $C_{ss}(L) = 0 \Rightarrow AL + C_0 = 0 \Rightarrow A = -\frac{C_0}{L}$.

Therefore,

$$C_{ss}(x) = C_0\left(1 - \frac{x}{L}\right).$$

This is a **linear concentration gradient** at steady state—a direct consequence of Fick's first law.

Step 2: Solve for the transient part

Let $C(x, t) = C_{ss}(x) + v(x, t)$, where $v(x, t)$ represents the deviation from the steady state. Substituting into the diffusion equation

$$\frac{\partial v}{\partial t} = D\frac{\partial^2 v}{\partial x^2}$$

with homogeneous boundary conditions:

- $v(0, t) = 0$,
- $v(L, t) = 0$,

and initial condition

$$v(x, 0) = C(x, 0) - C_{ss}(x) = 0 - C_0\left(1 - \frac{x}{L}\right) = -C_0\left(1 - \frac{x}{L}\right).$$

Step 3: Use separation of variables

As shown in the general treatment, the solution is

$$v(x, t) = \sum_{n=1}^{\infty} B_n e^{-D\left(\frac{n\pi}{L}\right)^2 t} \sin\left(\frac{n\pi x}{L}\right).$$

The coefficients are found from the initial condition

$$B_n = \frac{2}{L}\int_0^L \left[-C_0\left(1 - \frac{x}{L}\right)\right] \sin\left(\frac{n\pi x}{L}\right)dx.$$

Evaluating the integral:

$$B_n = -\frac{2C_0}{L}\int_0^L \left(1 - \frac{x}{L}\right) \sin\left(\frac{n\pi x}{L}\right)dx.$$

Using integration by parts:

$$B_n = -\frac{2C_0}{n\pi}.$$

Final Solution:

$$C(x,t) = C_0\left(1 - \frac{x}{L}\right) - \frac{2C_0}{\pi}\sum_{n=1}^{\infty}\frac{1}{n}e^{-D\left(\frac{n\pi}{L}\right)^2 t}\sin\left(\frac{n\pi x}{L}\right).$$

Physical and Chemical Interpretation:

1. **Steady-state flux:** At steady state, the flux through the membrane is: $J_{ss} = -D\frac{dC_{ss}}{dx}$. Since $C_{ss}(x) = C_0\left(1 - \frac{x}{L}\right)$, we have: $\frac{dC_{ss}}{dx} = C_0 \cdot \left(-\frac{1}{L}\right) = -\frac{C_0}{L}$. Therefore: $J_{ss} = -D\left(-\frac{C_0}{L}\right) = \frac{DC_0}{L}$. This is **Fick's first law**—flux is proportional to the concentration gradient. The positive value indicates flux in the positive x-direction (from high concentration at $x = 0$ to low concentration at $x = L$), as expected physically.
2. **Time to reach steady state:** The transient terms decay with a characteristic time: $\tau = \frac{L^2}{\pi^2 D}$. For a typical drug through a biological membrane with:
   - $L = 10\,\mu\text{m} = 10^{-5}\,\text{m}$,
   - $D = 10^{-10}\,\text{m}^2/\text{s}$,

   $$\tau = \frac{(10^{-5})^2}{\pi^2 \times 10^{-10}} \approx 0.10\,\text{s}.$$
3. **Pharmaceutical relevance:** This model describes:
   - Transdermal drug delivery patches.
   - Drug release from polymer matrices.
   - Permeation through skin or intestinal membranes.
   - Controlled-release formulations.
4. **Lag time:** Before steady-state flux is established, there is a "lag time" during which the membrane loads with drug. This is given by: $t_{\text{lag}} = \frac{L^2}{6D}$.

Simplified result for large times:
For $t > \tau$, only the first term is significant:

$$C(x,t) \approx C_0\left(1 - \frac{x}{L}\right) - \frac{2C_0}{\pi}e^{-\frac{\pi^2 Dt}{L^2}}\sin\left(\frac{\pi x}{L}\right).$$

Connection to Fick's Laws:
This solution beautifully illustrates both of Fick's laws:

- **Fick's First Law** (steady state): $J = -D\frac{dC}{dx}$ gives the steady-state flux $J_{ss} = DC_0/L$.
- **Fick's Second Law** (transient): $\frac{\partial C}{\partial t} = D\frac{\partial^2 C}{\partial x^2}$ describes how the system evolves from the initial state ($C = 0$ everywhere) to the steady state (linear gradient).

The complete solution shows the **transition** between these two regimes—initially dominated by Fick's second law (transient diffusion), eventually settling into the Fick's first law regime (steady flux). This is crucial in pharmaceutical formulation design, where both the time to reach steady release and the magnitude of steady-state flux determine drug efficacy.

14.10.2 Laplace's Equation

Laplace's equation describes the steady-state (time-independent) distribution of a field in a region containing no sources or sinks. It is the equilibrium solution to both the wave and heat equations. In two dimensions, it is given by

$$\nabla^2 u = \frac{\partial^2 u}{\partial x^2} + \frac{\partial^2 u}{\partial y^2}. \tag{14.36}$$

Solutions to Laplace's equation, known as harmonic functions, have the elegant property that their value at any point is precisely the average of their values in the immediate neighborhood. This "mean-value property" leads to solutions that are extremely smooth and cannot have local maxima or minima inside the domain—these must occur on the boundary. Laplace's equation is therefore an archetypal **boundary-value problem**. It is crucial in electrostatics for calculating potential fields and in fluid dynamics for modeling potential flow.

Example 14.3: Electrostatic Potential in an Electrochemical Cell

Objective: To determine the electric potential distribution in a simplified rectangular electrochemical cell using Laplace's equation.
Method Used: Solve Laplace's equation in two dimensions with boundary conditions representing electrode potentials.
Problem: Consider a simplified 2D cross-section of an electrochemical cell with dimensions $a \times b$. The bottom electrode (at $y = 0$) is held at potential V_0 (anode), the top electrode (at $y = b$) is grounded ($V = 0$, cathode), and the side walls are insulating (no current flow perpendicular to them). Find the potential distribution $V(x, y)$ in the electrolyte between the electrodes.

Boundary conditions:

- $V(x, 0) = V_0$ for $0 \leq x \leq a$ (anode);
- $V(x, b) = 0$ for $0 \leq x \leq a$ (cathode);
- $\frac{\partial V}{\partial x}|_{x=0} = 0$ for $0 \leq y \leq b$ (insulating wall);
- $\frac{\partial V}{\partial x}|_{x=a} = 0$ for $0 \leq y \leq b$ (insulating wall).

Solution:

Laplace's equation for the steady-state potential is

$$\nabla^2 V = \frac{\partial^2 V}{\partial x^2} + \frac{\partial^2 V}{\partial y^2} = 0.$$

Step 1: Recognize the simplified geometry
Because the electrodes are infinite in the x-direction (or the cell is much wider than it is tall), and the side boundary conditions are symmetric, the potential depends primarily on y and is approximately uniform in x.

For a cell where $a \gg b$, away from the side walls, the solution simplifies to a **one-dimensional problem**:

$$\frac{d^2 V}{dy^2} = 0.$$

B

Fourier Transform Pairs and Properties (Unitary Convention)

This appendix summarizes the transform pairs most commonly used in chemistry and physics, including signal analysis, spectroscopy, quantum mechanics, and differential equations. All results assume the **unitary Fourier transform**:

$$\mathcal{F}\{f(t)\} = \frac{1}{\sqrt{2\pi}} \int_{-\infty}^{\infty} f(t)\, e^{-i\omega t}\, dt, \quad \mathcal{F}^{-1}\{F(\omega)\} = \frac{1}{\sqrt{2\pi}} \int_{-\infty}^{\infty} F(\omega)\, e^{i\omega t}\, d\omega.$$

B.1 Basic Fourier Transform Pairs

1. **Dirac delta:**

$$f(t) = \delta(t) \Longleftrightarrow F(\omega) = \frac{1}{\sqrt{2\pi}},$$

$$f(t) = 1 \Longleftrightarrow F(\omega) = \sqrt{2\pi}\, \delta(\omega).$$

2. **Complex exponential:**

$$f(t) = e^{i\omega_0 t} \Longleftrightarrow F(\omega) = \sqrt{2\pi}\, \delta(\omega - \omega_0).$$

3. **Cosine and sine:**

$$\cos(\omega_0 t) \Longleftrightarrow F(\omega) = \frac{1}{\sqrt{\frac{\pi}{2}}}[\delta(\omega - \omega_0) + \delta(\omega + \omega_0)],$$

$$\sin(\omega_0 t) \Longleftrightarrow F(\omega) = \frac{1}{i\sqrt{\frac{\pi}{2}}}[\delta(\omega - \omega_0) - \delta(\omega + \omega_0)].$$

4. **Gaussian:**

Let $f(t) = e^{-at^2}$ with $a > 0$,

$$\mathcal{F}\left\{e^{-at^2}\right\} = \sqrt{\frac{1}{2a}}\, e^{-\omega^2/(4a)}.$$

Step 2: Solve the 1D Laplace equation

Integrating twice:

$$V(y) = Cy + D.$$

Applying boundary conditions:

- $V(0) = V_0 \Rightarrow D = V_0$.
- $V(b) = 0 \Rightarrow Cb + V_0 = 0 \Rightarrow C = -\frac{V_0}{b}$.

Final Solution (for the simplified case):

$$\boxed{V(y) = V_0\left(1 - \frac{y}{b}\right).}$$

This is a **linear potential drop** from anode to cathode.

Step 3: Calculate the electric field

The electric field is

$$\vec{E} = -\nabla V = -\frac{dV}{dy}\hat{j} = \frac{V_0}{b}\hat{j}.$$

The field is uniform and directed from anode (+) to cathode (−).

Step 4: Calculate current density

By Ohm's law in the electrolyte:

$$\vec{j} = \sigma\vec{E} = \frac{\sigma V_0}{b}\hat{j},$$

where σ is the conductivity of the electrolyte.
The total current through a cross-section of area A is

$$I = jA = \frac{\sigma V_0 A}{b}.$$

The cell resistance is

$$R = \frac{V_0}{I} = \frac{b}{\sigma A}.$$

This is the familiar relation: resistance is proportional to length and inversely proportional to cross-sectional area.

Physical and Chemical Interpretation:

1. **Electrochemical applications:**
   This model describes the potential distribution in:
   - simple parallel-plate electrochemical cells
   - electrolysis cells
   - battery electrodes
   - fuel cells

2. **Ohmic drop:** The linear potential profile represents the **ohmic drop** through the electrolyte. This is why:
   - Highly conductive electrolytes are desirable (minimize b/σ).
   - Thin electrolyte layers improve cell efficiency.
   - In real cells, concentration gradients near electrodes cause deviations.
3. **Current distribution:** In this simplified geometry, current density is uniform. In reality, near the edges and corners, current distribution becomes nonuniform (requiring the full 2D solution).
4. **Migration versus diffusion:** This solution assumes pure migration (charged species moving under the electric field). In real electrochemical systems, diffusion of electroactive species also occurs, requiring solution of the **Nernst–Planck equation** (which couples Laplace's equation with the diffusion equation).

Connection to Advanced Topics:

For cells with nonuniform boundary conditions or complex geometries, the full 2D solution using separation of variables is required. The method is identical, but the results show:

- nonuniform current distribution near edges;
- "edge effects" where current density is higher at corners;
- the need for supporting electrolytes to minimize migration effects in electroanalytical chemistry.

Practical Example:

For a typical aqueous electrochemical cell with:

- $b = 1\,\text{cm} = 10^{-2}\,\text{m}$,
- $\sigma = 10\,\text{S/m}$ (0.1 M KCl solution),
- $A = 10\,\text{cm}^2 = 10^{-3}\,\text{m}^2$,
- $V_0 = 1\,\text{V}$,

$$R = \frac{0.01}{10 \times 10^{-3}} = 1\,\Omega,$$

$$I = \frac{1\,\text{V}}{1\,\Omega} = 1\,\text{A}.$$

Comparison with Heat and Wave Equations:

Feature	Wave equation	Heat equation	Laplace's equation
Time behavior	Oscillatory	Exponential decay	None (static)
Spatial part	$\sin(n\pi x/L)$	$\sin(n\pi x/L)$	$\cos(n\pi x/a)$ ($n = 0, 1, 2, \ldots$)
y-dependence	N/A (1D)	N/A (1D)	$\sinh(n\pi y/a)$; linear in y for $n = 0$
Problem type	Initial-boundary	Initial-boundary	Pure boundary

A deep understanding of the one-dimensional wave equation and the method of separation of variables provides the essential toolkit for tackling these other equations. The procedure is analogous: assume a separable solution, apply boundary conditions to quantize the spatial part (eigenfunctions/eigenvalues), and then solve the resulting ordinary differential equation in time. The character of that time equation—oscillatory for waves, decaying for diffusion, or nonexistent for Laplace—is what defines the unique physics of each system.

Side Box: The Mathematics of Music

The wave equation isn't just for physicists—it is the reason music exists. A musical instrument, whether a guitar string, a flute's air column, or a drumhead, is a system that supports standing waves. The solutions we derived for the string,

$$u_n(x,t) \propto \cos(\omega_n t)\sin(n\pi x/L),$$

directly describe these standing waves. The fundamental frequency ($n = 1$) determines the perceived pitch of the note. The overtones ($n = 2, 3, \ldots$), which are integer multiples of the fundamental frequency for a string, determine the timbre or tone quality of the instrument. The unique blend of amplitudes (A_n) for these overtones is why a middle C on a piano sounds profoundly different from the same note on a violin. The wave equation thus provides a complete mathematical framework for understanding why instruments sound the way they do, linking physics, engineering, and art.

14.11 Summary and Conclusions

In this chapter, we have taken an important step toward quantum mechanics by connecting the **classical wave equation** with the **quantum-mechanical Schrödinger equation**.

Starting from the vibration of a stretched string, we introduced the **method of separation of variables**, discussed **eigenvalues**, **eigenfunctions**, and **boundary conditions**, and extended the treatment to three dimensions. Finally, by combining the classical expressions for energy and momentum with the de Broglie relations, we arrived at the **kinetic energy** and **Hamiltonian operators**, which form the mathematical foundation of quantum mechanics.

While the separation of variables and Fourier series provide exact solutions for simple, symmetric problems like the string and the particle-in-a-box, most real-world problems in engineering and science (e.g., drug diffusion in an irregularly shaped organ, airflow over a wing, the electronic structure of a complex molecule) require numerical methods for their solution. The Finite Difference Method and Finite Element Method are two powerful techniques that discretize PDEs like the ones studied here, allowing them to be solved on computers.

Looking Forward: From Analytical Solutions to Computational Science

The method of separation of variables and Fourier series provide powerful, exact solutions for the wave, heat, and Laplace equations on simple, symmetric domains like a string or a rectangle. However, many real-world problems in chemistry, physics, and engineering involve complex geometries, heterogeneous materials, and nonlinear phenomena that cannot be solved analytically.

For these challenges, scientists and engineers turn to numerical methods. Techniques like the Finite Difference Method (which approximates derivatives using values on a discrete grid) and the Finite Element Method (which breaks a complex domain into many small, simple pieces) are used to convert partial differential equations into systems of algebraic equations that can be solved by computers.

Whether simulating drug diffusion through an organ, modeling airflow over a wing, or calculating the electronic structure of a novel molecule, these computational tools solve the same fundamental PDEs you have studied here. The classical theory thus forms the essential foundation for the computational simulation tools that drive modern scientific discovery and technological innovation.

Further Reading

G. Arfken and H. Weber, *Mathematical Methods for Physicists* (7th Ed.), Elsevier, 2012.
E. Kreyszig, *Advanced Engineering Mathematics* (10th Ed.), Wiley, 2011.
D. J. Griffiths, *Introduction to Quantum Mechanics* (3d Ed.), Pearson, 2018.

Exercises

A. Conceptual Questions

1. Explain the difference between ordinary and PDEs with suitable examples.

2. Show that it is possible to satisfy either boundary condition (14.2) for the case $\lambda > 0$, but not both.

3. Why does quantization arise in the solution of the wave equation with fixed boundary conditions?

4. Discuss the physical significance of the separation constant and eigenvalues in the wave equation.

5. Explain why the principle of superposition applies to the solutions of the linear wave equation.

6. State and explain the role of boundary and initial conditions in determining the solution of a PDE.

7. What is meant by a normal mode of vibration? How is it related to the eigenfrequency?

8. Define the kinetic energy, momentum, and Hamiltonian operators in coordinate space.

9. Describe how the classical wave equation leads conceptually to the Schrödinger equation.

B. Problems

1. Solve the one-dimensional wave equation $\frac{\partial^2 u}{\partial t^2} = c^2 \frac{\partial^2 u}{\partial x^2}$ by the method of separation of variables:
   a. Discuss the nature of the solution for (i) $\lambda < 0$, (ii) $\lambda = 0$, and (iii) $\lambda > 0$.
   b. Identify which values satisfy the boundary conditions and find the allowed λ.

2. Show that the general solution of the wave equation can be written as $u(x,t) = \sum_n [A_n \cos(\omega_n t) + B_n \sin(\omega_n t)] \sin\left(\frac{n\pi x}{L}\right)$, and derive expressions for A_n and B_n using the initial conditions.

3. For a string with initial triangular deflection $u_0(x) = \begin{cases} \frac{2h}{L}x, & 0 \leq x \leq \frac{L}{2}, \\ \frac{2h}{L}(L-x), & \frac{L}{2} < x \leq L, \end{cases}$ and zero initial velocity, compute the coefficients A_n and determine $u(x,t)$.

4. Plot the first three normal modes ($n = 1, 2, 3$) of vibration for $0 \leq x \leq L$. How many nodes does each have?

5. Derive the three-dimensional wave equation and show that its form remains invariant under coordinate transformations.

6. Use the de Broglie relations to show how the kinetic energy operator can be written as $\hat{T} = -\frac{\hbar^2}{2m}\nabla^2$ and identify the momentum operator.

7. A rectangular container of dimensions $a \times b$ has three impermeable walls. The wall at $x = 0$ maintains a constant concentration C_0 of a dissolved species, while the concentration at the opposite wall ($x = a$) is maintained at zero. The other two walls ($y = 0$ and $y = b$) are impermeable:
   a. Write down the boundary conditions for this problem.
   b. Explain why the steady-state concentration profile depends only on x and not on y.
   c. Solve Laplace's equation $\nabla^2 C = 0$ to find $C(x)$.
   d. Calculate the steady-state flux through the container.

8. For the drug diffusion problem (Example 14.2), consider a membrane of thickness $L = 50\,\mu\text{m}$ with diffusion coefficient $D = 5 \times 10^{-11}\ \text{m}^2/\text{s}$:
   a. Calculate the characteristic time τ for the system to reach steady state.
   b. Calculate the lag time $t_{\text{lag}} = L^2/6D$.
   c. If the donor concentration is $C_0 = 10\,\text{mg/mL}$, what is the steady-state flux?

C. Advanced Exercises

1. Starting from the de Broglie relation and the classical wave equation, obtain the time-independent Schrödinger equation step by step.

2. For a vibrating rectangular membrane of sides a and b, derive the expression for the normal modes of vibration and discuss their degeneracies.

3. Show that the solution of the 3D wave equation can be expressed as a product of three orthogonal functions in x, y, and z.

4. Demonstrate that the Fourier series solution of the wave equation converges to the actual displacement function for piecewise continuous initial conditions.

5. For a particle confined in a one-dimensional box, compare the eigenfunctions of the Schrödinger equation with those of the classical stretched string.

6. Using the operator formalism, show that $[\hat{x}, \hat{p}] = i\hbar$ and discuss the physical meaning of this commutation relation.

7. Prove that the momentum eigenfunctions are of the form e^{ikx} and determine the corresponding eigenvalues.

8. Discuss how boundary conditions in quantum systems (e.g., infinite potential well) lead to discrete energy levels similar to the classical vibration modes of a string.

9. Solve the two-dimensional heat equation for a rectangular metal rod with cross-section $0 \leq x \leq a$, $0 \leq y \leq b$, infinite in the z-direction. The surfaces at $x = 0$, $x = a$, and $y = 0$ are held at zero temperature, while the surface at $y = b$ is held at temperature T_0:
   a. Use separation of variables to show that the solution has the form:

$$T(x, y) = \sum_{n=1}^{\infty} A_n \sinh\left(\frac{n\pi y}{a}\right) \sin\left(\frac{n\pi x}{a}\right).$$

   b. Determine the coefficients A_n and write the complete solution.
   c. Show that the temperature approaches zero as $y \to 0$ regardless of the value of T_0.

10. A thin rectangular metal plate occupies the region $0 \leq x \leq a$, $0 \leq y \leq b$. Three edges are held at temperature zero, while the edge at $y = b$ has a sinusoidal temperature distribution $u(x, b) = T_0 \sin(\pi x/a)$:
   a. Solve Laplace's equation with these boundary conditions.
   b. Show that only one term in the Fourier series is nonzero.
   c. Calculate the heat flux through the edge at $y = 0$.
   d. Sketch isotherms (curves of constant temperature) in the plate.

11. **Full 2D Drug Diffusion Problem**
   A drug-eluting polymer implant can be modeled as a rectangular slab with dimensions $a \times b \times L$. The thickness L is small compared to a and b. The bottom surface (at $z = 0$) releases drug at a controlled rate, maintaining concentration C_0. The top surface (at $z = L$) is exposed to blood flow, maintaining $C = 0$ (perfect sink). The four side surfaces are sealed (impermeable):

a. Write the steady-state diffusion equation and boundary conditions.
b. Explain why the problem reduces to one dimension in the z-direction away from the edges.
c. Solve for the concentration profile $C(z)$ and calculate the steady-state release rate per unit area.
d. If edge effects are considered, set up (but do not solve) the full 2D problem for the concentration near a corner.

12. **Electrochemical Cell with Nonuniform Electrode**
Repeat Example 14.3 but now assume the anode (at $y = 0$) has a nonuniform potential distribution:

$$V(x, 0) = V_0 \sin\left(\frac{\pi x}{a}\right),$$

while the cathode remains grounded: $V(x, b) = 0$. The side walls remain insulating:
a. Use separation of variables to solve for $V(x, y)$.
b. Calculate the electric field distribution $\vec{E}(x, y)$.
c. Determine where the current density is maximum.
d. Discuss the physical meaning: why would an electrode have a nonuniform potential?

A

Character Tables of Common Molecular Point Groups

Character tables summarize the symmetry properties of molecules and the transformation behavior of functions, vectors, and orbitals under the operations of a point group. They are indispensable in spectroscopy, quantum chemistry, selection rules, and molecular orbital analysis.

The following tables list:

- symmetry operations;
- irreducible representations (irreps);
- characters;
- typical basis functions (x, y, z; rotational functions; quadratic functions);
- notes relevant to spectroscopy (IR/Raman activity).

A.1 Point Group C_{2v}

C_{2v}	E	$C_2(z)$	$\sigma_v(xz)$	$\sigma'_v(yz)$	**Basis functions**
A_1	1	1	1	1	z, x^2, y^2, z^2
A_2	1	1	−1	−1	$R_z,\ xy$
B_1	1	−1	1	−1	x, R_y, xz
B_2	1	−1	−1	1	y, R_x, yz

Notes:

- IR active: $A_1(z)$, $B_1(x)$, $B_2(y)$.
- Raman active: all irreps (each has a quadratic basis function in this group).

A.2 Point Group C_{3v}

C_{3v}	E	$2C_3$	$3\sigma_v$	**Basis functions**
A_1	1	1	1	z, x^2+y^2, z^2
A_2	1	1	−1	R_z
E	2	−1	0	$(x, y), (R_x, R_y), (xz, yz)$

Notes:

- IR active: $A_1(z)$ and $E(x, y)$.
- Degenerate E representation is crucial for vibrational modes in trigonal pyramidal species (e.g., NH_3).

A.3 Point Group D_{3h}

D_{3h}	E	$2C_3$	$3C_2'$	σ_h	$2S_3$	$3\sigma_v$	Basis functions
A_1'	1	1	1	1	1	1	$z^2, x^2 + y^2$
A_2'	1	1	−1	1	1	−1	R_z
E'	2	−1	0	2	−1	0	$(x, y), (x^2 - y^2, xy)$
A_1''	1	1	1	−1	−1	−1	—
A_2''	1	1	−1	−1	−1	1	z
E''	2	−1	0	−2	1	0	$(R_x, R_y), (xz, yz)$

Note:
- Planar trigonal molecules (BF_3, SO_3): z transforms as A_2''.

A.4 Point Group D_{4h}

D_{4h}	E	$2C_4$	C_2	$2C_2'$	$2C_2''$	i	$2S_4$	σ_h	$2\sigma_v$	$2\sigma_d$	Basis functions
A_{1g}	1	1	1	1	1	1	1	1	1	1	$z^2, x^2 + y^2$
A_{2g}	1	1	1	−1	−1	1	1	1	−1	−1	R_z
B_{1g}	1	−1	1	1	−1	1	−1	1	1	−1	$x^2 - y^2$
B_{2g}	1	−1	1	−1	1	1	−1	1	−1	1	xy
E_g	2	0	−2	0	0	2	0	−2	0	0	$(R_x, R_y), (xz, yz)$
A_{1u}	1	1	1	1	1	−1	−1	−1	−1	−1	—
A_{2u}	1	1	1	−1	−1	−1	−1	−1	1	1	z
B_{1u}	1	−1	1	1	−1	−1	1	−1	−1	1	—
B_{2u}	1	−1	1	−1	1	−1	1	−1	1	−1	—
E_u	2	0	−2	0	0	−2	0	2	0	0	(x, y)

Note:
- Very important for square-planar complexes (D_{4h}).

A.5 Point Group D_{6h}

D_{6h}	E	$2C_6$	$2C_3$	C_2	$3C_2'$	$3C_2''$	i	$2S_3$	$2S_6$	σ_h	$3\sigma_d$	$3\sigma_v$	Basis functions
A_{1g}	1	1	1	1	1	1	1	1	1	1	1	1	z^2, x^2+y^2
A_{2g}	1	1	1	1	−1	−1	1	1	1	1	−1	−1	R_z
B_{1g}	1	−1	1	−1	1	−1	1	−1	1	−1	1	−1	—
B_{2g}	1	−1	1	−1	−1	1	1	−1	1	−1	−1	1	—
E_{1g}	2	1	−1	−2	0	0	2	1	−1	−2	0	0	$(xz, yz)(R_x, R_y)$
E_{2g}	2	−1	−1	2	0	0	2	−1	−1	2	0	0	(x^2-y^2, xy)
A_{1u}	1	1	1	1	1	1	−1	−1	−1	−1	−1	−1	—
A_{2u}	1	1	1	1	−1	−1	−1	−1	−1	−1	1	1	z
B_{1u}	1	−1	1	−1	1	−1	−1	1	−1	1	−1	1	—
B_{2u}	1	−1	1	−1	−1	1	−1	1	−1	1	1	−1	—
E_{1u}	2	1	−1	−2	0	0	−2	−1	1	2	0	0	(x, y)
E_{2u}	2	−1	−1	2	0	0	−2	1	1	−2	0	0	—

A.6 Point Group $D_{\infty h}$

$D_{\infty h}$	E	$2C_\infty$	C_2	σ_v	i	$2S_\infty$	σ_h	Functions/basis
Σ_g^+	1	1	1	1	1	1	1	x^2+y^2, z^2
Σ_g^-	1	1	−1	−1	1	1	−1	R_z
Σ_u^+	1	1	1	1	−1	−1	−1	z
Σ_u^-	1	1	−1	−1	−1	−1	1	
Π_g	2	$2\cos\varphi$	0	0	2	$2\cos\varphi$	0	$(R_x, R_y)(xz, yz)$
Π_u	2	$2\cos\varphi$	0	0	−2	$-2\cos\varphi$	0	$(x,\ y)$
Δ_g	2	$2\cos 2\varphi$	0	0	2	$2\cos 2\varphi$	0	$(x^2-y^2,\ xy)$
Δ_u	2	$2\cos 2\varphi$	0	0	−2	$-2\cos 2\varphi$	0	

A.7 Point Group T_d

T_d	E	$8C_3$	$3C_2$	$6S_4$	$6\sigma_d$	Basis functions
A_1	1	1	1	1	1	$x^2+y^2+z^2$
A_2	1	1	1	−1	−1	—
E	2	−1	2	0	0	$(2z^2-x^2-y^2, x^2-y^2)$
T_1	3	0	−1	1	−1	(R_x, R_y, R_z)
T_2	3	0	−1	−1	1	(x, y, z), (xy, yz, zx)

Notes:

- Basic group for tetrahedral complexes and CH_4 vibrations.
- p-orbitals → T_2.
- d-orbitals → $E + T_2$.

A.8 Point Group O_h

O_h	E	$8C_3$	$6C_2$	$6C_4$	$3C_2'$	i	$6S_4$	$8S_6$	$3\sigma_h$	$6\sigma_d$	Linear/ rotational	Quadratic
A_{1g}	1	1	1	1	1	1	1	1	1	1		$x^2+y^2+z^2$
A_{2g}	1	1	−1	−1	1	1	−1	1	1	−1		
E_g	2	−1	0	0	2	2	0	−1	2	0		$2z^2-x^2-y^2, x^2-y^2$
T_{1g}	3	0	−1	1	−1	3	1	0	−1	−1	(R_x, R_y, R_z)	
T_{2g}	3	0	1	−1	−1	3	−1	0	−1	1		(xz, yz, xy)
A_{1u}	1	1	1	1	1	−1	−1	−1	−1	−1		
A_{2u}	1	1	−1	−1	1	−1	1	−1	−1	1		
E_u	2	−1	0	0	2	−2	0	1	−2	0		
T_{1u}	3	0	−1	1	−1	−3	−1	0	1	1	(x, y, z)	
T_{2u}	3	0	1	−1	−1	−3	1	0	1	−1		

Notes:

- Octahedral complexes (O_h) dominate transition-metal chemistry.
- IR active: T_{1u} only.
- Raman active: A_{1g}, E_g, T_{2g}.

A.9 Direct Product Rules (Useful for Selection Rules)

C_{2v} (fully Abelian)

All irreps multiply like signs:

$A_1 \times \text{any} = \text{any}$
$A_2 \times A_2 = A_1$
$A_2 \times B_1 = B_2$
$A_2 \times B_2 = B_1$

$B_1 \times B_2 = A_2$
$B_1 \times B_1 = A_1$
$B_2 \times B_2 = A_1$

C_{3v}

$$A_1 \times A_1 = A_1$$
$$A_1 \times A_2 = A_2$$
$$A_2 \times A_2 = A_1$$
$$E \times E = A_1 + A_2 + E$$
$$E \times A_1 = E$$
$$E \times A_2 = E$$

T_d

$$T_2 \times T_2 = A_1 + E + T_1 + T_2$$
$$T_2 \times E = T_1 + T_2$$
$$E \times E = A_1 + A_2 + E$$
$$T_1 \times T_1 = A_1 + E + T_1 + T_2$$

A.10 Quick Spectroscopy Notes

- **IR activity:** irreps transforming as x, y, z.
- **Raman activity:** irreps transforming as quadratic functions.
- **Electronic transitions:** allowed if Γ(initial) $\otimes$ Γ(operator) $\otimes$ Γ(final) contains A_1 (or totally symmetric).

5. Exponential decay (damped function):

$$f(t) = e^{-\gamma|t|}, \gamma > 0,$$

$$F(\omega) = \sqrt{\frac{2}{\pi}} \frac{\gamma}{\gamma^2 + \omega^2}.$$

Note: Even function → real, even transform.

6. One-sided exponential (causal):

$$f(t) = e^{-\gamma t} u(t), \gamma > 0,$$

$$F(\omega) = \frac{1}{\sqrt{2\pi}} \frac{1}{\gamma + i\omega}.$$

7. Rectangular pulse:

$$f(t) = \mathrm{rect}\left(\frac{t}{2a}\right) = \begin{cases} 1, & |t| \leq a, \\ 0, & |t| > a, \end{cases}$$

$$F(\omega) = \sqrt{\frac{2}{\pi}}\, a\, \mathrm{sinc}(a\omega),$$

where

$$\mathrm{sinc}(x) = \frac{\sin x}{x}.$$

8. Triangular pulse:

$$f(t) = \mathrm{tri}\left(\frac{t}{a}\right) = \begin{cases} 1 - |t|/a, & |t| \leq a, \\ 0, & |t| > a, \end{cases}$$

$$F(\omega) = \sqrt{\frac{1}{2\pi}}\, a \sin \mathrm{c}^2\left(\frac{a\omega}{2}\right).$$

9. Sign function:

$$f(t) = \mathrm{sgn}(t) \Longleftrightarrow F(\omega) = -i\sqrt{\frac{2}{\pi}} \frac{1}{\omega}$$

(in the principal value sense).

B.2 Derivative and Multiplication Properties

All properties below follow directly from differentiating under the integral sign, consistent with the convention used in the book:

1. **Time derivative:**

$$\mathcal{F}\left\{\frac{df}{dt}\right\} = i\omega\, F(\omega),$$

$$\mathcal{F}\left\{\frac{d^n f}{dt^n}\right\} = (i\omega)^n\, F(\omega).$$

2. **Multiplication by *t*:**

$$\mathcal{F}\{t\,f(t)\} = i\,\frac{dF}{d\omega},$$

$$\mathcal{F}\{t^n f(t)\} = i^n\,\frac{d^n F}{d\omega^n}.$$

3. **Frequency-domain derivative:**

$$\mathcal{F}^{-1}\left\{\frac{dF}{d\omega}\right\} = -it\, f(t).$$

4. **Shift theorem:**
   Time shift:

$$\mathcal{F}\{f(t-t_0)\} = e^{-i\omega t_0} F(\omega).$$

   Frequency shift:

$$\mathcal{F}\{e^{i\omega_0 t} f(t)\} = F(\omega-\omega_0).$$

5. **Scaling theorem:**

$$\mathcal{F}\{f(at)\} = \frac{1}{|a|}F\left(\frac{\omega}{a}\right)$$

   (for real nonzero a).

B.3 Convolution and Product Theorems

1. **Convolution:**

$$(f*g)(t) = \int_{-\infty}^{\infty} f(\tau)\, g(t-\tau)\, d\tau,$$

$$\mathcal{F}\{f*g\} = \sqrt{2\pi}\, F(\omega)G(\omega).$$

2. **Product in the time domain:**

$$\mathcal{F}\{f(t)g(t)\} = \frac{1}{\sqrt{2\pi}}(F*G)(\omega).$$

   This matches exactly the conventions used in the book.

B.4 Symmetry Properties

1. **Even functions:**
   If $f(t) = f(-t)$ and $f(t)$ is real, then:

$$F(\omega) = F(-\omega) \quad \text{and} \quad F(\omega) \in \mathbb{R}.$$

2. **Odd functions:**
   If $f(t) = -f(-t)$ and $f(t)$ is real, then:

$$F(\omega) = -F(-\omega) \quad \text{and} \quad F(\omega) \in i\mathbb{R}.$$

These properties are extremely useful for spectroscopy lineshapes and real signal analysis.

B.5 Notes for Chemistry and Physics

- Real spectra imply Hermitian symmetry:

$$F(-\omega) = F \times (\omega).$$

- Exponential decays → Lorentzian frequency profiles.
- Rectangular pulses → sinc functions (diffraction patterns).
- Gaussians remain Gaussians (minimum-uncertainty states).

C

Laplace Transform Table (Compatible With Book Convention)

This appendix lists the most frequently used Laplace transform pairs and operational properties required for solving initial value problems, linear ODEs, and systems relevant to chemistry, physics, and engineering.

All transforms assume:

$$\mathscr{L}\{f(t)\} = \int_0^{\infty} f(t)\, e^{-st}\, dt.$$

C.1 Elementary Transforms

1. **Constant:**

$$\mathscr{L}\{1\} = \frac{1}{s}, \quad \mathbb{R}(s) > 0.$$

2. **Power functions:**

$$\mathscr{L}\{t^n\} = \frac{n!}{s^{n+1}}, \qquad n = 0, 1, 2, \ldots.$$

3. **Exponential:**

$$\mathscr{L}\{e^{at}\} = \frac{1}{s-a}, \quad \mathbb{R}(s) > \mathbb{R}(a).$$

4. **Damped exponential:**

$$\mathscr{L}\{e^{-\gamma t}\} = \frac{1}{s+\gamma}.$$

5. **Sine and cosine:**

$$\mathscr{L}\{\sin(bt)\} = \frac{b}{s^2 + b^2}, \quad \mathscr{L}\{\cos(bt)\} = \frac{s}{s^2 + b^2}.$$

6. **Exponentially damped sine/cosine:**

$$\mathscr{L}\{e^{-at}\sin(bt)\} = \frac{b}{(s+a)^2 + b^2},$$

$$\mathscr{L}\{e^{-at}\cos(bt)\} = \frac{s+a}{(s+a)^2 + b^2}.$$

C.2 Functions Used in Spectroscopy and Chemical Kinetics

1. **First-order decay:**

$$\mathcal{L}\{e^{-kt}\} = \frac{1}{s+k}.$$

2. **Rise-and-decay (difference of exponentials):**
Common in fluorescence, kinetic schemes, and relaxation:

$$\mathcal{L}\{e^{-at} - e^{-bt}\} = \frac{1}{s+a} - \frac{1}{s+b}.$$

3. **Gaussian (formal transform):**
Laplace transform exists in terms of the error function:

$$\mathcal{L}\left\{e^{-at^2}\right\} = \frac{1}{2}\sqrt{\frac{\pi}{a}}\, e^{s^2/(4a)}\, \mathrm{erfc}\left(\frac{s}{2\sqrt{a}}\right).$$

(Useful in diffusion problems.)

C.3 Hyperbolic Functions

$$\mathcal{L}\{\sinh(bt)\} = \frac{b}{s^2 - b^2}, \qquad \mathcal{L}\{\cosh(bt)\} = \frac{s}{s^2 - b^2},$$

$$\mathcal{L}\{e^{-at}\sinh(bt)\} = \frac{b}{(s+a)^2 - b^2}, \qquad \mathcal{L}\{e^{-at}\cosh(bt)\} = \frac{s+a}{(s+a)^2 - b^2}.$$

C.4 Derivative and Integral Properties

1. **First derivative:**

$$\mathcal{L}\{f'(t)\} = sF(s) - f(0).$$

2. **Second derivative:**

$$\mathcal{L}\{f''(t)\} = s^2F(s) - sf(0) - f'(0).$$

3. ***n*th derivative:**

$$\mathcal{L}\{f^{(n)}(t)\} = s^nF(s) - s^{n-1}f(0) - s^{n-2}f'(0) - \cdots - f^{(n-1)}(0).$$

4. **Integration:**

$$\mathcal{L}\left\{\int_0^t f(\tau)\,d\tau\right\} = \frac{1}{s}F(s).$$

C.5 Shifting Theorems

1. **Time shift with unit step:**

$$\mathscr{L}\{f(t-a)u(t-a)\} = e^{-as}F(s).$$

   (Used only where causality must be specified.)

2. **Frequency-domain shift:**

$$\mathscr{L}\{e^{at}f(t)\} = F(s-a).$$

C.6 Convolution Theorem

$$(f * g)(t) = \int_0^t f(\tau)\,g(t-\tau)\,d\tau,$$

$$\mathscr{L}\{f * g\} = F(s)\,G(s).$$

This theorem underlies Duhamel's principle and forced response solutions in ODEs.

C.7 Common Inverse Laplace Transforms

1. $\mathscr{L}^{-1}\left\{\dfrac{1}{s-a}\right\} = e^{at}.$
2. $\mathscr{L}^{-1}\left\{\dfrac{1}{s(s+a)}\right\} = \dfrac{1}{a}(1-e^{-at}).$
3. $\mathscr{L}^{-1}\left\{\dfrac{s}{(s^2+b^2)}\right\} = \cos bt.$
4. $\mathscr{L}^{-1}\left\{\dfrac{b}{s^2+b^2}\right\} = \sin bt.$
5. **Partial fraction form (very important for ODE problems):**
   If

$$F(s) = \frac{P(s)}{(s+a)(s+b)} \text{ with } \deg P(s) \leq 1,$$

   then

$$\mathscr{L}^{-1}\{F(s)\} = Ae^{-at} + Be^{-bt}.$$

C.8 Notes for Chemistry, Physics, and Engineering

- Laplace transforms convert **initial-value problems** into algebraic equations.
- Exponential decays, damped oscillations, first-order reactions, and RC/RL analogs map cleanly to rational functions of s.
- Shifts in s correspond to energetic or frequency shifts in physical systems.
- Convolution describes the response to forcing or relaxation phenomena.

D

Special Functions: Definitions, Orthogonality, and Recurrence Relations

Special functions arise naturally as solutions of second-order linear differential equations encountered in quantum mechanics, spectroscopy, molecular vibrations, and mathematical physics. This appendix summarizes essential definitions, differential equations, orthogonality relations, and recurrence formulas.

D.1 Legendre Polynomials $P_l(x)$

Differential Equation:

$$(1-x^2)\frac{d^2P_l}{dx^2} - 2x\frac{dP_l}{dx} + l(l+1)P_l = 0, l = 0, 1, 2, \dots .$$

Rodrigues's Formula:

$$P_l(x) = \frac{1}{2^l\, l!}\frac{d^l}{dx^l}(x^2-1)^l.$$

Orthogonality:

$$\int_{-1}^{1} P_l(x)P_k(x)dx = \frac{2}{2l+1}\delta_{lk}.$$

First Few Polynomials:

$$P_0(x) = 1, \quad P_1(x) = x, \quad P_2(x) = \frac{1}{2}(3x^2-1),$$

$$P_3(x) = \frac{1}{2}(5x^3-3x).$$

Recurrence Relations:

$$(2l+1)xP_l(x) = (l+1)P_{l+1}(x) + lP_{l-1}(x),$$

$$\frac{dP_l}{dx} = \frac{l}{x^2-1}(xP_l - P_{l-1}).$$

D.2 Associated Legendre Functions $P_l^m(x)$

Definition:

$$P_l^m(x) = (1-x^2)^{m/2}\frac{d^m}{dx^m}P_l(x), \quad m = 0, 1, \dots, l.$$

Orthogonality:

$$\int_{-1}^{1} P_l^m(x) P_{l'}^m(x) dx = \frac{2(l+m)!}{(2l+1)(l-m)!} \delta_{ll'}.$$

Parity:

$$P_l^m(-x) = (-1)^{l+m} P_l^m(x).$$

Spherical Harmonics:

$$Y_l^m(\theta, \phi) = (-1)^m \sqrt{\frac{2l+1}{4\pi} \frac{(l-m)!}{(l+m)!}} P_l^m(\cos\theta) e^{im\phi}.$$

D.3 Hermite Polynomials $H_n(x)$

Differential Equation:

$$\frac{d^2y}{dx^2} - 2x\frac{dy}{dx} + 2ny = 0.$$

Rodrigues's Formula:

$$H_n(x) = (-1)^n e^{x^2} \frac{d^n}{dx^n}\left(e^{-x^2}\right).$$

Orthogonality:

$$\int_{-\infty}^{\infty} e^{-x^2} H_n(x) H_m(x) dx = \sqrt{\pi}\, 2^n n!\, \delta_{nm}.$$

First Few Polynomials:

$$H_0(x) = 1, H_1(x) = 2x, H_2(x) = 4x^2 - 2.$$

Recurrence:

$$H_{n+1}(x) = 2xH_n(x) - 2nH_{n-1}(x).$$

Note (Quantum Mechanics):
Hermite polynomials appear in the eigenfunctions of the quantum harmonic oscillator.

D.4 Laguerre Polynomials $L_n(x)$

Differential Equation:

$$x\frac{d^2y}{dx^2} + (1-x)\frac{dy}{dx} + ny = 0.$$

Rodrigues's Formula:

$$L_n(x) = \frac{e^x}{n!} \frac{d^n}{dx^n}(e^{-x} x^n).$$

Orthogonality:

$$\int_0^\infty e^{-x} L_n(x)\, L_m(x) dx = \delta_{nm}.$$

First Few Polynomials:

$$L_0(x) = 1, L_1(x) = 1 - x, L_2(x) = \frac{1}{2}(x^2 - 4x + 2).$$

Recurrence:

$$(n+1)L_{n+1}(x) = (2n + 1 - x)L_n(x) - nL_{n-1}(x).$$

Note (Hydrogen Atom):
Radial wavefunctions contain associated Laguerre polynomials.

D.5 Associated Laguerre Polynomials $L_n^k(x)$

Definition:

$$L_n^k(x) = (-1)^k \frac{d^k}{dx^k} L_{n+k}(x).$$

Orthogonality:

$$\int_0^\infty x^k e^{-x}\, L_n^k(x) L_m^k(x) dx = \frac{(n+k)!}{n!} \delta_{nm}.$$

Note:
Important in hydrogen-like atomic orbitals:

$$R_{nl}(r) \propto e^{-r/na}\, r^l\, L_{n-l-1}^{2l+1}\left(\frac{2r}{na}\right).$$

D.6 Bessel Functions of the First Kind $J_\nu(x)$

Differential Equation:

$$x^2 \frac{d^2y}{dx^2} + x\frac{dy}{dx} + (x^2 - \nu^2)y = 0.$$

Series Definition:

$$J_\nu(x) = \sum_{k=0}^\infty \frac{(-1)^k}{k!\,\Gamma(k + \nu + 1)} \left(\frac{x}{2}\right)^{2k+\nu}.$$

Orthogonality:
For zeros $j_{\nu,n}$ of J_ν:

$$\int_0^1 J_\nu(j_{\nu,n}x)\, J_\nu(j_{\nu,m}x)\, x\, dx = \frac{\delta_{nm}}{2} \left[J_{\nu+1}(j_{\nu,n})\right]^2.$$

Recurrence:

$$J_{\nu-1}(x) + J_{\nu+1}(x) = \frac{2\nu}{x} J_\nu(x),$$

$$J_{\nu-1}(x) - J_{\nu+1}(x) = 2J'_\nu(x).$$

Note (Physics and Engineering):
Bessel functions arise in problems with cylindrical symmetry (heat conduction, vibrations, waveguides).

D.7 Modified Bessel Functions $I_\nu(x)$, $K_\nu(x)$

(For completeness)

$$I_\nu(x) = i^{-\nu} J_\nu(ix), K_\nu(x) = \frac{\pi}{2} \frac{I_{-\nu}(x) - I_\nu(x)}{\sin(\nu\pi)}.$$

Appear in diffusion and radial Schrödinger equations with imaginary arguments.

D.8 Gamma and Beta Functions

Gamma Function:

$$\Gamma(z) = \int_0^\infty t^{z-1} e^{-t} dt,$$

$$\Gamma(n+1) = n!, \Gamma(1/2) = \sqrt{\pi}.$$

Recurrence:

$$\Gamma(z+1) = z\,\Gamma(z).$$

Beta Function:

$$B(x, y) = \int_0^1 t^{x-1} (1-t)^{y-1} dt = \frac{\Gamma(x)\Gamma(y)}{\Gamma(x+y)}.$$

D.9 Notes for Chemistry and Physics

- Legendre, associated Legendre → angular momentum, spherical harmonics.
- Laguerre, associated Laguerre → hydrogen atom radial functions.
- Hermite → harmonic oscillator wavefunctions.
- Bessel → cylindrical wells, diffusion, and radial PDEs.
- Gamma/beta → normalization constants, integrals in statistics and quantum mechanics.

E

Differential Operators in Cartesian, Cylindrical, and Spherical Coordinates

Differential operators such as gradient, divergence, curl, and the Laplacian arise throughout chemistry, physics, and engineering. They appear in quantum mechanics, electrostatics, diffusion, heat flow, spectroscopy, and wave propagation. This appendix summarizes their standard forms in orthogonal coordinate systems.

E.1 Cartesian Coordinates x, y, z

The scale factors are:

$$h_x = h_y = h_z = 1.$$

Gradient:

$$\nabla f = \left(\frac{\partial f}{\partial x}, \frac{\partial f}{\partial y}, \frac{\partial f}{\partial z}\right).$$

Divergence:

$$\nabla \cdot \vec{A} = \frac{\partial A_x}{\partial x} + \frac{\partial A_y}{\partial y} + \frac{\partial A_z}{\partial z}.$$

Curl:

$$\nabla \times \vec{A} = \left(\frac{\partial A_z}{\partial y} - \frac{\partial A_y}{\partial z}, \frac{\partial A_x}{\partial z} - \frac{\partial A_z}{\partial x}, \frac{\partial A_y}{\partial x} - \frac{\partial A_x}{\partial y}\right).$$

Laplacian:

$$\nabla^2 f = \frac{\partial^2 f}{\partial x^2} + \frac{\partial^2 f}{\partial y^2} + \frac{\partial^2 f}{\partial z^2}.$$

E.2 Cylindrical Coordinates r, ϕ, z

Coordinate definitions:

$$x = r\cos\phi, y = r\sin\phi, \ z = z.$$

Scale factors:

$$h_r = 1, h_\phi = r, h_z = 1.$$

Gradient:

$$\nabla f = \frac{\partial f}{\partial r}\hat{e}_r + \frac{1}{r}\frac{\partial f}{\partial \phi}\hat{e}_\phi + \frac{\partial f}{\partial z}\hat{e}_z.$$

Divergence:

$$\nabla \cdot \vec{A} = \frac{1}{r}\frac{\partial}{\partial r}(rA_r) + \frac{1}{r}\frac{\partial A_\phi}{\partial \phi} + \frac{\partial A_z}{\partial z}.$$

Curl:

$$\nabla \times \vec{A} = \left[\frac{1}{r}\frac{\partial A_z}{\partial \phi} - \frac{\partial A_\phi}{\partial z}\right]\hat{e}_r + \left[\frac{\partial A_r}{\partial z} - \frac{\partial A_z}{\partial r}\right]\hat{e}_\phi + \frac{1}{r}\left[\frac{\partial}{\partial r}(rA_\phi) - \frac{\partial A_r}{\partial \phi}\right]\hat{e}_z.$$

Laplacian:
Scalar Laplacian:

$$\nabla^2 f = \frac{1}{r}\frac{\partial}{\partial r}\left(r\frac{\partial f}{\partial r}\right) + \frac{1}{r^2}\frac{\partial^2 f}{\partial \phi^2} + \frac{\partial^2 f}{\partial z^2}.$$

Note (Chemistry and Physics):
Cylindrical coordinates arise in:

- diffusion in cylindrical pores
- polymer chains
- vibrations of cylindrical rods
- rotational spectroscopy with cylindrical symmetry

E.3 Spherical Coordinates r, θ, ϕ

Definitions:

$$x = r\sin\theta\cos\phi,\ y = r\sin\theta\sin\phi,\ z = r\cos\theta.$$

Scale factors:

$$h_r = 1,\ h_\theta = r,\ h_\phi = r\sin\theta.$$

Gradient:

$$\nabla f = \frac{\partial f}{\partial r}\hat{e}_r + \frac{1}{r}\frac{\partial f}{\partial \theta}\hat{e}_\theta + \frac{1}{r\sin\theta}\frac{\partial f}{\partial \phi}\hat{e}_\phi.$$

Divergence:

$$\nabla \cdot \vec{A} = \frac{1}{r^2}\frac{\partial}{\partial r}(r^2 A_r) + \frac{1}{r\sin\theta}\frac{\partial}{\partial \theta}(\sin\theta\, A_\theta) + \frac{1}{r\sin\theta}\frac{\partial A_\phi}{\partial \phi}.$$

Curl:

$$\begin{aligned}\nabla \times \vec{A} = &\frac{1}{r\sin\theta}\left[\frac{\partial}{\partial \theta}(\sin\theta A_\phi) - \frac{\partial A_\theta}{\partial \phi}\right]\hat{e}_r \\ &+\frac{1}{r}\left[\frac{1}{\sin\theta}\frac{\partial A_r}{\partial \phi} - \frac{\partial}{\partial r}(rA_\phi)\right]\hat{e}_\theta \\ &+\frac{1}{r}\left[\frac{\partial}{\partial r}(rA_\theta) - \frac{\partial A_r}{\partial \theta}\right]\hat{e}_\phi.\end{aligned}$$

Laplacian:
Scalar Laplacian:

$$\nabla^2 f = \frac{1}{r^2}\frac{\partial}{\partial r}\left(r^2\frac{\partial f}{\partial r}\right) + \frac{1}{r^2 \sin\theta}\frac{\partial}{\partial\theta}\left(\sin\theta\frac{\partial f}{\partial\theta}\right) + \frac{1}{r^2\sin^2\theta}\frac{\partial^2 f}{\partial\phi^2}.$$

Note (Quantum Mechanics):
The Laplacian in spherical coordinates underlies the Schrödinger equation for central potentials:

$$\nabla^2\psi = \frac{1}{r^2}\frac{\partial}{\partial r}\left(r^2\frac{\partial\psi}{\partial r}\right) - \frac{\widehat{L}^2}{\hbar^2 r^2}\psi,$$

where $\widehat{L}^2$ is the angular momentum operator whose eigenfunctions are spherical harmonics.

E.4 Summary Table for Quick Reference

Operator	Cartesian	Cylindrical	Spherical
∇f	Simple partials	Extra $1/r$, $1/(r\sin\theta)$ factors	Radial and angular terms
$\nabla\cdot\vec{A}$	Sum of partials	$1/r$ factors	Radial + angular divergence
$\nabla\times\vec{A}$	Determinant form	Mixed derivatives	Most complex form
$\nabla^2 f$	Sum of second derivatives	Bessel-type forms appear	Separable into radial + spherical harmonics

F

Vector Calculus Identities

Vector calculus identities are used extensively in solving partial differential equations, deriving conservation laws, manipulating operator expressions in quantum mechanics, and simplifying electromagnetic and fluid-flow equations. This appendix lists the identities most frequently required in chemistry, physics, and engineering.

All vectors are assumed to be sufficiently smooth, and ∇ denotes the gradient operator.

F.1 Basic Identities

Gradient of a scalar product:
If f and g are scalar fields,

$$\nabla(fg) = f\,\nabla g + g\,\nabla f.$$

Divergence of a vector sum:

$$\nabla \cdot \left(\vec{A} + \vec{B}\right) = \nabla \cdot \vec{A} + \nabla \cdot \vec{B}.$$

Curl of a vector sum:

$$\nabla \times \left(\vec{A} + \vec{B}\right) = \nabla \times \vec{A} + \nabla \times \vec{B}.$$

Divergence of a scalar–vector product:

$$\nabla \cdot \left(f\vec{A}\right) = f\left(\nabla \cdot \vec{A}\right) + \vec{A} \cdot (\nabla f).$$

Curl of a scalar–vector product:

$$\nabla \times \left(f\vec{A}\right) = f\left(\nabla \times \vec{A}\right) + (\nabla f) \times \vec{A}.$$

F.2 Vector Identities Involving Cross and Dot Products

Curl of gradient:

$$\nabla \times (\nabla f) = \vec{0}.$$

Divergence of curl:

$$\nabla \cdot \left(\nabla \times \vec{A}\right) = 0.$$

Curl of curl:

$$\nabla \times \left(\nabla \times \vec{A}\right) = \nabla\left(\nabla \cdot \vec{A}\right) - \nabla^2 \vec{A}.$$

Gradient of a dot product:

$$\nabla\left(\vec{A} \cdot \vec{B}\right) = \left(\vec{A} \cdot \nabla\right)\vec{B} + (\vec{B} \cdot \nabla)\vec{A} + \vec{A} \times (\nabla \times \vec{B}) + \vec{B} \times \left(\nabla \times \vec{A}\right).$$

F.3 Product Identities for Divergence and Curl

Divergence of a cross product:

$$\nabla \cdot \left(\vec{A} \times \vec{B}\right) = \vec{B} \cdot \left(\nabla \times \vec{A}\right) - \vec{A} \cdot (\nabla \times \vec{B}).$$

Curl of a cross product:

$$\nabla \times \left(\vec{A} \times \vec{B}\right) = (\vec{B} \cdot \nabla)\vec{A} - \left(\vec{A} \cdot \nabla\right)\vec{B} + \vec{A}(\nabla \cdot \vec{B}) - \vec{B}\left(\nabla \cdot \vec{A}\right).$$

F.4 Laplacian Identities

Laplacian of a scalar product:

$$\nabla^2(fg) = f\nabla^2 g + g\nabla^2 f + 2\nabla f \cdot \nabla g.$$

Vector Laplacian:
In Cartesian coordinates, for a vector field $\vec{A}$,

$$\nabla^2 \vec{A} = (\nabla^2 A_x,\ \nabla^2 A_y,\ \nabla^2 A_z).$$

Combined with the curl identity,

$$\nabla^2 \vec{A} = \nabla\left(\nabla \cdot \vec{A}\right) - \nabla \times \left(\nabla \times \vec{A}\right).$$

F.5 Triple-product Identities

Scalar triple product:

$$\vec{A} \cdot \left(\vec{B} \times \vec{C}\right) = \vec{B} \cdot \left(\vec{C} \times \vec{A}\right) = \vec{C} \cdot \left(\vec{A} \times \vec{B}\right).$$

Vector triple product:

$$\vec{A} \times \left(\vec{B} \times \vec{C}\right) = \vec{B}\left(\vec{A} \cdot \vec{C}\right) - \vec{C}\left(\vec{A} \cdot \vec{B}\right),$$

$$\left(\vec{A} \times \vec{B}\right) \times \vec{C} = -\vec{C} \times \left(\vec{A} \times \vec{B}\right) = \vec{B}\left(\vec{A} \cdot \vec{C}\right) - \vec{A}\left(\vec{B} \cdot \vec{C}\right).$$

F.6 Integral Theorems (Summary)

Divergence theorem:

$$\int_V \left(\nabla \cdot \vec{A}\right) dV = \oint_{\partial V} \vec{A} \cdot d\vec{S}.$$

Stokes's theorem:

$$\int_S \left(\nabla \times \vec{A}\right) \cdot d\vec{S} = \oint_{\partial S} \vec{A} \cdot d\vec{r}.$$

Gradient fundamental theorem:

$$\int_A^B \nabla f \cdot d\vec{r} = f(B) - f(A).$$

F.7 Notes for Chemistry and Physics

- Curl and divergence identities appear in electrodynamics, fluid flow, and magnetic resonance.
- Laplacian identities are essential for the separation of variables in PDEs and in the Schrödinger equation.
- Triple-product identities are used in angular momentum algebra and vector operator manipulation.

G

Linear Algebra Identities and Matrix Decompositions

This appendix collects the most frequently used algebraic identities, operator properties, and matrix decompositions needed in quantum mechanics, spectroscopy, numerical analysis, and vector-space theory. All matrices are finite-dimensional unless otherwise stated.

G.1 Basic Matrix Identities

Transpose:

$$(AB)^T = B^T A^T.$$

Adjoint (Hermitian conjugate):

$$(AB)^\dagger = B^\dagger A^\dagger.$$

Inverse:

$$(AB)^{-1} = B^{-1} A^{-1}.$$

Determinant identities:

$$\det(AB) = \det(A)\det(B),$$

$$\det\left(A^T\right) = \det(A), \det\left(A^\dagger\right) = \det(A)^*.$$

Trace identities:

$$\mathrm{tr}(AB) = \mathrm{tr}(BA), \mathrm{tr}(ABC) = \mathrm{tr}(BCA) = \mathrm{tr}(CAB),$$

$$\mathrm{tr}\left(A^\dagger A\right) \geq 0 \text{ (positive semidefiniteness).}$$

Rank–Nullity theorem:
For a linear map $T\colon V \to V$,

$$\dim(\ker T) + \dim(\mathrm{im} T) = \dim(V).$$

G.2 Special Classes of Matrices

Hermitian matrices:

$$A^\dagger = A.$$

Properties:

- All eigenvalues are real.
- Eigenvectors belonging to distinct eigenvalues are orthogonal.
- Forms the basis of quantum mechanical observables.

Unitary matrices:

$$U^{\dagger}U = UU^{\dagger} = I.$$

Properties:

- Preserve norms: $|Ux| = |x|$.
- Eigenvalues lie on the unit circle.
- Represent symmetry operations and time evolution (finite-dimensional).

Orthogonal matrices:
(Real version of unitary.)

$$Q^T Q = I.$$

Normal matrices:

$$AA^{\dagger} = A^{\dagger}A.$$

Include Hermitian, unitary, and skew-Hermitian matrices.
Diagonalizable by a unitary matrix.

Positive definite matrices:

$$x^{\dagger}Ax > 0, \qquad \forall\, x \neq 0.$$

Equivalent characterizations:

- All eigenvalues positive.
- $A = B^{\dagger}B$ for some B of full column rank (e.g., invertible B).
- All leading principal minors positive.

G.3 Eigenvalue Identities

Similarity invariance:
If $B = S^{-1}AS$, then

$$\text{eigenvalues}(A) = \text{eigenvalues}(B).$$

Trace and determinant:

$$\text{tr}(A) = \sum_i \lambda_i,\ \det(A) = \prod_i \lambda_i.$$

Characteristic polynomial:

$$p(\lambda) = \det(A - \lambda I).$$

Spectral radius:

$$\rho(A) = \max_i |\lambda_i|.$$

Rayleigh quotient:
For Hermitian A,

$$R(x) = \frac{x^\dagger A x}{x^\dagger x},$$

satisfying

$$\lambda_{\min} \leq R(x) \leq \lambda_{\max}.$$

G.4 Matrix Norms (Frequently Used)

Frobenius norm:

$$|A|_F = \sqrt{\sum_{i,j} |a_{ij}|^2} = \sqrt{\mathrm{tr}(A^\dagger A)}.$$

Spectral norm:

$$\|A\|_2 = \sqrt{\lambda_{\max}(A^\dagger A)}.$$

G.5 Matrix Decompositions

These decompositions appear in numerical methods, stability analysis, quantum computing, and the study of linear operators.

1. **Eigenvalue decomposition:**
   (For diagonalizable matrices.)

   $$A = \mathrm{SDS}^{-1}, D = \mathrm{diag}(\lambda_1, \ldots, \lambda_n).$$

   If A is Hermitian,

   $$A = UDU^\dagger,$$

   with real λ_i.
2. **Spectral decomposition (Hermitian case):**

   $$A = \sum_i \lambda_i \,|\, v_i \rangle \langle v_i \,|.$$

3. **Singular Value Decomposition (SVD):**
   (Holds for all real/complex matrices.)

   $$A = U \Sigma V^\dagger,$$

where:
- U, V are unitary,
- $\Sigma = \text{diag}(\sigma_1, \ldots, \sigma_r)$,
- $\sigma_i \geq 0$ are singular values.

Chemical interpretation: SVD underlies PCA, low-rank approximations, data analysis, and regularization.

4. ***QR*** **Decomposition:**

$$A = QR,$$

with:
- Q orthogonal/unitary,
- R upper triangular.

Used in numerical eigenvalue algorithms.

5. ***LU*** **Decomposition:**

$$A = LU,$$

with:
- L lower triangular,
- U upper triangular.

Pivoting gives

$$PA = LU.$$

6. **Cholesky Decomposition:**
(For Hermitian positive definite matrices.)

$$A = LL^{\dagger}, L \text{ lower triangular.}$$

Very important in quantum chemistry, integral algorithms, and numerical optimization.

G.6 Outer Products and Projection Operators

Outer product:
For vectors u, v,

$$uv^{\dagger} \text{ is a matrix with rank} \leq 1.$$

Projection operator:
For normalized v,

$$P = |v\rangle\langle v|, P^2 = P.$$

Orthogonal projector onto a subspace U:

$$P_U = \sum_{i=1}^{k} |u_i\rangle\langle u_i|.$$

G.7 Notes for Chemistry, Physics, and Engineering

- Hermitian matrices correspond to observables in quantum mechanics.
- Unitary matrices represent symmetries (rotations, reflections) and time evolution.
- SVD is widely used for multidimensional spectroscopy and data compression.
- Cholesky factorization appears in variational calculations and integral evaluation.
- *QR* and *LU* decompositions form the backbone of numerical eigenvalue algorithms.

H

Numerical Methods Reference Sheet

This appendix summarizes the numerical techniques most frequently required for solving nonlinear equations, evaluating integrals, and approximating solutions of ordinary differential equations. Only the core formulas and algorithmic steps are included.

H.1 Root-finding Methods

1. **Bisection Method:**
   Given a continuous function $f(x)$ on $[a, b]$ with $f(a)f(b) < 0$:
   - Compute midpoint:
   $$c = \frac{a+b}{2}.$$
   Replace either a or b by c depending on the sign of $f(c)$.
   - Repeat until interval length is within tolerance.
   Guaranteed convergence, slow but robust.
2. **Fixed-point Iteration:**
   Rewrite equation $f(x) = 0$ as
   $$x = g(x).$$
   Iterate:
   $$x_{n+1} = g(x_n).$$
   Converges if $|g'(x)| < 1$ in a neighborhood of the solution.
3. **Newton–Raphson Method:**
   $$x_{n+1} = x_n - \frac{f(x_n)}{f'(x_n)}.$$
   - Quadratic convergence near the root
   - Requires derivative
   - Sensitive to initial guess
4. **Secant Method:**
   Derivative-free version of Newton's method:
   $$x_{n+1} = x_n - f(x_n)\frac{x_n - x_{n-1}}{f(x_n) - f(x_{n-1})}.$$
   Superlinear convergence; useful when $f'(x)$ is difficult to compute.

H.2 Numerical Integration (Quadrature)

1. **Trapezoidal Rule:**

$$\int_a^b f(x)\,dx \approx \frac{b-a}{2}[f(a)+f(b)].$$

Composite form (n subintervals):

$$\int_a^b f(x)dx \approx \frac{h}{2}\left[f(x_0)+2\sum_{i=1}^{n-1} f(x_i)+f(x_n)\right],$$

with $h=(b-a)/n$.

2. **Simpson's Rule:**
Requires even number of subintervals,

$$\int_a^b f(x)dx \approx \frac{h}{3}\left[f(x_0)+4\sum_{i=1,3,\ldots}^{n-1} f(x_i)+2\sum_{i=2,4,\ldots}^{n-2} f(x_i)+f(x_n)\right].$$

More accurate for smooth functions.

3. **Gaussian Quadrature (Legendre):**
Approximation:

$$\int_{-1}^{1} f(x)dx \approx \sum_{i=1}^{n} w_i\,f(x_i).$$

Here, x_i are roots of $P_n(x)$ and w_i are weights.
Used in quantum chemistry integrals and high-precision integration.

H.3 Numerical Differentiation

Forward-difference:

$$f'(x) \approx \frac{f(x+h)-f(x)}{h}.$$

Central-difference (more accurate):

$$f'(x) \approx \frac{f(x+h)-f(x-h)}{2h}.$$

Second derivative:

$$f''(x) \approx \frac{f(x+h)-2f(x)+f(x-h)}{h^2}.$$

H.4 Numerical Methods for ODEs

Consider $y'(t)=f(t,y)$ with $y(t_0)=y_0$.

1. **Euler's Method:**

$$y_{n+1}=y_n+hf(t_n,y_n).$$

First-order, simple, low accuracy.

2. **Improved Euler/Heun's Method:**
   Predictor:

$$y^{(p)} = y_n + hf(t_n, y_n).$$

   Corrector:

$$y_{n+1} = y_n + \frac{h}{2}\big[f(t_n, y_n) + f\big(t_{n+1}, y^{(p)}\big)\big].$$

3. **Midpoint (RK2) Method:**

$$k_1 = f(t_n, y_n), k_2 = f\left(t_n + \frac{h}{2}, y_n + \frac{hk_1}{2}\right),$$

$$y_{n+1} = y_n + hk_2.$$

4. **Runge–Kutta 4 (RK4):**

$$k_1 = f(t_n, y_n),$$

$$k_2 = f\left(t_n + \frac{h}{2}, y_n + \frac{hk_1}{2}\right),$$

$$k_3 = f\left(t_n + \frac{h}{2}, y_n + \frac{hk_2}{2}\right),$$

$$k_4 = f(t_n + h,\ y_n + hk_3),$$

$$y_{n+1} = y_n + \frac{h}{6}(k_1 + 2k_2 + 2k_3 + k_4).$$

   Fourth-order accuracy; standard in scientific computing.

5. **Stability Note (Important for PDEs and Stiff ODEs):**
   Explicitly Euler stable only if

$$|1 + h\lambda| \leq 1,$$

   where λ is eigenvalue of Jacobian.
   Implicit methods are preferred for stiff systems (diffusion and chemical kinetics).

H.5 Linear Algebra in Numerical Computations

Condition number:

$$\kappa(A) = |A||A^{-1}|.$$

Large $\kappa(A)\rightarrow$ ill-conditioning $\rightarrow$ sensitive to numerical error.

Iterative solvers (names only):

Conjugate gradient
Gauss–Seidel
Jacobi
(Useful in large-scale physical simulations.)

H.6 Notes for Chemistry, Physics, and Engineering

- Newton–Raphson is used for solving transcendental spectroscopy equations, rate constants, and energy minimization.
- Gaussian quadrature underlies integral evaluation in quantum chemistry.
- RK4 is widely used in time-dependent Schrödinger equations (approximate).
- Numerical stability is essential in diffusion, heat flow, and reaction–diffusion systems.

I

Physical Constants and Conversions

This appendix lists the physical constants and unit conversions most frequently required in spectroscopy, thermodynamics, quantum chemistry, molecular physics, and materials science.

Numerical values are rounded to appropriate significant figures for practical calculations.

I.1 Fundamental Physical Constants

Speed of light:

$$c = 2.99792458 \times 10^8 \text{ m s}^{-1} \text{ (exact).}$$

Planck's constant:

$$h = 6.62607015 \times 10^{-34} \text{ J s (exact by definition).}$$

Reduced Planck's constant:

$$\hbar = \frac{h}{2\pi} = 1.054571817 \times 10^{-34} \text{ J s.}$$

Avogadro constant:

$$N_A = 6.02214076 \times 10^{23} \text{ mol}^{-1} \text{ (exact).}$$

Boltzmann constant:

$$k_B = 1.380649 \times 10^{-23} \text{ J K}^{-1} \text{ (exact).}$$

Gas constant:

$$R = N_A k_B = 8.314462618 \text{ J mol}^{-1} \text{ K}^{-1}.$$

Electron charge:

$$e = 1.602176634 \times 10^{-19} \text{ C (exact).}$$

Electron mass:

$$m_e = 9.1093837015 \times 10^{-31} \quad \text{kg.}$$

Proton mass:

$$m_p = 1.6726219 \times 10^{-27} \text{ kg.}$$

Neutron mass:

$$m_n = 1.6749275 \times 10^{-27}\ \mathrm{kg}.$$

Atomic mass unit:

$$1\ \mathrm{u} = 1.66053906660 \times 10^{-27}\ \mathrm{kg}.$$

Vacuum permittivity:

$$\varepsilon_0 = 8.8541878128 \times 10^{-12}\ \mathrm{F\ m^{-1}}.$$

Vacuum permeability:

$$\mu_0 = 1.25663706212 \times 10^{-6}\ \mathrm{H\ m^{-1}}.$$

I.2 Constants Useful in Spectroscopy and Quantum Chemistry

Bohr radius:

$$a_0 = 5.291772109 \times 10^{-11}\ \mathrm{m}.$$

Hartree energy:

$$E_h = 4.359744722 \times 10^{-18}\ \mathrm{J} = 27.211386\ \mathrm{eV}.$$

Rydberg constant (for hydrogen):

$$R_\infty = 1.09737316 \times 10^{7}\ \mathrm{m^{-1}}.$$

Molar gas constant in different units:

$$R = 8.314462618\ \mathrm{J\ mol^{-1}\ K^{-1}} = 0.082057\ \mathrm{L\ atm\ mol^{-1}\ K^{-1}}.$$

Faraday's constant:

$$F = N_A e = 96485.33212\ \mathrm{C\ mol^{-1}}.$$

Debye unit:

$$1\ \mathrm{D} = 3.33564 \times 10^{-30}\ \mathrm{C\ m}.$$

I.3 Unit Conversions

Energy:

$$1\ \mathrm{eV} = 1.602176634 \times 10^{-19}\ \mathrm{J},$$

$$1\ \mathrm{hartree} = 27.211386\ \mathrm{eV},$$

$$1 \text{ kcal mol}^{-1} = 4.184 \text{ kJ mol}^{-1},$$

$$1 \text{ cm}^{-1} = 100 \text{ hc} = 1.98644586 \times 10^{-23} \text{ J}.$$

Frequency and wavelength:

$$\nu = \frac{c}{\lambda}, \ \widetilde{\nu} = \frac{1}{\lambda}, \ E = h\nu = hc\widetilde{\nu}.$$

Pressure:

$$1 \text{ atm} = 101{,}325 \text{ Pa},$$

$$1 \text{ bar} = 10^5 \text{ Pa},$$

$$1 \text{ torr} = 133.322 \text{ Pa}.$$

Temperature:

$$T\ (\text{K}) = T\ (^\circ\text{C}) + 273.15.$$

I.4 Constants for Materials Science and Solid-state Physics

Electron mobility constant:

$$\mu_e \approx 0.14 \text{ m}^2 \text{ V}^{-1} \text{ s}^{-1} \text{ (in Si)}.$$

Permittivity of common media (approx.):

Water (25 °C):

$$\varepsilon_r \approx 78.3.$$

Air:

$$\varepsilon_r \approx 1.0006.$$

I.5 Notes for Chemistry, Physics, and Engineering

- Exact constants (c, h, k_B, N_A, e) simplify modern metrology and dimensional analysis.
- Wavenumber–energy conversions are essential for IR, Raman, UV–V is spectroscopy.
- Debye is critical for dipole moment calculations.
- Conversions among eV, Hartree, and J are routinely used in quantum chemistry and materials calculations.

J

Extended Proofs and Derivations

J.1 Orthogonal Curvilinear Coordinates

Let us consider an arbitrary orthogonal coordinate system with coordinates (u_1, u_2, u_3). In terms of these, a small displacement $d\vec{r}$ can be expressed as:

$$d\vec{r} = \left(\frac{\partial \vec{r}}{\partial u_1}\right)du_1 + \left(\frac{\partial \vec{r}}{\partial u_2}\right)du_2 + \left(\frac{\partial \vec{r}}{\partial u_3}\right)du_3. \tag{J.1}$$

Let $\hat{e}_1, \hat{e}_2, \hat{e}_3$ be the unit vectors in the direction of $\vec{u}_1, \vec{u}_2, \vec{u}_3$. Then:

$$\hat{e}_i = \frac{\partial \vec{r}/\partial u_i}{|\partial \vec{r}/\partial u_i|}, \quad i = 1,\ 2,\ 3. \tag{J.2}$$

As this is an orthogonal coordinate system, $\hat{e}_1, \hat{e}_2, \hat{e}_3$ are mutually orthogonal. The quantities in the denominator are called the scale factors h_1, h_2, h_3, and are expressed as:

$$h_i = |\partial \vec{r}/\partial u_i|, \quad i = 1,\ 2,\ 3, \tag{J.3}$$

so that

$$\partial \vec{r}/\partial u_i = h_i \hat{e}_i, i = 1,\ 2,\ 3, \tag{J.4}$$

and the infinitesimal displacement $d\vec{r}$ can be expressed as:

$$d\vec{r} = h_1\hat{e}_1 du_1 + h_2\hat{e}_2 du_2 + h_3\hat{e}_3 du_3. \tag{J.5}$$

The square of the magnitude of the displacement is:

$$(ds)^2 = (d\vec{r} \cdot d\vec{r}) = h_1^2(du_1)^2 + h_2^2(du_2)^2 + h_3^2(du_3)^2. \tag{J.6}$$

The volume element $d\tau$ is given by:

$$d\tau = |h_1\hat{e}_1 \cdot h_2\hat{e}_2 \times h_3\hat{e}_3| du_1 du_2 du_3 = h_1 h_2 h_3 du_1 du_2 du_3. \tag{J.7}$$

In view of Equation (J.4), the volume element can also be expressed as:

$$d\tau = \left|\frac{\partial \vec{r}}{\partial u_1} \cdot \frac{\partial \vec{r}}{\partial u_2} \times \frac{\partial \vec{r}}{\partial u_3}\right| du_1 du_2 du_3. \tag{J.8}$$

Since $\vec{r} = \hat{i}x + \hat{j}y + \hat{k}z$,

$$\frac{\partial \vec{r}}{\partial u_1} = \hat{i}\frac{\partial x}{\partial u_1} + \hat{j}\frac{\partial y}{\partial u_1} + \hat{k}\frac{\partial z}{\partial u_1},$$
$$\frac{\partial \vec{r}}{\partial u_2} = \hat{i}\frac{\partial x}{\partial u_2} + \hat{j}\frac{\partial y}{\partial u_2} + \hat{k}\frac{\partial z}{\partial u_2},$$
$$\frac{\partial \vec{r}}{\partial u_3} = \hat{i}\frac{\partial x}{\partial u_3} + \hat{j}\frac{\partial y}{\partial u_3} + \hat{k}\frac{\partial z}{\partial u_3},$$

and, using the expression for triple product in Equation (J.8),

$$d\tau = \begin{vmatrix} \frac{\partial x}{\partial u_1} & \frac{\partial y}{\partial u_1} & \frac{\partial z}{\partial u_1} \\ \frac{\partial x}{\partial u_2} & \frac{\partial y}{\partial u_2} & \frac{\partial z}{\partial u_2} \\ \frac{\partial x}{\partial u_3} & \frac{\partial y}{\partial u_3} & \frac{\partial z}{\partial u_3} \end{vmatrix} du_1 du_2 du_3.$$

The determinant in the above equation is just the transpose of the **Jacobian** of the transformation, $J(u_1, u_2, u_3)$, given by:

$$J(u_1, u_2, u_3) = \left|\frac{\partial(x, y, z)}{\partial(u_1, u_2, u_3)}\right| = \begin{vmatrix} \frac{\partial x}{\partial u_1} & \frac{\partial x}{\partial u_2} & \frac{\partial x}{\partial u_3} \\ \frac{\partial y}{\partial u_1} & \frac{\partial y}{\partial u_2} & \frac{\partial y}{\partial u_3} \\ \frac{\partial z}{\partial u_1} & \frac{\partial z}{\partial u_2} & \frac{\partial z}{\partial u_3} \end{vmatrix}. \tag{J.9}$$

Note that all rows in the Jacobian have a common numerator, all columns have a common denominator, and the magnitude of the determinant remains unchanged by taking the transpose.

Therefore, in terms of the Jacobian, the volume element is given by:

$$d\tau = \left|\frac{\partial(x, y, z)}{\partial(u_1, u_2, u_3)}\right| du_1 du_2 du_3. \tag{J.10}$$

Gradient

The total differential of a scalar field φ in terms of these coordinates is given by:

$$d\varphi = \left(\frac{\partial \varphi}{\partial u_1}\right)du_1 + \left(\frac{\partial \varphi}{\partial u_2}\right)du_2 + \left(\frac{\partial \varphi}{\partial u_3}\right)du_3. \tag{J.11}$$

For orthogonal coordinates, this can be written as:

$$d\varphi = \nabla\varphi \cdot d\vec{r}. \tag{J.12}$$

Substitution of Equation (J.5) in (J.11) gives:

$$\nabla\varphi = \frac{\hat{e}_1}{h_1}\frac{\partial \varphi}{\partial u_1} + \frac{\hat{e}_2}{h_2}\frac{\partial \varphi}{\partial u_2} + \frac{\hat{e}_3}{h_3}\frac{\partial \varphi}{\partial u_3}. \tag{J.13}$$

Divergence:

Let $\vec{A}$ be an arbitrary vector field. It can be expressed in terms of the base vectors as:

$$\vec{A} = A_1\hat{e}_1 + A_2\hat{e}_2 + A_3\hat{e}_3. \tag{J.14}$$

Then

$$\nabla.\vec{A} = \left(\frac{\widehat{e}_1}{h_1}\frac{\partial}{\partial u_1} + \frac{\widehat{e}_2}{h_2}\frac{\partial}{\partial u_2} + \frac{\widehat{e}_3}{h_3}\frac{\partial}{\partial u_3}\right).(A_1\widehat{e}_1 + A_2\widehat{e}_2 + A_3\widehat{e}_3)$$

(accounting for the u_i-dependence of $\widehat{e}_1, \widehat{e}_2, \widehat{e}_3$, this reduces to)

$$= \frac{1}{h_1h_2h_3}\left[\frac{\partial}{\partial u_1}(h_2h_3A_1) + \frac{\partial}{\partial u_2}(h_3h_1A_2) + \frac{\partial}{\partial u_3}(h_1h_2A_3)\right]. \tag{J.15}$$

Laplacian:
Let the vector field $\vec{A}$ be the derivative of a scalar field. Then:

$$\vec{A} = \nabla\varphi = \frac{\widehat{e}_1}{h_1}\left(\frac{\partial\varphi}{\partial u_1}\right) + \frac{\widehat{e}_2}{h_2}\left(\frac{\partial\varphi}{\partial u_2}\right) + \frac{\widehat{e}_3}{h_3}\left(\frac{\partial\varphi}{\partial u_3}\right) = A_1\widehat{e}_1 + A_2\widehat{e}_2 + A_3\widehat{e}_3. \tag{J.16}$$

Thus,

$$A_i = \frac{1}{h_i}\frac{\partial\varphi}{\partial u_i}. \tag{J.17}$$

The Laplacian is given by $\nabla \cdot \nabla\varphi$. Substituting the above expression for the A_i in Equation (J.15), we have

$$\nabla^2\varphi = \nabla \cdot \nabla\varphi = \frac{1}{h_1h_2h_3}\left[\frac{\partial}{\partial u_1}\left(\frac{h_2h_3}{h_1}\frac{\partial\varphi}{\partial u_1}\right) + \frac{\partial}{\partial u_2}\left(\frac{h_3h_1}{h_2}\frac{\partial\varphi}{\partial u_2}\right) + \frac{\partial}{\partial u_3}\left(\frac{h_1h_2}{h_3}\frac{\partial\varphi}{\partial u_3}\right)\right]. \tag{J.18}$$

Curl:
For the vector field $\vec{A} = A_1\widehat{e}_1 + A_2\widehat{e}_2 + A_3\widehat{e}_3$, the curl is given by

$$\nabla \times \vec{A} = \frac{1}{h_1h_2h_3}\begin{vmatrix} h_1\widehat{e}_1 & h_2\widehat{e}_2 & h_3\widehat{e}_3 \\ \dfrac{\partial}{\partial u_1} & \dfrac{\partial}{\partial u_2} & \dfrac{\partial}{\partial u_3} \\ h_1A_1 & h_2A_2 & h_3A_3 \end{vmatrix}. \tag{J.19}$$

Table J.1.1 summarizes the various expressions obtained for vector operations.

Table J.1.1 Vector operators in orthogonal curvilinear coordinates u_1, u_2, u_3. In these expressions, φ is a scalar field and $\widehat{A}$ is a vector field.

$\nabla\varphi = \frac{1}{h_1}\frac{\partial\varphi}{\partial u_1}\widehat{e}_1 + \frac{1}{h_2}\frac{\partial\varphi}{\partial u_2}\widehat{e}_2 + \frac{1}{h_3}\frac{\partial\varphi}{\partial u_3}\widehat{e}_3$
$\nabla \cdot \vec{A} = \frac{1}{h_1h_2h_3}\left[\frac{\partial}{\partial u_1}(h_2h_3A_1) + \frac{\partial}{\partial u_2}(h_3h_1A_2) + \frac{\partial}{\partial u_3}(h_1h_2A_3)\right]$
$\nabla \times \vec{A} = \frac{1}{h_1h_2h_3}\begin{vmatrix} h_1\widehat{e}_1 & h_2\widehat{e}_2 & h_3\widehat{e}_3 \\ \frac{\partial}{\partial u_1} & \frac{\partial}{\partial u_2} & \frac{\partial}{\partial u_3} \\ h_1A_1 & h_2A_2 & h_3A_3 \end{vmatrix}$
$\nabla^2\varphi = \frac{1}{h_1h_2h_3}\left[\frac{\partial}{\partial u_1}\left(\frac{h_2h_3}{h_1}\frac{\partial\varphi}{\partial u_1}\right) + \frac{\partial}{\partial u_2}\left(\frac{h_3h_1}{h_2}\frac{\partial\varphi}{\partial u_2}\right) + \frac{\partial}{\partial u_3}\left(\frac{h_1h_2}{h_3}\frac{\partial\varphi}{\partial u_3}\right)\right]$

We also need expressions for the partial derivatives with respect to the coordinates in terms of Cartesian coordinates. Writing the various partial derivatives of scalar $\varphi(x, y, z)$ using the chain rule:

$$\begin{aligned}
\frac{\partial\varphi}{\partial u_1} &= \left(\frac{\partial\varphi}{\partial x}\right)\left(\frac{\partial x}{\partial u_1}\right) + \left(\frac{\partial\varphi}{\partial y}\right)\left(\frac{\partial y}{\partial u_1}\right) + \left(\frac{\partial\varphi}{\partial z}\right)\left(\frac{\partial z}{\partial u_1}\right),\\
\frac{\partial\varphi}{\partial u_2} &= \left(\frac{\partial\varphi}{\partial x}\right)\left(\frac{\partial x}{\partial u_2}\right) + \left(\frac{\partial\varphi}{\partial y}\right)\left(\frac{\partial y}{\partial u_2}\right) + \left(\frac{\partial\varphi}{\partial z}\right)\left(\frac{\partial z}{\partial u_2}\right),\\
\frac{\partial\varphi}{\partial u_3} &= \left(\frac{\partial\varphi}{\partial x}\right)\left(\frac{\partial x}{\partial u_3}\right) + \left(\frac{\partial\varphi}{\partial y}\right)\left(\frac{\partial y}{\partial u_3}\right) + \left(\frac{\partial\varphi}{\partial z}\right)\left(\frac{\partial z}{\partial u_3}\right).
\end{aligned} \tag{J.20}$$

In terms of the Jacobian, these equations can be written as:

$$\begin{bmatrix} \frac{\partial\varphi}{\partial u_1} & \frac{\partial\varphi}{\partial u_2} & \frac{\partial\varphi}{\partial u_3} \end{bmatrix} = \begin{bmatrix} \frac{\partial\varphi}{\partial x} & \frac{\partial\varphi}{\partial y} & \frac{\partial\varphi}{\partial z} \end{bmatrix} \begin{bmatrix} \frac{\partial x}{\partial u_1} & \frac{\partial x}{\partial u_2} & \frac{\partial x}{\partial u_3} \\ \frac{\partial y}{\partial u_1} & \frac{\partial y}{\partial u_2} & \frac{\partial y}{\partial u_3} \\ \frac{\partial z}{\partial u_1} & \frac{\partial z}{\partial u_2} & \frac{\partial z}{\partial u_3} \end{bmatrix}. \tag{J.21}$$

(where we transpose the vectors to maintain the determinant in Jacobian form). Cramer's rule can be used to solve for various partial derivatives.

We now derive specific expressions for the spherical polar coordinates, which we use to solve for the rigid rotator and hydrogen atom, both of which have spherical boundary conditions. From Table J.1.1, it is clear that we must express displacement in terms of u_i, from which we can evaluate h_i given by $h_i = |\partial\vec{r}/\partial u_i|$, $i = 1, 2, 3$.

J.1.1 Spherical Polar Coordinates

Figure J.1.1 shows spherical polar coordinates in a Cartesian frame.

In this system, vector $\vec{r}$ describes the position of point P with respect to the origin. Its length is r $(0 \le r \le \infty)$, θ $(0 \le \theta \le \pi)$ is the angle this vector makes with the z-axis, and ϕ $(0 \le \phi \le 2\pi)$ is the angle its projection in the xy-plane makes with the x-axis. From Figure J.1.1, it is clear that, in terms of the spherical polar coordinates,

$$\begin{aligned}
x &= r \sin\theta \cos\phi,\\
y &= r \sin\theta \sin\phi,\\
z &= r \cos\theta.
\end{aligned} \tag{J.22}$$

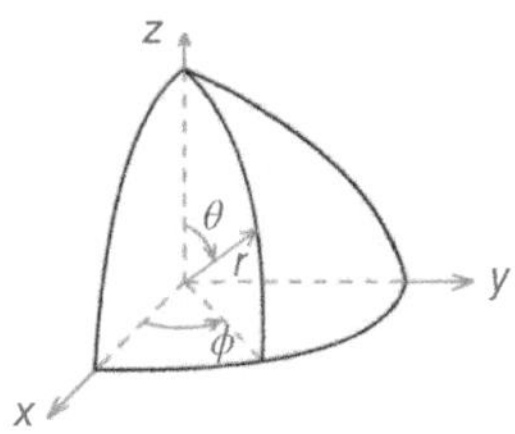

Figure J.1.1 The spherical polar coordinate system.

The spherical polar coordinates in terms of Cartesian coordinates are:

$$r=\sqrt{x^2+y^2+z^2},$$
$$\theta=\cos^{-1}\left(\frac{z}{r}\right)=\cos^{-1}\left(\frac{z}{\sqrt{x^2+y^2+z^2}}\right),\tag{J.23}$$
$$\phi=\tan^{-1}\left(\frac{y}{x}\right).$$

Hence, the position vector of point P may be written as:

$$\vec{r}=r\sin\theta\cos\phi\hat{i}+r\sin\theta\sin\phi\hat{j}+r\cos\theta\hat{k}.\tag{J.24}$$

Let $\hat{e}_r, \hat{e}_\theta, \hat{e}_\phi$ be the unit vectors in three coordinate directions. Then:

$$\hat{e}_r=\frac{\partial\vec{r}/\partial r}{|\partial\vec{r}/\partial r|}=\frac{\sin\theta\cos\phi\hat{i}+\sin\theta\sin\phi\hat{j}+\cos\theta\hat{k}}{\left(\sin^2\theta\cos^2\phi+\sin^2\theta\sin^2\phi+\cos^2\theta\right)^{1/2}}.\tag{J.25}$$

Hence,

$$\begin{aligned}h_r^2&=\sin^2\theta\cos^2\phi+\sin^2\theta\sin^2\phi+\cos^2\theta\\&=\sin^2\theta\left(\cos^2\phi+\sin^2\phi\right)+\cos^2\theta\\&=\sin^2\theta+\cos^2\theta=1.\end{aligned}\tag{J.26}$$

Therefore, $h_r=1$.
Similarly,

$$\hat{e}_\theta=\frac{\partial\vec{r}/\partial\theta}{|\partial\vec{r}/\partial\theta|}=\frac{r\cos\theta\cos\phi\hat{i}+r\cos\theta\sin\phi\hat{j}-r\sin\theta\hat{k}}{\left(r^2\cos^2\theta\cos^2\phi+r^2\cos^2\theta\sin^2\phi+r^2\sin^2\theta\right)^{1/2}}.\tag{J.27}$$

Hence,

$$\begin{aligned}h_\theta^2&=r^2\cos^2\theta\cos^2\phi+r^2\cos^2\theta\sin^2\phi+r^2\sin^2\theta\\&=r^2\cos^2\theta\left(\cos^2\phi+\sin^2\phi\right)+r^2\sin^2\theta\\&=r^2\cos^2\theta+r^2\sin^2\theta=r^2.\end{aligned}\tag{J.28}$$

Therefore, $h_\theta=r$.
For ϕ, the expression is

$$\hat{e}_\phi=\frac{\partial\vec{r}/\partial\phi}{|\partial\vec{r}/\partial\phi|}=\frac{-r\sin\theta\sin\phi\hat{i}+r\sin\theta\cos\phi\hat{j}}{\left(r^2\sin^2\theta\sin^2\phi+r^2\sin^2\theta\cos^2\phi\right)^{1/2}}.\tag{J.29}$$

Therefore,

$$h_\phi^2=r^2\sin^2\theta\sin^2\phi+r^2\sin^2\theta\cos^2\phi=r^2\sin^2\theta\left(\sin^2\phi+\cos^2\phi\right)=r^2\sin^2\theta\tag{J.30}$$

Hence, $h_\phi=r\sin\theta$.
Therefore, the unit vectors in the direction of increasing r, θ, ϕ are

$$\begin{aligned}\hat{e}_r&=\hat{i}\sin\theta\cos\phi+\hat{j}\sin\theta\sin\phi+\hat{k}\cos\theta,\\\hat{e}_\theta&=\hat{i}\cos\theta\cos\phi+\hat{j}\cos\theta\sin\phi-\hat{k}\sin\theta,\\\hat{e}_\phi&=-\hat{i}\sin\phi+\hat{j}\cos\phi.\end{aligned}\tag{J.31}$$

Applying Equation (J.24), a displacement $d\vec{r}$ can be expressed in spherical polar coordinates as:

$$d\vec{r} = \hat{e}_r dr + r\hat{e}_\theta d\theta + r \sin\theta \hat{e}_\phi d\phi.$$

The volume element is given by Equation (J.7):

$$d\tau = r^2 \sin\theta dr d\theta d\phi. \tag{J.32}$$

Alternatively, using Equations (J.8) and (J.9), in terms of the Jacobian and substituting the various partial derivatives in Equation (J.9), using the definitions in Equation (J.20), we have:

$$J(r,\theta,\phi) = \left|\frac{\partial(x,y,z)}{\partial(r,\theta,\phi)}\right| = \begin{vmatrix} \sin\theta\cos\phi & r\cos\theta\cos\phi & -r\sin\theta\sin\phi \\ \sin\theta\sin\phi & r\cos\theta\sin\phi & r\sin\theta\cos\phi \\ \cos\theta & -r\sin\theta & 0 \end{vmatrix}. \tag{J.33}$$

The determinant can be expanded using the last row, giving:

$$\begin{aligned} J(r,\theta,\phi) &= \cos\theta\left(r^2\cos\theta\sin\theta\cos^2\phi + r^2\sin\theta\cos\theta\sin^2\phi\right) \\ &\quad + r\sin\theta\left(r\sin^2\theta\cos^2\phi + r\sin^2\theta\sin^2\phi\right) \\ &= r^2\sin\theta\cos^2\theta\left(\cos^2\phi + \sin^2\phi\right) + r^2\sin^3\theta\left(\cos^2\phi + \sin^2\phi\right) \\ &= r^2\sin\theta\cos^2\theta + r^2\sin^3\theta \\ &= r^2\sin\theta\left(\cos^2\theta + \sin^2\theta\right) = r^2\sin\theta. \end{aligned} \tag{J.34}$$

Therefore, the volume element is given by $d\tau = r^2 \sin\theta\, dr d\theta d\phi$, as expressed in Equation (J.32).

Using the Jacobian for finding the volume element may seem more complicated than using Equation (J.7), but the Jacobian has other uses. It can be used to obtain expressions for the partial derivatives in the new coordinate system with respect to the old one. Substituting the various terms in Equation (J.21), we get:

$$\begin{bmatrix} \frac{\partial\varphi}{\partial r} & \frac{\partial\varphi}{\partial\theta} & \frac{\partial\varphi}{\partial\phi} \end{bmatrix} = \begin{bmatrix} \frac{\partial\varphi}{\partial x} & \frac{\partial\varphi}{\partial y} & \frac{\partial\varphi}{\partial z} \end{bmatrix} \begin{bmatrix} \sin\theta\cos\phi & r\cos\theta\cos\phi & -r\sin\theta\sin\phi \\ \sin\theta\sin\phi & r\cos\theta\sin\phi & r\sin\theta\cos\phi \\ \cos\theta & -r\sin\theta & 0 \end{bmatrix}. \tag{J.35}$$

For example, to solve for $\partial/\partial x$, using Cramer's rule, replace the first *row* with the vector on the left of this equation and find the determinant. Therefore,

$$\begin{aligned} \frac{\partial\varphi}{\partial x} = \frac{\frac{\partial(\varphi,y,z)}{\partial(r,\theta,\phi)}}{\frac{\partial(x,y,z)}{\partial(r,\theta,\phi)}} &= \frac{\begin{vmatrix} \partial\varphi/\partial r & \partial\varphi/\partial\theta & \partial\varphi/\partial\phi \\ \sin\theta\sin\phi & r\cos\theta\sin\phi & r\sin\theta\cos\phi \\ \cos\theta & -r\sin\theta & 0 \end{vmatrix}}{|J(r,\theta,\phi)|} \\ &= \left(r^2\sin^2\theta\cos\phi\frac{\partial\varphi}{\partial r} + r\sin\theta\cos\theta\cos\phi\frac{\partial\varphi}{\partial\theta} - r\sin\phi\frac{\partial\varphi}{\partial\phi}\right)/|J(r,\theta,\phi)|. \end{aligned} \tag{J.36}$$

The determinant in the denominator is the Jacobian, that is, $r^2\sin\theta$. Hence, the expression for the partial derivative is

$$\frac{\partial\varphi}{\partial x} = \sin\theta\cos\phi\frac{\partial\varphi}{\partial r} + \frac{\cos\theta\cos\phi}{r}\frac{\partial\varphi}{\partial\theta} - \frac{\sin\phi}{r\sin\theta}\frac{\partial\varphi}{\partial\phi}. \tag{J.37}$$

Similarly, by substituting into the second row, we obtain $\left|\frac{\partial(x, \varphi, z)}{\partial(r, \theta, \phi)}\right|$, and use this to obtain the derivative with respect to y. The final expressions for the other two derivatives are

$$\frac{\partial \varphi}{\partial y} = \sin\theta \sin\phi \frac{\partial \varphi}{\partial r} + \frac{\cos\theta \sin\phi}{r}\frac{\partial \varphi}{\partial \theta} + \frac{\cos\phi}{r\sin\theta}\frac{\partial \varphi}{\partial \phi}, \tag{J.38}$$

$$\frac{\partial \varphi}{\partial z} = \cos\theta \frac{\partial \varphi}{\partial r} - \frac{\sin\theta}{r}\frac{\partial \varphi}{\partial \theta}. \tag{J.39}$$

Applying the equations given in Table J.1.1, the gradient of the scalar field φ is given by

$$\nabla\varphi = \frac{\partial \varphi}{\partial r}\hat{e}_r + \frac{1}{r}\frac{\partial \varphi}{\partial \theta}\hat{e}_\theta + \frac{1}{r\sin\theta}\frac{\partial \varphi}{\partial \phi}\hat{e}_\phi. \tag{J.40}$$

Therefore, the del operator in spherical polar coordinates is

$$\nabla = \frac{\partial}{\partial r}\hat{e}_r + \frac{1}{r}\frac{\partial}{\partial \theta}\hat{e}_\theta + \frac{1}{r\sin\theta}\frac{\partial}{\partial \phi}\hat{e}_\phi. \tag{J.41}$$

For the vector field $\vec{A}$, the div operation yields

$$\begin{aligned}\nabla\cdot\vec{A} &= \frac{1}{r^2\sin\theta}\left[\frac{\partial}{\partial r}(r^2\sin\theta A_r) + \frac{\partial}{\partial \theta}(r\sin\theta A_\theta) + \frac{\partial}{\partial \phi}(rA_\phi)\right]\\ &= \frac{1}{r^2\sin\theta}\left[\sin\theta\frac{\partial}{\partial r}(r^2A_r) + r\frac{\partial}{\partial \theta}(\sin\theta A_\theta) + r\frac{\partial}{\partial \phi}(A_\phi)\right]\\ &= \left[\frac{1}{r^2}\frac{\partial}{\partial r}(r^2A_r) + \frac{1}{r\sin\theta}\frac{\partial}{\partial \theta}(\sin\theta A_\theta) + \frac{1}{r\sin\theta}\frac{\partial}{\partial \phi}(A_\phi)\right].\end{aligned} \tag{J.42}$$

The curl is given by:

$$\nabla\times\vec{A} = \frac{1}{r^2\sin\theta}\begin{vmatrix}\hat{e}_r & r\hat{e}_\theta & r\sin\theta\hat{e}_\phi\\ \frac{\partial}{\partial r} & \frac{\partial}{\partial \theta} & \frac{\partial}{\partial \phi}\\ A_r & rA_\theta & r\sin\theta A_\phi\end{vmatrix}. \tag{J.43}$$

Finally, the Laplacian is given by:

$$\begin{aligned}\nabla^2\varphi &= \frac{1}{r^2\sin\theta}\left[\frac{\partial}{\partial r}\left(r^2\sin\theta\frac{\partial \varphi}{\partial r}\right) + \frac{\partial}{\partial \theta}\left(\sin\theta\frac{\partial \varphi}{\partial \theta}\right) + \frac{\partial}{\partial \varphi}\left(\frac{1}{\sin\theta}\frac{\partial \varphi}{\partial \phi}\right)\right]\\ &= \frac{1}{r^2}\frac{\partial}{\partial r}\left(r^2\frac{\partial \varphi}{\partial r}\right) + \frac{1}{r^2\sin\theta}\frac{\partial}{\partial \theta}\left(\sin\theta\frac{\partial \varphi}{\partial \theta}\right) + \frac{1}{r^2\sin^2\theta}\frac{\partial^2 \varphi}{\partial \phi^2}.\end{aligned} \tag{J.44}$$

J.2 Simultaneous Diagonalization

In the text, we referred to the simultaneous diagonalization of two matrices, provided that one of them is **positive definite**. The following example illustrates the procedure for the simultaneous diagonalization of the potential energy and kinetic energy matrices for a linear triatomic molecule.

Example J.2.1

a. Consider the vibrations of a linear triatomic molecule. Let x_1, x_2, and x_3 be the displacements of the three atoms of masses $m_1 = m,\ \ m_2 = 2m,\ \ m_3 = m$.

b. Evaluate the kinetic energy T.

c. Evaluate the potential energy of the system and show that it reduces to:

$$V = \frac{1}{2}k\left[(x_2 - x_1)^2 + (x_3 - x_2)^2\right].$$

d. Show that the two quadratic forms are:

$$T = \frac{1}{2}\left[m\dot{x}_1^2 + 2m\dot{x}_2^2 + m\dot{x}_3^2\right],$$

$$F = \frac{1}{2}k\left[x_1^2 + 2x_2^2 + x_3^2 - 2x_1x_2 - 2x_2x_3\right].$$

e. Show that T is positive definite.

f. Find the eigenvalues λ and eigenvectors y of $(F - \lambda T)y = 0$.

g. Determine the transformation matrix that simultaneously diagonalizes T and F.

h. Write down the diagonal forms of T and F.

i. Determine the characteristic frequencies and normal modes and sketch them.

Solution:

a. The kinetic energy is given by:

$$T = \frac{1}{2}\left[m\dot{x}_1^2 + 2m\dot{x}_2^2 + m\dot{x}_3^2\right].$$

b. Because there are two bonds, and assuming both have the same force constant k, the potential energy is the sum of the energies stored in the two bonds:

$$V = \frac{1}{2}k\left[(x_2 - x_1)^2 + (x_3 - x_2)^2\right].$$

c. These can be written in matrix form as:

$$T = \frac{1}{2}\dot{\boldsymbol{x}}^T\begin{bmatrix} m & 0 & 0 \\ 0 & 2m & 0 \\ 0 & 0 & m \end{bmatrix}\dot{\boldsymbol{x}},\quad F = \frac{1}{2}k\boldsymbol{x}^T\begin{bmatrix} 1 & -1 & 0 \\ -1 & 2 & -1 \\ 0 & -1 & 1 \end{bmatrix}\boldsymbol{x}.$$

d. For simplicity, we factor out the common mass m upfront:

$$T = m\begin{bmatrix} 1 & 0 & 0 \\ 0 & 2 & 0 \\ 0 & 0 & 1 \end{bmatrix},$$

and then calculate minors on this dimensionless matrix:

$$M_1 = 1,\quad M_2 = \det\begin{bmatrix} 1 & 0 \\ 0 & 2 \end{bmatrix} = 2,\quad M_3 = \det\begin{bmatrix} 1 & 0 & 0 \\ 0 & 2 & 0 \\ 0 & 0 & 1 \end{bmatrix} = 2.$$

The principal minors of T are:

$$M_1 = 1,\ M_2 = 2,\ M_3 = 2,$$

all positive, hence T is **positive definite**.

e. For simplicity, we omit constant factors.
The eigenvalues are found from:

$$\det(F - \lambda T) = 0,$$

which yields $\lambda = 0,\ \lambda = 1,\ \lambda = 2$.
Substituting the eigenvalues, we obtain the eigenvectors:

$$\begin{aligned} \lambda = 0&: \ y_1 = [1, 1, 1]^T, \\ \lambda = 1&: \ y_2 = [1, 0, -1]^T, \\ \lambda = 2&: \ y_3 = [1, -1, 1]^T. \end{aligned}$$

For $\lambda = 1$ and 2, the linear momentum is conserved. For $\lambda = 2$, for example, $m(1) + 2m(-1) + m(1) = 0$. (Center moves opposite to ends with same amplitude.)

f. To normalize the eigenvectors, we multiply them by factors $s,\ t$, and u, respectively.
To obtain the identity matrix for T, $s = u = 1/2$ and $t = 1/\sqrt{2}$.
Hence, the transformation matrix is:

$$S = \begin{bmatrix} 1/2 & 1/\sqrt{2} & 1/2 \\ 1/2 & 0 & -1/2 \\ 1/2 & -1/\sqrt{2} & 1/2 \end{bmatrix}.$$

g. Since $S^T T S = I$, both matrices are diagonalized simultaneously.
The transformations give:

$$T' = S^T T S = I, V' = S^T V S = \begin{bmatrix} 0 & 0 & 0 \\ 0 & 1 & 0 \\ 0 & 0 & 2 \end{bmatrix}.$$

h. We ignored k and m in the derivation.

Including them, the eigenvalues are $0,\ k/m,\ 2k/m$.
Since $\lambda = \omega^2$, the characteristic frequencies are:

$$\omega_1 = 0, \omega_2 = \sqrt{k/m}, \omega_3 = \sqrt{2k/m}.$$

The corresponding normal modes are:

$$\begin{aligned} \lambda = 0&: \quad [1, 1, 1]^T \qquad \text{(translation)}, \\ \lambda = 1&: \quad [1, 0, -1]^T \quad \text{(symmetric stretch)}, \\ \lambda = 2&: \quad [1, -1, 1]^T \quad \text{(antisymmetric stretch)}. \end{aligned}$$

These are illustrated schematically below:

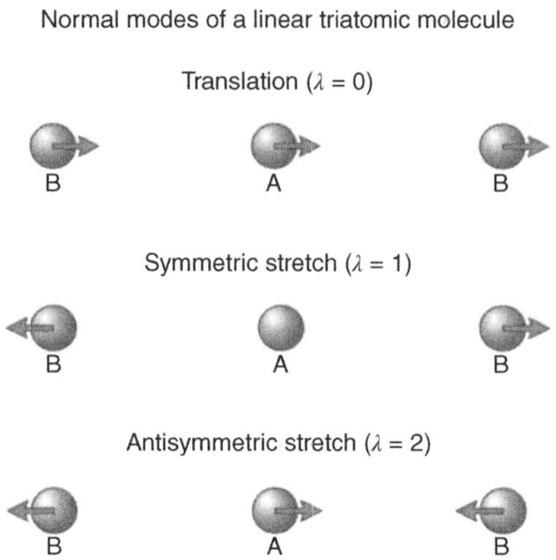

J.3 The Wilson–Decius–Cross *GF* Method (Optional)

This appendix outlines the traditional *GF* matrix formulation of molecular vibrations, included for completeness and for readers familiar with spectroscopic notation. The main text develops normal coordinate analysis through simultaneous diagonalization of kinetic and potential energy quadratic forms and does not rely on the *GF* method.

J.3.1 Harmonic Approximation and Internal Coordinates

Let $\{s_i\}$ denote a set of internal displacement coordinates (bond stretches, angle bends). Under the harmonic approximation, the potential energy is written as

$$V = \frac{1}{2}\sum_{i,j} F_{ij} s_i s_j,$$

where F_{ij} are the **force constants**, forming the **force constant matrix** F.

The kinetic energy can be expressed in the form

$$T = \frac{1}{2}\sum_{i,j} G_{ij}\, \dot{s}_i \dot{s}_j,$$

where the elements G_{ij} depend on atomic masses and molecular geometry. The matrix G is called the **kinetic energy matrix**.

J.3.2 Lagrange's Equations and the *GF* Eigenvalue Problem

Applying Lagrange's equations to the quadratic forms T and V yields:

$$G\,\ddot{\mathbf{s}} + F\,\mathbf{s} = 0.$$

Seeking solutions of the form $\mathbf{s}(t) = \mathbf{a}\, e^{i\omega t}$, one obtains:

$$\left(F - \omega^2 G\right)\mathbf{a} = 0.$$

Nontrivial solutions exist only if:

$$\det\left|F - \omega^2 G\right| = 0,$$

which constitutes the *GF* **secular equation**. The eigenvalues ω^2 give the squared normal-mode frequencies.

J.3.3 Example: Linear Triatomic Molecule (Stretching Modes Only)

Consider a linear symmetric triatomic molecule $X—Y—X$ such as CO_2 with two equivalent bond stretches s_1, s_2.

Potential energy

Assuming equal force constants k,

$$V = \frac{1}{2}k\left(s_1^2 + s_2^2\right), \quad F = \begin{pmatrix} k & 0 \\ 0 & k \end{pmatrix}.$$

Kinetic energy

The kinetic energy couples the two stretching motions through the central atom, giving:

$$G = \begin{pmatrix} g & g' \\ g' & g \end{pmatrix},$$

where g and g' depend on the masses of X and Y.

J.3.4 Normal Coordinates and Decoupling

Solving the secular equation leads to two normal modes:

- **Symmetric stretch:**

$$Q_s = \frac{1}{\sqrt{2}}(s_1 + s_2).$$

- **Antisymmetric stretch:**

$$Q_a = \frac{1}{\sqrt{2}}(s_1 - s_2).$$

In these coordinates, both T and V become diagonal, and the motions are dynamically independent.

J.3.5 Relation to the Main Text

The *GF* method provides a **coordinate-specific realization** of the more general mathematical procedure used in the main text:

- T and V are quadratic forms.
- Normal coordinates arise from a **simultaneous diagonalization**.
- The transformation involved is a **congruent transformation**, not a similarity transformation.

Thus, the *GF* method is a special case of the general normal coordinate analysis developed earlier.

Pedagogical note:
Readers interested primarily in the mathematics of normal modes may skip this appendix without loss of continuity.

J.4 Geometric Proof

J.4.1 Conjugacy of the Three σ_v-planes

Let's take σ_v as the reference operation (say, the plane passing through one vertex of a trigonal pyramid like NH_3).

Let us apply conjugation by C_3:

$$C_3^{-1}\,\sigma_v\,C_3 = ?.$$

Step 1: Interpret geometrically:

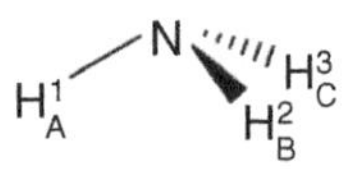

The operation C_3 rotates the molecule counterclockwise by 120° around the principal axis. σ_v is a plane that contains this axis and one vertex (H_A).

When you perform C_3, the vertex H_A moves to H_B, H_B moves to H_C, and H_C to H_A.

The plane that contained H_A now contains H_C, so H_A and H_B exchange positions on $\sigma_v$reflection.

$C_3^{-1} \equiv C_3^2$ rotates the atoms in the opposite (clockwise) direction.

Therefore, $C_3^{-1}\,\sigma_v\,C_3$ corresponds to reflection through that new plane—that is σ_v''

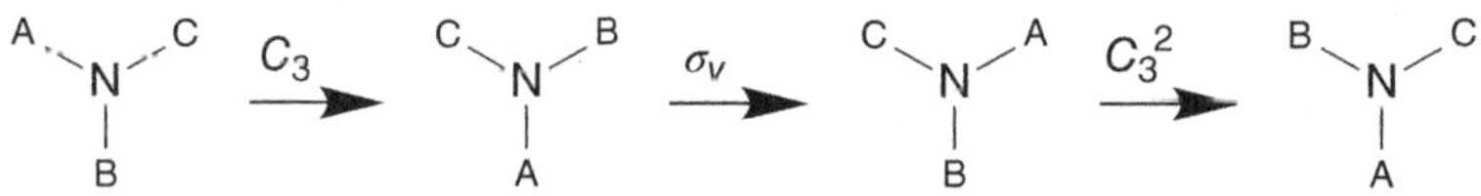

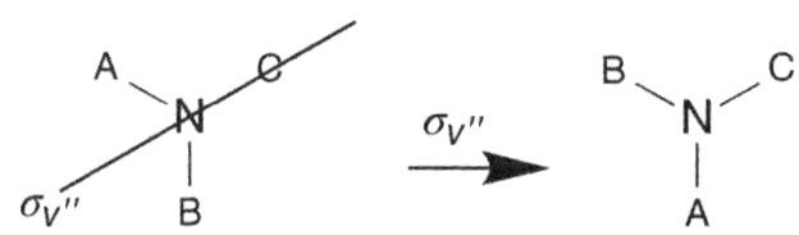

Step 2: Write the result

$$C_3^{-1}\,\sigma_v\,C_3 = \sigma_v''.$$

If we repeat the operation again:

$$C_3^{-1}\,\sigma_v''\,C_3 = \sigma_v',$$

and once more:

$$C_3^{-1}\,\sigma_v'\,C_3 = \sigma_v.$$

Thus, the three σ_v-type planes are related by conjugation with C_3 and therefore form one class,

$$\boxed{\text{Class 1: } (\sigma_v, \sigma_v', \sigma_v'').}$$

J.4.2 Conjugacy of C_3 and C_3^2

Now test C_3 and C_3^2.

Let $A = C_3$. We must find C such that:

$$C^{-1}\, C_3\, C = C_3^2.$$

Let's choose $C = \sigma_v$. Then:

$$\sigma_v^{-1}\, C_3\, \sigma_v = ?.$$

Step 1: Use geometric reasoning

Reflect the molecule in σ_v → this inverts the sense of rotation.

Then perform C_3 → rotate 120°.

Reflect back → reverses the direction again.

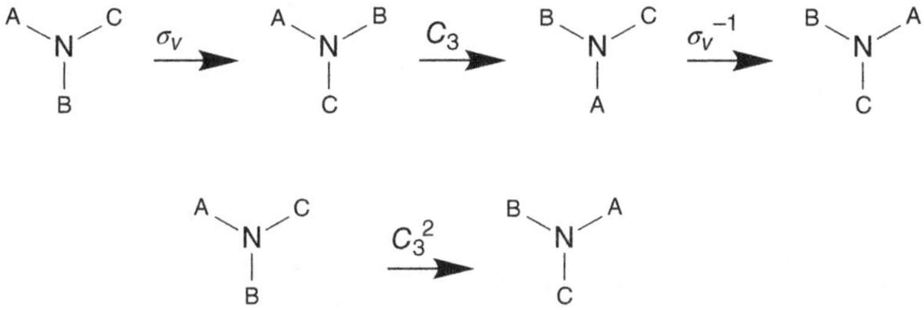

The net effect is equivalent to rotating by −120°, i.e., C_3^2,

$$\sigma_v^{-1}\, C_3\, \sigma_v = C_3^2.$$

Hence, C_3 and C_3^2 are conjugates,

$$\boxed{\text{Class 2: } (C_3, C_3^2).}$$

J.4.3 Identity Element

The identity element E always forms a class by itself:

$$\boxed{\text{Class 3: } (E).}$$

Final Classification of Classes in C_{3v}

Class	Operations	Number of elements
I	E	1
II	$C_3,\ C_3^2$	2
III	$\sigma_v, \sigma_v', \sigma_v''$	3

Therefore:
$\sigma_v, \sigma_v', \sigma_v''$ are mutually conjugate because rotation (C_3) cyclically permutes them.
C_3 and C_3^2 are conjugate because reflection (σ_v) interchanges them.
Hence, both statements are proved.

J.5 Frobenius Expansion for Legendre's Equation at $x = 0$: Detailed Derivation

Legendre's equation is

$$(1 - x^2)y'' - 2xy' + \lambda y = 0.$$

Although $x = 0$ is an **ordinary point**, we begin with a **general Frobenius form** to test whether a non-zero exponent r is required:

$$y = x^r \sum_{n=0}^{\infty} a_n x^n, \; a_0 \neq 0. \tag{J.45}$$

Step 1: Compute derivatives

$$y = \sum_{n=0}^{\infty} a_n x^{n+r}, \quad y' = \sum_{n=0}^{\infty} (n+r) a_n x^{n+r-1}, \quad y'' = \sum_{n=0}^{\infty} (n+r)(n+r-1) a_n x^{n+r-2}.$$

Step 2: Substitute into the equation

We substitute into

$$(1 - x^2)y'' - 2xy' + \lambda y.$$

a. **Expand $(1 - x^2)y''$**

$$(1 - x^2)y'' = y'' - x^2 y''.$$

First part:

$$y'' = \sum_{n=0}^{\infty} (n+r)(n+r-1) a_n x^{n+r-2}. \tag{J.46}$$

Second part:

$$-x^2 y'' = -\sum_{n=0}^{\infty} (n+r)(n+r-1) a_n x^{n+r}. \tag{J.47}$$

b. **Expand $-2xy'$**

$$-2xy' = -2 \sum_{n=0}^{\infty} (n+r) a_n x^{n+r}. \tag{J.48}$$

c. **Expand λy**

$$\lambda y = \lambda \sum_{n=0}^{\infty} a_n x^{n+r}. \tag{J.49}$$

Step 3: Identify the *lowest* power of x

The indicial equation comes from the **lowest exponent** in the entire expression.
Look at the powers of x:

- In (J.46): exponent is $n + r - 2 \rightarrow$ **lowest** is $r - 2$ when $n = 0$.
- In (J.47), (J.48), (J.49): lowest exponent is x^r.

Since $r - 2 < r$ for all real r, the **lowest power** comes only from (J.46).
We therefore isolate the coefficient of x^{r-2}.
From (J.46), the term with $n = 0$ is

$$(n + r)(n + r - 1)a_n x^{n+r-2} = (r)(r - 1)a_0 x^{r-2}.$$

No other term from the substitution produces x^{r-2}.
Thus, the coefficient of x^{r-2} must vanish:

$$\boxed{r(r - 1)a_0 = 0.} \tag{J.50}$$

Since $a_0 \neq 0$, this gives the **indicial equation**:

$$\boxed{r(r - 1) = 0}$$

$$\Rightarrow r = 0,\ 1.$$

J.6 Proofs

J.6.1 Proof 1: Orthogonality of Legendre Polynomials

Strategy:

1. Start with the differential equations for P_l and $P_{k.}$
2. Multiply each equation by the other polynomial, integrate over $[-1, 1]$, and subtract the two resulting equations.
3. Use integration by parts to show that the boundary terms vanish due to the $(1 - x^2)$ factor, forcing the orthogonality integral to be zero when $l \neq k$.

Proof

Since the $P_l(x)$ satisfy Legendre's equation, we may write from Equation (2) in Math Box 13.1:

$$\left[(1 - x^2)P_l'(x)\right]' + l(l + 1)P_l(x) = 0,$$

where $P_l' = dP_l/dx$. Multiplying throughout by $P_k(x)$ and integrating from $x = -1$ to $x = +1$, we obtain:

$$\int_{-1}^{1} P_k(x)\left[(1 - x^2)P_l'(x)\right]' dx + \int_{-1}^{1} P_k(x)l(l + 1)P_l(x)dx = 0.$$

Integrating the first term by parts and noting that the boundary contribution vanishes at both limits because of the factor $(1 - x^2)$, we find:

$$\int_{-1}^{1} P_k'(x)(1 - x^2)P_l'(x)dx + \int_{-1}^{1} P_k(x)l(l + 1)P_l(x)dx = 0.$$

Now, if we reverse the roles of l and k and subtract one expression from the other, we find that:

$$[k(k+1) - l(l+1)] \int_{-1}^{1} P_k(x)P_l(x)dx = 0,$$

and, therefore, since $k \neq l$, the term in the square brackets is nonzero, so the integral must be zero, and we must have the result (13.1).

J.6.2 Proof 2: Rodrigues's Formula for Legendre Polynomials

Strategy:

1. Define a candidate function $y = \frac{d^l}{dx^l}(x^2-1)^l$.
2. Show that this y satisfies a modified version of Legendre's equation by differentiating the expression $(x^2-1)\frac{dy}{dx} = 2lxy(l+1)$ times using Leibniz's theorem for higher derivatives of a product.
3. Compare the resulting equation with the standard Legendre equation to confirm y is a solution, up to a multiplicative constant c_l.
4. Determine c_l by evaluating $\frac{d^l}{dx^l}(x^2-1)^l$ at $x = 1$ and using the standardization condition $P_l(1) = 1$.

Proof

$$\text{Let } y = (x^2-1)^l. \tag{J.51}$$

Then $y' = 2lx(x^2-1)^{l-1}$,
and, hence, multiplying both sides by (x^2-1), we obtain:

$$(x^2-1)y' = 2lx(x^2-1)^l = 2lxy \Rightarrow (x^2-1)y' - 2lxy = 0.$$

In order to show the equivalence of this equation to Equation (J.51), we need to differentiate this expression $(l+1)$ times. We use *Leibniz's theorem*, which states that

If u and v are two functions of x such that all their desired differential coefficients exist, then the nth differential coefficient of their product is given by:

$$D^{(n)}(uv) = \left(D^{(n)}u\right) \cdot v + {}^nC_1 D^{(n-1)}u \cdot Dv + {}^nC_2 D^{(n-2)}u \cdot D^2 v + \ldots + {}^nC_r D^{(n-r)}u \cdot D^r v + \ldots + u \cdot D^n v.$$

Applying this theorem to differentiate the above equation $(l+1)$ times with
$u = y'$; $v = (x^2-1)$ for the first term, and
$u = y$; $v = x$ for the second
results in the expression:

$$y^{(l+2)}(x^2-1) + (l+1)y^{(l+1)} \cdot 2x + \frac{l(l+1)}{2}y^{(l)} \cdot 2 - 2l\left[y^{(l+1)}x + (l+1)y^l\right] = 0,$$

which simplifies to:

$$(1-x^2)y^{(l+2)} - 2xy^{(l+1)} + l(l+1)y^{(l)} = 0$$

after a change of sign throughout.

On comparison of this equation with Legendre's Equation (J.51), we can see that:

$$y^{(l)}(x) \equiv c_l P_l(x),$$

where c_l is a constant dependent on l. To determine this constant, we differentiate y, as given by (J.51) l times, again using Leibniz's theorem, as follows:

$$y(x) = (x^2 - 1)^l = (x+1)^l(x-1)^l,$$
$$y^{(l)}(x) = (x+1)^l D^{(l)}(x-1)^l + lD(x+1)^l D^{(l-1)}(x-1)^l + \cdots.$$

Each derivative of $(x - 1)^l$ is a polynomial in $(x - 1)$ and hence vanishes at $x = 1$, except the first term, which is equal to $(x+1)^l l!(x-1)^{l-l} = (x+1)^l l!$. When $x = 1$, this expression equals $2^l l!$, and since $P_l(1) = 1$ for all l, from Equation (13.6), we have $c_l = 2^l l!$, and hence Equation (13.4) is proved.

J.6.3 Proof 3: Normalization Integral for Legendre Polynomials

Strategy:

1. Substitute Rodrigues's formula into the normalization integral $\int_{-1}^{1} |P_l(x)|^2 dx$.
2. Perform integration by parts l times. Each time, the boundary terms vanish because they still contain a factor of $(x^2 - 1)^l$.
3. This process transforms the integral into $\int_{-1}^{1} (x^2-1)^l \frac{d^{2l}}{dx^{2l}}(x^2-1)^l dx$. The $(2l)$th derivative of $(x^2 - 1)^l$ is simply $(2l)!$.
4. The remaining integral $K_l = \int_{-1}^{1} (1 - x^2)^l dx$ is evaluated by finding a recurrence relation for K_l through integration by parts, eventually reducing it to a known base case.

Proof

For $l = 0$, this is obviously true. For $l \geq 1$, by Rodrigues's formula:

$$I_l = \frac{1}{2^{2l}(l!)^2} \int_{-1}^{1} \left[\frac{d^l(x^2-1)^l}{dx^l}\right]\left[\frac{d^l(x^2-1)^l}{dx^l}\right] dx.$$

Integrating by parts, we obtain:

$$I_l = \frac{1}{2^{2l}(l!)^2}\left[\frac{d^l(x^2-1)^l}{dx^l}\frac{d^{l-1}(x^2-1)^l}{dx^{l-1}}\Bigg|_{-1}^{1} - \int_{-1}^{1} \frac{d^{l+1}(x^2-1)^l}{dx^{l+1}}\frac{d^{l-1}(x^2-1)^l}{dx^{l-1}} dx\right].$$

The first term vanishes at both limits because of the presence of $(x^2 - 1)$ in the second differential, since differentiating $(x^2 - 1)^l$ less than l times leaves an expression that has $(x^2 - 1)$ in it.

Repeating the integration by parts l times, with all boundary conditions vanishing, results in:

$$I_l = \frac{(-1)^l}{2^{2l}(l!)^2}\int_{-1}^{1} (x^2-1)^l \left[\frac{d^{2l}(x^2-1)^l}{dx^{2l}}\right] dx = \frac{(-1)^l(2l)!}{2^{2l}(l!)^2}\int_{-1}^{1} (x^2-1)^l dx$$
$$= \frac{(2l)!}{2^{2l}(l!)^2}\int_{-1}^{1} (1-x^2)^l dx,$$

since the $(2l)$th derivative of the polynomial $\left(x^2-1\right)^l$, which has degree $2l$, is $(2l)!$.

If we now write $K_l = \int_{-1}^{1} (1-x^2)^l dx$ and integrate by parts, taking a factor 1 as the second part, we get

$$K_l = \left[(1-x^2)^l x\right]_{-1}^{1} - \int_{-1}^{1} \frac{d}{dx}(1-x^2)^l x dx = 2l \int_{-1}^{1} (1-x^2)^{l-1} x^2 dx.$$

The first term vanishes because of the presence of $(1-x^2)$.
Let $x^2 = -(1-x^2)+1$. Then:

$$\begin{aligned} K_l &= 2l \int_{-1}^{1} (1-x^2)^{l-1}(-(1-x^2)+1)dx \\ &= -2l \int_{-1}^{1} (1-x^2)^l dx + 2l \int_{-1}^{1} (1-x^2)^{l-1} dx \\ &= -2lK_l + 2lK_{l-1}, \end{aligned}$$

and hence, we have the recurrence relation:

$$(2l+1)K_l = 2lK_{l-1}.$$

We therefore find:

$$K_l = \frac{2l}{2l+1}K_{l-1} = \frac{2l}{2l+1}\frac{2(l-1)}{2l-1}K_{l-2} = \ldots$$

until the last term becomes K_0. There are l such terms,

$$K_l = \frac{2l}{2l+1}\frac{2l-2}{2l-1}\cdots\frac{2}{3}K_0 = 2^l l!\frac{2^l l!}{(2l+1)!}2 = \frac{2^{2l+1}(l!)^2}{(2l+1)!}.$$

Since

$$I_l = \frac{(2l)!}{2^{2l}(l!)^2}K_l,$$

this gives

$$I_l = \frac{(2l)!}{2^{2l}(l!)^2}\frac{2^{2l+1}(l!)^2}{(2l+1)!} = \frac{2}{2l+1}.$$

J.6.4 Proof 4: Generating Function Produces Legendre Polynomials

Strategy:

1. Assume the generating function $G(x,t) = \dfrac{1}{\sqrt{1-2xt+t^2}} = \sum_{n=0}^{\infty} P_n(x)t^n$ defines the polynomials $P_n(x)$.
2. Differentiate $G(x,t)$ with respect to t and x to derive two different recurrence relations for the $P_n(x)$ (Equations J.53 and J.54).
3. Combine and manipulate these recurrence relations to ultimately show that the $P_n(x)$ defined by the generating function must satisfy Legendre's differential equation. This confirms they are the genuine Legendre polynomials.

Proof

To do this, we first differentiate both sides of this equation with respect to x. We get

$$t(1-2xt+t^2)^{-3/2} = \sum_{n=0}^{\infty} P_n'(x)t^n. \tag{J.52}$$

Using Equation (13.9), we can rewrite this equation as:

$$\begin{aligned} t\sum_{n=0}^{\infty} P_n(x)t^n &= (1-2xt+t^2)\sum_{n=0}^{\infty} P_n'(x)t^n \\ \Rightarrow \sum_{n=0}^{\infty} P_n(x)t^{n+1} &= \sum_{n=0}^{\infty} P_n'(x)t^n \\ &\quad -2\sum_{n=0}^{\infty} xP_n'(x)t^{n+1} + \sum_{n=0}^{\infty} P_n'(x)t^{n+2}. \end{aligned}$$

Equating the coefficients of t^{n+1}, we obtain the recurrence relation:

$$P_n(x) = P_{n+1}'(x) - 2xP_n'(x) + P_{n-1}'(x). \tag{J.53}$$

We now differentiate Equation (13.9) with respect to t and find

$$(x-t)(1-2xt+t^2)^{-3/2} = \sum_{n=0}^{\infty} nP_n(x)t^{n-1}. \tag{J.54}$$

Substituting from Equation (J.52), we get:

$$\begin{aligned} (x-t)\sum_{n=0}^{\infty} P_n'(x)t^{n-1} &= \sum_{n=0}^{\infty} nP_n(x)t^{n-1} \\ &\Rightarrow \sum_{n=0}^{\infty} xP_n'(x)t^{n-1} \\ &\quad - \sum_{n=0}^{\infty} P_n'(x)t^n = \sum_{n=0}^{\infty} nP_n(x)t^{n-1} \end{aligned} \tag{J.55}$$

Again, equating the coefficients for t^{n-1}, we get a second recurrence relation:

$$xP_n'(x) - P_{n-1}'(x) = nP_n(x). \tag{J.56}$$

Substituting for $P_{n-1}'(x)$ from Equation (13.11), we get:

$$\begin{aligned} &xP_n'(x) - P_n(x) + P_{n+1}'(x) - 2xP_n'(x) = nP_n(x) \\ &\Rightarrow (n+1)P_n(x) = P_{n+1}'(x) - xP_n'(x). \end{aligned} \tag{J.57}$$

In Equation (J.57), if we replace n by $n-1$ and add x times Equation (J.56) to it, we get:

$$(1-x^2)P_n' = n(P_{n-1}(x) - xP_n(x)).$$

Differentiate both sides of this equation with respect to x,

$$(1-x^2)P_n'' - 2xP_n'(x) = n(P_{n-1}'(x) - xP_n'(x) - P_n(x)).$$

Substitute for $P_{n-1}'(x)$ from Equation (J.56) to get:

$$(1-x^2)P_n'' - 2xP_n'(x) + n(n+1)P_n(x) = 0,$$

and thus, the $P_n(x)$ defined by the generating function satisfy Legendre's Equation (13.1).

K

Notations and Conventions

K.1 Chapters 2–6

Because these chapters move between **matrix algebra**, **geometric vector analysis**, and **quantum-mechanical operator language**, the mix of notations (boldface, arrows, and Dirac's bra-ket notation) can be confusing.

Symbol/notation	Meaning	Context/remarks
A, B, S, I	Matrices (linear operators)	Italic uppercase letters denote matrices. When elements are shown explicitly, square brackets [ ] are used
$a, b, c, x, y, z, \lambda$	Scalars, matrix elements, eigenvalues	Italic lowercase letters denote real or complex numbers
$\mathbf{x}, \mathbf{c}$	Component column vectors	Bold lowercase letters denote column vectors of components of a vector or state in a given basis Example: $\mathbf{c}' = S^{-1}\mathbf{c}$
$\vec{v}, \vec{u}, \vec{x}$	Geometric vectors (geometric form)	Used in coordinate geometry, change-of-basis, and visualization of quadratic surfaces
$\vert v\rangle, \langle v\vert$	Ket and bra vectors	Used for algebraic manipulations and in contexts leading toward operator formalism. Equivalent to column and row vectors, respectively
$\langle u\vert v\rangle$	Inner product	Equals $\vec{u}\cdot\vec{v}$ or $u_1^* v_1 + u_2^* v_2 + \ldots$; provides the metric for orthogonality and normalization
$A\vert v\rangle = \lambda\vert v\rangle$	Eigenvalue equation in operator form	Equivalent to $A\vec{v} = \lambda\vec{v}$ or $A\mathbf{v} = \lambda\mathbf{v}$ in matrix–vector form
$S^{-1} A\, S$	Similarity transformation	Expresses the same linear operator in a new basis
$S^{\dagger} A\, S$	Unitary transformation	Applies when the basis vectors are orthonormal $(S^{\dagger} = S^{-1})$
$\mathrm{Tr}(A)$	Trace of a matrix	Sum of diagonal elements; invariant under similarity transformation
$\det(A)$	Determinant of a matrix	Product of eigenvalues; invariant under similarity transformation
δ_{ij}	Kronecker delta	1 if $i = j$, 0 otherwise; used for orthonormality

(*Continued*)

(Continued)

Symbol/notation	Meaning	Context/remarks
$x^T A\,x, x^\dagger\;A\,x$	Quadratic (or Hermitian) form	Used for real or complex matrices, respectively; represents physical quantities such as energy
Λ	Diagonal matrix of eigenvalues	Appears in diagonalization: $A = S\Lambda S^{-1}$ or $A = S\Lambda S^\dagger$
V, T, F	Special matrices in physical contexts	V: potential-energy matrix; T: kinetic-energy matrix; F: force-constant matrix (preferred in spectroscopy and quantum chemistry)
ω, μ, k	Angular frequency, reduced mass, force constant	Used in normal-mode analysis and molecular vibration examples
$\vec{r} = (x_i, y_i, z_i)$	Cartesian coordinates of the ith atom	Used in moment-of-inertia and polarizability tensors
$A^\dagger$	Hermitian (conjugate) transpose	$(A^\dagger)_{ij} = (A_{ji})^*$; used for complex matrices
Commuting matrices: $[A, B] = 0$	Commutator definition	When $[A, B] = 0$, the matrices share the same eigenvectors and can be simultaneously diagonalized

Notes:

- Vectors are represented by **arrows** ($\vec{v}$) in geometric or coordinate-based discussions and by **Dirac notation** ($|v\rangle$) in operator form. Bold roman ($\mathbf{v}$) denotes component column vectors.
- Matrices remain **italic** throughout (no boldface).
- Physical matrices follow chemistry convention: T for kinetic energy, F (or occasionally V) for the force constant or potential matrix.
- The three notations are equivalent:

$$A|\,v\rangle = \lambda|\,v\rangle \Longleftrightarrow A\vec{v} = \lambda\vec{v} \Longleftrightarrow A\mathbf{v} = \lambda\mathbf{v}.$$

K.2 Chapter 12

- $y(x)$: solution of the differential equation;
- $y_1(x), y_2(x)$: two linearly independent solutions;
- $W(y_1, y_2)$: Wronskian of the pair of solutions;
- a_n: coefficient of x^n in a power-series expansion;
- x_0: point about which the series is expanded;
- r: exponent in a Frobenius solution of the form $x^r \sum_{n=0}^{\infty} a_n x^n$;
- l: degree (order) of a Legendre polynomial $P_l(x)$;
- $P_l(x)$: Legendre polynomial of degree l;
- $[-1, 1]$: domain on which Legendre polynomials are defined and orthogonal;
- "Ordinary point": point where the coefficients of the ODE are analytic;
- "Regular singular point": a singular point x_0 at which $(x - x_0)\,P(x)$ and $(x - x_0)^2\,Q(x)$ remain analytic, guaranteeing a Frobenius series solution;
- "Recurrence relation": formula linking a_{n+2} to a_n.

Index

Wiley The manufacturer's authorized representative according to the EU General Product Safety Regulation is Wiley-VCH GmbH, Boschstr. 12, 69469 Weinheim, Germany, e-mail: Product_Safety@wiley.com.

Printed and bound by CPI Group (UK) Ltd, Croydon, CR0 4YY
14/09/2026
02216811-0002